T0269088

LONDON MATHEMATICAL SOCIETY LECTURE NOTE SERIES

Managing Editor: Professor J.W.S. Cassels, Department of Pure Mathematics and Mathematical Statistics, University of Cambridge, 16 Mill Lane, Cambridge CB2 1SB, England

The titles below are available from booksellers, or, in case of difficulty, from Cambridge University Press.

34 Representation theory of Lie groups, M.F. ATIYAH *et al*
46 p-adic analysis: a short course on recent work, N. KOBLITZ
50 Commutator calculus and groups of homotopy classes, H.J. BAUES
59 Applicable differential geometry, M. CRAMPIN & F.A.E. PIRANI
66 Several complex variables and complex manifolds II, M.J. FIELD
69 Representation theory, I.M. GELFAND *et al*
76 Spectral theory of linear differential operators and comparison algebras, H.O. CORDES
77 Isolated singular points on complete intersections, E.J.N. LOOIJENGA
83 Homogeneous structures on Riemannian manifolds, F. TRICERRI & L. VANHECKE
86 Topological topics, I.M. JAMES (ed)
87 Surveys in set theory, A.R.D. MATHIAS (ed)
88 FPF ring theory, C. FAITH & S. PAGE
89 An F-space sampler, N.J. KALTON, N.T. PECK & J.W. ROBERTS
90 Polytopes and symmetry, S.A. ROBERTSON
92 Representation of rings over skew fields, A.H. SCHOFIELD
93 Aspects of topology, I.M. JAMES & E.H. KRONHEIMER (eds)
94 Representations of general linear groups, G.D. JAMES
95 Low-dimensional topology 1982, R.A. FENN (ed)
96 Diophantine equations over function fields, R.C. MASON
97 Varieties of constructive mathematics, D.S. BRIDGES & F. RICHMAN
98 Localization in Noetherian rings, A.V. JATEGAONKAR
99 Methods of differential geometry in algebraic topology, M. KAROUBI & C. LERUSTE
100 Stopping time techniques for analysts and probabilists, L. EGGHE
104 Elliptic structures on 3-manifolds, C.B. THOMAS
105 A local spectral theory for closed operators, I. ERDELYI & WANG SHENGWANG
107 Compactification of Siegel moduli schemes, C-L. CHAI
108 Some topics in graph theory, H.P. YAP
109 Diophantine analysis, J. LOXTON & A. VAN DER POORTEN (eds)
110 An introduction to surreal numbers, H. GONSHOR
113 Lectures on the asymptotic theory of ideals, D. REES
114 Lectures on Bochner-Riesz means, K.M. DAVIS & Y-C. CHANG
115 An introduction to independence for analysts, H.G. DALES & W.H. WOODIN
116 Representations of algebras, P.J. WEBB (ed)
118 Skew linear groups, M. SHIRVANI & B. WEHRFRITZ
119 Triangulated categories in the representation theory of finite-dimensional algebras, D. HAPPEL
121 Proceedings of *Groups - St Andrews 1985*, E. ROBERTSON & C. CAMPBELL (eds)
122 Non-classical continuum mechanics, R.J. KNOPS & A.A. LACEY (eds)
125 Commutator theory for congruence modular varieties, R. FREESE & R. MCKENZIE
126 Van der Corput's method of exponential sums, S.W. GRAHAM & G. KOLESNIK
128 Descriptive set theory and the structure of sets of uniqueness, A.S. KECHRIS & A. LOUVEAU
129 The subgroup structure of the finite classical groups, P.B. KLEIDMAN & M.W. LIEBECK
130 Model theory and modules, M. PREST
131 Algebraic, extremal & metric combinatorics, M-M. DEZA, P. FRANKL & I.G. ROSENBERG (eds)
132 Whitehead groups of finite groups, ROBERT OLIVER
133 Linear algebraic monoids, MOHAN S. PUTCHA
134 Number theory and dynamical systems, M. DODSON & J. VICKERS (eds)
135 Operator algebras and applications, 1, D. EVANS & M. TAKESAKI (eds)
136 Operator algebras and applications, 2, D. EVANS & M. TAKESAKI (eds)
137 Analysis at Urbana, I, E. BERKSON, T. PECK, & J. UHL (eds)
138 Analysis at Urbana, II, E. BERKSON, T. PECK, & J. UHL (eds)
139 Advances in homotopy theory, S. SALAMON, B. STEER & W. SUTHERLAND (eds)
140 Geometric aspects of Banach spaces, E.M. PEINADOR & A. RODES (eds)
141 Surveys in combinatorics 1989, J. SIEMONS (ed)
144 Introduction to uniform spaces, I.M. JAMES
145 Homological questions in local algebra, JAN R. STROOKER
146 Cohen-Macaulay modules over Cohen-Macaulay rings, Y. YOSHINO
147 Continuous and discrete modules, S.H. MOHAMED & B.J. MÜLLER
148 Helices and vector bundles, A.N. RUDAKOV *et al*
149 Solitons, nonlinear evolution equations and inverse scattering, M. ABLOWITZ & P. CLARKSON
150 Geometry of low-dimensional manifolds 1, S. DONALDSON & C.B. THOMAS (eds)
151 Geometry of low-dimensional manifolds 2, S. DONALDSON & C.B. THOMAS (eds)
152 Oligomorphic permutation groups, P. CAMERON

153 L-functions and arithmetic, J. COATES & M.J. TAYLOR (eds)
154 Number theory and cryptography, J. LOXTON (ed
155 Classification theories of polarized varieties, TAKAO FUJITA
156 Twistors in mathematics and physics, T.N. BAILEY & R.J. BASTON (eds)
157 Analytic pro-*p* groups, J.D. DIXON, M.P.F. DU SAUTOY, A. MANN & D. SEGAL
158 Geometry of Banach spaces, P.F.X. MÜLLER & W. SCHACHERMAYER (eds)
159 Groups St Andrews 1989 volume 1, C.M. CAMPBELL & E.F. ROBERTSON (eds)
160 Groups St Andrews 1989 volume 2, C.M. CAMPBELL & E.F. ROBERTSON (eds)
161 Lectures on block theory, BURKHARD KÜLSHAMMER
162 Harmonic analysis and representation theory, A. FIGA-TALAMANCA & C. NEBBIA
163 Topics in varieties of group representations, S.M. VOVSI
164 Quasi-symmetric designs, M.S. SHRIKANDE & S.S. SANE
165 Groups, combinatorics & geometry, M.W. LIEBECK & J. SAXL (eds)
166 Surveys in combinatorics, 1991, A.D. KEEDWELL (ed)
167 Stochastic analysis, M.T. BARLOW & N.H. BINGHAM (eds)
168 Representations of algebras, H. TACHIKAWA & S. BRENNER (eds)
169 Boolean function complexity, M.S. PATERSON (ed)
170 Manifolds with singularities and the Adams-Novikov spectral sequence, B. BOTVINNIK
171 Squares, A.R. RAJWADE
172 Algebraic varieties, GEORGE R. KEMPF
173 Discrete groups and geometry, W.J. HARVEY & C. MACLACHLAN (eds)
174 Lectures on mechanics, J.E. MARSDEN
175 Adams memorial symposium on algebraic topology 1, N. RAY & G. WALKER (eds)
176 Adams memorial symposium on algebraic topology 2, N. RAY & G. WALKER (eds)
177 Applications of categories in computer science, M. FOURMAN, P. JOHNSTONE, & A. PITTS (eds)
178 Lower K- and L-theory, A. RANICKI
179 Complex projective geometry, G. ELLINGSRUD *et al*
180 Lectures on ergodic theory and Pesin theory on compact manifolds, M. POLLICOTT
181 Geometric group theory I, G.A. NIBLO & M.A. ROLLER (eds)
182 Geometric group theory II, G.A. NIBLO & M.A. ROLLER (eds)
183 Shintani zeta functions, A. YUKIE
184 Arithmetical functions, W. SCHWARZ & J. SPILKER
185 Representations of solvable groups, O. MANZ & T.R. WOLF
186 Complexity: knots, colourings and counting, D.J.A. WELSH
187 Surveys in combinatorics, 1993, K. WALKER (ed)
188 Local analysis for the odd order theorem, H. BENDER & G. GLAUBERMAN
189 Locally presentable and accessible categories, J. ADAMEK & J. ROSICKY
190 Polynomial invariants of finite groups, D.J. BENSON
191 Finite geometry and combinatorics, F. DE CLERCK *et al*
192 Symplectic geometry, D. SALAMON (ed)
193 Computer algebra and differential equations, E. TOURNIER (ed)
194 Independent random variables and rearrangement invariant spaces, M. BRAVERMAN
195 Arithmetic of blowup algebras, WOLMER VASCONCELOS
196 Microlocal analysis for differential operators, A. GRIGIS & J. SJÖSTRAND
197 Two-dimensional homotopy and combinatorial group theory, C. HOG-ANGELONI, W. METZLER & A.J. SIERADSKI (eds)
198 The algebraic characterization of geometric 4-manifolds, J.A. HILLMAN
199 Invariant potential theory in the unit ball of C^n, MANFRED STOLL
200 The Grothendieck theory of dessins d'enfant, L. SCHNEPS (ed)
201 Singularities, JEAN-PAUL BRASSELET (ed)
202 The technique of pseudodifferential operators, H.O. CORDES
203 Hochschild cohomology of von Neumann algebras, A. SINCLAIR & R. SMITH
204 Combinatorial and geometric group theory, A.J. DUNCAN, N.D. GILBERT & J. HOWIE (eds)
205 Ergodic theory and its connections with harmonic analysis, K. PETERSEN & I. SALAMA (eds)
206 An introduction to noncommutative differential geometry and its physical applications, J. MADORE
207 Groups of Lie type and their geometries, W.M. KANTOR & L. DI MARTINO (eds)
208 Vector bundles in algebraic geometry, N.J. HITCHIN, P. NEWSTEAD & W.M. OXBURY (eds)
209 Arithmetic of diagonal hypersurfaces over finite fields, F.Q. GOUVÊA & N. YUI
210 Hilbert C*-modules, E.C. LANCE
211 Groups 93 Galway / St Andrews I, C.M. CAMPBELL *et al*
212 Groups 93 Galway / St Andrews II, C.M. CAMPBELL *et al*
214 Generalised Euler-Jacobi inversion formula and asymptotics beyond all orders, V. KOWALENKO, N.E. FRANKEL, M.L. GLASSER & T. TAUCHER
215 Number theory, S. DAVID (ed)
216 Stochastic partial differential equations, A. ETHERIDGE (ed)
217 Quadratic forms with applications to algebraic geometry and topology, A. PFISTER
218 Surveys in combinatorics, 1995, PETER ROWLINSON (ed)
220 Algebraic set theory, A. JOYAL & I. MOERDIJK
221 Harmonic approximation, S.J. GARDINER
222 Advances in linear logic, J.-Y. GIRARD, Y. LAFONT & L. REGNIER (eds)
223 Analytic semigroups and semilinear initial boundary value problems, KAZUAKI TAIRA
226 Novikov conjectures, index theorems and rigidity I, S. FERRY, A. RANICKI & J. ROSENBERG (eds)
227 Novikov conjectures, index theorems and rigidity II, S. FERRY, A. RANICKI & J. ROSENBERG (eds)

London Mathematical Society Lecture Note Series. 227

Novikov Conjectures, Index Theorems and Rigidity

Volume 2

Oberwolfach 1993

Edited by

Steven C. Ferry
State University of New York, Binghampton

Andrew Ranicki
University of Edinburgh

Jonathan Rosenberg
University of Maryland

CAMBRIDGE
UNIVERSITY PRESS

Published by the Press Syndicate of the University of Cambridge
The Pitt Building, Trumpington Street, Cambridge CB2 1RP
40 West 20th Street, New York, NY 10011–4211, USA
10 Stamford Road, Oakleigh, Melbourne 3166, Australia

© Cambridge University Press 1995

First published 1995

Library of Congress cataloguing in publication data available

British Library cataloguing in publication data available

ISBN 0 521 49795 7 paperback

Transferred to digital printing 2003

The papers in this volume were typeset by the editors using the TEX typesetting program and the AMS-TEX, LATEX, LAMS-TEX, AMS-LATEX, and XY-pic macro packages. Style files were prepared by the editors using templates prepared by Cambridge University Press. Some of the figures were prepared in PostScript. TEX and AMS-TEX are trademarks of the American Mathematical Society; PostScript is a registered trademark of Adobe Systems Incorporated. LATEX is copyright (1992) by Leslie Lamport, LAMS-TEX is copyright (1989) by the TEXplorators Corporation, and AMS-LATEX is copyright (1991) by the American Mathematical Society. The XY-pic package is copyright 1991–1994 by Kristoffer H. Rose.

To Sergei P. Novikov
with respect and admiration

Contents

Contents, Volume 2

Contents, Volume 1

Preface

These volumes grew out of the conference which we organized at the Mathematisches Forschungsinstitut Oberwolfach in September, 1993, on the subject of "Novikov conjectures, index theorems and rigidity." The aim of the meeting was to examine the Novikov conjecture, one of the central problems of the topology of manifolds, along with the vast assortment of refinements, generalizations, and analogues of the conjecture which have proliferated over the last 25 years. There were 38 participants, coming from Australia, Canada, France, Germany, Great Britain, Hong Kong, Poland, Russia, Switzerland, and the United States, with interests in topology, analysis, and geometry. What made the meeting unusual were both its interdisciplinary scope and the lively and constructive interaction of experts from very different fields of mathematics. The success of the meeting led us to try to capture its spirit in print, and these two volumes are the result.

It was not our intention to produce the usual sort of conference proceedings volume consisting of research announcements by the participants. There are enough such tomes gathering dust on library shelves. Instead, we have hoped to capture a snapshot of the status of work on the Novikov conjecture and related topics, now that the subject is about 25 years old. We have also tried to produce volumes which will be helpful to beginners in the area (especially graduate students), and also to those working in some aspect of the subject who want to understand the connection between what they are doing and what is going on in other fields. Accordingly, we have included here :

(a) a fairly detailed historical survey of the Novikov conjecture, including an annotated reprint of the original statement (both in the original Russian and in English translation), and a reasonably complete bibliography of the subsequent developments;

(b) the texts of hitherto unpublished classic papers by Milnor, Browder, and Kasparov relevant to the Novikov conjecture, which are known to the experts but hard for the uninitiated to locate;

(c) several papers (Ferry, Ferry-Weinberger, Ranicki, Rosenberg) which, while they present some new work, also attempt to survey aspects of the subject; and

(d) research papers which reflect the wide range of current techniques used to attack the Novikov conjecture: geometry, analysis, topology, algebra, ...

All the research papers have been refereed.

We hope that the reader will find the two volumes worthwhile, not merely as a technical reference tool, but also as stimulating reading to be browsed through at leisure.

We should like to thank the Director and staff of the Mathematisches Institut Oberwolfach for their expert logistical help, for their financial support, and for the marvellous working environment that made possible the 1993 conference that got this project started. Thanks are due as well to all the participants at the meeting, to the contributors to the two volumes, to the referees of the research papers, to the contributors to the problem list and bibliography, and to Roger Astley and David Tranah of Cambridge University Press. We thank the National Science Foundation of the U. S. for its support under grants DMS 90-03746, DMS-93-05758 and DMS 92-25063, the European Union for its support via the K-theory Initiative under Science Plan SCI–CT91–0756, as well as the Centenary Fund of the Edinburgh Mathematical Society.

Steve Ferry
(Binghamton, NY)
Andrew Ranicki
(Edinburgh, Scotland)
Jonathan Rosenberg
(College Park, MD)
June, 1995

Proper affine isometric actions of amenable groups

M. E. B. Bekka, P.-A. Cherix and A. Valette

A property of a (countable) group Γ of relevance both in harmonic analysis and operator algebras is the so-called ***Haagerup's approximation property*** (see [Cho], [JV], [Ro]): one possible definition is to say that the abelian C^*-algebra $c_0(\Gamma)$ has an approximate unit consisting of positive definite functions on Γ.

On the other hand, in §§7.A and 7.E of his book [Gr], M. Gromov introduced the following definition (and dubious pun): Γ is ***a-T-menable*** if Γ admits a proper affine isometric action on some Hilbert space $\mathcal{H}$, where "proper action" means that, for any bounded subsets B, C in $\mathcal{H}$, the set of elements $g \in \Gamma$ such that $\alpha(g)B$ meets C is finite. (This non-standard sense of properness is relevant for actions on general metric spaces.)

Our first result is that the two concepts are actually equivalent.

Lemma. *For a group Γ, the following are equivalent:*

(i) Γ has the Haagerup approximation property;

(ii) Γ admits a proper function of conditionally negative type;

(iii) Γ is a-T-menable.

Proof. (i) $\Leftrightarrow$ (ii) is due to Akemann and Walter (Theorem 10 in [AW]).

(ii) $\Rightarrow$ (iii). Let ψ be a proper function of conditionally negative type on Γ. By Proposition 14 of [HV], there exists an affine isometric action α of Γ on a Hilbert space $\mathcal{H}$ such that, for any $g \in \Gamma$,

$$\psi(g) = \|\alpha(g)(0)\|^2.$$

We claim that α is a proper action. To see this, it is enough to check that, for any $R > 0$, the set $F_R = \{g \in \Gamma : \alpha(g)B_R \cap B_R \neq \emptyset\}$ is finite (here B_R denotes the closed ball with radius R centered at 0).

For $g \in G$, denote by $\pi(g)$ the linear part of $\alpha(g)$, so that $\alpha(g)(0)$ is its translation part, i.e.,

$$\alpha(g)\xi = \pi(g)\xi + \alpha(g)(0)$$

for any $\xi \in \mathcal{H}$.

If $g \in F_R$, we find $\xi \in B_R$ such that $\|\alpha(g)(\xi)\| \leq R$, which implies $\|\alpha(g)(0)\| \leq R + \|\pi(g)(\xi)\| \leq 2R$ or $\psi(g) \leq 4R^2$. Thus F_R is contained in $\{g \in \Gamma : \psi(g) \leq 4R^2\}$, a finite set by assumption.

(iii) $\Rightarrow$ (ii). If α is a proper isometric action of Γ on $\mathcal{H}$, then the function $\psi : \Gamma \to \mathbb{R}; g \mapsto \|\alpha(g)(0)\|^2$ is of conditionally negative type (see *n° 13* of Chapter 5 in [HV]). On the other hand, it is clear that ψ is proper. □

During the problem session at the Oberwolfach Conference on "Novikov conjectures, index theorems and rigidity,"[1] (Sept. 5–11, 1993), Gromov asked whether any amenable group is a-T-menable; doing so, he advertised a Question in §7.E of [Gr]. Our main result is that Gromov's question has an affirmative answer.

Let us fix more notation. We denote by λ_Γ the left regular representation of Γ on $\ell^2(\Gamma)$, and by π_Γ the direct sum of countably many copies of λ_Γ, acting on

$$\mathcal{H}_\Gamma = \ell^2(\Gamma) \oplus \ell^2(\Gamma) \oplus \ell^2(\Gamma) \oplus \cdots$$

(countably many summands).

Proposition. *Let Γ be a countable amenable group. Then Γ admits a proper affine isometric action α on $\mathcal{H}_\Gamma$, such that the linear part of α is π_Γ.*

Proof. Let $(F_k)_{k\geq 1}$ be an increasing family of finite subsets of Γ, such that $\bigcup_{k=1}^{\infty} F_k = \Gamma$. By Følner's property, we find for any $k \geq 1$ a finite subset U_k of Γ, such that for any $g \in F_k$,

$$\frac{|gU_k \,\Delta\, U_k|}{|U_k|} < 2^{-k}.$$

Let ξ_k be the normalized characteristic function of U_k, i.e.,

$$\xi_k(x) = \begin{cases} |U_k|^{-\frac{1}{2}}, & \text{if } x \in U_k, \\ 0, & \text{otherwise.} \end{cases}$$

Then

$$\|\lambda_\Gamma(g)\xi_k - \xi_k\|^2 = \frac{|gU_k \,\Delta\, U_k|}{|U_k|}$$

for any $g \in \Gamma$.

---

[1] The third author thanks the organizers for inviting him to that stimulating week!

For $g \in \Gamma$, set now $b(g) = \bigoplus_{k=0}^{\infty} k(\lambda_\Gamma(g)\xi_k - \xi_k)$. This series converges in $\mathcal{H}_\Gamma$ because, for $g \in F_n$, we have

$$\left\| \bigoplus_{k=n}^{\infty} k(\lambda_\Gamma(g)\xi_k - \xi_k) \right\|^2 = \sum_{k \geq n} k^2 \|\lambda_\Gamma(g)\xi_k - \xi_k\|^2 \leq \sum_{k \geq n} k^2 2^{-k} < \infty.$$

It is immediate to check that b is a 1-cocycle with respect to π_Γ, i.e., for any $g, h \in \Gamma$, $b(gh) = \pi_\Gamma(g)(b(h)) + b(g)$. Thus, defining $\alpha(g) : \mathcal{H}_\Gamma \to \mathcal{H}_\Gamma$ by $\alpha(g)\xi = \pi_\Gamma(g)\xi + b(g)$, we see that α is an affine isometric action of Γ on $\mathcal{H}_\Gamma$, with linear part π_Γ.

Now, we define an function ψ conditionally of negative type on Γ by $\psi(g) = \|b(g)\|^2$, and we claim that ψ is a proper function. This amounts to proving that, for $R \geq 0$, the set $C_R = \{g \in \Gamma : \|b(g)\| \leq R\}$ is finite. To see this, we fix $N \in \mathbb{N}$ such that $R \leq N$. Then, for $g \in C_R$:

$$N^2 \|\lambda_\Gamma(g)\xi_N - \xi_N\|^2 \leq \|b(g)\|^2 \leq R^2,$$

hence $|gU_N \triangle U_N| \leq |U_N|$ or $\frac{|U_N|}{2} \leq |U_N \cap gU_N|$. But the set of h's in Γ such that $\frac{|U_N|}{2} \leq |U_N \cap hU_N|$ is clearly finite.

From the fact that ψ is a proper function conditionally of negative type, one deduces the fact that α is a proper action as in the implication (ii) $\Rightarrow$ (iii) of the Lemma. □

Remark. It is known (see e.g. [HV, Chapter 4]) that a countable group Γ does *not* have Kazhdan's property (T) if and only if Γ admits an affine isometric action with *unbounded* orbits on some Hilbert space. For Γ a countably infinite amenable group, the action α just constructed appeared in [Che] to give an explicit example of such an affine isometric action with unbounded orbits. Guichardet also proved that any such group admits an affine isometric action with unbounded orbits on $\ell^2(G)$, the linear part of which is λ_Γ (see [Gu], Cor. 2.4 in Chapter III; the proof, appealing to the closed graph theorem, is not constructive, so we do not know whether such an action is proper or not).

References

[AW] C. A. Akemann and M. E. Walter, *Unbounded negative definite functions*, Canad. J. Math. **33** (1981), 862–871.

[Che] P.-A. Cherix, *Propriétés de groupes héritées par commensurabilité*, Master's thesis, Univ. de Neuchâtel, Neuchâtel, Switzerland, 1991.

[Cho] M. Choda, *Group factors of the Haagerup type*, Proc. Japan Acad. **59** (1983), 174–177.

[HV] P. de la Harpe and A. Valette, *La propriété (T) de Kazhdan pour les groupes localement compacts*, Astérisque, vol. 175, Soc. Math. de France, Paris, 1989.

[Gr] M. Gromov, *Asymptotic invariants of infinite groups*, Geometric Group Theory (G. A. Niblo and M. A. Roller, eds.), London Math. Soc. Lect. Notes, vol. 182, Cambridge Univ. Press, Cambridge, 1993, pp. 1–295.

[Gu] A. Guichardet, *Cohomologie des groupes topologiques et des algèbres de Lie*, Cedic–F. Nathan, Paris, 1980.

[JV] P. Jolissaint and A. Valette, *Normes de Sobolev et convoluteurs bornés sur $L^2(G)$*, Ann. Institut Fourier (Grenoble) **41** (1991), 797–822.

[RFo] A. G. Robertson, *Property (T) for II_1- factors and unitary representations of Kazhdan groups*, Math. Ann. **296** (1993), 547–555.

Département de Mathématiques, Université de Metz, Île du Saulcy, F-57045 Metz, France

email: bekka@dmi.univ-metz.fr

Institut de Mathématiques, Université de Neuchâtel, Chantemerle 20, CH-2007 Neuchâtel, Switzerland

email: cherix@maths.unine.ch

Institut de Mathématiques, Université de Neuchâtel, Chantemerle 20, CH-2007 Neuchâtel, Switzerland

email: valette@maths.unine.ch

Bounded K–theory and the Assembly Map in Algebraic K–theory

Gunnar Carlsson[1]

Introduction

Since their introduction 20 years ago, the evaluation of Quillen's higher algebraic K–groups of rings [35] has remained a difficult problem in homotopy theory. One does have Quillen's explicit evaluation of the algebraic K–theory of finite fields [36] and Suslin's theorem about the K–theory of algebraically closed fields [41]. In addition, Quillen's original paper [35] gave a number of useful formal properties of the algebraic K–groups: localization sequences, homotopy property for the K–theory of polynomial rings, reduction by resolution, and reduction by "devissage". These tools provide, for a large class of commutative rings, a fairly effective procedure which reduces the description of the K–theory of the commutative ring to that of fields. The K–theory of a general field remains an intractable problem due to the lack of a good Galois descent spectral sequence, although by work of Thomason [45] one can understand its so–called Bott–periodic localization. In the case of non–commutative rings, the formal properties of Quillen are not nearly as successful as in the commutative case.

A central theme in the subject has been the relationship with properties of manifolds which are not homotopy invariant; here the ring in question is usually the group ring $\mathbb{Z}[\Gamma]$, with Γ the fundamental group of a manifold. Examples include Wall's finiteness obstruction [49], the s–cobordism theorem of Barden–Mazur–Stallings [31], Hatcher–Wagoner's work [24] on the connection between K_2 and the homotopy type of the pseudoisotopy space, and finally Waldhausen's description [48] of the pseudoisotopy space in terms of the K–theory of "rings up to homotopy". Since the rings $\mathbb{Z}[\Gamma]$ are typically not commutative, the reduction methods of [35] are not adequate for the description of the K–theory of $\mathbb{Z}[\Gamma]$.

Fortunately, in the case of group rings, there is a reasonable conjecture concerning the structure of the K–theory of $\mathbb{Z}[\Gamma]$, or more generally $A[\Gamma]$, where A is a commutative ring. Let Γ be any group, let $B\Gamma_+$ denote

[1]Supported in part by NSF DMS 8907771.A01

its classifying space with a disjoint basepoint added, and let $\underset{\sim}{K}A$ denote the connective spectrum associated to the symmetric monoidal category of finitely generated projective A–modules. (We will allow ourselves to use the language of spectra freely, in this introduction and throughout the paper; see [1] or [30] for a discussion.) The homotopy groups of $\underset{\sim}{K}A$ are precisely the higher algebraic K–groups of Quillen. Then we have the "assembly map"

$$\alpha \;:\; B\Gamma_+ \wedge \underset{\sim}{K}A \longrightarrow KA[\Gamma]$$

(see [27] or [47]), which is constructed out of the group homomorphism

$$\Gamma \times GL_nA \to GL_1A[\Gamma] \times GL_nA \xrightarrow{\otimes} GL_nA[\Gamma] \; .$$

Here, Γ is included in $GL_1A[\Gamma]$ by recognizing that any $\gamma \in \Gamma$ can be viewed as a unit in the ring $A[\Gamma]$, and $\otimes$ is the homomorphism

$$GL_1(A[\Gamma]) \times GL_n(A[\Gamma] \underset{A}{\otimes} A) \to GL_n(A[\Gamma]) \;;\; (M,N) \to M \otimes N \; .$$

A preliminary conjecture about $\underset{\sim}{K}A[\Gamma]$ is that α is an equivalence. This would provide a complete description of $\underset{\sim}{K}A[\Gamma]$ in terms of its constituent parts $\widetilde{K}A$ and $B\Gamma_+$, and would in particular yield a spectral sequence with $E^2_{p,q} = H_p(\Gamma, K_qA)$ converging to $K_{p+q}A[\Gamma]$. The conjecture is however known to fail in general. If Γ is finite, then even with the coefficient ring $\mathbb{C}$, one can show by direct calculation that the map fails to be an isomorphism. On the other hand, if A contains nilpotent elements, one can show that even for the group $\Gamma = \mathbb{Z}$ (the conjecture holds for A regular and $\Gamma = \mathbb{Z}$ by the methods of Quillen [35]), α fails to be an equivalence due to the presence of Nil–groups (see [5]), in the K–theory of the polynomial ring over A. However, there are no known counter–examples to the conjecture with A regular and Γ a group which admits a finite classifying space. In [47], Waldhausen proves that α is an equivalence, when A is regular, for a large class of groups built from $\mathbb{Z}$ by processes of amalgamated product and extension by $\mathbb{Z}$. The rationalized form of the conjecture ($\pi_i(\alpha) \underset{\mathbb{Z}}{\otimes} \mathrm{id}_{\mathbb{Q}}$ is an isomorphism) is much more approachable, and has been studied by many people. Quinn for instance, has shown that this rationalized form holds when Γ is a torsion free Bieberbach group [37], and Bökstedt–Hsiang–Madsen [6] have shown that the $\pi_i(\alpha) \underset{\mathbb{Z}}{\otimes} \mathrm{id}_{\mathbb{Q}}$ is injective for any group with finitely generated homology. Low dimensional integral information has also been obtained by Farrell–Hsiang [16,17]. More recently, the startling results of Farrell–Jones [18,19] have given a complete description of the pseudo–isotopy space of a closed compact manifold admitting a Riemannian metric with negative curvature; one consequence of this is that $\pi_i(\alpha) \underset{\mathbb{Z}}{\otimes} \mathrm{id}_{\mathbb{Q}}$ is an isomorphism for the fundamental groups of these manifolds. The method here relies on the properties of geodesic flow on the tangent bundle to the manifold. Their method will generalize to a larger class of groups, including fundamental groups of manifolds admitting a Riemannian metric with non–positive curvature and cocompact discrete torsion-free subgroups of Lie groups, but always relies on fairly precise control on the geometry (i.e., metric structure) of the manifold.

In this paper, we will study the map α directly, without rationalization. We describe our method. Suppose G is a finite group. (Note that our results will certainly not apply to finite groups; it is included here for motivational purposes.) Then the group ring $A[G]$ can be viewed as the algebra of G–invariant A–linear transformations from $A[G]$ to $A[G]$, where G acts on $\mathrm{Hom}_A(A[G], A[G])$ by the conjugation action $(gf)(x) = gf(g^{-1}x)$. Thus, if $|G| = n$, $A[G]$ is the invariant subring of a group action of G on the $n \times n$ matrices over A, given by conjugating by a subgroup of the permutation matrices in $GL_n(A)$. From this fact, one derives that $\underset{\sim}{K}A[G]$ can be obtained as the fixed point spectrum $(\underset{\sim}{K}A)^G$ of an action of G on $\underset{\sim}{K}A$. We now recall the notion of the *homotopy fixed point set* of an action of a group G on a space (or spectrum) X. For any group G, let EG denote a contractible space on which G acts freely. Then we can equip the function space (or spectrum) $F(EG, X)$ with the conjugation G–action $(gf)(z) = gf(g^{-1}z)$. There is of course the equivariant map $EG \to$ point, which induces an equivariant map

$$\varepsilon : X \to F(\text{point}, X) \to F(EG, X) ,$$

and because of the contractibility of EG, this map is a homotopy equivalence, although not necessarily a G–homotopy equivalence. We denote $F(EG, X)^G$ by X^{hG}, and refer to it as the homotopy fixed point set of X; ε^G is now a map from X^G to X^{hG}. The advantage of X^{hG} over X^G is that X^{hG} is computable from BG and X in an explicit way; for instance, if G acts trivially on X, $X^{hG} = F(BG_+, X)$. More generally, there is a spectral sequence with $E_2^{p,q} = H^{-p}(G, \pi_q X)$ converging to $\pi_{p+q}(X^{hG})$. We informally say that X^{hG} can be constructed in a "homotopy theoretic" way out of the constituent pieces BG_+ and X, as $BG_+ \wedge \underset{\sim}{K}A$ is built in a homotopy–theoretic way out of BG_+ and $\underset{\sim}{K}A$. We now obtain a map

$$\underset{\sim}{K}A[G] = (\underset{\sim}{K}A)^G \to (\underset{\sim}{K}A)^{hG} ,$$

which we use as a detecting device for the assembly map α. Atiyah [2] has shown that it is an effective such device when $A = \mathbb{C}$. In our work, however, we will be exclusively interested in groups Γ which admit a finite classifying space; in particular, they are torsion–free. In the case we can still describe

$A[\Gamma]$ as $\operatorname{Hom}_A(A[\Gamma], A[\Gamma])^\Gamma$, i.e., the fixed point set of an action of Γ on a ring of infinite matrices, which we write $M_\infty(A) = \operatorname{Hom}_A(A[\Gamma], A[\Gamma])$. Unfortunately, the K–theory spectrum of $M_\infty = \operatorname{Hom}_A(A[\Gamma], A[\Gamma])$ can easily be shown to be contractible; elementary properties of the $(-)^{h\Gamma}$ construction (see, e.g., [12]) show that $(\underset{\sim}{K} M_\infty A)^{h\Gamma}$ is therefore contractible. To remedy this, one might attempt to replace $M_\infty(A)$ by a Γ–invariant subring containing $A[\Gamma]$, and whose K–theory spectrum allows one to detect more. Roughly speaking, this is the procedure we use in this paper.

Precisely what we do is sketched as follows. We recall that E. Pedersen and C. Weibel [33], [34], have introduced, for a ring R and metric space X, the *bounded K–theory* $\underset{\sim}{K}(X, R)$; $\underset{\sim}{K}(X, R)$ is a spectrum, has appropriate covariant functoriality properties in both X and R, and when applied to the Euclidean space E^k produces a k–fold developing of the K–theory spectrum of R, equivalent to the Gersten–Wagoner k–fold delooping [22], [46]. When a group Γ acts on a metric space, we introduce a related *equivariant bounded* K–theory spectrum, $\underset{\sim}{K}^\Gamma(X; R)$; it is a spectrum with Γ–action. Viewed as a spectrum without Γ-action, $\underset{\sim}{K}^\Gamma(X; R)$ is equivalent to the original Pedersen–Weibel construction $\underset{\sim}{K}(X; R)$. In general, its fixed point set is difficult to describe. However, suppose X is a compact Riemannian manifold; with $\pi_1(X) = \Gamma$. Then the universal cover of X becomes a Riemannian manifold with free, isometric Γ–action, and the Γ–fixed point set of $\underset{\sim}{K}^\Gamma(X, R)$ is equivalent to the K–theory spectrum $\underset{\sim}{K}(R[\Gamma])$. We have thus achieved the construction of a spectrum with Γ–action with $\underset{\sim}{K}(R[\Gamma])$ as fixed point set, and hence obtain a map

$$\varepsilon^\Gamma : \underset{\sim}{K}(R[\Gamma]) \to \underset{\sim}{K}^\Gamma(X, R)^{h\Gamma}.$$

An elementary (but not as convenient for our purpose as the one given in the paper) Γ–homotopy equivalent version of the $\underset{\sim}{K}^\Gamma(X; R)$–construction can be described as follows. Let Γ be a finitely generated group, and Ω a finite generating set for Γ. For any $\gamma \in \Gamma$, let

$$\ell(\gamma) = \min\{n \mid \text{there is a word } w_1^{\pm 1} \dots w_n^{\pm 1} \text{ which is equal to } \alpha\}.$$

For any $\gamma \in \Gamma$ and $\ell \geq 0$, set $N_\ell(\gamma) = \operatorname{span}\{\bar{\gamma} \mid \ell(\gamma^{-1}\bar{\gamma}) \leq \ell\}$. Define $M^b(A) \subseteq \operatorname{Hom}(A[\Gamma], A[\Gamma])$ to be the subgroup of all $f : A[\Gamma] \to A[\Gamma]$ so that there exists some ℓ so that $f(\gamma) \in N_\ell(\gamma)$ for all $\gamma \in \Gamma$; $M^b(A)$ is of course a subring, closed under the Γ–action and it contains $A[\Gamma]$. One can show that when $\Gamma = \pi_1(X)$ as above, $\underset{\sim}{K}^\Gamma(X, R)$ is equivariantly equivalent to $\underset{\sim}{K} M^b(A)$ with the Γ action induced from the one on $M^b(A)$.

Having obtained our detecting map ε^Γ, we must find some way to evaluate it on the assembly map. We choose to do this by realizing the assembly map as the induced map on fixed point sets of an equivariant map with target $\underset{\sim}{K}^\Gamma(X;R)$. We define, for any spectrum A, a functor ${}^b\underset{\sim}{h}^{\ell f}(-,A)$ defined on an appropriate category of metric spaces, and a natural transformation ${}^b\underset{\sim}{h}^{\ell f}(-,\underset{\sim}{K}(R)) \to \underset{\sim}{K}(-,R)$. When $\underset{\sim}{A}$ is an Eilenberg–MacLane spectrum ${}^b\underset{\sim}{h}^{\ell f}(-,A)$ is closely related to Borel–Moore homology [7]. When X is equipped with an isometric Γ–action ${}^b\underset{\sim}{h}^{\ell f}(-,\underset{\sim}{K}(R))$ becomes a spectrum with Γ–action, and we obtain an equivariant natural transformation ${}^b\underset{\sim}{h}^{\ell f}(-,\underset{\sim}{K}(R)) \to \underset{\sim}{K}^\Gamma(-,R)$. The functor ${}^b\underset{\sim}{h}^{\ell f}(-,\underset{\sim}{K}(R))$ on a large (including all cases of interest to us) category of metric spaces with Γ–action has two crucial properties. The first is that when the Γ–action is free and properly discontinuous, ${}^b\underset{\sim}{h}^{\ell f}(X,\underset{\sim}{K}(R))^\Gamma \cong {}^b\underset{\sim}{h}^{\ell f}(\Gamma\backslash X,\underset{\sim}{K}(R))$. In particular, when $\Gamma\backslash X$ is compact, ${}^b\underset{\sim}{h}^{\ell f}(\Gamma\backslash X,\underset{\sim}{K}(R))$ reduces to ordinary homology of $\Gamma\backslash X$ with coefficients in $\underset{\sim}{K}(R)$, i.e., $(\Gamma\backslash X)_+ \wedge \underset{\sim}{K}(R)$. The second crucial property is that the natural map $\varepsilon^\Gamma : {}^b\underset{\sim}{h}^{\ell f}(X,\underset{\sim}{K}(R))^\Gamma \to {}^b\underset{\sim}{h}^{\ell f}(X,\underset{\sim}{K}(R))^{h\Gamma}$ is an equivalence. Now suppose that X is a compact Riemannian manifold, $\pi_1(X)=\Gamma$, and that $\tilde{X}$ is contractible. Then the natural transformation ${}^b\underset{\sim}{h}^{\ell f}(-,\underset{\sim}{K}(R)) \to \underset{\sim}{K}^\Gamma(-,R)$, by restricting to fixed point sets, gives rise to a map $X_+ \wedge \underset{\sim}{K}(R) \to \underset{\sim}{K}(R[\Gamma])$. Since $\tilde{X}$ is contractible, X is a model for $B\Gamma$, and this map can be identified with the assembly map. We now have the commutative diagram

$$\begin{array}{ccc}
B\Gamma_+ \wedge \underset{\sim}{K}(R) & \longrightarrow & \underset{\sim}{K}(R[\Gamma]) \\
\wr\downarrow & & \downarrow \\
{}^b\underset{\sim}{h}^{\ell f}(\tilde{X},\underset{\sim}{K}(R))^\Gamma & \longrightarrow & \underset{\sim}{K}^\Gamma(\tilde{X},R)^\Gamma \\
\varepsilon^\Gamma\downarrow & & \downarrow \\
{}^b\underset{\sim}{h}^{\ell f}(\tilde{X},\underset{\sim}{K}(R))^{h\Gamma} & \longrightarrow & \underset{\sim}{K}^\Gamma(\tilde{X},R)^{h\Gamma}
\end{array}$$

The lower left hand vertical arrow is an equivalence by the second of the properties of ${}^b\underset{\sim}{h}^{\ell f}(-,\underset{\sim}{K}(R))$. We recall that if $X \xrightarrow{f} Y$ is an equivariant map, which is an equivalence without reference to the Γ–action, then $X^{h\Gamma} \to Y^{h\Gamma}$ is an equivalence; this is a general property of homotopy fixed point sets. Therefore, if we show that the map ${}^b\underset{\sim}{h}^{\ell f}(-,\underset{\sim}{K}(R)) \to \underset{\sim}{K}(\tilde{X},R)$ is an equivalence, the lowest horizontal arrow above will also be an equivalence.

Therefore, so is the composite

$$B\Gamma_+ \wedge \underset{\sim}{K}(R) \to \underset{\sim}{K}(R[\Gamma]) \to \underset{\sim}{K}^{\Gamma}(\tilde{X}, R)^{h\Gamma},$$

and the assembly map can be identified up to homotopy with the inclusion of a wedge summand of spectra. We now state our two main theorems. Let $\underset{\sim}{\mathcal{K}}(R)$ denote the non-connective Gersten–Wagoner spectrum of R; its homotopy groups agree with those of $\underset{\sim}{K}(R)$ in nonnegative dimensions, and in all dimensions if R is regular (see [22] or [46]). In III.20 a natural transformation ${}^b\underset{\sim}{h}^{\ell f}(-, \underset{\sim}{\mathcal{K}}(R)) \to \underset{\sim}{\mathcal{K}}(X, R)$ is constructed where $\underset{\sim}{\mathcal{K}}(X, R)$ denotes a non-connective version of $\underset{\sim}{K}(X, R)$; it also agrees with $\underset{\sim}{K}(X, R)$ in non-negative dimensions.

Theorem A. *Suppose X is a Riemannian manifold with $\pi_1(X) \cong \Gamma$, and $\tilde{X}$ is contractible. Equip $\tilde{X}$ with the Riemannian metric induced from X. If the map ${}^b\underset{\sim}{h}^{\ell f}(\tilde{X}, \underset{\sim}{\mathcal{K}}(R)) \to \underset{\sim}{\mathcal{K}}(\tilde{X}, R)$ is an equivalence, then the assembly map can be identified up to homotopy with the inclusion of a wedge summand of the spectrum $\underset{\sim}{\mathcal{K}}(R[\Gamma])$.*

The condition that ${}^b\underset{\sim}{h}^{\ell f}(\tilde{X}, \underset{\sim}{\mathcal{K}}(R)) \to \underset{\sim}{\mathcal{K}}(\tilde{X}, R)$ is an equivalence depends only on the behavior of the metric on $\tilde{X}$ "in the large." After some contemplation, one can see that it only depends on the structure (again in the large) of the metric space whose points are elements of Γ and where the metric is the one associated to a length function for some finite generating set for Γ. It does not depend on the fine algebraic structure of Γ; for instance, all torsion–free cocompact subgroups of the groups of isometries of Euclidean k–space give the same result.

A second result that we prove in this paper is that if G is a connected Lie group, K is a maximal compact subgroup, and G/K is equipped with a left–invariant Riemannian metric, then ${}^b\underset{\sim}{h}^{\ell f}(G/K, \underset{\sim}{\mathcal{K}}(R)) \to \underset{\sim}{\mathcal{K}}(G/K, R)$ is an equivalence. Since $\Gamma\backslash G/K$ is a Riemannian manifold, with G/K contractible, the following now is a consequence of Theorem A.

Theorem B. *Suppose Γ is a discrete, cocompact, torsion–free subgroup of a connected Lie group G. Then the assembly map*

$$B\Gamma_+ \wedge \underset{\sim}{\mathcal{K}}(R) \to \underset{\sim}{\mathcal{K}}(R[\Gamma])$$

may be identified up to homotopy with the inclusion of a wedge summand of spectra.

Our methods are more homotopy theoretic than previous work in the area, with the exception of Waldhausen's work [47]. We feel that there are three important advantages. The first is that our methods produce integral results, i.e., not results after rationalizing. This is crucial for future geometric applications. The second is that they are applicable to situations where the group Γ acts freely not on a Riemannian manifold, but on some other space with metric. An interesting example of this is the case of cocompact, discrete, torsion–free subgroups Γ of reductive p–groups G, which act on the so–called Bruhat–Tits buildings [8]. These buildings carry metrics but are definitely not manifolds; P. Mostad in his thesis [32] has applied the methods of this paper to them to conclude in the case $G = SL_n(\hat{\mathbb{Q}}_p)$ that the assembly map can be identified up to homotopy with the inclusion of a wedge summand in this case as well. The third advantage is that via the work of Ranicki [39], the Pedersen–Weibel theory has an L–theoretic analogue, and one would expect that the methods of this paper and the sequels should extend to give results about the L–theory analogue of the assembly map. This map plays a key role in attempts to prove the "Borel conjecture" that any two closed compact $K(\Gamma, 1)$–manifolds are homeomorphic.

In a later paper, we plan to study the surjectivity of the assembly map on homotopy groups; in cases where it has been shown to be an injection, this shows it to be an equivalence. We also plan to develop a "simplicial" version of the theory, in order to obtain a more homotopic theoretic method to study the assembly map ${}^b\underset{\sim}{h}^{\ell f}(\tilde{X}, \underset{\sim}{K}(R)) \to \underset{\sim}{K}(\tilde{X}, R)$, less dependent on strong geometric hypotheses on X.

The paper is arranged as follows. § I contains homotopy theoretic material required later in the paper. It need not be read carefully by the expert, although some nonstandard terminology is introduced. § II gives the construction of ${}^b\underset{\sim}{h}^{\ell f}$ and related analogues to Borel–Moore homology. We prove excision results and discuss equivariant properties; the latter should have independent interest for the study of classifying spaces of infinite groups. § III defines bounded K–theory, develops some of its elementary properties, and constructs the natural transformation ${}^b\underset{\sim}{h}^{\ell f}(-, \underset{\sim}{K}(R)) \to \underset{\sim}{K}(-, R)$. § IV proves some useful excision properties of the bounded K–theory spectra. Specifically, we prove an excision theorem for finite coverings and for certain families of infinite coverings, whose nerves have dimension 1. § V then applies the results of § IV to prove that the transformation ${}^b\underset{\sim}{h}^{\ell f}(G/K, \underset{\sim}{K}(R)) \to \underset{\sim}{K}(G/K, R)$ is an equivalence, where G is a connected Lie group, $K \subseteq G$ is a maximal compact, and G/K is equipped with a left invariant Riemannian metric. § VI defines the equivariant bounded K–theory and proves the main theorems.

The author wishes to thank a number of mathematicians for helpful discussions on various aspects of this paper. The list includes W. Dwyer, T. Farrell, W.C. Hsiang, L. Jones, M. Karoubi, I. Madsen, A. Nicas, E. Pedersen, F. Waldhausen, and C. Weibel. This paper was written during 1991 while the author was at Princeton University and visiting the University of Minnesota.

I. Homotopy Theoretic Preliminaries

We refer the reader to [11] or [28] for background material on simplicial homotopy theory. Throughout this paper, the word "space" will mean simplicial set, and "map" will mean simplicial map. A map $f : X. \to Y.$ is said to be a *weak equivalence* if the induced map on geometric realization is a weak equivalence. A *simplicial space* will thus mean a bisimplicial set, and we use the notation $|\ |$ for the functor which assigns to a simplicial space its diagonal simplicial set. Recall that a map of simplicial spaces $f : X. \to Y.$ is a weak equivalence if $f_k : X_k \to Y_k$ is a weak equivalence for all k. Similarly, recall that a cosimplicial space is a covariant functor from $\underline{\Delta}$ to spaces. We assume that the reader is familiar with the notation "Kan complex", "Kan fibration", and "fibrant cosimplicial space". A map of cosimplicial spaces is said to be a weak equivalence if it is an equivalence in each level. Spaces, maps, and cosimplicial spaces can be converted in a functorial way to Kan complexes, Kan fibrations, and fibrant cosimplicial spaces. Recall the notion of total space $\mathrm{Tot}(X.)$ of a cosimplicial space, and that if $f : X. \to Y.$ is a weak equivalence of fibrant cosimplicial spaces, then $\mathrm{Tot}(f)$ is a weak equivalence. Let $X. \xrightarrow{f} Y. \xrightarrow{g} Z.$ be a diagram of based spaces, and suppose that gf is constant. Let $\hat{g} : \hat{Y}. \to \hat{Z}.$ denote the result of the functorial conversion of g to a Kan fibration. There is a canonically induced map $\hat{f}$ from $X.$ into the inverse image under $\hat{g}$ of the basepoint $\hat{Z}.$, and we say that the sequence $X. \xrightarrow{f} Y. \xrightarrow{g} Z.$ is a fibration up to homotopy if $\hat{f}$ is a weak equivalence. Finally, if $\underline{C}$ is any category, we let $\underline{S.C}$ denote the category of simplicial objects in $\underline{C}$. For $\underline{C}$ any category, we let $N.\underline{C}$ denote the nerve of $\underline{C}$, the simplicial set whose k–simplices are given by diagrams

$$X_0 \xrightarrow{f_0} X_1 \xrightarrow{f_1} X_2 \xrightarrow{f_2} \cdots \xrightarrow{f_{k-1}} X_k$$

in $\underline{C}$, and whose face and degeneracy maps are given by compositions and insertions of the identity as in [35]. We recall from [35] that a functor $F : \underline{C} \to \underline{D}$ induces a map $N.F : N.\underline{C} \to N.\underline{C}$, and that if $F, G : \underline{C} \to \underline{D}$ are functors and $N : F \to G$ is a natural transformation, then N induces a simplicial homotopy from $N.F$ to $N.G$. We permit ourselves the abuse of notation $x \in \underline{C}$ for "x is an object of $\underline{C}$." We say a functor F is a weak equivalence if $N.F$ is, and that a category is *contractible* if $N.\underline{C}$ is weakly equivalent to a point. For any functor $F : \underline{C} \to \underline{C}$ and object $x \in \underline{D}$, we let $F \downarrow x$ denote the category whose objects are pairs (ξ, θ), where ξ is an object of $\underline{C}$ and $\theta : F(\xi) \to x$ is a morphism in $\underline{D}$, which we call the reference map, and where a morphism from (ξ, θ) to (ξ', θ') is given by a morphism $\varphi : \xi \to \xi'$ in $\underline{C}$ making the diagram

$$\begin{array}{ccc} \xi & \xrightarrow{F(\phi)} & \xi' \\ & \searrow^{\theta} \quad \swarrow^{\theta'} & \\ & x & \end{array}$$

commute. The category $x \downarrow F$ is defined in a dual fashion, as in [35]. We recall that Quillen's Theorem A [35] asserts that F is a weak equivalence if for all $x \in \underline{D}$, $F \downarrow x$ is a contractible category. Similarly, if $x \downarrow F$ is contractible for all x, F is a weak equivalence.

We will be dealing with spectra throughout this paper. We will adopt the definitions of [10], i.e., that a spectrum $\underset{\sim}{X}$ is a family of based spaces $\{X_i\}_{i=0}$ with maps $\sigma_i : S^1 \wedge X_i \to X_{i+1}$, where S^1 denotes the simplicial circle. A map of spectra $f : X \to Y$ is a family of maps $f_i : X_i \to Y_i$, making the diagrams

$$\begin{array}{ccc} S^1 \wedge X_i & \xrightarrow{\mathrm{Id} \wedge f_i} & S^1 \wedge Y_i \\ \sigma_i \downarrow & & \downarrow \sigma_i \\ X_{i+1} & \xrightarrow{f_{i+1}} & Y_{i+1} \end{array}$$

commute. Note that if we applied geometric realization, we would obtain what are usually called prespectra and maps of prespectra. For a simplicial set, let $\|X.\|$ denote its geometric realization, a topological space. If $\underset{\sim}{X}$ is a spectrum, we obtain continuous maps $\|\sigma_i\| : \|S^1\| \wedge \|X_i\| \to \|X_{i+1}\|$, and hence suspension homomorphisms $\pi_{k+1}(\|X_i\|) \to \pi_{k+i+1}(\|X_{i+1}\|)$. We define $\pi_k(\underset{\sim}{X})$ to be $\varinjlim_i \pi_{k+i}(\|X_i\|)$, and note that π_k is a functor from the category of spectra $\underline{S}$ to $\underline{Ab}$. We say a map $f : \underset{\sim}{X} \to \underset{\sim}{Y}$ of spectra is a weak equivalence if $\pi_k(f)$ is an equivalence for all k.

A spectrum $\underset{\sim}{X}$ is said to be an *Ω-spectrum* if for each i, the map $\|X_i\| \to \Omega\|X_{i+1}\|$ adjoint to the realization of $\|\sigma_i\|$ is a weak equivalence of topological spaces. Note that for an Ω-spectrum $\underset{\sim}{X}$, $\pi_k(\underset{\sim}{X}) \cong \pi_{k+i}(\|X_i\|)$, whenever the right hand side is defined. Let $\underline{\omega S} \subset \underline{S}$ denote the full subcategory of Ω-spectra. Then every $\underset{\sim}{X} \in \underline{S}$ is weakly equivalent to an Ω-spectrum. Indeed, there is a functor $Q : \underline{S} \to \underline{\omega S}$ and a natural transformation $\mathrm{Id} \to Q$ which is a weak equivalence for all $\underset{\sim}{X} \in \underline{S}$. Note that a map $f : \underset{\sim}{X} \to \underset{\sim}{Y}$ between Ω-spectra $\underset{\sim}{X}$ and $\underset{\sim}{Y}$ is a weak equivalence if and only if all the homomorphisms $\pi_i(\|f_k\|) : \pi_i(\|X_k\|) \to \pi_i(\|\tilde{Y}_k\|)$ are isomorphisms.

We say a map of spectra $f : \underset{\sim}{X} \to \underset{\sim}{Y}$ is a cofibration if each f_i is. In this case, we can form the cofibre spectrum $\underset{\sim}{Y \backslash X}$ by setting $(Y/X)_i = Y_i/X_i$. Note that if $\underset{\sim}{X}$ and $\underset{\sim}{Y}$ are Ω–spectra, $\underset{\sim}{Y \backslash X}$ is generally not. We say a spectrum $\underset{\sim}{X}$ is *Kan* if each X_i is a Kan complex, and that a map of spectra $f : \underset{\sim}{X} \to \underset{\sim}{Y}$ is a *Kan fibration* if each f_i is a Kan fibration. Spectra and maps of spectra can be replaced by Kan spectra and Kan fibrations of spectra in a functorial way. Suppose we have a sequence $\underset{\sim}{X} \xrightarrow{f} \underset{\sim}{Y} \xrightarrow{g} \underset{\sim}{Z}$ of maps of spectra, and that $g \cdot f$ is the constant map of the spectra. Then the sequence $\underset{\sim}{X} \xrightarrow{f} \underset{\sim}{Y} \xrightarrow{g} \underset{\sim}{Z}$ is said to be a *fibration up to homotopy* if each $X_k \xrightarrow{f_k} Y_k \xrightarrow{g_k} Z_k$ is. Cofibrations and fibrations up to homotopy both induce long exact sequences on homotopy groups of spectra.

Now recall from [11] the definitions of homotopy colimits and homotopy limits of functions from a category $\underline{C}$ to *s–sets*. The homotopy colimit is the diagonal space of a certain simplicial space, and the homotopy limit is the total space of a certain cosimplicial space, which is fibrant if the value of the functor is a Kan complex for every object of $\underline{C}$. We adopt the based versions of both constructions.

We summarize the relevant properties of these constructions.

Proposition 1.1.

(a) *Let* $\Phi : \underline{C} \to \underline{D}$ *and* $F : \underline{D} \to \underline{s\text{–}sets}$ *be functors. Then we obtain maps* $\Phi_! : \underset{\overrightarrow{\underline{C}}}{\operatorname{hocolim}} F \circ \Phi \to \underset{\overrightarrow{\underline{D}}}{\operatorname{hocolim}} F$ *and* $\Phi^! : \underset{\overleftarrow{\underline{D}}}{\operatorname{holim}} F \to \underset{\overleftarrow{\underline{C}}}{\operatorname{holim}} F \circ \Phi$. *Further,* $\Psi_! \cdot \Phi_! = (\Psi \cdot \Phi)_!$ *and* $\Phi^! \cdot \Psi^! = (\Phi \cdot \Psi)^!$.

(b) *Let* $F, G : \underline{C} \to \underline{s\text{–}sets}$ *be functors and let* $N : F \to G$ *be a natural transformation. Then* N *induces maps* $\underline{N} : \underset{\overrightarrow{\underline{C}}}{\operatorname{hocolim}} F \to \underset{\overrightarrow{\underline{C}}}{\operatorname{hocolim}} G$ *and* $\bar{N} : \underset{\overleftarrow{\underline{C}}}{\operatorname{holim}} F \to \underset{\overleftarrow{\underline{C}}}{\operatorname{holim}} G$. $\overline{N_1}\,\overline{N_2} = \overline{N_1 N_2}$ *and* $\underline{N_1}\,\underline{N_2} = \underline{N_1 N_2}$.

(c) *Suppose* $F, G : \underline{C} \to \underline{s\text{–}sets}$ *are functors, and* $N : F \to G$ *is a natural transformation, so that* $N(x) : F(x) \to G(x)$ *is a weak equivalence for all* $x \in \underline{C}$. *Then* $\underline{N}$ *is a weak equivalence. If* $F(x)$ *and* $G(x)$ *are Kan for all* $x \in \underline{C}$ *then* $\bar{N}$ *is a weak equivalence.*

(d) *There are natural maps*

$$\underset{\overrightarrow{\underline{C}}}{\operatorname{hocolim}} F \to \underset{\overrightarrow{\underline{C}}}{\operatorname{colim}} F$$

and

$$\lim_{\overleftarrow{\underline{C}}} F \to \operatorname{holim}_{\overleftarrow{\underline{C}}} F.$$

(e) *Suppose $N : F \to G$ is a natural transformation of functors from $\underline{C}$ to $\underline{s\text{-sets}}$ and that $N(x)$ is a cofibration for all $x \in \underline{C}$. Then $\underline{N}$ is a cofibration. Similarly, if $N_1 : F \to G$ and $N_2 : G \to H$ are natural transformations of functors from $\underline{C}$ to $\underline{s\text{-sets}}$, $F(x), G(x)$, and $H(x)$ are Kan for all $x \in \underline{C}$, so that for each $x \in \underline{C}$, $F(x) \to G(x) \to H(x)$ is a fibration up to homotopy of spaces, then the sequence*

$$\operatorname{holim}_{\overleftarrow{\underline{C}}} F \xrightarrow{\tilde{N}_1} \operatorname{holim}_{\overleftarrow{\underline{C}}} G \xrightarrow{\tilde{N}_2} \operatorname{holim}_{\overleftarrow{\underline{C}}} H$$

is a fibration up to homotopy.

(f) *Let $X.$ be a simplicial space, i.e., a functor $F : \underline{\Delta}^{\mathrm{op}} \to \underline{s\text{-sets}}$. Then $\operatorname{hocolim}_{\overrightarrow{\underline{\Delta}^{\mathrm{op}}}} F$ is naturally equivalent to $|X|$.*

(g) *(See [14].) Let $\Phi : \underline{C} \to \underline{D}$ and $F : \underline{D} \to \underline{s\text{-sets}}$ be functors. if the categories $x \downarrow \Phi$ are contractible for all $x \in \underline{D}$, then $\Phi_!$ is an equivalence. If the categories $\Phi \downarrow x$ are contractible for all $x \in \underline{D}$, and $F(x)$ is a Kan complex for all $x \in \underline{D}$, then $\Phi^!$ is an equivalence.*

Now suppose $F : \underline{C} \to \underline{S}$ is a functor. For each k, let $F_k : \underline{C} \to \underline{s\text{-sets}}$ denote the "k–th space" functor of F. The maps σ_i induce maps $\bar{\sigma}_i : S^1 \wedge \operatorname{hocolim}_{\overrightarrow{\underline{C}}} F_k \to \operatorname{hocolim}_{\overrightarrow{\underline{C}}} F_{k+1}$, and therefore defines a spectrum which we write as $\operatorname{hocolim}_{\overrightarrow{\underline{C}}} F$. Similarly, we obtain maps $S^1 \wedge \operatorname{holim}_{\overleftarrow{\underline{C}}} F_k \to \operatorname{holim}_{\overleftarrow{\underline{C}}} F_{k+1}$, and hence a homotopy inverse limit spectrum which we denote $\operatorname{holim}_{\overleftarrow{\underline{C}}} F$. $\operatorname{hocolim}_{\overrightarrow{\underline{C}}} F$ is generally not Kan, even if each spectrum $F(x)$ is. It also is not an Ω–spectrum, even if $F(x)$ is for all $x \in \underline{C}$. However, if $F(x)$ is Kan and is an Ω–spectrum for each $x \in \underline{C}$, then $\operatorname{holim}_{\overleftarrow{\underline{C}}} F$ is a Kan Ω–spectrum. We now record the analogue to I.1 for spectra. If $\underset{\sim}{X}.$ is a simplicial object in $\underline{S}.$, we let $|\underset{\sim}{X}.|$ denote the spectrum obtained by applying $|\ |$ termwise.

Proposition I.2. *I.1 holds as stated for functors with values in $\underline{S}$.*

We will refer to a functor $F : \underline{C} \to \underline{S}$ as a (spectrum valued) $\underline{C}$–diagram. A natural transformation $N \colon F \to G$ of functors is referred to as a map of $\underline{C}$–diagrams, and is called a weak–equivalence if $N(x)$ is a weak equivalence of spectra for all $x \in \underline{C}$. A *homotopy natural transformation* from a $\underline{C}$–diagram G is a family of $\underline{C}$–diagrams $\{H_i\}_{i=0}^{n}$ with $H_0 = F$ and $H_n = G$, together with natural transformations $\theta_i : H_i \to H_{i-1}$ and $\eta_i : H_i \to H_{i+1}$ for all $1 \le i \le n-1$ so that each θ_i is a weak equivalence. Homotopy natural transformations may be composed in the evident way, and we say that the homotopy natural transformation is a *homotopy natural equivalence* if in addition each η_i is a weak equivalence. Homotopy natural transformations are just morphisms in in the homotopy category obtained from the category of $\underline{C}$–diagrams by inverting the weak equivalences. We say two $\underline{C}$–diagrams are weakly equivalent if there is a homotopy weak equivalence connecting them. Weakly equivalent diagrams yield weakly equivalent homotopy colimits. If $F(x)$ and $G(x)$ are Kan Ω–spectra for all $x \in \underline{C}$, and are weakly equivalent, then $\underset{\underline{C}}{\underleftarrow{\operatorname{holim}}}\, F$ and $\underset{\underline{C}}{\underleftarrow{\operatorname{holim}}}\, G$ are weakly equivalent spaces. Let $K^\infty : \underline{S} \to \underline{S}$ denote the functorial replacement of $\underset{\sim}{X}$ by a weakly equivalent Kan spectrum. Then if F is any $\underline{C}$–diagram, $K^\infty QF$ is a weakly equivalent diagram to F, and $K^\infty QF(x)$ is a Kan Ω–spectrum for all $x \in \underline{C}$. We will frequently have diagrams F whose values are not Kan Ω–spectra, and we will also wish to consider $\underset{\underline{C}}{\underleftarrow{\operatorname{holim}}}\, K^\infty QF$ instead of $\underset{\underline{C}}{\underleftarrow{\operatorname{holim}}}\, F$, since $\underset{\underline{C}}{\underleftarrow{\operatorname{holim}}}\, F$ does not have good homotopy invariance properties. For this reason, it will be understood in the remainder of this paper that if $F : \underline{C} \to \underline{S}$ is a diagram whose values are not Kan Ω–spectra, then the notation $\underset{\underline{C}}{\underleftarrow{\operatorname{holim}}}\, F$ will mean $\underset{\underline{C}}{\underleftarrow{\operatorname{holim}}}\, K^\infty QF$. We allow ourselves this abuse of notation in the interest of streamlining the notation; it should not create any real confusion.

We wish to introduce a tool for analyzing the homotopy type of homotopy colimits and limits. Let $\underline{C}$ be any category and let $\underline{\mathcal{F}}(\underline{C}, \underline{\mathrm{Ab}})$ denote the category of Abelian group valued functors on $\underline{C}$. $\underline{\mathcal{F}}(\underline{C}, \underline{\mathrm{Ab}})$ has enough injectives and projectives, and the functors $\underset{\underline{C}}{\underleftarrow{\lim}}$ and $\underset{\underline{C}}{\underrightarrow{\operatorname{colim}}}$ are left and right exact, respectively. Consequently, one may define right derived functions $\underleftarrow{\lim}^s$ and left derived functors $\underrightarrow{\operatorname{colim}}^s$.

Proposition I.3. *Let $F : \underline{C} \to \underline{S}$ be any functor, and let $\pi_i F$ denote the Abelian group valued functor obtained by composing F with π_i. Then there is a spectral sequence with $E^2_{p \cdot q} = \underset{\underline{C}}{\underrightarrow{\operatorname{colim}}}^p \pi_q F$, converging to $\pi_*(\underset{\underline{C}}{\underrightarrow{\operatorname{hocolim}}}\, F)$.*

Similarly, there is a spectral sequence with $E_2^{p,q} = \varprojlim^p_{\underline{C}} \pi_q F$, *converging to* $\pi_*(\underset{\overleftarrow{\underline{C}}}{\text{holim}}\, F)$.

Proof: This follows directly by applying homology and homotopy spectral sequences to the simplicial and cosimplicial spaces defining $\underset{\overrightarrow{\underline{C}}}{\text{hocolim}}\, F$ and $\underset{\overleftarrow{\underline{C}}}{\text{holim}}\, F$. Q.E.D.

We wish to discuss the functoriality of the $\underset{\overrightarrow{\underline{C}}}{\text{hocolim}}$ and $\underset{\overleftarrow{\underline{C}}}{\text{holim}}$ constructions. Let $\underline{\mathcal{E}}$ be any subcategory of the category of small categories, and let $\underline{C}$ denote any category. We define a category $\underline{C}^{\mathcal{E}}$ as follows. The objects of $\underline{C}^{\mathcal{E}}$ are pairs $(\underline{E}, \varphi)$, where $\underline{E} \in \underline{\mathcal{E}}$, and $\varphi : \underline{E} \to \underline{C}$ is a functor. If $(\underline{E}, \varphi)$ and $(\underline{E}', \varphi')$ are objects in $\underline{C}^{\mathcal{E}}$, then a morphism from $(\underline{E}, \varphi)$ to $(\underline{E}', \varphi')$ is a pair (F, ν), where $F : \underline{E} \to \underline{E}'$ is a functor and ν is a natural transformation from φ to $\varphi' \cdot F$. These morphisms are composed in the evident way. Similarly, $\underline{C}_0^{\mathcal{E}}$ has the same objects as $\underline{C}^{\mathcal{E}}$, but a morphism from $(\underline{E}, \varphi)$ to $(\underline{E}', \varphi')$ in $\underline{C}_0^{\mathcal{E}}$ is a pair (F, ν), where $F : \underline{E}' \to \underline{E}$ is a functor, and $\nu : \varphi' \cdot F \to \varphi$ is a natural transformation.

Proposition I.4. *Let* $\underline{CAT}$ *denote the category of small categories. Then* $\overrightarrow{\text{hocolim}}$ *and* $\overleftarrow{\text{holim}}$ *define functors from* $\underline{S}^{\varepsilon}$ *and* $\underline{S}_0^{\varepsilon}$, *respectively, to* $\underline{S}$.

Proof: Follows directly from I.2 (a) and (b). Q.E.D.

Let $\underline{C}$ be a category, and let $F : \underline{C} \to \underline{CAT}$ be a functor. Then as in [42], we define a category $\underline{C} \wr F$ as follows. The objects of $\underline{C} \wr F$ are pairs (x, ξ), where $x \in \underline{C}$ and $\xi \in F(x)$. A morphism from (x, ξ) to (x', ξ') is a pair (f, θ), where $f : x \to x'$ is a morphism in $\underline{C}$ and θ is a morphism in $F(x')$ from $F(f)(\xi)$ to ξ'. Once again, the composition is given by an obvious formula. Let $\underline{A}$ be any category and let $\pi : \underline{A}^{CAT} \to \underline{CAT}$ be the functor defined by $\pi(\underline{E}, \varphi) = \underline{E}$, $\pi(F, \nu) = F$. If $\Phi : \underline{C} \to \underline{A}^{CAT}$ is any functor, we obtain an associated functor $\Phi^v : \underline{C} \wr (\pi \cdot \Phi) \to \underline{A}$ by $\Phi^v(x, \xi) = \phi_x(\xi)$, where $\Phi(x) = (\underline{E}_x, \varphi_x)$. This sets up a natural bijection between functors $\underline{C} \xrightarrow{\Phi} \underline{A}^{CAT}$ with $\pi \cdot \Phi = F$ and functors from $\underline{C} \wr F$ to $\underline{A}$.

Proposition I.5. *Let* $\Phi : \underline{C} \to \underline{S}^{CAT}$ *(or* $\Phi : \underline{C} \to \underline{s\text{–}sets}^{CAT}$*) be a functor. Then there is a natural equivalence*

$$\underset{\overleftarrow{\underline{C} \wr \pi \cdot \Phi}}{\text{holim}}\, \Phi^v \to \underset{\overleftarrow{\underline{C}}}{\text{holim}}\, (\overleftarrow{\text{holim}} \cdot \Phi)\,.$$

Proof: This is proved in [42] for the case where Φ^v is constant, in the case of s–sets. The extension to this generality is direct and we leave it to the reader. Q.E.D.

We will also need symmetric monoidal category-theoretic models for spectrum homotopy colimits. We recall from [43] the notion of a symmetric monoidal category; it is a category equipped with sum operations and associativity and commutativity isomorphisms satisfying certain coherence conditions; we refer to [15] for the explicit definitions. A symmetric monoidal category is *permutative* if the associativity isomorphism is actually an identity. We adopt Thomason's terminology "lax symmetric monoidal functors," "strict symmetric monoidal functors," and use "symmetric monoidal functor" for Thomason's "strong symmetric monoidal functor." SymMon will denote the category of small symmetric monoidal categories and lax symmetric monoidal functors. We will also use the term *unital* in the same sense as Thomason does. Thomason constructs a functor $Spt : \underline{\text{SymMon}} \to \underline{S}$, whose properties we summarize below. For a unital symmetric monoidal category $\underline{C}$, let $\pi_0(\underline{C})$ denote $\pi_0(N\underline{C})$. $\pi_0(\underline{C})$ is a commutative monoid, and can be identified with the equivalence classes of objects of $\underline{C}$ under the equivalence relation generated by the relation $\{x \sim y \text{ if and only if } \exists f : x \to y \text{ in } \underline{C}\}$. A submonoid $M \subset \pi_0(\underline{C})$ is said to be *cofinal* if for all $x \in \pi_0(\underline{C})$, there is a $y \in \pi_0(\underline{C})$ so that $x + y \in M$. For any $M \subset \pi(\underline{C})$, let $\underline{C}[M]$ denote the symmetric monoidal subcategory of $\underline{C}$ on objects which lie in an equivalence class belonging to M. We also recall from [29] the notion of a *symmetric monoidal pairing* $\underline{A} \times \underline{B} \to \underline{C}$. The following theorem summarizes the information we need concerning infinite loop space machines; parts (a)–(c) are in Thomason [43], and part (d) is contained in [29].

Theorem I.6. *There is a functor* Spt : $\underline{\text{SymMon}} \to \underline{S}$ *satisfying the following conditions.*

(a) *If $f : \underline{C} \to \underline{D}$ is a lax symmetric monoidal functor and $N.f$ is a weak equivalence of simplicial sets then* Spt(f) *is an equivalence of spectra.*

(b) *For any symmetric monoidal category $\underline{C}$, let* $\text{Spt}_0(\underline{C})$ *denote the zeroth space of* $\text{Spt}(\underline{C})$. *There is a natural map $N\underline{C} \to \text{Spt}_0(\underline{C})$, which induces an isomorphism*

$$(\pi_0 N\underline{C})^{-1} H_*(N\underline{C}) \xrightarrow{\sim} H_*(\text{Spt}_0(\underline{C})).$$

(c) *Let $f : \underline{C} \to \underline{D}$ be a unital symmetric monoidal functor between unital symmetric monoidal categories $\underline{C}$ and $\underline{D}$, and suppose $\pi_0(\underline{C})$ contains a cofinal submonoid M so that $\pi_0(f)(M)$ is also a cofinal submonoidal of $\pi_0(\underline{D})$. Suppose further that for every object $x \in \underline{D}$ lying in an* *equivalence class belonging to $\pi_0(f)(M)$, $x \downarrow f$ (or $f \downarrow x$) is a contractible category. The $\pi_i(\mathrm{Spt}(f))$ is an isomorphism for $i > 0$.*

(d) *If $\mu : \underline{A} \times \underline{B} \to \underline{C}$ is a symmetric monoidal pairing, then there is an induced map*

$$\mathrm{Spt}(\mu) : \mathrm{Spt}(\underline{A}) \wedge \mathrm{Spt}\,(\underline{B}) \to \mathrm{Spt}(\underline{C})$$

so that the composite

$$N.\underline{A} \times N.\underline{B} \to \mathrm{Spt}_0(\underline{A}) \wedge \mathrm{Spt}_0(\underline{B}) \to (\mathrm{Spt}\,\underline{A} \wedge \mathrm{Spt}\,\underline{B})_0 \to \mathrm{Spt}_0\,\underline{C}$$

is equal to the composite

$$N.\underline{A} \times N.\underline{B} \xrightarrow{N.\mu} N.\underline{C} \to \mathrm{Spt}_0(\underline{C}).$$

Proof: Parts (a) and (b) are conditions 2.1 and 2.2 of [43]. Part (c) follows directly from (a) and (b) directly together with Quillen's theorem A. Q.E.D.

We now wish to understand how homotopy colimits work in this setting. Thomason [43] accomplishes this as follows. Let $\underline{C}$ be a category, and let $F : \underline{C} \to \underline{\mathrm{SymMon}}$ be a functor. Thus $\mathrm{Spt} \cdot F$ is a functor from $\underline{C}$ to $\underline{S}$, and in [43], Thomason produces a symmetric monoidal category $\mathrm{Perm}\underset{\overrightarrow{\underline{C}}}{\mathrm{hocolim}}\, F$ and a natural equivalence of spectra

$$\mathrm{Spt}\,(\mathrm{Perm}\underset{\overrightarrow{\underline{C}}}{\mathrm{hocolim}}\, F) \to \underset{\overrightarrow{\underline{C}}}{\mathrm{hocolim}}(\mathrm{Spt} \cdot F).$$

The objects of $\mathrm{Perm}\underset{\overrightarrow{\underline{C}}}{\mathrm{hocolim}}\, F$ are of the form $n[(L_1, X_1), \ldots, (L_n, X_n)]$, where n is a positive integer, L_i is an object of $\underline{C}$, and X_i is an object of $F(L_i)$. A morphism

$$n[(L_1, X_1), \ldots, (L_n, X_n)] \to m[(L_1', X_1'), \ldots, (L_n', X_n')]$$

consists of data (ℓ_i, Ψ, x_j), where

(1) $\Psi : \{1, \ldots, n\} \to \{1, \ldots, m\}$ is a surjection of sets,

(2) $\ell_i : L_i \to L'_{\Psi(i)}$ is a morphism in $\underline{C}$ for $i = 1, \ldots, n$, and

(3) $x_j : \oplus_{\Psi(i)=j} F(\ell_i(x_j)) \to x_j'$ is a morphism in $F(L_j')$ for $j = 1, \ldots, m$. (The order of summation in $\oplus_{\Psi(i)=j} F(\ell_i)(X_j)$ is given by the ordering

$\{1, \ldots, m\}$, and the brackets are "piled up on the left.") The composition law for morphisms is defined in a straightforward way; see [43] for details. The sum in $\operatorname{Permhocolim}_{\overrightarrow{\underline{C}}} F$ is given on objects by

$$n[(L_1, X_1), \ldots, (L_n, X_n)] \bigoplus m[(L'_1, X'_1), \ldots, (L'_n, X'_n)]$$
$$= (n+m)[(L_1, X_1), \ldots, (L_n, X_n), (L'_1, X'_1), \ldots, (L'_n, X'_n)].$$

The definition on morphisms is easily worked out by the reader. For simplicity, we will use the notation $\underline{C} \wr F$ for $\operatorname{Permhocolim}_{\overrightarrow{\underline{C}}} F$.

Proposition I.7. *The rule* $(\underline{C}, F) \to \underline{C} \wr F$ *gives a functor from* $\underline{\text{SymMon}}^{CAT}$ *to* $\underline{\text{SymMon}}$*; i.e.,* $\underline{C} \wr F$ *has the same functoriality properties as the spectrum level homotopy colimit.*

Thomason proves the following universal mapping property for $\underline{C} \wr F$.

Proposition I.8 ([43], Proposition 3.1.) *Let* $F : \underline{C} \to \underline{\text{SymMon}}$ *be a functor, and let* $\underline{S}$ *be a permutative category. For each object* $x \in \underline{C}$*, let* $i_x : F(x) \to \underline{C} \wr F$ *be the functor given by* $i_x(\xi) = 1[(x, \xi)]$*, where* i_x *is symmetric monoidal. Moreover, for each morphism* $f : x \to y$ *in* $\underline{C}$*, we obtain a symmetric monoidal natural transformation* $N_f : i_x \to i_y \cdot F(f)$*, and the equation* $N_f \cdot N_g = N_{f \cdot g}$ *holds for all composable pairs of morphisms* f *and* g*. Given lax symmetric monoidal functors* $f_x : F(x) \to \underline{S}$ *for all* $x \in \underline{C}$*, and symmetric monoidal natural transformations* $M(g) : f_x \to f_y \cdot F(g)$ *for every morphism* $g : x \to y$ *in* $\underline{C}$ *satisfying* $M(g_1)\, M(g_2) = M(g_1 g_2)$*, then there is a unique strict symmetric monoidal functor* $\mu : \underline{C} \wr F \to \underline{S}$ *such that* $\mu \cdot i_x = f_x$ *and* $\mu \cdot N(g) = M(g)$.

The construction $\underline{C} \wr F$ is a bit unwieldy; fortunately, in two situations where we will require it, much more economical models are available.

Suppose first that $\underline{L}$ is the category of proper subsets of the set $\{0, 1\}$; $\underline{C}$ has three objects, $\emptyset, \{0\}$ and $\{1\}$, with unique morphisms $\emptyset \to \{0\}$ and $\emptyset \to \{1\}$. A functor from $\underline{L}$ into a category $\underline{C}$ is just a diagram of the form

$$y \leftarrow x \to z$$

in $\underline{C}$. In the case of spectra, the homotopy colimit of such a diagram is the double mapping cylinder of the diagram. Suppose we have a functor F from L to $\underline{\text{SymMon}}$, say $\underline{B} \xleftarrow{u} \underline{A} \xrightarrow{v} \underline{C}$, so that $\underline{A}, \underline{B}$, and $\underline{C}$ are unital and permutative monoidal functors. Then we define as in [43], the simplified double mapping cylinder of this diagram, $\operatorname{Cyl}(\underline{B} \leftarrow \underline{A} \to \underline{C})$, to be the following symmetric monoidal category. The objects are triples (b, a, c), with $b \in \underline{B}$, $a \in \underline{A}$, and $c \in \underline{C}$. A morphism $(b,a,c) \to (b',a',c')$ is an equivalence class of data, a datum consisting of objects U and V of $\underline{A}$, a morphism $\Psi : a \to U \oplus a' \oplus V$ and morphisms $\Psi_1 : b \oplus uU \to b'$ and $\Psi_2 : vV \oplus c \to c'$ in $\underline{B}$ and $\underline{C}$, respectively. If $\mu : U' \xrightarrow{\sim} U$ and $\nu : V' \xrightarrow{\sim} V$ are isomorphisms in $\underline{A}$, the datum (U,V,Ψ,Ψ_1,Ψ_2) is equivalent to

$$(U',V',(\mu^{-1} \oplus \mathrm{Id}'_a \oplus \nu^{-1}) \cdot \Psi, \Psi_1 \cdot (\mathrm{Id}_b \oplus v\mu), \Psi_2 \cdot (v\nu \oplus \mathrm{Id}_c)) .$$

The composite of the two morphisms (U,V,Ψ,Ψ_1,Ψ_2) and $(U',V',\Psi',\Psi'_1,\Psi'_2)$ is given by $(U \oplus U', V' \oplus V, \Psi'', \Psi''_1, \Psi''_2)$, where Ψ'' is the composite

$$a \xrightarrow{\Psi} U \oplus a' \oplus V \xrightarrow{\mathrm{Id}\oplus\Psi\oplus\mathrm{Id}} U \oplus U' \oplus a \oplus V' \oplus V,$$

Ψ''_1 is the composite

$$b \oplus u(U \oplus U') \cong b \oplus uU \oplus uU' \xrightarrow{\Psi_1 \oplus \mathrm{Id}} b' \oplus uU' \xrightarrow{\Psi'_1} b'' ,$$

and Ψ''_2 is the composite

$$v(V' \oplus V) \oplus c \xrightarrow{\sim} vV' \oplus vV \oplus c \xrightarrow{\mathrm{Id}\oplus\Psi_2} vV' \oplus c' \xrightarrow{\Psi'_2} c'' .$$

The symmetric monoidal structure is given by $(b,a,c) \oplus (b',a',c') = (b \oplus b', a \oplus a', c \oplus c')$, note that $\mathrm{Cyl}\,(\underline{B} \xleftarrow{u} \underline{A} \xrightarrow{b} \underline{C})$ is permutative. Note that we have inclusion functors $i_A : \underline{A} \to \mathrm{Cyl}\,(\underline{B} \xleftarrow{u} \underline{A} \xrightarrow{b} \underline{C})$, $i_B : \underline{B} \to v\mathrm{Cyl}\,(\underline{B} \xleftarrow{u} \underline{A} \xrightarrow{b} \underline{C})$, and $i_C : \underline{C} \to \mathrm{Cyl}\,(\underline{B} \xleftarrow{u} \underline{A} \xrightarrow{b} \underline{C})$, which are symmetric monoidal, and we have symmetric monoidal natural transformations $i_A \to i_B \circ u$ and $i_A \to i_C \circ v$. This data is precisely what is required in the universal mapping principle I.8 to give a (strict) symmetric monoidal functor θ from $\underline{L} \wr F$ to $\mathrm{Cyl}\,(\underline{B} \leftarrow \underline{A} \to \underline{C})$. We observe that the construction Cyl also has a universal mapping property. Let $\underline{X}$ be a permutative category, and suppose $\beta : \underline{B} \to \underline{X}$ and $\gamma : \underline{C} \to \underline{X}$ are symmetric monoidal functors so that $\beta u = \gamma v$. Then we define a symmetric monoidal functor $\alpha_x : \mathrm{Cyl}(\underline{B} \xleftarrow{u} \underline{A} \xrightarrow{v} \underline{C}) \to \underline{X}$ on objects by $\alpha_x(b,a,c) = \beta b \oplus \beta u a \oplus \gamma c$, and on morphisms by letting $\alpha_x(U,V,\Psi,\Psi_1,\Psi_2)$ be the following composite

$$\beta b \oplus \beta u(a) \oplus \gamma c \xrightarrow{\mathrm{Id}\oplus\beta u(\Psi)\oplus\mathrm{Id}} \beta b \oplus \beta u(U \oplus a' \oplus V) \oplus \gamma c \xrightarrow{\sim}$$

$$\beta b \oplus \beta uU \oplus \beta u(a') \oplus \beta uV \oplus \gamma c = \beta b \oplus \beta uU \oplus \beta u(a') \oplus \gamma vV \oplus \gamma c \xrightarrow{\sim}$$

$$\beta(b \oplus \beta uU) \oplus \beta u(a') \oplus \gamma(vV \oplus c) \xrightarrow{\beta(\Psi_1)\oplus\mathrm{Id}\oplus\beta(\Psi_2)} \beta(b') \oplus \beta u(a') \oplus \gamma(c')$$

The second arrow is the isomorphism arising from the fact that βu is symmetric monoidal, and the fourth arises from the fact that β and γ are symmetric monoidal. Thomason's theorem now reads as follows.

Theorem I.9. ([43], Theorem 5.2.) Spt (θ) *is an equivalence of spectra. Further, note that the functors* β *and* γ *constitute data to which the universal mapping principle I.8 applies (the natural transformations are identities), and we denote the associated symmetric monoidal functor by* ν*. Then the diagram*

$$\begin{array}{ccc} \mathrm{Spt}\,(\underline{L}\wr\wr F) & \xrightarrow{\mathrm{Spt}\,(\nu)} & \mathrm{Spt}(\underline{X}) \\ \searrow_{\mathrm{Spt}\,(\theta)} & & \nearrow_{\mathrm{Spt}\,(\alpha_x)} \\ & \mathrm{Spt}\,(\mathrm{Cyl}\,\underline{B}\overset{u}{\leftarrow}\underline{A}\overset{v}{\rightarrow}\underline{C}) & \end{array}$$

is functorial and naturally homotopy commutative for $\underline{L}$*–diagrams with* $\underline{A}$*,* $\underline{B}$*, and* $\underline{C}$ *permutative.*

Proof: The first statement is stated in [43]. The rest are immediate consequences of the universal mapping principle; $\alpha_x \cdot \theta$ and ν differ by a canonical choice of commutativity isomorphism. Q.E.D.

We consider one variant on this construction. Let $\underline{M}$ be the category with two objects x and y, with $\mathrm{Mor}_{\underline{M}}(x,x) = \{\mathrm{Id}_x\}$, $\mathrm{Mor}_{\underline{M}}(y,y) = \{\mathrm{Id}_y\}$, and $\mathrm{Mor}_{\underline{M}}(x,y) = \{f,g\}$, and $\mathrm{Mor}_{\underline{M}}(y,x) = \emptyset$. A morphism from $\underline{M}$ to a category $\underline{C}$ is just a diagram of the form

$$X \underset{g}{\overset{f}{\rightrightarrows}} Y$$

in $\underline{C}$. If $X \underset{g}{\overset{f}{\rightrightarrows}} Y$ is an $\underline{M}$–diagram in the category of based spaces, then the homotopy colimit of this diagram is the quotient $(X \times I) \coprod Y/\cong$, where $\cong$ is the equivalence relation generated by $(x,0) \cong f(x) \in Y$, $(x,1) \cong g(x) \in Y$, and $(*,t) \cong * \in Y$. The spectrum version is obtained by applying this construction termwise to the spectra involved. There is a cofibration sequence of spectra $\underset{\sim}{Y} \to \underset{\underset{\underline{M}}{\longrightarrow}}{\mathrm{hocolim}}\, F \to \Sigma\underset{\sim}{X}$, and the spectrum homotopy colimit can be described as the cofibre of the stable map $\underset{\sim}{X} \xrightarrow{f-g} \underset{\sim}{Y}$. Consider an $\underline{M}$–diagram in $\underline{\mathrm{SymMon}}$, say $u,v\colon \underline{A} \longrightarrow \underline{B}$ so that $\underline{A}$ and $\underline{B}$ are unital and permutative, so that u and v are unital symmetric monoidal functors, and so that the morphisms in $\underline{A}$ and $\underline{B}$ are all isomorphisms. We defined a category $\mathrm{Tor}(u,v\colon \underline{A} \longrightarrow \underline{B}) = \underline{T}$ as follows. The objects of $\underline{T}$ are pairs (a,b), where a and b are objects of $\underline{A}$ and $\underline{B}$ respectively. A morphism from (a,b) to (a',b') consists of equivalence classes of data, where a datum is given by (U,V,φ,ψ), where U and V are objects of $\underline{A}$, $\varphi : a \to U \oplus a' \oplus V$ is an isomorphism in $\underline{A}$, and $\psi : uU \oplus b \oplus vV \to b'$ is an isomorphism in $\underline{B}$. Two sets of data $(U_1,V_1,\varphi_1,\Psi_1)$ and $(U_2,V_2,\varphi_2,\psi_2)$ are equivalent if there are isomorphisms $\theta : U_1 \to U_2$ and $\eta : V_1 \to V_2$ in $\underline{A}$ so that $\varphi_2 = (\theta \oplus \mathrm{Id}_{a'} \oplus \eta) \cdot \varphi$ and $\psi_1 = \psi_2 \cdot (u\theta \oplus \mathrm{Id}_b \oplus u\eta)$. If $(U_1,V_1,\varphi_1,\psi_1)$ and $(U_2,V_2,\varphi_2,\psi_2)$ are morphisms $(a,b) \to (a',b')$ and $(a',b') \to (a'',b'')$ respectively, then their composite is given by

$$(U_1 \oplus U_2, v_2 \oplus V_1, (\mathrm{Id}_{U_1} \oplus \varphi_2 \oplus \mathrm{Id}_{V_1}) \cdot \varphi_1, \psi_2 \cdot (\mathrm{Id}_{u_2} \oplus \psi_1 \oplus \mathrm{Id}_{V_2}) \cdot (T_U \oplus \mathrm{Id}_b \oplus T_V)),$$

where $T_U : U_1 \oplus U_2 \to U_2 \oplus U_1$ and $T_V : V_1 \oplus V_2 \to V_2 \oplus V_1$ are the commutativity isomorphisms. Also, if $\underline{E}$ is a permutative category and $\alpha : \underline{A} \to \underline{E}$ and $\beta : \underline{B} \to \underline{E}$ are symmetric monoidal functors so that $\beta \cdot u = \alpha = \beta \cdot v$, we define an associated symmetric monoidal functor $\nu : \underline{T} \to \underline{E}$ on objects by $\nu(a,b) = \alpha(a) \oplus \beta(b)$, and on morphisms by the requirement that $\nu(U, V, \varphi, \psi)$ be the composite

$$\alpha(a) \oplus \beta(b) \xrightarrow{\alpha(\varphi)\oplus \mathrm{Id}} \alpha(U \oplus a' \oplus V) \oplus \beta(b) \xrightarrow{\sim} \alpha(U) \oplus \alpha(a') \oplus \alpha(V) \oplus \beta(b) \xrightarrow{\sim}$$

$$\alpha(a') \oplus \beta u(U) \oplus \beta(b) \oplus \beta v(V) \longrightarrow \alpha(a') \oplus \beta(uU \oplus b \oplus V) \overset{\mathrm{Id}\oplus\beta(\psi)}{\alpha} (a') \oplus \beta(b')$$

By the universal mapping principle I.8, α and β determine a symmetric monoidal functor $\xi : \underline{M} \wr F \to \underline{E}$. We also have inclusion functors $j_A : \underline{A} \to \underline{T}$ and $j_B : \underline{B} \to \underline{T}$, together with natural transformations N_f and N_g from j_A to $j_A \cdot u$ and $j_A \cdot v$, respectively. The universal mapping principle gives a symmetric monoidal functor $\theta : \underline{M} \wr F \to \underline{T}$.

Proposition I.10. Spt (θ) *is an equivalence of spectra. Moreover, the diagram*

$$\begin{array}{ccc} \mathrm{Spt}\,(\underline{M} \wr F) & \xrightarrow{\mathrm{Spt}\,(\xi)} & \mathrm{Spt}(\underline{E}) \\ & {\searrow}_{\mathrm{Spt}\,(\theta)} \quad \nearrow_{\mathrm{Spt}\,(\nu)} & \\ & \mathrm{Spt}\,(\underline{T}) & \end{array}$$

is functorial and naturally homotopy commutative for $\underline{M}$*–diagrams with* $\underline{A}$ *and* $\underline{B}$ *permutative and unital and* u *and* v *unital.*

Proof: The naturality and homotopy commutativity statements are immediate as in I.9. It remains to show that Spt (θ) is an equivalence. The proof of this fact closely parallels the proof of Theorem 5.2 of [43]. Let $\underline{\tilde{P}} = \underline{M} \wr F$. We define a symmetric monoidal functor $\rho : \underline{\tilde{P}} \to \underline{T}$. For a given object $n[P_1, \ldots, P_n]$ with P_i an object of either $\underline{A}$ or $\underline{B}$, let $a = a(P_1, \ldots, P_n)$ be the sum in $\underline{A}$ of the objects belonging to $\underline{A}$, and let $b = b(P_1, \ldots, P_n)$ be the sum of objects belonging to $\underline{B}$. The ordering on the sum is the one inherited from the ordering of the P_i's. ρ is now defined on objects by $\rho(n[P_1, \ldots, P_n]) = (a(P_1, \ldots, P_n), b(P_1, \ldots, P_n))$. Suppose we have a morphism in $\underline{\tilde{P}}$, from $n[P_1, \ldots, P_n]$ to $m[P'_1, \ldots, P'_m]$. Let U denote the direct sum of all the P_i's for which $\ell_i = f$, and let V denote the direct sum of all the P_i's for which $\ell_i = g$. Then there is determined an isomorphism from $a(P_1, \ldots, P_n)$ to $U \oplus a(P'_1, \ldots, P'_m) \oplus V$, and an isomorphism $uU \oplus b(P_1, \ldots, P_n) \oplus vV \to b(P'_1, \ldots, P'_m)$. This defines a morphism in $\underline{T}$, which is defined to be the image under ρ of the given morphism in $\underline{\tilde{P}}$. Observe that ρ and θ are naturally isomorphic symmetric monoidal functors;

the isomorphism involves only repeated application of the isomorphic symmetric $O_A \oplus a \cong a$ and $O_B \oplus b \cong b$ coming from the unital structure on $\underline{A}$ and $\underline{B}$, where O_A and O_B denote the zero objects of $\underline{A}$ and $\underline{B}$, respectively. Consequently, it will suffice to prove that $\mathrm{Spt}\,(\rho)$ is an equivalence of spectra. Let $\underline{\tilde{P}'} \subset \underline{\tilde{P}}$ be the full subcategory on objects $n[P_1, \ldots, P_n]$, so that at least one object P_i is in $\underline{B}$, i.e., so that at least one L_i is equal to y. $\underline{\tilde{P}'}$ is a symmetric monoidal subcategory of $\underline{\tilde{P}}$. Let $\underline{P'}$ be the full subcategory of $\underline{\tilde{P}'}$ consisting of objects $n[P_1, \ldots, P_n]$ so that P_n is an object of $\underline{B}$, and all the other P_i''s are objects of $\underline{A}$. The inclusion $\underline{P'} \hookrightarrow \underline{\tilde{P}'}$ has a left adjoint, sending $m[P_1', \ldots, P_m']$ in $\underline{\tilde{P}'}$ to $n[P_1, \ldots, P_n]$ in $\underline{P'}$, where P_n is the sum of all the P_i's belonging to $\underline{B}$ (ordered in the ordering inherited from the ordering $[P_1', \ldots, P_m']$), and the objects $P_1, \ldots, P_{n-1}$ are simply the ordered list of the objects P_i' belonging to $\underline{A}$. Consequently, $\underline{P'} \to \underline{\tilde{P}'}$ is a weak equivalence.

We now claim that $\underline{\tilde{P}'} \to \underline{\tilde{P}}$ induces a weak homotopy equivalence $\mathrm{Spt}\,(\underline{\tilde{P}'}) \to \mathrm{Spt}\,(\underline{\tilde{P}})$. For, consider the functor $\underline{\tilde{P}} \to \underline{\tilde{P}'}$, given on objects $n[P_1, \ldots, P_n] \to (n+1)[P_1, \ldots, P_n, O_B]$. We have an isomorphism $O_B \oplus O_B \sim O_B$ which shows that our functor is lax symmetric monoidal. As in [43], proof of Theorem 5.2, we now see that the composite $\underline{\tilde{P}} \to \underline{\tilde{P}'} \to \underline{\tilde{P}}$ induces multiplication by an idempotent element of $\pi_0(\mathrm{Spt}\,(\underline{\tilde{P}}))$, hence that $\mathrm{Spt}\,(\underline{\tilde{P}}) \to \mathrm{Spt}\,(\underline{\tilde{P}'}) \to \mathrm{Spt}\,(\underline{\tilde{P}})$ is homotopic to the identity map. Since $\underline{P'} \to \underline{\tilde{P}'}$ induces a weak equivalence on nerves, it suffices by I.6 (a) to show that the composite functor sending $n[P_1, \ldots, P_n]$ in $\underline{P'}$ to $(n+1)[P_1, \ldots, P_n, O_B]$ in $\underline{\tilde{P}'}$ is homotopic to the inclusion. But there is an evident natural transformation $(n+1)[P_1, \ldots, P_n, O_B] \to n[P_1, \ldots, P_n]$, arising from the isomorphism $P_n \oplus O_B \cong P_n$. This shows that $\mathrm{Spt}\,(\underline{\tilde{P}'}) \to \mathrm{Spt}\,(\underline{\tilde{P}})$ is a weak equivalence.

Finally, we must check that the composite $\mathrm{Spt}\,(\underline{\tilde{P}'}) \to \mathrm{Spt}\,(\underline{\tilde{P}}) \to \mathrm{Spt}\,(\underline{T})$ is an equivalence, this will follow if $\underline{\tilde{P}'} \to \underline{T}$ is weak equivalence, which in turn will follow if the composite $\underline{P'} \xrightarrow{i} \underline{\tilde{P}} \xrightarrow{\rho} \underline{T}$ induces an equivalence on nerves, again by I.6 (a). We must show that $(a,b) \downarrow \rho \cdot i$ is contractible for all objects (a,b) of $\underline{T}$. An object in $(a,b) \downarrow \rho \cdot i$ consists of an object $n[P_1, \ldots, P_n]$ of $\underline{P'}$ and a morphism $(U, V, \varphi, \Psi) : (a,b) \to (\oplus_{i=1}^{n-1} P_i, P_n)$ in $\underline{T}$. Consider the full subcategory of $(a,b) \downarrow \rho \cdot i$ consisting of those objects for which the reference map is an isomorphism. The inclusion of this subcategory has a right adjoint, sending $(U, V, \varphi, \Psi) : (a,b) \to (\oplus_{i=1}^{n-1} P_i, P_n)$ to the object $(O_A, O_B, \varphi, \mathrm{Id}) : (a,b) \to (U \oplus \oplus_{i=1}^{n-1} P_i \oplus B, b) = (\oplus_{i=1}^{n-1} P_i, P_n)$ so $N.((a,b) \downarrow \rho \cdot i)$ is equivalent to the nerve of this subcategory. But the subcategory has the terminal object $(2[a,b], \mathrm{Id} : (a,b) \to \rho \cdot i(2[a,b]))$, hence we have the result. Q.E.D.

We will also need to understand the symmetric monoidal category–theoretic versions of mapping telescopes of spectra. Let $\underline{\mathbb{N}}$ denote the category whose objects are the non–negative integers, with a unique morphism from i to j when $i \leq j$ and $\mathrm{Hom}_{\underline{\mathbb{N}}}(i,j) = \emptyset$ when $i < j$. Let $\underline{\mathbb{N}}^s$ denote the category whose objects are the non–empty subsets S of $\mathbb{N}$, so that any two elements of S are adjacent, and with a unique morphism $S \to T$ if and only if $T \subset S$. Of course, $ob\,\underline{\mathbb{N}}^s = \mathbb{N} \coprod \{\{i, i+1\},\ i \in \mathbb{N}\}$, and the nerve of $\underline{\mathbb{N}}^s$ is the barycentric subdivision of the nerve of $\underline{\mathbb{N}}$, whose geometric realization is homeomorphic to the non–negative half–line $[0, +\infty)$. We have a functor $\pi : \underline{\mathbb{N}}^s \to \underline{\mathbb{N}}$ given by $\pi(S) = \min(S)$. Let $\Phi : \underline{\mathbb{N}} \to \underline{CAT}$ and $\Psi : \underline{\mathbb{N}}^s \to \underline{CAT}$ be any functors. Then we have, as above, the Grothendieck constructions $\underline{\mathbb{N}} \wr \Phi$ and $\underline{\mathbb{N}}^s \wr \Psi$.

Proposition I.11. *There are equivalences* $N.(\underline{\mathbb{N}} \wr \Phi)_+ \to \underset{\underline{\mathbb{N}}}{\underrightarrow{\mathrm{hocolim}}}(N. \cdot \Phi)_+$ *and* $N.(\underline{\mathbb{N}} \wr \Psi)_+ \to \underset{\underline{\mathbb{N}}^s}{\underrightarrow{\mathrm{hocolim}}}(N. \cdot \Psi)_+$ *of simplicial sets, where* $X._+$ *denotes* $X.$ *with a disjoint base point added, which are natural on the categories of* $\underline{\mathbb{N}}$*–diagrams and* $\underline{\mathbb{N}}^s$*–diagrams in* $\underline{CAT}$*, respectively. In particular,* $N.(N \wr \Phi)$ *has the weak homotopy type of the mapping telescope of* $N.\Phi(0) \to N.\Phi(1) \to \cdots$.

Proof: Clear from I.5. Q.E.D.

Proposition I.12. *Let* $\Phi : \underline{\mathbb{N}} \to \underline{CAT}$ *and* $\Psi : \underline{\mathbb{N}}^s \to \underline{CAT}$ *be functors, and let* $\underline{E}$ *be a category*

(a) Functors from $\underline{\mathbb{N}} \wr \Phi$ *to* $\underline{E}$ *are in bijective correspondence with collections of data* $\{F_i, N_i :\ i \geq 0\}$*, where* $F_i : \Phi(i) \to \underline{E}$ *is a functor, and* N_i *is a natural transformation from* F_i *to* $F_{i+1} \cdot \Phi(i \to i+1)$*. Functors from* $\underline{\mathbb{N}}^s \wr \Psi$ *to* $\underline{E}$ *are in bijective correspondence with collections of data* $\{F_i, G_i, N_i, N_i^! :\ i \geq 0\}$*, where* $F_i : \Psi(\{i\}) \to \underline{E}$ *and* $G_i : \Psi(\{i, i+1\}) \to \underline{E}$ *are functors, and where* N_i *(respectively* $N_i^!$*) are natural transformations from* G_i *to* $F_i : \Psi(\{i, i+1\} \to \{i\})$ *respectively to* $F_{i+1} \cdot \Psi(\{i, i+1\} \to \{i+1\})$.

Proof: Immediate from the paragraphs preceding I.5. Q.E.D.

Proposition I.13. *Let* Φ *and* Φ' *be the functors from* $\underline{\mathbb{N}}$ *to* $\underline{CAT}$*. Suppose we have functors* $F_i : \Phi(i) \to \Phi'(i)$ *and natural transformations* N_i *from* $\Phi'(i < i+1) \cdot F_i$ *to* $F_{i+1} \cdot \Phi(i < i+1)$*. Then the data* $\{F_i, N_i;\ i \geq 0\}$ *determine a functor* $\underline{\mathbb{N}} \wr \Phi \to \underline{\mathbb{N}} \wr \Phi'$*, which restricts to* F_n *on the subcategory* $(n, \Phi(n)) \cong \Phi(n)$*, and which induces an equivalence on nerves if each* F_n *does. Similarly, if* Ψ *and* Ψ' *are functors from* $\underline{\mathbb{N}}^s$ *to* $\underline{CAT}$*, and we have functors* $F_i : \Psi(i) \to \Psi'(i)$ *and* $G_k : \Psi(\{i, i+1\}) \to \Psi'(\{i, i+1\})$*, together with natural transformations* N_i *(respectively* $N_i^!$*) from* $\Psi'(\{i, i+1\} \to \{i\}) \cdot G_i$ *to* $F_i \cdot \Psi(\{i, i+1\} \to \{i\})$ *(respectively from* $\Psi'(\{i, i+1\} \to \{i+1\}) \cdot G_i$ *to* $F_{i+1} \cdot \Psi(\{i, i+1\} \to \{i+1\})$*), we obtain a functor* $\underline{\mathbb{N}}^s \wr \Psi \to \underline{\mathbb{N}}^s \wr \Psi'$*, which restricts to* F_i *on* $(\{i\}, \Psi(i))$ *and to* G_i *on* $(\{i, i+1\}, \Psi(\{i, i+1\}))$*, and which induces an equivalence on nerves if the* F_i*'s and* G_i*'s all do.*

Proof: Let $\mathcal{F}_i : \Phi(i) \to \underline{\mathbb{N}} \wr \Phi'$ be given by $x \to (i, F_i(x))$. Let $\mathcal{F}_i^! :$

$\Phi(i) \to \underline{\mathbb{N}} \wr \Phi'$ be given by $x \to (i+1, \Phi'(i < i+1) \cdot F_i(x))$. There is a natural transformation $\mathcal{F}_i \to \mathcal{F}_i^!$ (on an object x, it is the morphism $(i, \xi) \to (i+1, \Phi'(i < i+1)(\xi))$ associated to the identity map of $\Phi'(i < i+1)(\xi)$, and N_i gives a natural transformation from $\mathcal{F}_i^!$ to the functor $x \to (i+1, F_{i+1} \cdot \Phi(i < i+1)(x)) = \mathcal{F}_{i+1} \cdot \Phi(i < i+1)$. The composite of these two natural transformations give the required functor via I.12, which we call $\mathcal{F}$. To see that it is an equivalence if the F_i's all are, we note that if $\underline{\mathbb{N}}^{(i)}$ denotes the full subcategory of $\underline{\mathbb{N}}$ consisting of integers less than or equal to i, then we have a commutative diagram

$$\begin{array}{ccc} \Phi(i) & \xrightarrow{F_i} & \Phi'(i) \\ f \downarrow & & \downarrow g \\ \underline{\mathbb{N}}^{(i)} \wr \Phi & \xrightarrow{\mathcal{F} \mid \underline{\mathbb{N}}^{(i)} \wr \Phi} & \underline{\mathbb{N}}^{(i)} \wr \Phi' \end{array}$$

and one easily checks that the vertical arrows are equivalences. The result now follows easily from the well known properties of mapping telescopes. The case of $\underline{\mathbb{N}}^s$ is similar. Q.E.D.

Proposition I.14. *Let $\Phi : \underline{\mathbb{N}} \to \underline{CAT}$ be a functor. Then the natural map $\underline{\mathbb{N}}^s \wr (\Phi \cdot \pi) \to \underline{\mathbb{N}} \wr \Phi$ is a weak equivalence.*

Proof: Direct from I.1 (g); the categories $i \downarrow \pi$ are easily seen to be contractible. Q.E.D.

We use Quillen's Theorem A to obtain a nice criterion for a functor $\underline{\mathbb{N}} \wr \Phi \to \underline{E}$ to be a weak equivalence. Thus, let $\Phi : \underline{\mathbb{N}} \to \underline{CAT}$ be a functor, and let G be a functor from $\underline{\mathbb{N}} \wr \Phi$ to a category $\underline{E}$, determined by data $\{F_i, N_i\}$. For any $x \in \underline{E}$, we may consider the categories $x \downarrow F_i$. We define a functor $l_i : x \downarrow F_i \to x \downarrow F_{i+1}$ on objects by $l_i(y, \theta) = (\Phi(i < i+1)(y), \theta)$, which makes sense since $F_{i+1} \cdot \Phi(i < i+1)(y) = F_i(y)$. On morphisms, $l_i(\eta) = \Phi(i < i+1)(\eta)$. Thus, for every object x of $\underline{E}$, we obtain a functor $G_x : \underline{\mathbb{N}} \to \underline{CAT}$.

Proposition I.15. *If for each $x \in \underline{E}$, the mapping telescope associated to $N. \circ G_x$ is weakly contractible, then G induces a weak equivalence on nerves.*

Proof: One sees that $\underline{\mathbb{N}} \wr G_x$ is isomorphic to the category $x \downarrow G$, and applies Quillen's Theorem A and I.11. Q.E.D.

When we have functors $\Phi : \underline{\mathbb{N}} \to \underline{\text{SymMon}}$, we will see that $\underline{\mathbb{N}} \wr \Phi$ and sometimes $\underline{\mathbb{N}}^s \wr \Psi$ give small models for the spectrum homotopy colimits of Spt $\cdot\, \Phi$ and Spt $\cdot\, \Psi$. For $\Phi : \underline{\mathbb{N}} \to \underline{\text{SymMon}}$, we define $\underline{\mathbb{N}} \wr \Phi \xrightarrow{l} \underline{\mathbb{N}} \wr\wr \Phi$ as follows. From I.12, we are required to produce functors $\Phi(i) \xrightarrow{F_i} \underline{\mathbb{N}} \wr\wr \Phi$ and natural transformations N_i from F_i to $F_{i+1} \cdot \Phi(i \to \imath + 1)$. F_i will be the functor from $\Phi(i) \to \underline{\mathbb{N}} \wr\wr \Phi$ given by $\xi \to 1[(i,\xi)]$ and $N_i(\xi)$ will be the morphism $1[(i,\xi)] \to 1[(i+1, \Phi(i < i+1)(\xi))]$ corresponding to the identity map of $\Phi(i < i+1)(\xi)$. Similarly, we defined $l^s : \underline{\mathbb{N}}^s \wr \Psi \to \underline{\mathbb{N}}^s \wr\wr \Psi$ for $\Psi : \underline{\mathbb{N}}^s \to \underline{\text{SymMon}}$.

Proposition I.16. *Let Φ, Ψ, l, and l^s be defined as above.*

(a) *l is a weak equivalence*

(b) *If $\Psi(\{i, i+1\} \to \{i\})$ is a weak equivalence for all i, then l^s is a weak equivalence.*

Proof: In the proof of his spectrum homotopy colimit theorem [43], Thomason shows that $N.(\underline{C} \wr\wr F)$ is weakly equivalent to the diagonal simplicial set obtained by taking the nerve of a simplicial category, which in every level k is $\underline{C} \wr\wr F^{(k)}$, where $F^{(k)}$ is of the form $S \cdot G$, for $G : \underline{C} \to \underline{CAT}$ a functor and S denoting the "free symmetric monoidal category functor." See [43] for details of this construction. Consequently, since homotopy colimits commute with $|\ |$, it suffices to consider the case where Φ and Ψ are of the form $S \cdot \Phi^0$ and $S \cdot \Psi^0$, $\Phi^0 : \underline{\mathbb{N}} \to \underline{CAT}$ and $\Psi^0 : \underline{\mathbb{N}} \to \underline{CAT}$. Moreover, Thomason [43], p. 1637–1638] shows that there is a commutative diagram of functors

$$\begin{array}{ccc} \underline{C} \wr S \circ F & \longrightarrow & S(\underline{C} \wr F) \\ & \searrow & \downarrow \\ & & \underline{C} \wr\wr S \circ F \end{array}$$

where the right hand vertical arrow is a weak equivalence and the diagonal arrow is l (respectively l^s) in the case $\underline{C} = \underline{\mathbb{N}}$ (respectively $\underline{C} = \underline{\mathbb{N}}^s$). We describe the horizontal functor. Let $\underline{\Sigma_n}$ denote the symmetric group on n letters, viewed as a category with on object, and let $\underline{E}$ be a category. The permutation action of Σ_n on the category $\underline{E}^n$ can be viewed as a functor $P_n^E : \underline{\Sigma_n} \to \underline{CAT}$, and $S\underline{E}$ can be identified with $\coprod_{n \geq 1} v\underline{\Sigma_n} \wr P_n^E$. If $\underline{C} \xrightarrow{G} \underline{CAT}$ is any functor, let $F_n : \underline{C} \times \underline{\Sigma_n} \to \underline{CAT}$ be the functor $(x, e) \to F(x)^n$, with evident Σ_n action. Then $\underline{C} \wr S \cdot F$ can be identified with $\coprod_n (\underline{C} \times \underline{\Sigma_n}) \wr F_n$, and $S(\underline{C} \wr F)$ can be identified with $\prod_n \underline{\Sigma_n} \wr P_n^{\underline{C} \wr F}$, and there is an evident inclusion $\underline{C} \wr S \cdot F \to S(\underline{C} \wr F)$, involving the diagonal maps $\underline{C} \wr F_n \to (\underline{C} \wr F)^n$. This is the horizontal arrow in Thomason's diagram above. It will thus suffice to show that this horizontal arrow is a weak equivalence in our situation. Since this functor breaks up as a disjoint union of functors over n, it will suffice to check that $(\underline{C} \times \underline{\Sigma_n}) \wr F_n \to \underline{\Sigma_n} \wr P_n^{\underline{C} \wr F}$ is a weak equivalence for each n. But it is easy to see that $N.(\underline{\Sigma_n} \wr P_n^E) \cong E\Sigma_n \times_{\Sigma_n} (N.\underline{E})^n$, where $E\Sigma_n$ denotes a contractible simplicial set on which Σ_n acts freely. Part (a) of our proposition now amounts to showing that mapping telescopes commute with the n–adic construction $X \to E\Sigma_n \times_{\Sigma_n} X^n$, which is clear.

For the second case, let $\underline{\mathbb{N}}^s(k)$ denote the full subcategory on the objects $0, 1, \ldots, k, \{0,1\}, \ldots, \{k, k+1\}$. $(\mathbb{N}^s \times \Sigma_n) \wr \Psi$ can be written as $\varinjlim_k (\underline{\mathbb{N}}^s(k) \times \Sigma_n) \wr \Psi_n$. But $(\underline{\mathbb{N}}^s(k) \times \underline{\Sigma_n}) \wr \Psi_n$ has nerve equivalent to $\underline{\Sigma_n} \wr P_n^{\mathbb{N}^s(k) \wr \Psi}$, since both have nerves weakly equivalent to $E\,\Sigma_n \times_{\Sigma_n} \Psi(\{k, k+1\})^n$, as one easily checks using the hypothesis that $\Psi(\{i, i+1\} \to \{i\})$ induces an equivalence on nerves for all i. Q.E.D.

We finally need to examine group actions on inverse limits. Let $\underline{C}$ be a category with an action by group Γ. We may view this action as a functor $\underline{\Gamma} \to \underline{CAT}$, where $\underline{\Gamma}$ denotes Γ viewed as a category with one object. We write $\underline{\Gamma} \wr \underline{C}$ for the Grothendieck construction on this functor. Suppose $F : \underline{\Gamma} \wr \underline{C} \to \underline{s\text{-sets}}$ is a functor. Then by the discussion preceding I.5, F corresponds to a functor $\underline{\Gamma} \to \underline{s\text{-sets}}^{CAT}$.

Recall from [11, Ch. XI, § 5] that if $F : \underline{C} \to \underline{s\text{-sets}}$ is any functor, the cosimplicial space X defining $\underset{\underline{C}}{\overleftarrow{\text{holim}}} F$ is given in codimension k by $X^k = \prod_{x_0 \to \cdots x_k} F(x_k)$. We say a category is *discrete* if its only morphisms are identity morphisms; discrete subcategories are the same things as sets. We may now view $N.\underline{C}$ as a simplicial discrete category, and if $F : \underline{C} \to \underline{s\text{-sets}}$ is a functor, we define functors $F_k : N_k \underline{C} \to \underline{s\text{-sets}}$ by $F_k(x_0 \to \cdots x_k) = F(x_k)$. One readily checks that the usual definitions of cofaces and codegeneracy maps in $X.$ make $\underline{k} \to (N_k \underline{C}, F_k)$ into a functor $\underline{s} \to \underline{s\text{-set}}^{CAT_\delta}$, where $CAT_\delta \subset CAT$ is the full subcategory of discrete categories. Consequently, $\underset{\underline{C}}{\overleftarrow{\text{holim}}}\, F$ can be described as the total space of a cosimplicial space $X.$, in which each X^k is itself the homotopy inverse limit of a functor over a discrete category. If $\underline{C}$ is acted on by a group Γ, and we are given a functor $F : \underline{\Gamma} \wr \underline{C} \to \underline{s\text{-sets}}$, then the cosimplicial space defining $\underset{\underline{C}}{\overleftarrow{\text{holim}}}\, F \,|\, C$ is a cosimplicial Γ–space, which in every level k is the homotopy inverse limit over $N_k \underline{C}$ of $(F \,|\, \underline{C})_k$ associated to a functor $\underline{\Gamma} \wr N_k \underline{C} \to \underline{s\text{-sets}}$, then the cosimplicial space defining $\underset{\underline{C}}{\overleftarrow{\text{holim}}} F \,|\, C$ is a cosimplicial Γ–space, which in every level k is the homotopy inverse limit over $N_k \underline{C}$ of $(F \,|\, \underline{C})_k$ associated to a functor $\underline{\Gamma} \wr N_k \underline{C} \to \underline{s\text{-sets}}$. If $X.$ is any Γ–space, we let $X^{h\Gamma}$ denote the homotopy inverse limit $\underset{\underline{\Gamma}}{\overleftarrow{\text{holim}}} X.$, where $X.$ is viewed as a functor from $\underline{\Gamma}$ to $\underline{s\text{-sets}}$. Alternatively, one can take $X^{h\Gamma}$ to be the fixed point set of the conjugation action of Γ on the function complex $F(W\Gamma, X)$, where $W\Gamma$ is any free contractible Γ–space. The map $W\Gamma \to *$ gives a map $X^{\Gamma}. \to X^{h\Gamma}..$

Proposition I.17. *Let a group* Γ *act on a category* $\underline{C}$, *and suppose* F :

$\underline{\Gamma} \wr \underline{C} \to \underline{s\text{-}sets}$ is a functor. Suppose further that Γ acts freely on the objects of $\underline{C}$, and that $F(\xi)$ is a Kan complex for all $\xi \in \underline{\Gamma} \wr \underline{C}$. Let $X.$ denote the cosimplicial Γ–space defining $\underset{\overleftarrow{\underline{C}}}{\text{holim}}\, F \mid \underline{C}$.

(a) *$(X.)^{\Gamma}$ and $(X.)^{h\Gamma}$ are fibrant cosimplicial spaces.*

(b) *The natural map of cosimplicial spaces $(X.)^{\Gamma} \to (X\cdot)^{h\Gamma}$ is a weak equivalence, hence by (a), $\left(\underset{\overleftarrow{\underline{C}}}{\text{holim}}\, F \mid \underline{C}\right)^{\Gamma} \to \left(\underset{\overleftarrow{\underline{C}}}{\text{holim}}\, F \mid \underline{C}\right)^{h\Gamma}$ is a weak equivalence.*

Proof: (a) is direct, entirely analogous to the proof that the cosimplicial space defining $\underset{\overleftarrow{\underline{C}}}{\text{holim}}$ is fibrant if $F(x)$ is a Kan complex for all $x \in \underline{C}$. We leave it to the reader. To prove (b), we must show that if $\underline{C}$ is discrete category with free Γ–action, and $\underline{\Gamma} \wr \underline{C} \to \underline{s\text{-}sets}$ is a functor, then the natural map $\left(\underset{\overleftarrow{\underline{C}}}{\text{holim}}\, F \mid \underline{C}\right)^{\Gamma} \to \left(\underset{\overleftarrow{\underline{C}}}{\text{holim}}\, F \mid \underline{C}\right)^{h\Gamma}$ is a weak equivalence. But to give a functor $\underline{\Gamma} \wr \underline{C} \to \underline{s\text{-}sets}$ is the same as to specify simplicial sets E_x for every $x \in \underline{C}$, and maps $F_{\gamma,x} : E_x \to E_{\gamma x}$ for all $\gamma \in \Gamma$, $x \in \underline{C}$, such that $F_{\gamma_2,\gamma_1 x} \cdot F_{\gamma_1,x} = F_{\gamma_2\gamma_1,x}$ for γ_1, γ_2 and x. The homotopy inverse limit $\underset{\overleftarrow{\underline{C}}}{\text{holim}} F \mid \underline{C}$ is now the product $\prod_{x \in \underline{C}} E_x$, and the Γ–action is given by the equation $\gamma \cdot \{e_x\} = \{F_{\gamma} e_{\gamma^{-1}x}\}$. Let A denote the set of orbits of the Γ–action on the objects of $\underline{C}$, and let $\underline{C}_{\alpha}$ denote the subcategory on objects belonging to the orbit α. Then the above description shows that $\underset{\overleftarrow{\underline{C}}}{\text{holim}}\, F \mid \underline{C} \xrightarrow{\sim} \prod_{\alpha \in A} \underset{\overleftarrow{\underline{C}_{\alpha}}}{\text{holim}}\, F \mid \underline{C}_{\alpha}$, and the isomorphism is equivariant, so it suffices to deal with the case where the Γ–action is transitive on the objects of $\underline{C}$. In this case, $\underset{\overleftarrow{\underline{C}}}{\text{holim}}\, F \mid \underline{C}$ is Γ–isomorphic to the Γ–space $\prod_{\gamma \in \Gamma} X_e$, with Γ acting by permuting factors, and this space is in turn Γ–isomorphic to the space of functions $F(\Gamma, X_e)$. But for a space of the form, the statement is obvious. Q.E.D.

For Γ a group, a Γ–spectrum is just a spectrum with action by the group Γ. Fixed point sets and homotopy fixed point sets are defined termwise.

Corollary I.18. *Let $\underline{C}$ be any category with Γ–action, and let $F : \underline{\Gamma} \wr \underline{C} \to \underline{S}$ be a functor. Then $\underset{\overleftarrow{\underline{C}}}{\text{holim}}\, F \mid \underline{C}$ is equipped with a Γ–action from F.*

Suppose that Γ *acts freely on the objects of* $\underline{C}$. *Then the natural map* $\left(\underset{\overleftarrow{\underline{C}}}{\mathrm{holim}}\, F \mid \underline{C}\right)^{\Gamma} \to \left(\underset{\overleftarrow{\underline{C}}}{\mathrm{holim}}\, F \mid \underline{C}\right)^{h\Gamma}$ *is a weak equivalence of spectra.*

Proof: Follows directly from I.17; recall that conventionally, all spectrum homotopy inverse limits are taken only after converting the functor F into one where the values are all Kan Ω–spectra. Q.E.D.

II. Locally Finite Homology

In this section, we describe the theory of locally finite "Borel–Moore" [7] homology with coefficients in a spectrum, and clarify its equivariant properties.

Definition II.1. *A map* $f : X \to Y$, *where* X *and* Y *are sets, is said to be proper if for all finite sets* $U \subseteq Y$, $f^{-1}U \subseteq X$ *is finite. In particular, a one to one map is always proper.*

Let $A[-]$ denote the "free Abelian group" functor. Applied to a set X, it assigns to X all formal linear combinations $\sum_{x \in X} n_x x$, where $n_x = 0$ except for a finitely many x. Let $\hat{A}[X]$ denote the group of *all* linear combinations $\sum_{x \in X} n_x x$, without any constraints on the n_x's. $\hat{A}[-]$ cannot be made functorial for all maps of sets; only proper maps will induce homomorphisms on $\hat{A}$. Let $\underline{\mathrm{sets}}^p$ denote the category of sets and proper maps; $\hat{A}$ defines a functor from $\underline{\mathrm{sets}}^p$ to $\underline{\mathrm{Ab}}$. We give an alternate version of the definition of $\hat{A}$. For a set X, let $\mathcal{F}(X)$ denote the category of finite subsets of X under inclusions. We define functor $\Phi^X : \mathcal{F}(X)^{\mathrm{op}} \to \underline{\mathrm{Ab}}$ by $A[X]/A[X-U] \cong A[U]$; note that $A[\Phi]$ is interpreted as the zero group, to cover the case where X itself is finite. If $U \subseteq V$, $X - V \subseteq X - U$, and we obtain an evident projection $\Phi^X(V) \to \Phi^X(U)$. Note that if $f : X \to Y$ is a proper map of sets, we obtain a functor $f_! : \mathcal{F}(Y)^{\mathrm{op}} \to \mathcal{F}(X)^{\mathrm{op}}$, defined on objects by $f_!(U) = f^{-1}U$. Also associated to f is a natural transformation $\nu_f : \Phi^\circ f_! \to \Phi^Y$ of functors on $\mathcal{F}(Y)^{\mathrm{op}}$, given by the homomorphism $A[X]/A[X - f^{-1}U] \to A[Y]/A[Y-U]$ induced by f. We obtain a functor $\hat{A}^0 : \underline{\mathrm{sets}}^p \to \underline{\mathrm{Ab}}_0^{CAT}$ by setting $\hat{A}^0[X] = (\mathcal{F}(X)^{\mathrm{op}}, \Phi^X)$ and $\hat{A}^0[f] = (f_!, \nu_f)$.

Proposition II.2. *The functors* $\hat{A}[\;]$ *and* $\underleftarrow{\lim} \circ \hat{A}^0$ *are isomorphic as functors from* $\underline{\mathrm{sets}}^p$ *to* $\underline{\mathrm{Ab}}$.

Proof: We must produce a natural isomorphism $\hat{A} \to \underleftarrow{\lim} \circ \hat{A}^0$. To produce a homomorphism from $\hat{A}[X]$ to

$$(\underleftarrow{\lim} \circ \hat{A}^0)[X] = \underleftarrow{\lim}_{U \in \mathcal{F}(X)^{\text{op}}} A[X]/A[X-U],$$

it suffices to produce homomorphisms to $A[X]/A[X-U]$ for each U satisfying the obvious compatibility conditions. The projections

$$\pi_U : \hat{A}[X] \to \hat{A}[X]/\hat{A}[X-U] \cong A[X]/A[X-U]$$

give such a compatibility, and it follows directly from the definition of inverse limits that the map is an isomorphism. Naturality is also clear. Q.E.D.

We may view this construction a bit more homotopy theoretically as follows. First, for any set, let X_+ denote X with a disjoint basepoint added. Then $A[X]$ can be identified with $\pi_0(\underset{\sim}{K} \wedge X_+)$, where $\underset{\sim}{K}$ denotes the integral Eilenberg–MacLane spectrum $K(\mathbb{Z}, 0)$. Moreover, $\pi_i(\underset{\sim}{K} \wedge X_+) = 0$ for $i > 0$, so the (discrete) space $A[X]$ is homotopy equivalent to the zero–th space of the spectrum $\underset{\sim}{K} \wedge X_+$. We now wish to construct a version of $\hat{A}$ that allows an arbitrary spectrum $\underset{\sim}{W}$ to replace $\underset{\sim}{K}$. When one is dealing with spaces or spectra, one must replace inverse limits by homotopy inverse limits. Let $\underline{\text{sets}}^p$ be as above. For any set X and spectrum $\underset{\sim}{W}$, we define a functor $\Phi^X_W : \mathcal{F}(X)^{\text{op}} \to \underline{\mathcal{S}}$ by $\Phi^X_W(U) = \underset{\sim}{W} \wedge (X/X-U)$; as before, when $U \subseteq V$, we have the evident collapse map $X/X-V \to X/X-U$. Note that the image of $X-U$ in $X/X-U$ is taken as the basepoint when applying the smash product. Also, when $f : X \to Y$ is a proper map, we obtain a natural transformation $\nu_f : \Phi^X_W \circ f_! \to \Phi^Y_W$ exactly as above, using the set map $X/X - f^{-1}U \to Y/Y-U$ induced by f. $H_W : \underline{\text{sets}}^p \to \underline{\mathcal{S}}_0^{CAT}$ is now defined by $H_W(X) = (\mathcal{F}(X)^{\text{op}}, \Phi^X_W)$ and $H_W(f) = (f_!, \nu_f)$.

Definition II.3. *Let $\underset{\sim}{W}$ be any spectrum. We define a functor $\underset{\sim}{h}^{\ell f}(-;\underset{\sim}{W})$ from $\underline{\text{sets}}^p$ to spectra by $\underset{\sim}{h}^{\ell f}(-;\underset{\sim}{W}) = \underleftarrow{\text{holim}} \circ H_W$.*

For any set X and Abelian group G, let $\hat{G}[X]$ denote $\underleftarrow{\lim}_{\mathcal{F}(X)^{\text{op}}} G \otimes_{\mathbb{Z}} A[X/X-U]$; $\hat{G}$ is a functor from $\underline{\text{sets}}^p$ to $\underline{\text{Ab}}$, and can be identified with the group of all infinite formal linear combinations $\sum_{x \in X} g_x x$, $g_x \in G$.

Proposition II.4. *For any set X and spectrum $\underset{\sim}{W}$, we have*

$$\pi_i(\underset{\sim}{h}^{\ell f}(X;\underset{\sim}{W})) \cong \widehat{\pi_i(\underset{\sim}{W})}[X],$$

the isomorphism being natural with respect to morphisms in $\underline{\text{sets}}^p$. *In particular, if* $\underset{\sim}{W}$ *is the Eilenberg–MacLane spectrum* $K(G,0)$ *for an Abelian group* G, *then we have*

$$\begin{cases} \pi_0(\underset{\sim}{h}^{\ell f}(X;\underset{\sim}{W})) \cong \hat{G}[X] & \text{and} \\ \pi_i(\underset{\sim}{h}^{\ell f}(X;\underset{\sim}{W})) \cong 0 & \text{for } i \neq 0, \end{cases}$$

so $\underset{\sim}{h}^{\ell f}(X;\underset{\sim}{W})$ *is naturally equivalent to the spectrum* $K(\hat{G}[X],0)$. *(Here* $K(-,0)$ *is viewed as a functor which assigns to an Abelian group* G *the Eilenberg–MacLane spectrum* $K(G,0)$.*)*

Proof: There is by I.3 a spectral sequence with $E_2^{-p,q}$–term $\underset{\mathcal{F}(X)^{\text{op}}}{\overleftarrow{\lim}}{}^p \pi_q \circ \Phi_W^X$, converging to $\pi_{p+q}(\underset{\mathcal{F}(X)^{\text{op}}}{\overleftarrow{\text{holim}}}\,\Phi_W^X)$. The result will follow if we can demonstrate the vanishing of these derived functors for $p > 0$, since $\underset{\mathcal{F}(X)^{\text{op}}}{\overleftarrow{\lim}} \pi_q \circ \Phi_W^X$ is readily seen to be isomorphic to $\widehat{\pi_q(\underset{\sim}{W})}[X]$. Let A be any Abelian group, and define a functor $\mathcal{F}(X)^{\text{op}} \xrightarrow{G_A} \underline{\text{Ab}}$ by $G_A(U) = F(U,A)$, the set of functions from U to A under pointwise addition. Note that if $A = \pi_q(\underset{\sim}{W})$, then the functors $\pi_q \circ \Phi_W^X$ and G_A are isomorphic; we thus establish the vanishing of the higher derived functors for G_A. $\underset{\mathcal{F}^{\text{op}}}{\overleftarrow{\lim}}{}^* G_A$ is by [11, Ch. XI] computable as the cohomotopy of the cosimplicial Abelian group

$$\underline{k} \to \prod_{U_0 \supseteq U_1 \supseteq \cdots \supseteq U_k} G_A(U_k),$$

with the coface and codegeneracy maps given by the formulas defining homotopy inverse limits. Let $H : \mathcal{F}(X) \to \underline{\text{sets}}$ be the functor $H(U) = U$; then the cosimplicial Abelian group $\underline{k} \to \prod_{U_0 \supseteq U_1 \supseteq \cdots \supseteq U_k} G_A(U_k)$ is obtained by applying $F(-,A)$ to the simplicial space $\underline{k} \to \coprod_{U_0 \subseteq U_1 \subseteq \cdots \subseteq U_k} H(U_k)$, i.e., to the homotopy colimit $\underset{\mathcal{F}(X)}{\overrightarrow{\text{hocolim}}}\, H$. If we can show that $H_p(\underset{\mathcal{F}(X)}{\overrightarrow{\text{hocolim}}}\, H)$ vanishes for $p > 0$ and is free for $p = 0$, we can conclude the result by the universal coefficient theorem. But by Thomason's homotopy colimit theorem [43], Theorem 5.2, $\underset{\mathcal{F}(X)}{\overrightarrow{\text{hocolim}}}\, H$ has the same weak homotopy type as the nerve of the category $\mathcal{F}(X) \wr H$, where H is viewed as a functor from $\mathcal{F}(X)$ to discrete categories. The objects of $\mathcal{F}(X) \wr H$ are pairs (U,x), where $U \in \mathcal{F}(X)$ and $x \in U$. There is a unique morphism $(U,x) \to (U',x')$ if $U \subseteq U'$ and

$x = x'$. Let $\mathcal{F}(X)^\delta \subseteq \mathcal{F}(X)$ denote the subcategory of $\mathcal{F}(X)$ of sets U with exactly one element. We define an inclusion $i : \mathcal{F}(X)^\delta \to \mathcal{F}(X) \wr H$ by $\{x\} \xrightarrow{i} (\{x\}, x)$, and a functor $s : \mathcal{F}(X) \wr H \to \mathcal{F}(X)^\delta$ by $s((U, x)) = \{x\}$. $s \circ i$ is the identity functor on $\mathcal{F}(X)^\delta$, and the morphism $(\{x\}, x) \to (U, x)$ gives a natural transformation from $i \circ s$ to the identity functor on $\mathcal{F}(X) \wr H$, so the nerve of $\mathcal{F}(X) \wr H$ is weakly equivalent to the nerve of the discrete category $\mathcal{F}(X)^\delta(X)$. This gives the isomorphism stated in the theorem. The naturality of the isomorphism follows from the naturality of the spectral sequence with respect to maps in $\underline{\mathcal{S}}_0^{CAT}$. Q.E.D.

We now extend the definition of $\underset{\sim}{h}^{\ell f}(-; W)$ to simplicial sets, and finally to spaces.

Definition II.5. *By a locally finite simplicial set, we mean a simplicial object in* $\underline{\text{sets}}^p$. *If* $X.$ *is a locally finite simplicial set, then applying* $\underset{\sim}{h}^{\ell f}(-; \underset{\sim}{W})$ *gives a simplicial spectrum for any spectrum* $\underset{\sim}{W}$. *We define the simplicial locally finite homology of* $X.$ *to be the diagonalization of this simplicial spectrum, and denote it by* ${}^s\underset{\sim}{h}^{\ell f}(X., \underset{\sim}{W})$.

We now extend the definition to a theory on topological spaces. For $\underset{\sim}{W}$ an Eilenberg–MacLane spectrum, Borel–Moore homology [7] gives a way to construct this theory. However, this approach seems too sheaf–theoretic to extend to the case of coefficients in a spectrum, and for our purposes, it is preferable to have a version more closely related to singular homology.

If X is a topological space, let $S.X$ denote its "singular complex," see [28]. This is a simplicial set, but it is not locally finite, so we cannot apply $\underset{\sim}{h}^{\ell f}(-; \underset{\sim}{W})$ directly. We proceed as follows. For a singular k–simplex $\sigma : \Delta^k \to X$, we let $\mathrm{im}(\sigma)$ denote the image of σ, a compact subset of X. A subset $\mathcal{A} \subseteq S_k X$ is said to be *locally finite* if, for every point $x \in X$, there is a neighborhood U of x, so that $U \cap \mathrm{im}(\sigma)$ is non–empty for only finitely many $\sigma \in \mathcal{A}$. It is clear that if $\mathcal{A}$ is locally finite, then so are $d_i\mathcal{A} = \{d_i\sigma, \ \sigma \in \mathcal{A}\}$ and $s_i\mathcal{A} = \{s_i\sigma, \ \sigma \in \mathcal{A}\}$, and that $d_i \,|\, \mathcal{A}$ and $s_i \,|\, \mathcal{A}$ are proper maps of sets. For a topological space X, let $\mathcal{L}_k X$ denote the partially ordered set of locally finite subsets of $S_k X$. The face maps d_i and degeneracy maps s_i induce maps of partially ordered sets $\mathcal{L}d_i$ and $\mathcal{L}s_i$. Define $\mathcal{J}_k(X; \underset{\sim}{W})$ to be $\varinjlim_{\mathcal{A} \in \mathcal{L}_k X} \underset{\sim}{h}^{\ell f}(\mathcal{A}; \underset{\sim}{W})$. This makes sense since for $\mathcal{A}_1 \subseteq \mathcal{A}_2$ locally finite subsets of S_k, the inclusion is proper (any inclusion of sets is a proper map). Therefore, it induces a map on $\underset{\sim}{h}^{\ell f}(-; \underset{\sim}{W})$. Let $\mathcal{J}.(X; \underset{\sim}{W})$ denote the resulting simplicial spectrum.

Definition II.6. *For any topological space X and spectrum $\underset{\sim}{W}$, we define $\underset{\sim}{h}^{\ell f}(X;\underset{\sim}{W})$ to be $|\mathcal{J}.(X;\underset{\sim}{W})|$.*

We examine the degree of functoriality of this construction.

Lemma II.7. *Suppose $X \xrightarrow{f} Y$ is a proper (in the usual topological sense) map, and Y is locally compact. Suppose $\mathcal{A} \subseteq S_k X$ is a locally finite collection of singular k–simplices. Then the collection $f \circ \mathcal{A} = \{f \circ \sigma, \sigma \in \mathcal{A}\} \subseteq S_k Y$ is locally finite.*

Proof: Let $y \in Y$ be any point, and let U be a neighborhood of y, with $\bar{U}$ compact. Then

$$\{f \circ \sigma \mid \operatorname{im}(f \circ \sigma) \cap U \neq \emptyset\} \subseteq \{f \circ \sigma \mid \operatorname{im}(f \circ \sigma) \cap \bar{U} \neq \emptyset\}.$$

But the set $\{f \circ \sigma \mid \operatorname{im}(f \circ \sigma) \cap \bar{U} \neq \emptyset\}$ is the surjective image of the set $\{\sigma \mid \operatorname{im}(f \circ \sigma) \cap f^{-1}\bar{U} \neq \emptyset\}$. By the properness of f, $f^{-1}\bar{U}$ is compact. For each $x \in \bar{U}$, let B_x be an open neighborhood of x, so that $\{\sigma \in \mathcal{A} \mid \operatorname{im}(\sigma) \cap B_x \neq \emptyset\}$ is finite. The covering $\{B_x \cap \bar{U}\}_{x \in \bar{U}}$ of $\bar{U}$ has finite subcovering, say $B_{x_1}, \ldots, B_{x_s}$. Now

$$\{\sigma \mid \operatorname{im}(\sigma) \cap f^{-1}\bar{U} \neq \emptyset\} \subseteq \bigcup_{i=1}^{s} \{\sigma \mid \operatorname{im}(\sigma) \cap B_{x_i} \neq \emptyset\},$$

and the right hand side is a finite set. This gives the result. Q.E.D.

Corollary II.8. *The construction $X \to \underset{\sim}{h}^{\ell f}(X;\underset{\sim}{W})$ is functorial for proper maps of locally compact topological spaces.*

Proof: Let $f : X \to Y$ be a proper map between locally compact spaces and let $f_k^! : \mathcal{L}_k X \to \mathcal{L}_k Y$ be the order preserving map $\mathcal{A} \to f \circ \mathcal{A}$. Then we have the natural map

$$\varinjlim_{\mathcal{A} \in \mathcal{L}_k X} \underset{\sim}{h}^{\ell f}(\mathcal{A};\underset{\sim}{W}) \to \varinjlim_{\mathcal{B} \in \mathcal{L}_k Y} \underset{\sim}{h}^{\ell f}(\mathcal{B};\underset{\sim}{W})$$

induced by $f_k^!$, and hence an induced map of simplicial spectra $\mathcal{J}.(X;\underset{\sim}{W}) \to \mathcal{J}.(Y,\underset{\sim}{W})$. This clearly gives the required functoriality. Q.E.D.

We now turn to homotopy invariance and excision. We will first deal with the case where $\underset{\sim}{W}$ is an Eilenberg–MacLane spectrum $K(G,0)$. Let $\hat{G}[-]$ be defined as in Proposition II.4. If X is a topological space, we define $\hat{C}_*(X;G)$ to be the chain complex associated to the simplicial Abelian group

$\underline{k} \to \varinjlim_{\mathcal{A}\in\mathcal{L}_k X} \hat{G}[\mathcal{A}]$. Thus, $\hat{C}_*(X;G)$ is the chain complex of locally finite sums of singular simplices in X with coefficients in G. Let $\underset{\sim}{K}(-;0)$ denote the functor which assigns to an Abelian group G the Eilenberg–MacLane spectrum $K(G,0)$ corresponding to that group.

Proposition II.9. *There is a natural (with respect to proper maps of locally compact topological spaces) isomorphism*

$$H_i(\hat{C}_*(X;G)) \to \pi_i(\underset{\sim}{h}^{\ell f}(X;\underset{\sim}{K}(G,0))),$$

for all i. (For $i<0$, this just means that $\pi_i(\underset{\sim}{h}^{\ell f}(X;\underset{\sim}{K}(G,0)))=0$ for $i<0$, which follows directly from II.4.)

Proof: Consider the simplicial Abelian group $\underline{k} \to \varinjlim_{\mathcal{A}\in\mathcal{L}_k X} \hat{G}[\mathcal{A}]$. It is the levelwise zero–th space of the simplicial spectrum

$$\underline{k} \to \underset{\sim}{K}(\varinjlim_{\mathcal{A}\in\mathcal{L}_k X} \hat{G}[\mathcal{A}],0) \cong \varinjlim_{\mathcal{A}\in\mathcal{L}_k X} \hat{G}[\mathcal{A}],$$

since $\underset{\sim}{K}(-,0)$ commutes with filtered direct limits. Also, $\underset{\sim}{K}(-,0)$ commutes with inverse limits in the following sense. Let $B^n\underset{\sim}{K}(-,0)$ denote the simplicial Abelian group–valued functor which assigns to G the n–th delooping in $\underset{\sim}{K}(-,0)$. If $\underset{\sim}{A}$ is a pro–Abelian group, then the natural map

$$B^n\underset{\sim}{K}(\varprojlim \underset{\sim}{A},0) \to \varprojlim B^n(\underset{\sim}{K}(\underset{\sim}{A},0))$$

is an isomorphism of simplicial Abelian groups. We therefore have a map of spectra

$$\underset{\sim}{K}(\varprojlim \underset{\sim}{A},0) \to \underset{\underline{C}}{\operatorname{holim}}\, \underset{\sim}{K}(\underset{\sim}{A},0),$$

where $\underline{C}$ denotes the parameter category for the pro–Abelian group $\underset{\sim}{A}$. Moreover, this map is natural for morphisms of pro–Abelian groups. The simplicial Abelian group $\underline{k} \to \varinjlim_{\mathcal{A}\in\mathcal{L}_k X} \hat{G}\mathcal{A}$ is delooped n times by applying $B^n\underset{\sim}{K}(-,0)$ levelwise to obtain a bisimplicial Abelian group and taking the diagonal simplicial group. Since all face and degeneracy maps and the maps in the directed system defining $\varinjlim_{\mathcal{A}\in\mathcal{L}_k X} \hat{G}[\mathcal{A}]$ are induced by maps of pro–Abelian groups, we obtain for any locally compact space X a morphism of simplicial spectra

$$[\underline{k} \to \underset{\sim}{K}(\varinjlim_{\mathcal{A}\in\mathcal{L}_k X} \hat{G}[\mathcal{A}],0)] \to [\underline{k} \to \varinjlim_{\mathcal{A}\in\mathcal{L}_k X} \underset{U\in\mathcal{F}(\mathcal{A})^{op}}{\text{holim}} \underset{\sim}{K}(G\otimes(A[\mathcal{A}]/A[\mathcal{A}-U],0))],$$

and furthermore this morphism of simplicial spectra induces an equivalence after applying $|\ |$ since it is an equivalence in each level by II.4. On the other hand, there is a canonical equivalence of functors from based sets to spectra, $\underset{\sim}{K}(G,0)\wedge X \to \underset{\sim}{K}(G\otimes A[X],0)$, and hence an equivalence

$$\underset{U\in\mathcal{F}(\mathcal{A})^{op}}{\text{holim}} \underset{\sim}{K}(G,0)\wedge[\mathcal{A}/\mathcal{A}-U] \to \underset{U\in\mathcal{F}(\mathcal{A})^{op}}{\text{holim}} \underset{\sim}{K}(G\otimes A[\mathcal{A}]/G\otimes A[\mathcal{A}-U],0)$$

of spectra. This together with I.2. (c), (f) gives that the realization of the simplicial spectrum $\underline{k} \to \underset{\sim}{K}(\varinjlim_{\mathcal{A}\in\mathcal{L}_k X} \hat{G}[\mathcal{A}],0)$ is naturally equivalent to $\underset{\sim}{h}^{\ell f}(X;\underset{\sim}{K}(G,0))$, and since the homotopy groups of the simplicial Abelian group $\underline{k} \to \varinjlim_{\mathcal{A}\in\mathcal{L}_k X} \hat{G}[\mathcal{A}]$ are isomorphic to the homotopy groups of $\hat{C}_*(X;G)$, we obtain the result. Q.E.D.

Corollary II.10. *Let X be a locally compact space, and let I denote the closed unit interval. Then the inclusions $i_0, i_1 : X \to X\times I$ induce equivalences of spectra*

$$\underset{\sim}{h}^{\ell f}(X;\underset{\sim}{K}(G,0)) \to \underset{\sim}{h}^{\ell f}(X\times I;\underset{\sim}{K}(G,0)).$$

Proof: II.9 reduces one to showing that the induced maps $H_i(\hat{C}_*(X;G)) \to H_i(\hat{C}_*(X\times I;G))$. But one easily checks that the prism operators used to prove the corresponding result for $C_*(X,G)$ in, say [23], extend to $\hat{C}_*(X;G)$ to produce the required chain homotopies. Q.E.D.

Now, let X be a locally compact space, and let $\mathcal{U} = \{U_\alpha\}_{\alpha\in A}$ be a covering of X by subsets of X. A singular simplex $\sigma : \Delta^k \to X$ is said to be *small of order* $\mathcal{U}$ if $\sigma(\Delta^k) \subseteq U_\alpha$, for some $\alpha \in A$. An element $\sum_{\sigma\in\mathcal{L}_k x} g_\sigma \sigma = \xi$ of $\hat{C}_k(X;G)$ is said to be small of order $\mathcal{U}$ if $g_\sigma \neq 0$ implies that σ is small of order $\mathcal{U}$, i.e., if ξ is an infinite sum of singular simplices each of which is small of order $\mathcal{U}$. Let $\mathcal{M}_k X \subseteq \mathcal{L}_k X$ be the collection of locally finite subsets of singular simplices which are small of order $\mathcal{U}$, and define $\hat{C}_k^{\mathcal{U}}(X,G) \subseteq \hat{C}_k(X,G)$ by

$$\hat{C}_k^{\mathcal{U}}(X,G) = \varinjlim_{\mathcal{A}\in\mathcal{M}_k X} \varprojlim_{U\in\mathcal{F}(\mathcal{A})^{op}} G\otimes(A[\mathcal{A}]/A[\mathcal{A}-U]).$$

The $\hat{C}^{\mathcal{U}}_k(X,G)$'s fit together into a subcomplex $\hat{C}^{\mathcal{U}}_k(X,G) \subseteq \hat{C}^{\mathcal{U}}_*(X,G)$. We say the covering $\mathcal{U}$ is *excisive* if for every singular simplex $\sigma : \Delta^k \to X$, the covering $\sigma^{-1}(U_\alpha)$, $\alpha \in A$, admits a Lebesgue number. This condition holds, for instance, if $\mathcal{U}$ is an open covering or more generally if $\mathcal{U}$ is any covering so that the family of subsets $\{U^0_\alpha\}_{\alpha \in A}$ (U^0 denotes the interior of U) also covers X

Proposition II.11. *Suppose that there is an increasing union $X = \cup_{i=0}^\infty U_i$, where each U_i is compact, and $U_i \subseteq U^0_{i+1}$. This holds, for instance, in a metric space where all closed balls are compact or more generally in a countable disjoint union of such. Suppose further that $\mathcal{U}$ is an excisive covering of X. Then the inclusion $\hat{C}^{\mathcal{U}}_*(X,G) \to \hat{C}_*(X,G)$ induces an isomorphism on homology.*

Proof: We recall first from [23] that one has a natural chain transformation $Sd(X) : C_*(X,G) \to C_*(X,G)$ and a natural chain homotopy $H(X) : C_*(X,G) \to C_{*+1}(X,G)$, with $\partial H(X) + H(X)\partial = \mathrm{Id} - Sd(X)$. We show that for a space X satisfying the hypotheses of the proposition, $Sd(X)$ and $H(X)$ extend in a natural way to $\hat{C}_*(X,G)$. For any subspace $U \subseteq X$, we write $\hat{C}_*(X,U,G)$ for the quotient complex $\hat{C}_*(X;G)/\hat{C}_*(U,G)$. Let the U_n's be as in the statement of the theorem. Then it is easy to verify that $\hat{C}_*(X,G) \cong \varprojlim \hat{C}_*(X, X - U^0_n; G)$ and that

$$\hat{C}_*(X, X - U^0_n; G) \cong C_*(X, X - U^0_n; G) \cong C_*(X,G)/C_*(X - U^0_n; G);$$

the second statement requires the compactness of the U_n's. The naturality of the operators Sd and H with respect to maps of spaces shows that they extend to $\varprojlim_n C_*(X, X - U_n; G)$. Similarly, for any subspace $W \subseteq X$, let $C^{\mathcal{U}}_*(W;G)$ denote the subcomplex of all singular chains on W which are small of order $\mathcal{U}$ when viewed as chains on X, or equivalently which are small of order $\mathcal{U}_W$, where $\mathcal{U}_W$ denotes the covering $\{U_\alpha \cap W\}_{\alpha \in A}$. Also, let $C^{\mathcal{U}}_*(X,W;G)$ denote the quotient complex $C^{\mathcal{U}}_*(X;G)/C^{\mathcal{U}}_*(W,G)$. We observe, as we did for $\hat{C}_*$, that $\hat{C}^{\mathcal{U}}_*(X,G)$ is isomorphic to

$$\varprojlim_n C^{\mathcal{U}}_*(X, X - U^0_n; G).$$

We wish to show that $\hat{C}^{\mathcal{U}}_*(X;G) \to \hat{C}_*(X;G)$ induces an isomorphism on homology. We first prove surjectivity, so let ξ be an n–cycle in $\hat{C}_n(X;G)$. We will construct elements $C^k_0 \in C^{\mathcal{U}}_*(U^0_1;G)$, $C^k_1 \in C^{\mathcal{U}}_*(U^0_2 - U_0;G)$, $\ldots$, $C^k_{k-1} \in C^{\mathcal{U}}_*(U^0_k - U_{k-2};G)$, $C^k_k \in \hat{C}_*(X - U_{k-1},G)$, and $H^k \in \hat{C}_*(X - U_{k-2};G)$ so that the following conditions hold:

(i) $C^k_i = C^{k-1}_i$ for $i \le k-2$,

(ii) $C^{k-1}_{k-1} - (C^k_{k-1} + C^k_k) = \partial H^k$,

(iii) $C^0_0 = \xi$. (Note that C^0_0 is only required to lie in $\hat{C}_*(X;G)$; we conventionally set $U_{-1} = \emptyset$.)

Let $\alpha_k = \sum_{i=0}^k C^k_i$; the sum is taken by viewing all the given chains in $\hat{C}_*(X;G)$ via inclusions. Since $\alpha_{k+1} - \alpha_k = C^{k+1}_{k+1} + C^{k+1}_k - C^k_k \in \hat{C}_*(X - U_{k-1};G)$, we see that the sequence $\{\alpha_k\}_{k=0}^{\infty}$ is convergent in the inverse limit topology, and yields an element $\lim_{k\to\infty} \alpha_k = \alpha$ in $\hat{C}_*(X;G)$. Since $H^k \in \hat{C}_*(X - U_{k-1};G)$, the infinite sum $\sum_{k=1}^{\infty} H^k$ is convergent, yielding an element $\mathcal{H}$ in $\hat{C}_*(X;G)$, $\partial\mathcal{H} = \sum_{i=1}^{\infty} \alpha_i - \alpha_{i-1} = (\lim_{i\to\infty} \alpha_i) - \alpha_0$, so ξ and α differ by a boundary. Note that the projection of α_k in $\hat{C}_*(X, X - U_{k-1};G)$ lies in the subcomplex $\hat{C}^{\mathcal{U}}_*(X, X - U_{k-1};G)$, since it is represented by the sum $C^k_0 + \cdots + C^k_{k-1}$. Consequently, α lies in $\underleftarrow{\lim} C^{\mathcal{U}}_*(X, X - U_n;G) \cong \hat{C}^{\mathcal{U}}_*(X;G)$. Consequently, the homology class of ξ is in the image of the homology of $\hat{C}^{\mathcal{U}}_*(X;G)$, so the construction of the C^k_i's and H^k's would imply the desired surjectivity. We proceed with the construction of the C^k_i's and H^k's.

To perform this construction, we note that for any set X, there is an isomorphism of Abelian groups $\hat{G}[X] \cong F(X;G)$, where F denotes the Abelian group of G–valued functions on X under pointwise addition. The isomorphism is given by $\sum_{x\in X} g_x x \to \{x \to g_x\}$. The *support* of $f \in F(X,G)$, $\mathrm{Supp}\,(f)$, is the set of elements at which f does not vanish. For any $f \in F(X,G)$, and partition $\mathrm{Supp}\,(f) = A \coprod B$, we have a corresponding decomposition $f = f_A + f_B$ with $\mathrm{Supp}\,(f_A) = A$, $\mathrm{Supp}\,(f_B) = B$. Under this identification, $\underset{\mathcal{A}\in\mathcal{L}_k X}{\underleftarrow{\lim}} \hat{G}[\mathcal{A}]$ corresponds to the subgroup of $F(S_k X;G)$ of functions whose support is a locally finite subset of $S_k X$. Since $\hat{C}_n(X;G) \cong \underset{\mathcal{A}\in\mathcal{L}_n X}{\underrightarrow{\lim}} \hat{G}[\mathcal{A}]$, we now have the notion of support of an element of $\hat{C}_n(X;G)$, and if $\xi \in \hat{C}_n(X,G)$, with $\mathrm{Supp}\,(\xi) = A \coprod B$, then ξ can be written uniquely as $\xi_A + \xi_B$, with $\mathrm{Supp}(\xi_A) = A$, $\mathrm{Supp}(\xi_B) = B$. First, we set $C^0_0 = \xi$. Next, we decompose the locally finite set $\mathrm{Supp}(\xi) = A \coprod B$, where $A = \{\sigma \in \mathrm{Supp}(\xi) \mid \mathrm{im}(\sigma) \cap U_0 \neq \emptyset\}$ and $B = \{\sigma \in \mathrm{Supp}(\xi) \mid \mathrm{im}(\sigma) \cap U_0 \neq \emptyset\}$, and write $\xi = \xi_A + \xi_B$. The set A is finite by the compactness of U_0 and local compactness of X, so ξ_A is a finite sum and can be viewed as an element in $C_n(X,G)$. Using the argument for proving the usual excision theorem for singular homology, we find that there is a k so that $Sd^k(\xi_A)$ is small of order $\mathcal{U}$. In particular, $Sd^k(\xi_A) = \xi'_A + \xi''_A$, where $\xi'_A \in C_n(U^0_1;G)$ and $\xi''_A \in C_n(X - U_0;G)$. We now define $C'_0 = \xi'_A$ and $C'_1 = \xi''_A + Sd^k \xi_B$. Note that $\mathrm{Supp}(\xi_B)$ consists entirely of simplices with image disjoint from U_0, so

ξ_B, and hence $Sd^k\xi_B$, is in $\hat{C}_n(X - U_0; G)$. Consequently, $\xi''_A + Sd^k\xi_B \in \hat{C}_n(X - U_0; G)$. We must define H^1. Note that $H + SdH + \cdots Sd^{k-1}H = \tilde{H}$ satisfies $\partial\tilde{H} + \tilde{H}\partial = \mathrm{Id} - Sd^k$, so if we set $H^1 = \tilde{H}\xi$, we see that $\partial(\tilde{H}\xi) = \xi - Sd^k\xi = C_0^0 - (C'_0 + C'_1)$. H^1 is of course in $\hat{C}_n(X; G)$, which is all that was required of it. This completes the case $k = 1$, and we proceed by induction on k. Suppose $C_0^k, \ldots, C_k^k, H^k$, have been constructed as above. Set $C_i^{k+1} = C_i$ for $i \leq k-1$. Consider $C_k^k \in \hat{C}_n(X - U_{k-1}; G)$. $\mathrm{Supp}(C_k^k)$ of course consists entirely of simplices with image in $X - U_{k-1}$. Write $\mathrm{Supp}(C_k^k) = A \coprod B$, where $A = \{\sigma \in \mathrm{Supp}(C_k^k) \mid \mathrm{im}(\sigma) \cap U_k \neq \emptyset\}$ and $B = \{\sigma \in \mathrm{Supp}(C_k^k) \mid \mathrm{im}(\sigma) \cap U_k \neq \emptyset\}$. Again, by local compactness of X and compactness of U_k, the set A is finite, and we can write $C_k^k = C_A = C_B$, where $C_A \in C_n(X - U_{k-1}; G)$ and $C_B \in \hat{C}_n(X - U_k; G)$. Since C_A is a finite chain, there is an ℓ so that $Sd^\ell(C_A)$ is small of order $\mathcal{U}$, so we write $Sd^\ell(C_A) = C'_A + C''_A$, where $C'_A \in C_n(U_{k+1}^0 - U_{k-1}; G)$ and $C''_A \in C_n(X - U_k; G)$. We note that $\partial C_k^k \in C_{n-1}(U_k^0 - U_{k-1}; G)$. As before, let $\tilde{H} = H + Sd\,H + \cdots + Sd^{\ell-1}H$; $\partial\tilde{H} + \tilde{H}\partial = \mathrm{Id} - Sd^\ell$. $\tilde{H}\partial C_k^k$ is a chain in $C_n^{\mathcal{U}}(U_k^0 - U_{k-1}; G)$, by the naturality of $\tilde{H}$, and since $\partial C_k^k = -\sum_{i=0}^{k-1} C_i^k$, with C_i^k small of order $\mathcal{U}$ for all $i < k$. Now, set $C_i^{k+1} = C'_A + \tilde{H}\partial C_k^k$ and $C_{k+1}^{k+1} = C''_A + Sd^\ell C_B$. C'_A is small of order $\mathcal{U}$, so $C_k^{k+1} \in C_n^{\mathcal{U}}(U_{k+1}^0 - U_{k-1}; G)$, and C_{k+1}^{k+1} is clearly in $\hat{C}_n(X - U_k; G)$, so C_k^{k+1} and C_{k+1}^{k+1} belong to the right groups. Finally, let $H^{k+1} = \tilde{H}C_k^k$. Then H^{k+1} is clearly a chain in $\hat{C}_{n+1}(X - U_{k-1}; G)$. We have

$$\partial H^{k+1} = \partial\tilde{H}C_k^k = -\tilde{H}\partial C_k^k + C_k^k - Sd^\ell C_k^k$$
$$= C_k^k - (\tilde{H}\partial C - k^k + Sd^\ell C_k^k) = C_k^k - (\tilde{H}\partial C_k^k + C'_A + C''_A + Sd^\ell(C_B))$$
$$= C_k^k - ((\tilde{H}\partial C_k^k + C'_A) + (C''_A + Sd^\ell C_B)) = C_k^k - (C_k^{k+1} + C_{k+1}^{k+1}).$$

This gives the C_i^k's, and hence that the map $\hat{C}_*^{\mathcal{U}}(X; G) \to \hat{C}_*(X; G)$ induces a surjection on homology. The injectivity statement is proved in an entirely similar way, and leave the proof to the reader. Q.E.D.

We now use these results to obtain homotopy invariance and excision results for the functor $\underset{\sim}{h}^{\ell f}(X, \underset{\sim}{A})$, where $\underset{\sim}{A}$ is an arbitrary spectrum.

Lemma II.12. *Let $\underset{\sim}{A} \to \underset{\sim}{B} \to \underset{\sim}{C}$ be a homotopy fibre sequence of spectra, all of which are n–connected for some (possibly negative) integer n. Then the induced sequence $\underset{\sim}{h}^{\ell f}(X, \underset{\sim}{A}) \to \underset{\sim}{h}^{\ell f}(X, \underset{\sim}{B}) \to \underset{\sim}{h}^{\ell f}(X, \underset{\sim}{C})$ is a homotopy fibre sequence of spectra.*

Proof: Direct from the definitions, using I.2 (c) and I.2 (f). Q.E.D.

Lemma II.13. *If $\underset{\sim}{A}$ is an n–connected spectrum and X is a locally compact space, then $\underset{\sim}{h}^{\ell f}(X; \underset{\sim}{A})$ is n–connected.*

Proof: Direct from the definition of $\underset{\sim}{h}^{\ell f}$ and Proposition II.4. Q.E.D.

We are now in a position to prove the proper homotopy invariance of $\underset{\sim}{h}^{\ell f}(-;\underset{\sim}{A})$.

Proposition II.14 *Let X be a locally compact space. Then the inclusions $i_0, i_1 : X \to X \times [0,1]$ induce isomorphisms $\underset{\sim}{h}^{\ell f}(X;\underset{\sim}{A}) \to \underset{\sim}{h}^{\ell f}(X \times [0,1];\underset{\sim}{A})$. Consequently, if $f, g : X \to Y$ are properly homotopic maps between locally compact spaces X and Y, then $\underset{\sim}{h}^{\ell f}(f;\underset{\sim}{A}) \cong \underset{\sim}{h}^{\ell f}(g;\underset{\sim}{A})$.*

Proof: By II.10, the result holds for the Eilenberg–MacLane spectrum $\underset{\sim}{K}(G,0)$. A direct application of II.12 shows that it holds for $\underset{\sim}{K}(G,n)$, and for any spectrum with finitely many non–zero homotopy groups. Consider now any n–connected spectrum $\underset{\sim}{A}$, and consider any map $\underset{\sim}{A} \to \underset{\sim}{B}$, where the maps $\pi_i(\underset{\sim}{A}) \to \pi_i(\underset{\sim}{B})$ are isomorphisms for $i \leq n+k$, and $\underset{\sim}{B}$ has only finitely many non–zero homotopy groups. For instance, use the Postnikov tower to obtain such a sequence. Consider the diagram

$$\begin{array}{ccc}
\pi_j(\underset{\sim}{h}^{\ell f}(X,\underset{\sim}{A})) & \longrightarrow & \pi_j(\underset{\sim}{h}^{\ell f}(X \times I;\underset{\sim}{A})) \\
\big\downarrow & & \big\downarrow \\
\pi_j(\underset{\sim}{h}^{\ell f}(X,\underset{\sim}{B})) & \longrightarrow & \pi_j(\underset{\sim}{h}^{\ell f}(X \times I;\underset{\sim}{B}))
\end{array}$$

for $j < n+k$. The two vertical arrows are isomorphisms by II.12 and II.13, and the lower horizontal arrow is an isomorphism since $\underset{\sim}{B}$ has only finitely many non–zero homotopy groups. Consequently, the result holds for spectra which are n–connected for some n. The general result now follows by passing to direct limits over n. Q.E.D.

We now deal with excision. Let X be a locally compact space, and suppose $X = \cup_{i=0}^{\infty} U_i$, where U_i is compact and $U_{i+1}^0 \supset U_i$. Suppose $X = U \cup V$, where U and V are closed subsets of X, so that $X = U^0 \cup V^0$. The covering $\mathcal{U} = \{U, V\}$ is excisive. Let $\underset{\sim}{A}$ be any spectrum; then we have a diagram of spectra

$$\begin{array}{ccc}
\underset{\sim}{h}^{\ell f}(U \cap V,\underset{\sim}{A}) & \longrightarrow & \underset{\sim}{h}^{\ell f}(V,\underset{\sim}{A}) \\
\big\downarrow & & \big\downarrow \\
\underset{\sim}{h}^{\ell f}(U,\underset{\sim}{A}) & \longrightarrow & \underset{\sim}{h}^{\ell f}(X,\underset{\sim}{A})
\end{array}$$

Let $\mathcal{P}(U, V, \underset{\sim}{A})$ denote the homotopy pushout of the diagram below.

$$\begin{array}{ccc} \underset{\sim}{h}^{\ell f}(U \cap V, \underset{\sim}{A}) & \longrightarrow & \underset{\sim}{h}^{\ell f}(V, \underset{\sim}{A}) \\ \downarrow & & \\ \underset{\sim}{h}^{\ell f}(U, \underset{\sim}{A}) & & \end{array}$$

Then there is a natural map $\mathcal{P}(U, V, \underset{\sim}{A}) \xrightarrow{\alpha} \underset{\sim}{h}^{\ell f}(X, \underset{\sim}{A})$.

Proposition II.15. *α is a weak equivalence of spectra.*

Proof: We first deal with the case $\underset{\sim}{A} = \underset{\sim}{K}(G, 0)$, where G is an Abelian group. From the definition of homotopy pushouts, there is a long exact Mayer–Vietoris sequence

$$\cdots \to \pi_i(\underset{\sim}{h}^{\ell f}(U \cap V, \underset{\sim}{A})) \to \pi_i(\underset{\sim}{h}^{\ell f}(U, \underset{\sim}{A})) \oplus \pi_i(\underset{\sim}{h}^{\ell f}(V, \underset{\sim}{A}))$$

$$\cdots \to \pi_i(\mathcal{P}(U, V, \underset{\sim}{A})) \to \cdots$$

On the other hand, since U and V are closed, $\hat{C}^{\mathcal{U}}_*(X, G)$ is the pushout in the category of chain complexes of the diagram

$$\begin{array}{ccc} \hat{C}_*(U \cap V; G) & \longrightarrow & \hat{C}_*(V; G) \\ \downarrow & & \\ \hat{C}(U; G) & & \end{array}$$

so we have a long exact sequence

$$\cdots \to H_i(\hat{C}_*(U \cap V, G)) \to H_i(\hat{C}_*(U, G)) \oplus H_i(\hat{C}_*(V, G))$$
$$\cdots \to H_i(\hat{C}^{\mathcal{U}}_*(X; G)) \to \cdots$$

and it follows easily from II.9 that these long exact sequences are identified, and that the map $\pi_i(\mathcal{P}(U, V, \underset{\sim}{A})) \to \pi_i(\underset{\sim}{h}^{\ell f}(X, \underset{\sim}{A}))$ is identified with the map $H_i(\hat{C}^{\mathcal{U}}_*(X; G)) \to H_i(\hat{C}_*(X; G))$. The result for $\underset{\sim}{A} = \underset{\sim}{K}(G, 0)$ now follows from II.11.

If $\underset{\sim}{A} = \underset{\sim}{K}(G, n)$, the result follows from II.12 using the homotopy fibre sequences $\underset{\sim}{K}(G, n) \to * \to \underset{\sim}{K}(G, n+1)$ of spectra, and I.2 (c), which applies since a homotopy pushout can be viewed as a homotopy colimit. We also apply I.2 (c) to obtain a proof in the case where $\underset{\sim}{A}$ has only finitely many non–zero homotopy groups by induction on the number of non–zero groups.

An argument similar to that in the proof of II.14 gives the result for an n–connected spectrum, where n is any integer. Here one uses the fact that a homotopy pushout of k–connected spectra is again k–connected.

Finally, the case of a general spectrum follows by a passage to direct limits over n. Q.E.D.

We now prove some miscellaneous results about $\underset{\sim}{h}^{\ell f}$ which we will find useful later. Let $\underset{\sim}{Z}_\alpha$, $\alpha \in A$ denote any indexed family of Ω–spectra. Let $\prod_{\alpha\in A} \underset{\sim}{Z}_\alpha$ denote the spectrum whose k–th space is the product over $\alpha \in A$ of the k–th spaces of the spectra $\underset{\sim}{Z}_\alpha$. If X is a locally compact space, and $X = \coprod_{\alpha\in A} X_\alpha$, one checks the following Proposition directly from the definitions.

Proposition II.16 *There is a natural (for families of proper maps indexed over A between families of locally compact spaces indexed over A) equivalence of spectra*

$$\underset{\sim}{h}^{\ell f}\left(\coprod_{\alpha\in A} X_\alpha; \underset{\sim}{W}\right) \to \prod_{\alpha\in A} \underset{\sim}{h}^{\ell f}(X_\alpha; \underset{\sim}{W}).$$

Recall from the beginning of this section that if $X.$ is a locally finite simplicial set, then there is a simplicial spectrum obtained by applying $\underset{\sim}{h}^{\ell f}(-; \underset{\sim}{A})$ levelwise, and we denote it now by ${}^s\underset{\sim}{h}^{\ell f}(X.; A)$. This ${}^s\underset{\sim}{h}^{\ell f}(X.; A)$ is much "smaller" than the locally finite homology spectrum of the geometric realization of $X.$ with coefficients in $\underset{\sim}{A}$; it bears the same relationship to that spectrum as the homology of the complex of simplicial chains does to singular homology. There is of course a natural map ${}^s\underset{\sim}{h}^{\ell f}(X.; \underset{\sim}{A}) \to \underset{\sim}{h}^{\ell f}(\|X.\|; \underset{\sim}{A})$.

Corollary II.17 *If $X.$ is a finite dimensional locally finite simplicial set, then $\|X.\|$ is a locally compact space and the map*

$${}^s\underset{\sim}{h}^{\ell f}(X.; A) \to \underset{\sim}{h}^{\ell f}(\|X.\|; \underset{\sim}{A})$$

is an equivalence.

Proof: This is an easy consequence of II.1, since $\|X.\|$ can be expressed as a finite iterated pushout, and since the result holds for arbitrary disjoint unions of simplices of a fixed dimension by II.14 or II.16. Q.E.D.

We also wish to look at a version of this theory associated with metric spaces. It will be convenient to introduce a slightly more general notion of metric space than is usual.

Definition II.18. *A metric space is a set X together with a function $d: X \times X \to [0, +\infty) \cup \{+\infty\}$, satisfying the following conditions.*

(a.) $d(x,y) = d(y,x)$

(b.) $d(x,y) = 0 \Leftrightarrow x = y$

(c.) $d(x,z) \le d(x,y) + d(y,z)$

Condition (c) is interpreted in the obvious way when some value is infinite. If $\{X_\alpha\}_{\alpha \in A}$ is an indexed family of metric spaces we let $\coprod_{\alpha \in A} X_\alpha$ denote the metric space whose underlying set is $\coprod_{\alpha \in A} X_\alpha$ and where the metric is defined by $d(x_1, x_2) = d_\alpha(x_1, x_2)$ if $x_1, x_2 \in X_\alpha$ and d_α denotes the metric on X_α. $d(x_1, x_2) = +\infty$ if $x_1 \in X_\alpha$, $x_2 \in X_\beta$, $\alpha \neq \beta$. As in the case of an ordinary metric space, i.e., where $d(x,y)$ never takes the value $+\infty$, we associate a topology on X to the metric. Let $\sim$ denote the equivalence relation on X given by $x \sim y \Leftrightarrow d(x,y) < +\infty$, and let $\pi_0(X) = X/\sim$. For each $\alpha \in \pi_0(X)$, we let X_α denote the subspace of X consisting of the equivalence class α. Then X is isometric to $\coprod_{\alpha \in \pi_0 X} X_\alpha$, and the underlying spaces are homeomorphic.

Recall that $\mathcal{L}_k X$ denotes the collection of all locally finite subsets of $S_k X$. We define $\mathcal{B}_k X \subseteq \mathcal{L}_k X$ to be the collection of all locally finite subsets $\mathcal{A}$ of $S_k X$ so that there exists a number N for which $\operatorname{diam}(\operatorname{image}(\sigma)) < N$ for all $\sigma \in \mathcal{A}$, and so that for every ball $B_r(x)$ in X, the set

$$\{\sigma \in \mathcal{A} \mid \operatorname{image}(\sigma) \cap B_r(x) \neq \emptyset\}$$

is finite. We define an associated subcomplex ${}^b\hat{C}_*(X;G) \subseteq \hat{C}_*(X;G)$ by letting

$${}^b\hat{C}_k(X,G) = \{\xi \in \hat{C}_k(X,G) \mid \operatorname{Supp}(\xi) \in \mathcal{B}_k X\}.$$

We are thinking of this as the complex of locally finite chains on simplices of uniformly bounded diameter.

Proposition II.19. *Let X be a metric space for which d takes only finite values and in which all closed balls are compact, or more generally a countable disjoint union of such. Then the inclusion ${}^b\hat{C}_*(X;G) \to \hat{C}_*(X;G)$ induces an isomorphism on homology.*

Proof: Let $\mathcal{U}_r$ denote the covering of X by balls of radius r. We claim that a chain $\xi \in \hat{C}_*(X;G)$ lies in ${}^b\hat{C}_*(X;G)$ if and only if it is small of order $\mathcal{U}_r$ for some r. For if $\xi \in \hat{C}_*(X;G)$ is a chain so that the image of any singular simplex in Supp(ξ) has diameter less than N, then the chain is clearly small of order $\mathcal{U}_N$. Conversely, if ξ is small of order $\mathcal{U}_r$, then the diameter of any singular simplex in Supp(ξ) is less than or equal to $2r$. Moreover, since by hypothesis, any closed ball is compact, we see that if ξ is locally finite, the set $\{\sigma \in \text{supp}(\xi) \mid \text{image}(\sigma) \cap B_R(x)\}$ is finite for all R and x. Consequently, if ξ is small of order $\mathcal{U}_r$ for some r, it lies in ${}^b\hat{C}_*(X;G)$. Let $\hat{C}_*^{\mathcal{U}_r}(X;G)$ denote the subcomplex of $\hat{C}_*(X;G)$ consisting of chains small of order $\mathcal{U}_r$. We have shown that

$$ {}^b\hat{C}_*(X;G) = \bigcup_r \hat{C}_*^{\mathcal{U}_r}(X;G) \subseteq \hat{C}_*(X;G). $$

If we can show that each inclusion $\hat{C}_*^{\mathcal{U}_r}(X;G) \to \hat{C}(X;G)$ induces an isomorphism on homology, then the result will follow. But $\mathcal{U}_r$ is clearly excisive, so it remains to show that X satisfies the hypotheses of II.11. We write $X = \coprod_{i=0}^{\infty} X_i$, where each X_i is a metric space for which d takes only finite values, and in which all closed balls are compact. For each i, fix a point $x_i \in X_i$. Let $U_n = \coprod_{i=0}^{n} B_n(x_i)$; then the U_i's are an increasing family of subsets satisfying the hypotheses of II.11. This gives the result. Q.E.D.

One now defines the analogous theory ${}^b\underset{\sim}{h}^{\ell f}(-,\underset{\sim}{A})$, for $\underset{\sim}{A}$ any spectrum, to be the realization of the simplicial spectrum $\underline{k} \to \varinjlim_{\mathcal{A}\in B_k} \underset{\sim}{h}^{\ell f}(\mathcal{A};\underset{\sim}{A})$. One has the following results about this theory, analogous to the results about $\underset{\sim}{h}^{\ell f}(X;\underset{\sim}{A})$.

Proposition II.20

(a) *Let $\underset{\sim}{A} = \underset{\sim}{K}(G,0)$ be an Eilenberg–MacLane spectrum. Then there is a natural isomorphism*

$$ \pi_i({}^b\underset{\sim}{h}^{\ell f}(X;\underset{\sim}{A})) \cong H_i({}^b\hat{C}_*(X;G)) $$

compatible with the analogous isomorphism

$$ \pi_i(\underset{\sim}{h}^{\ell f}(X;\underset{\sim}{A})) \cong H_i(\hat{C}_*(X;G)). $$

(b) *Let $\underset{\sim}{A} \to \underset{\sim}{B} \to \underset{\sim}{C}$ be a fibre sequence of spectra up to homotopy. Then the induced sequence*

$$ {}^b\underset{\sim}{h}^{\ell f}(X;\underset{\sim}{A}) \to {}^b\underset{\sim}{h}^{\ell f}(X;\underset{\sim}{B}) \to {}^b\underset{\sim}{h}^{\ell f}(X;\underset{\sim}{C}) $$

is a fibre sequence of spectra up to homotopy.

(c) *Let $\underset{\sim}{A}$ be an n–connected spectrum. Then ${}^b\underset{\sim}{h}^{\ell f}(X;\underset{\sim}{A})$ is also n–connected.*

Corollary II.21 *Let X be a countable disjoint union of metric spaces X_i, so that in each X_i, d takes only finite values, and so that all closed balls in X_i are compact. Then the natural map ${}^b\underset{\sim}{h}^{\ell f}(X;\underset{\sim}{A}) \to \underset{\sim}{h}^{\ell f}(X;\underset{\sim}{A})$ is an equivalence of spectra.*

Proof: Follows from II.19, via an induction similar to that used in the proof of II.14, using the results of II.20. Q.E.D.

Finally, we discuss the equivariant situation. Let X be any left Γ–set, where Γ is a group. Then, because of the functoriality of the $\underset{\sim}{h}^{\ell f}(-;A)$ construction, $\underset{\sim}{h}^{\ell f}(X;A)$ becomes a spectrum with left Γ–action. On the other hand, $\mathcal{F}(X)^{op}$ is a category with left Γ–action. Let $\underset{\sim}{A}$ be any spectrum. We define a functor $\alpha_A : \underline{\Gamma} \wr \mathcal{F}(X)^{op} \to \underline{\mathcal{S}}$ on objects by $\alpha_A(e,U) = \underset{\sim}{A} \wedge (X/(X - U))$, and on morphisms by $\alpha(\gamma, e) = \mathrm{Id}_{\underset{\sim}{A}} \wedge (\gamma\cdot : X/(X-U) \to X/(X-\gamma U))$ and $\alpha(e, U \subseteq V) = \mathrm{Id}_{\underset{\sim}{A}} \wedge (X/(X-V) \to X/(X-U))$. By the discussion preceding I.17, we obtain a right Γ–action on $\underset{\overleftarrow{\mathcal{F}(X)^{op}}}{\mathrm{holim}}\, \Phi^X_{\underset{\sim}{A}} = \underset{\sim}{h}^{\ell f}(X;\underset{\sim}{A})$, and one checks that one obtains the first left Γ–action on $\underset{\sim}{h}^{\ell f}(X;\underset{\sim}{A})$ from the second right action via the the anti automorphism $\gamma \to \gamma^{-1}$ of Γ.

Proposition II.22 *Let X be a free left Γ–set, where Γ is a torsion–free group. Let $\underset{\sim}{h}^{\ell f}(X,\underset{\sim}{A})$ be given the above described left Γ–action. Then the natural map $\underset{\sim}{h}^{\ell f}(X,\underset{\sim}{A}) \to \underset{\sim}{h}^{\ell f}(X,A)^{h\Gamma}$ is an equivalence of spectra.*

Proof: The torsion–freeness of Γ implies that the action of Γ on the objects of $\mathcal{F}(X)^{op}$ is free, so I.18 applies. Q.E.D.

Corollary II.23 *Let Γ be a group which acts freely on a contractible finite dimensional simplicial complex E. Let $X.$ denote any locally finite simplicial free Γ–set. Let ${}^s\underset{\sim}{h}^{\ell f}(X.,\underset{\sim}{A})$ be as in II.17; ${}^s\underset{\sim}{h}^{\ell f}(X.;\underset{\sim}{A})$ is obtained by applying the functor $\underset{\sim}{h}^{\ell f}(-;\underset{\sim}{A})$: $\underline{\text{sets}^p} \to \underline{\mathcal{S}}$ levelwise to $X.$. ${}^s\underset{\sim}{h}^{\ell f}(X.,\underset{\sim}{A})$ is of course also a spectrum with Γ–action. The natural map ${}^s\underset{\sim}{h}^{\ell f}(X.,\underset{\sim}{A})^{\Gamma} \to {}^s\underset{\sim}{h}^{\ell f}(X.,\underset{\sim}{A})^{h\Gamma}$ is a weak equivalence of spectra.*

Proof: Since E is finite dimensional, the functor $(-)^{h\Gamma} = F(E,-)^{\Gamma}$ commutes with $|\cdot|$, so the result follows from II.22. Q.E.D.

We also wish to describe ${}^{s}\underset{\sim}{h}^{\ell f}(X;\underset{\sim}{A})^{\Gamma}$ explicitly. To get an idea of what the fixed set should be, we note that if X is any left Γ–set, then $\widehat{\mathbb{Z}}[X]^{\Gamma}$ is the product of copies of $\mathbb{Z}$, one for each Γ– orbit in X, or equivalently $\widehat{\mathbb{Z}}[\Gamma\backslash X]$. In fact, on the category of free left Γ– sets, the abelian group valued functors $X \to \widehat{\mathbb{Z}}[X]^{\Gamma}$ and $X \to \widehat{\mathbb{Z}}[\Gamma\backslash X]$ are isomorphic. The analogous result for spectra, which we now prove, is a bit complicated technically, but the intuition should be clear. Let X be a free left Γ–set, where Γ is any torsion-free group. Let $\mathcal{F}_0(X)^{op}$ denote the full subcategory of $\mathcal{F}(X)^{op}$ consisting of those finite sets which intersect each Γ–orbit of X in at most one element. $\mathcal{F}_0(X)^{op}$ is closed under the Γ–action, and we obtain an equivariant restriction map

$$\underset{\mathcal{F}(X)^{op}}{\underleftarrow{\mathrm{holim}}}\, \Phi^X_{\underset{\sim}{A}} \to \underset{\mathcal{F}_0(X)^{op}}{\underleftarrow{\mathrm{holim}}}\, \Phi^X_{\underset{\sim}{A}} \mid \mathcal{F}_0(X)^{op}\,.$$

Proposition II.24 *The construction* $X \to \underset{\mathcal{F}_0(X)^{op}}{\underleftarrow{\mathrm{holim}}}\, \Phi^X_{\underset{\sim}{A}} \mid \mathcal{F}_0(X)^{op}$ *is functorial for proper equivariant maps of free* Γ*–sets. The induced map on fixed point sets*

$$\left(\underset{\mathcal{F}(X)^{op}}{\underleftarrow{\mathrm{holim}}}\, \Phi^X_{\underset{\sim}{A}}\right)^{\Gamma} \to \left(\underset{\mathcal{F}_0(X)^{op}}{\underleftarrow{\mathrm{holim}}}\, \Phi^X_{\underset{\sim}{A}} \mid \mathcal{F}_0(X)^{op}\right)^{\Gamma}$$

of the above mentioned restriction map is a natural weak equivalence of functors from the category of free Γ*–sets and proper equivariant maps to* $\underset{\sim}{\mathcal{S}}$.

Proof: The functoriality for proper equivariant maps follows immediately from the observation that if $f : X \to Y$ is a proper equivariant map, where X and Y are free Γ–sets, and $U \in \mathcal{F}_0(Y)$, then $f^{-1}U \in \mathcal{F}_0(X)$. This gives an equivariant functor $f_0^! : \mathcal{F}_0(Y) \to \mathcal{F}_0(X)$, and the functoriality is now shown exactly as in the case of $\mathcal{F}(X)$. The naturality of the transformation is direct. To see that it is an equivalence, we observe that the Γ actions on $\mathcal{F}(X)^{op}$ and $\mathcal{F}_0(X)^{op}$ are both free on objects and hence that we have a commutative diagram

$$\begin{array}{ccc}
\left(\underset{\mathcal{F}(X)^{op}}{\underleftarrow{\mathrm{holim}}}\, \Phi^X_{\underset{\sim}{A}}\right)^{\Gamma} & \longrightarrow & \left(\underset{\mathcal{F}(X)^{op}}{\underleftarrow{\mathrm{holim}}}\, \Phi^X_{\underset{\sim}{A}}\right)^{h\Gamma} \\
\downarrow & & \downarrow \\
\left(\underset{\mathcal{F}(X)^{op}}{\underleftarrow{\mathrm{holim}}}\, (\Phi^X_{\underset{\sim}{A}} \mid \mathcal{F}_0(X)^{op})\right)^{\Gamma} & \longrightarrow & \left(\underset{\mathcal{F}(X)^{op}}{\underleftarrow{\mathrm{holim}}}\, (\Phi^X_{\underset{\sim}{A}} \mid \mathcal{F}_0(X)^{op})\right)^{h\Gamma}
\end{array}$$

where the horizontal arrows are equivalences by I.18, so it suffices to prove that the right hand vertical arrow is an equivalence. By I.2 (c), it suffices to show that the restriction

$$\underset{\mathcal{F}(X)^{op}}{\underleftarrow{\text{holim}}}\, \Phi^X_{\underset{\sim}{A}} \to \underset{\mathcal{F}_0(X)^{op}}{\underleftarrow{\text{holim}}}\, \Phi^X_{\underset{\sim}{A}} \mid \mathcal{F}_0(X)^{op}$$

(on ambient spectra, not fixed point spectra) is an equivalence. As in the proof of II.4, it will suffice to prove that the derived functors $\underset{\mathcal{F}_0(X)^{op}}{\underleftarrow{\lim}}{}^p - G_A$ vanish for $p > 0$, where $G_A : \mathcal{F}_0(X)^{op} \to \underline{\text{Ab}}$ is the functor $U \to F(U, A)$, for any Abelian group A. It is direct to verify that

$$\underset{\mathcal{F}(X)^{op}}{\underleftarrow{\lim}}\ \pi_q \circ \Phi^X_{\underset{\sim}{A}} \to \underset{\mathcal{F}_0(X)^{op}}{\underleftarrow{\lim}}\ \pi_q \circ (\Phi^X_{\underset{\sim}{A}} \mid \mathcal{F}_0(X)^{op})$$

is an isomorphism. Precisely as in the proof of II.4, the derived functors are computed by applying the contravariant functor $U \to F(U, A)$ from $\underline{\text{sets}}$ to $\underline{\text{Ab}}$ to the simplicial space $\underline{k} \to \coprod_{U_0 \subseteq \cdots \subseteq U_k} H(U_0)$, where $U_0 \subseteq \cdots \subseteq U_k$ is an increasing chain of subsets in $\mathcal{F}_0(X)$ and $H : \mathcal{F}_0(X) \to \underline{\text{sets}}$ denotes the functor $H(U) = U$. As in the proof of II.4, we note that the diagonal of this simplicial space has the homotopy type of $\mathcal{F}_0(X) \wr H$, and the result follows from the observation that the functor $i : \mathcal{F}(X)^\delta \to \mathcal{F}(X) \wr H$ actually has image in $\mathcal{F}_0(X) \wr H$. Q.E.D.

Let $U \in \mathcal{F}_0(X)^{op}$; let U_Γ denote the image of U in $\Gamma\backslash X$. Note that the natural projection $U \to U_\Gamma$ is a bijection for $U \in \mathcal{F}_0(X)^{op}$, and similarly that $X/(X - U) \to (\Gamma\backslash X)/(\Gamma\backslash X - U_\Gamma)$ is a bijection of based sets. Therefore, if we set

$$\Psi^X_{\underset{\sim}{A}} = [U \to \underset{\sim}{A} \wedge ((\Gamma\backslash X)/(\Gamma\backslash X) - U_\Gamma)]$$

we obtain an equivalence of functors

$$\Phi^X_{\underset{\sim}{A}} \mid \mathcal{F}_0(X)^{op} \to \Psi^X_{\underset{\sim}{A}}.$$

on $\mathcal{F}_0(X)^{op}$. Moreover, this equivalence is functorial in X for proper, Γ–equivariant maps between free Γ–sets. Consequently, we obtain an equivalence of functors

$$\text{(I)} \qquad \left[X \to \underset{\mathcal{F}_0(X)^{op}}{\underleftarrow{\text{holim}}}\, \Phi^X_{\underset{\sim}{A}} \mid \mathcal{F}_0(X)^{op} \right] \to \left[X \to \underset{\mathcal{F}_0(X)^{op}}{\underleftarrow{\text{holim}}}\, \Psi^X_{\underset{\sim}{A}} \right]$$

on the category of free Γ–sets and proper equivariant maps. $\underset{U \in \mathcal{F}_0(X)^{op}}{\underleftarrow{\text{holim}}}\ \Phi^X_{\underset{\sim}{A}}$ is given a Γ–action by restricting the functor $\underline{\Gamma} \wr \mathcal{F}(X)^{op} \xrightarrow{\alpha_A} \underline{S}$ used in defining the group action on $\underset{\sim}{h}^{\ell f}(X;\underset{\sim}{W})$ to the subcategory $\Gamma \wr \mathcal{F}_0(X)^{op}$. Similarly, we define an action of Γ on $\underset{U\in\mathcal{F}_0(X)^{op}}{\underset{\longleftarrow}{\text{holim}}}\ \Psi^X_{\underset{\sim}{A}}$ by defining a second functor β_A : $\underline{\Gamma}\wr\mathcal{F}_0(X)^{op} \to \underline{S}$ as follows. On objects, $\beta_A(e,U) = \underset{\sim}{A}\wedge((\Gamma\backslash X)/(\Gamma\backslash X - U_\Gamma))$, and on morphisms we let

$$\beta_A(e, U \subseteq V) = \underset{\sim}{A} \wedge ((\Gamma\backslash X)/(\Gamma\backslash X - V_\Gamma)) \to \underset{\sim}{A} \wedge ((\Gamma\backslash X)/(\Gamma\backslash X - U_\Gamma))$$

be the projection and $\beta_A(\gamma, \text{Id}_U)$ be the identity map on $\underset{\sim}{A}\wedge((\Gamma\backslash X)/(\Gamma\backslash X - U_\Gamma))$. There is a natural transformation $\alpha_A \to \beta_A$, given on objects (e,U) by the projection $\underset{\sim}{A} \wedge (X/(X-U)) \to \underset{\sim}{A} \wedge ((\Gamma\backslash X)/(\Gamma\backslash X - U_\Gamma))$, which is an equivalence since $U \in \mathcal{F}_0(X)^{op}$. Consequently, the natural transformation (I) constructed above is Γ–equivariant and is a weak equivalence of functors. Since Γ acts freely on the objects of $\mathcal{F}_0(X)^{op}$, a straightforward argument using I.18 shows that the map

$$\left[\underset{U\in\mathcal{F}_0(X)^{op}}{\underset{\longleftarrow}{\text{holim}}}\ \Phi^X_{\underset{\sim}{A}}\right]^\Gamma \to \left[\underset{U\in\mathcal{F}_0(X)^{op}}{\underset{\longleftarrow}{\text{holim}}}\ \Psi^X_{\underset{\sim}{A}}\right]^\Gamma$$

is an equivalence.

Now, let $\pi : \mathcal{F}_0(X)^{op} \to \mathcal{F}(\Gamma\backslash X)^{op}$ denote the functor $\pi(U) = U_\Gamma$. Then it is easy to see that $\Psi^X_{\underset{\sim}{A}} = \Phi^{\Gamma\backslash X}_{\underset{\sim}{A}} \circ \pi$. We obtain an equivariant natural transformation (again on the category of free Γ–sets and proper Γ–equivariant maps)

$$\left[X \to \underset{\mathcal{F}(\Gamma\backslash X)^{op}}{\underset{\longleftarrow}{\text{holim}}}\ \Phi^{\Gamma\backslash X}_{\underset{\sim}{A}}\right] \to \left[X \to \underset{\mathcal{F}_0(X)}{\underset{\longleftarrow}{\text{holim}}}\, \Psi^X_{\underset{\sim}{A}}\right],$$

where the action on the left hand side is trivial.

Proposition II.25 *The map*

$$\left[X \to \underset{\mathcal{F}(\Gamma\backslash X)^{op}}{\underset{\longleftarrow}{\text{holim}}}\ \Phi^{\Gamma\backslash X}_{\underset{\sim}{A}}\right] \to \left[X \to \underset{\mathcal{F}_0(X)}{\underset{\longleftarrow}{\text{holim}}} \left(\Psi^X_{\underset{\sim}{A}}\right)\right]^\Gamma$$

is a weak equivalence of functors on the category of free Γ–sets and proper equivariant maps.

Proof: Let $\mathcal{F}^\delta(\Gamma\backslash X)^{op}$ and $\mathcal{F}_0^\delta(X)$ denote the full subcategories of subsets with exactly one element. Then using argument similar to that in the proofs of II.4 and II.24, we find that we have a commutative diagram

$$\text{(I)}\qquad \begin{array}{ccc} \underset{\mathcal{F}(\Gamma\backslash X)^{op}}{\underleftarrow{\text{holim}}}\ \Phi^{\Gamma\backslash X}_{\underset{\sim}{A}} & \longrightarrow & (\underset{\mathcal{F}_0(X)^{op}}{\underleftarrow{\text{holim}}}\ \Psi^X_{\underset{\sim}{A}})^\Gamma \\ \downarrow & & \downarrow \\ \underset{\mathcal{F}^\delta(\Gamma\backslash X)^{op}}{\underleftarrow{\text{holim}}}\ \Phi^{\Gamma\backslash X}_{\underset{\sim}{A}} & \longrightarrow & (\underset{\mathcal{F}^\delta_0(X)^{op}}{\underleftarrow{\text{holim}}}\ \Psi^X_{\underset{\sim}{A}})^\Gamma \end{array}$$

where the left hand vertical arrow is an equivalence. We claim the right hand arrow is an equivalence; to see this, consider the commutative diagram

$$\begin{array}{ccc} \left(\underset{\mathcal{F}_0(X)^{op}}{\underleftarrow{\text{holim}}}\ \Phi^X_{\underset{\sim}{A}}\right)^\Gamma & \xrightarrow{\sim} & \left(\underset{\mathcal{F}_0(X)^{op}}{\underleftarrow{\text{holim}}}\ \Psi^X_{\underset{\sim}{A}}\right)^{h\Gamma} \\ \downarrow & & \downarrow \\ \left(\underset{\mathcal{F}^\delta_0(X)^{op}}{\underleftarrow{\text{holim}}}\ \Phi^X_{\underset{\sim}{A}}\right)^\Gamma & \xrightarrow{\sim} & \left(\underset{\mathcal{F}^\delta_0(X)^{op}}{\underleftarrow{\text{holim}}}\ \Psi^X_{\underset{\sim}{A}}\right)^{h\Gamma} \end{array}$$

The horizontal arrows are equivalences by I.18 since Γ acts freely on the objects of $\mathcal{F}^\delta_0(X)^{op}$ and $\mathcal{F}_0(X)^{op}$; thus, to prove that the left hand arrow is an equivalence, it will suffice to prove that the right hand arrow is. But the right hand arrow will be an equivalence if $\underset{\mathcal{F}_0(X)^{op}}{\underleftarrow{\text{holim}}}\ \Psi^X_{\underset{\sim}{A}} \to \underset{\mathcal{F}^\delta_0(X)^{op}}{\underleftarrow{\text{holim}}}\ \Psi^X_{\underset{\sim}{A}}$ is. But the functor $\Psi^X_{\underset{\sim}{A}}$ is naturally equivalent to $\Phi^X_{\underset{\sim}{A}}$, and the result now follows immediately as in the proof of II.4. Thus, both vertical arrows in (I) are weak equivalences. Consequently, the proposition would follow if we could show that the lower horizontal arrow in (I) is an equivalence. But $\underset{\mathcal{F}^\delta(\Gamma\backslash X)^{op}}{\underleftarrow{\text{holim}}}\ \Phi^{\Gamma\backslash X}_{\underset{\sim}{A}}$ is just $\prod_{z\in\Gamma\backslash X} \underset{\sim}{A} \wedge \{z\}_+$, and $\underset{\mathcal{F}^\delta_0(X)^{op}}{\underleftarrow{\text{holim}}}\ \Psi^X_{\underset{\sim}{A}}$ is just $\prod_{x\in X} \underset{\sim}{A} \wedge \{x_\Gamma\}_+$, where x_Γ is the image of x in $\Gamma\backslash X$. The Γ action is give by permuting the factors in the wedge sum, since $x_\Gamma = (\gamma x)_\Gamma$ for all γ. Consequently, $(\underset{\mathcal{F}^\delta_0(X)^{op}}{\underleftarrow{\text{holim}}}\ \Psi^X_{\underset{\sim}{A}})^\Gamma$ may also be identified with $\prod_{x\in\Gamma\backslash X} \underset{\sim}{A} \wedge \{x\}_+$, and we leave it to the reader to verify that the lower vertical arrow in (I) corresponds to the identity map when the identifications are made. Q.E.D.

Corollary II.26 *The functors* $X \to \underset{\sim}{h}^{\ell f}(X; \underset{\sim}{A})^\Gamma$ *and* $X \to \underset{\sim}{h}^{\ell f}(\Gamma\backslash X; \underset{\sim}{A})$ *on the category of free* Γ*–sets and proper equivariant maps are naturally weakly equivalent.*

Corollary II.27 *Let* Γ *be any torsion free group, and let* $X.$ *be any locally finite simplicial set with free* Γ*–action. Then there is a natural equivalence* *of functors*

$$\underset{\sim}{h}^{\ell f}(\Gamma\backslash X.;\underset{\sim}{A}) \to \underset{\sim}{h}^{\ell f}(X.,\underset{\sim}{A})^{\Gamma}$$

on the category of locally finite free simplicial Γ*–sets and proper equivariant simplicial maps.*

We remark that if X is any locally compact space with Γ–action, then $\underset{\sim}{h}^{\ell f}(X;\underset{\sim}{A})$ becomes a spectrum with Γ–action. Moreover, if X is locally compact, then the orbit space $\Gamma\backslash X$ is also locally compact. The following is now a direct consequence of II.25.

Corollary II.28 *Let* Γ *be any torsion free group. Then there is a natural equivalence of functors* $\underset{\sim}{h}^{\ell f}(\Gamma\backslash X.;\underset{\sim}{A}) \to \underset{\sim}{h}^{\ell f}(X.,\underset{\sim}{A})^{\Gamma}$ *on the category of locally compact properly discontinuous free* Γ*–spaces and proper equivariant maps.*

Proof: Let $\mathcal{L}_k^{\Gamma}X$ denote the collection of Γ–invariant locally finite collections of singular simplices in X. Then the hypotheses of the corollary show that elements of $\mathcal{L}_k^{\Gamma}X$ are in bijective correspondence with elements of $\mathcal{L}_k(\Gamma\backslash X)$, and we see that $\underset{\sim}{h}^{\ell f}(X;\underset{\sim}{A})^{\Gamma}$ is the diagonalization $|\;|$ of the simplicial spectrum $\underline{k} \to \varinjlim_{\mathcal{A}\in\mathcal{L}_k(\Gamma\backslash X)} \underset{\sim}{h}^{\ell f}(\pi^{-1}\mathcal{A},\underset{\sim}{A})^{\Gamma}$, where $\pi : S.(X) \to S.(\Gamma\backslash X)$ is the projection. II.26 now shows that this is equivalent to the diagonalization of the simplicial spectrum $\underline{k} \to \varinjlim_{\mathcal{A}\in\mathcal{L}_k(\Gamma\backslash X)} \underset{\sim}{h}^{\ell f}(\mathcal{A};\underset{\sim}{A})$, and this is $\underset{\sim}{h}^{\ell f}(\Gamma\backslash X;\underset{\sim}{A})$. Q.E.D.

Similarly, suppose (X,d) is a locally compact metric space, with Γ acting on X by isometries. Then ${}^{b}\underset{\sim}{h}^{\ell f}(X;\underset{\sim}{A})$ becomes a spectrum with Γ–action; indeed, ${}^{b}\underset{\sim}{h}^{\ell f}(-;\underset{\sim}{A})$ defines a functor from the category of locally compact metric spaces with isometric Γ–action and proper, uniformly continuous, equivariant maps to the category of spectra with Γ–action. Suppose now that the action is properly discontinuous, and equip $\Gamma\backslash X$ with the orbit space metric $\Delta([x],[y]) = \min_{\gamma\in\Gamma} d(x,\gamma y)$. Let $\mathcal{B}_k^{\Gamma}X$ denote the collection of Γ–invariant elements of $\mathcal{B}_kX$. Each element of $\mathcal{B}_k^{\Gamma}(X)$ corresponds (by applying π) to an element of $\mathcal{B}_k(\Gamma\backslash X)$, but not every element in $\mathcal{B}_k(\Gamma\backslash X)$ is in the image of $\mathcal{B}_k^{\Gamma}(X)$. The point is that although all elements of $\mathcal{B}_k(\Gamma\backslash X)$ correspond to elements of $\mathcal{L}_k^{\Gamma}(X)$, these elements do not necessarily lie in $\mathcal{B}_k^{\Gamma}(X) \subseteq \mathcal{L}_k^{\Gamma}(X)$. The inclusion $\mathcal{B}_k^{\Gamma}(X) \to \mathcal{B}_k(\Gamma\backslash X)$ induces a map of simplicial spectra

$$\left[\underline{k} \to \lim_{\substack{\longrightarrow \\ \mathcal{A}\in\mathcal{B}_k^\Gamma X}} \underset{\sim}{h}^{\ell f}(\mathcal{A}; \underset{\sim}{A})^\Gamma\right] \to \left[\underline{k} \to \lim_{\substack{\longrightarrow \\ \mathcal{A}\in\mathcal{B}_k(\Gamma\backslash X)}} \underset{\sim}{h}^{\ell f}(\pi^{-1}\mathcal{A}; \underset{\sim}{A})^\Gamma\right].$$

Applying $|-|$ and II.26, we obtain a natural transformation $\nu : {}^b\underset{\sim}{h}^{\ell f}(X, \underset{\sim}{A}) \to {}^b\underset{\sim}{h}^{\ell f}(\Gamma\backslash X; \underset{\sim}{A})$ on the category of locally compact metric spaces in which all closed balls are compact, with free properly discontinuous Γ–actions and equivariant, proper, uniformly continuous maps.

Proposition II.29 *$\nu(X)$ is a weak equivalence if $\Gamma\backslash X$ is paracompact.*

Proof: We consider first the case where $\underset{\sim}{A} = \underset{\sim}{K}(G, 0)$, the Eilenberg–MacLane spectrum for G. Let ${}^b\hat{C}_*(\Gamma\backslash X) \subseteq \hat{C}_*(\Gamma\backslash; G)$ be defined as above. Let ${}^b\hat{C}'_*(\Gamma\backslash X; G) \subseteq {}^b\hat{C}_*(\Gamma\backslash X; G)$ be the subcomplex of chains $\xi \in \hat{C}_*(\Gamma\backslash X; G)$ so that $\mathrm{Supp}(\xi) \subseteq \mathrm{image}\,(\mathcal{B}_k^\Gamma(X) \to \mathcal{B}_k(X))$. An argument very similar to the proof of II.9 now shows that

$$H_*({}^b\hat{C}'_*(\Gamma\backslash X); G) \cong \pi_*({}^b\underset{\sim}{h}^{\ell f}(X; \underset{\sim}{A})^\Gamma),\ \ H_*({}^b\hat{C}_*(\Gamma\backslash X); G) \cong \pi_*({}^b\underset{\sim}{h}^{\ell f}(X; \underset{\sim}{A})),$$

and hence that it suffices to show that the inclusion ${}^b\hat{C}'_*(\Gamma\backslash X; G) \subseteq {}^b\hat{C}_*(\Gamma\backslash X; G)$ induces an isomorphism on homology. To do this, we construct a certain covering $\mathcal{U}$ of $\Gamma\backslash X$. Note that $p : X \to \Gamma\backslash X$ is an open mapping. Fix a number $R > 0$ and for every x choose an open ball $B_x \subseteq X$ so that the diameter of B_x is less than R and so that $p(B_x)$ is an open neighborhood of $p(x)$ which is evenly covered. The sets $\{p(B_x)\}_{x\in X}$ form an open covering of $\Gamma\backslash X$; by paracompactness, there is a locally finite refinement, $\{U_\alpha\}_{\alpha\in A}$, which we call $\mathcal{U}$. Note that if $U \in \mathcal{U}$, $p^{-1}U$ is a disjoint union of sets whose diameter is less than R. Let $\{\sigma_\beta\}_{\beta\in B}$ be any locally finite family of singular simplices of $\Gamma\backslash X$, which is small of order $\mathcal{U}$. Then the associated Γ–invariant element of $\mathcal{L}_k^\Gamma X$, obtained by taking all possible lifts of singular simplices σ_β to X, is in $\mathcal{B}_k^\Gamma(X)$, since any lift of σ_β clearly has diameter less than R. Consequently, we have an inclusion $\hat{C}_*^{\mathcal{U}}(\Gamma\backslash X; G) \to {}^b\hat{C}'_*(\Gamma\backslash X; G)$. The composite $\hat{C}_*^{\mathcal{U}}(\Gamma\backslash X; G) \to {}^b\hat{C}'_*(\Gamma\backslash X; G) \to {}^b\hat{C}_*(\Gamma\backslash X; G)$ induces an isomorphism on homology, since the chain maps $\hat{C}_*^{\mathcal{U}}(\Gamma\backslash X; G) \to \hat{C}_*(\Gamma\backslash X; G)$ and ${}^b\hat{C}_*(\Gamma\backslash X; G) \to \hat{C}_*(\Gamma\backslash X; G)$ do by II.11 and II.19, respectively. Consequently, ${}^b\hat{C}'_*(\Gamma\backslash X; G) \to {}^b\hat{C}_*(\Gamma\backslash X; G)$ induces a surjection on homology. To prove it induces an injection, it will suffice to prove that $\hat{C}_*^{\mathcal{U}}(\Gamma\backslash X; G) \to {}^b\hat{C}'_*(\Gamma\backslash X; G)$ induces a surjection on homology. We recall that the proof that $\hat{C}_*^{\mathcal{U}}(\Gamma\backslash X; G) \to \hat{C}_*(\Gamma\backslash X; G)$ is surjective proceeds by constructing for each $\xi \in \hat{C}_*(X; G)$ a family of elements C_i^k, $i \leq j$, and H^k, satisfying certain compatibility conditions, and showing that $\partial\left(\sum_{i=0}^\infty H^k\right) = \alpha - \xi$, where α

is small of order $\mathcal{U}$. The proof of the surjectivity of $H_*(\hat{C}_*^{\mathcal{U}}(\Gamma\backslash X;G)) \to H_*(\hat{C}_*(\Gamma\backslash X;G))$ is now obtained by noting that all the elements C_i^k and H^k can be taken to lie in ${}^b\hat{C}'_*(X;G)$ if $\xi \in {}^b\hat{C}'_*(X;G)$. For a general spectrum, the proof proceeds by induction on the Postnikov tower as in II.14. Q.E.D.

Corollary II.30 *Suppose X is a locally finite simplicial complex, with free Γ–action, where Γ is a torsion free group acting freely on a finite dimensional simplicial complex E. Then the natural map $\underset{\sim}{h}^{\ell f}(X;\underset{\sim}{A})^{\Gamma} \to \underset{\sim}{h}^{\ell f}(X;\underset{\sim}{A})^{h\Gamma}$ is an equivalence. If X is further equipped with a metric, so that all the simplices of X have uniformly bounded diameter, then ${}^b\underset{\sim}{h}^{\ell f}(X;\underset{\sim}{A}) \to {}^b\underset{\sim}{h}^{\ell f}(X;\underset{\sim}{A})^{h\Gamma}$ is an equivalence.*

Proof: It follows from II.27, II.28, and II.29 that the maps ${}^s\underset{\sim}{h}^{\ell f}(X;\underset{\sim}{A}) \to \underset{\sim}{h}^{\ell f}(X;\underset{\sim}{A})$ and ${}^s\underset{\sim}{h}^{\ell f}(X;\underset{\sim}{A}) \to {}^b\underset{\sim}{h}^{\ell f}(X;\underset{\sim}{A})$ are actually equivariant equivalences. The result now follows from II.23. Q.E.D.

III. Bounded K–theory

We will be working with the Pedersen–Weibel "bounded K–theory." See [33,34] for a complete discussion of this construction. We adopt the definition given in §II (Definition II.18) as our definition of a metric space.

Let R be a ring, and let X be a metric space. We define a category $\mathcal{C}_X(R)$ as follows.

Definition III.1

(a) *The objects of $\mathcal{C}_X(R)$ are triples (F, B, φ), where R is a free left R–module (perhaps not finitely generated), B is a basis for F, and $\varphi : B \to X$ is a function, the "labeling function", so that for any ball $B_r(x)$, $r \in R$, $x \in X$, $\varphi^{-1}(B_r(x))$ is finite.*

(b) *Let (F, B, φ) and (F', B', φ') be objects of $\mathcal{C}_X(R)$, and let $L : F \to F'$ be any R–linear homomorphism. For $\beta \in B$, $\beta' \in B'$, let elements $r_L(\beta', \beta) \in R$ be defined by the equation*

$$L(\beta) = \sum_{\beta' \in B'} r_L(\beta', \beta)\,\beta' .$$

This uniquely defines $r_L(\beta', \beta)$ since F is generated by B. We say L is bounded by N (or is bounded of filtration N) if $d(\varphi'(\beta'), \varphi(\beta)) > N$ implies $r_L(\beta', \beta) = 0$. The morphisms in $\mathcal{C}_X(R)$ are now defined to consist of all L for which there exists an N so that L is bounded by N.

Note that $\mathcal{C}_X(R)$ is an additive category. For any category $\underline{A}$, let $\widehat{\underline{A}}$ denote the *idempotent completion* of $\underline{A}$ (see [19]). The objects of $\widehat{\underline{A}}$ are pairs (a, e), where $a \in \underline{A}$ and $e : a \to a$ is a morphism so that $e^2 = e$. A morphism from (a_1, e_1) to (a_2, e_2) is a morphism $f : a_1 \to a_2$ so that $e_2 f e_1 = f$. We write $\widehat{\mathcal{C}}_X(R)$ for $\widehat{\mathcal{C}_X(R)}$. Let $f : (a_1, e_1) \to (a_2, e_2)$ be a morphism in $\widehat{\mathcal{C}}_X(R)$; say f is bounded by N (or has filtration N) if there is a morphism $f' : a_1 \to a_2$ in $\mathcal{C}_X(R)$ which is bounded by N, so that $e_2 f' = f$.

Proposition III.2 *Let $f : x \to y$ and $g : y \to z$ be morphisms in $\mathcal{C}_X(R)$ or $\widehat{\mathcal{C}}_X(R)$. If f is bounded by M and g is bounded by N, then $g \circ f$ is bounded by $M + N$.*

Proof: Clear in the case of $\mathcal{C}_X(R)$. If we are in $\widehat{\mathcal{C}_X(R)}$, let $x = (a_1, e_1)$, $y = (a_2, e_2)$, $z = (a_3, e_3)$, let $f' : a_1 \to a_2$ have a filtration M and $g' : a_2 \to a_3$ have a filtration N, and $e_2 f' = f$, $e_3 g' = g$. Then $g' \circ g'$ has filtration $M + N$, and

$$e_3 g' f' = g f f' = e_3 g e_2 f' = e_3 g f = e_3 e_3 g f e_2 = e_3 f g e_2 = g f .$$

Q.E.D.

We write $i\,\mathcal{C}_X(R)$ and $i\widehat{\mathcal{C}}_X(R)$ for the categories of isomorphisms of $\mathcal{C}_X(R)$ and $\widehat{\mathcal{C}}_X(R)$, respectively. Any object of $\mathcal{C}_X(R)$ has an underlying free R–module. Similarly, an object of $\widehat{\mathcal{C}}_X(R)$ has an underlying free R–module and an idempotent of that R–module; this corresponds to a projective R–module. A morphism $\mathcal{C}_X(R)$ (respectively $\widehat{\mathcal{C}}_X(R)$) is an isomorphism if it is an isomorphism on the underlying free R–module (or the underlying projective R–module) and if both f and f^{-1} are bounded of some filtration n. Note that it is possible for an isomorphism of underlying free R–modules to be bounded without its inverse being bounded. Choose symmetric monoidal structures on the categories of left R–modules, R–mod, and sets, so that in each case the sum of two objects is the categorical sum. Any such sum structures give rise to a symmetric monoidal structure on $i\mathcal{C}_X(R)$ and $i\widehat{\mathcal{C}}_X(R)$; for instance, in the case of $i\mathcal{C}_X(R)$, $(F, B, \varphi) \oplus (F', B', \varphi') = (F \oplus F', B \coprod B', \varphi \coprod \varphi')$, where $\coprod$ denotes the sum in sets. If the symmetric monoidal structures on $\underline{R\text{–mod}}$ and $\underline{\text{sets}}$ are permutative, then so are the structures on $i\mathcal{C}_X(R)$ and $i\widehat{\mathcal{C}}_X(R)$, and we assume permutative structures chosen once and for all, and $i\mathcal{C}_X(R)$ and $i\widehat{\mathcal{C}}_X(R)$ will denote the corresponding permutative categories.

Definition III.3 *Let* Spt : $\underline{\text{SymMon}} \to \underline{\mathcal{S}}$ *be the functor described in I.6. Then we define* $\underset{\sim}{K}(X; R) = \text{Spt}(i\mathcal{C}_X(R))$ *and* $\widetilde{\underset{\sim}{K}}(X; R) = \text{Spt}(i\widehat{\mathcal{C}}_X(R))$.

This construction is evidently covariantly functorial in the variable R. We discuss its functoriality in X.

Definition III.4 *Let (X,d) and (X',d') be metric spaces, and let $f: X \to X'$ be any function. f is said to be proper if the inverse images of sets of finite diameter are finite unions of sets of finite diameter, and f is eventually continuous if there is a function $\Phi : \mathbb{R} \to \mathbb{R}$ such that $d'(fx, fy) \leq \Phi(d(x,y))$, if $d(x,y) < +\infty$. Given a proper, eventually continuous map of metric spaces $f : X \to X'$, we obtain functors $\mathcal{C}_f : i\mathcal{C}_X(R) \to i\mathcal{C}_{X'}(R)$ and $\widehat{\mathcal{C}}_f : i\widehat{\mathcal{C}}_X(R) \to i\widehat{\mathcal{C}}_{X'}(R)$ of symmetric monoidal categories by*

$$\mathcal{C}_f(F, B, \varphi) = (F, B, f \circ \varphi)$$

and

$$\widehat{\mathcal{C}}_f(F, B, \varphi, e) = (F, B, f \circ \varphi, e).$$

On morphisms, $\mathcal{C}_f(L) = L$ and $\widehat{\mathcal{C}}_f(L) = L$. The properness condition assures that $\mathcal{C}_f(F, B, \varphi)$ is an object of $\mathcal{C}_{X'}(R)$, and the eventual continuity assures that morphisms in $i\mathcal{C}_X(R)$ are sent to morphisms in $i\mathcal{C}_{X'}(R)$. We obtain corresponding maps of spectra

$$\underset{\sim}{K}(X,R) \xrightarrow{\underset{\sim}{K}(f,R)} \underset{\sim}{K}(X',R)$$

and

$$\underset{\sim}{\widetilde{K}}(X,R) \xrightarrow{\underset{\sim}{\widetilde{K}}(f,R)} \underset{\sim}{\widetilde{K}}(X',R).$$

Let $\underline{\mu}$ denote the category of metric spaces and proper, eventually continuous maps; then the above constructions make $\underset{\sim}{K}(-;R)$ and $\underset{\sim}{\widetilde{K}}(-;R)$ into functors from $\underline{\mu}$ to $\underline{S}$.

We record some straightforward properties of this construction.

Proposition III.5 *Let (X,d) be a bounded metric space, i.e., so that $d(x,y) < N$ for all x, y and a fixed real number N. Then $\underset{\sim}{K}(X,R)$ is equivariant to the K–theory spectrum of the symmetric monoidal category of free finitely generated left R-modules, and $\underset{\sim}{\widehat{K}}(X;R)$ is equivalent to the K–theory spectrum of the category of finitely generated projective left R–modules.*

Proposition III.6 *Let d and d' be two metrics on a set X, and suppose that there are functions $\Phi : \mathbb{R} \to \mathbb{R}$ and $\Psi : \mathbb{R} \to \mathbb{R}$ so that $d(x,y) \leq \Phi(d'(x,y))$ and $d'(x,y) \leq \Psi(d(x,y))$, whenever either $d(x,y)$ or $d'(x,y)$ is finite. Then the identity map on X is eventually continuous and proper when viewed as a map $(X,d) \to (X,d')$ or $(X,d') \to (X,d)$, and induces equivalences $\underset{\sim}{K}((X,d),R) \to \underset{\sim}{K}((X,d'),R)$ and $\underset{\sim}{\widehat{K}}((X,d),R) \to \underset{\sim}{\widehat{K}}((X,d'),R)$.*

Proposition III.7 *Let X be a metric space, and let $Y \subseteq X$ be a submetric space. Suppose that there is a real number N so that for every $x \in X$, there is a point $y \in Y$ so that $d(x,y) \leq N$. Then the inclusion maps $\underset{\sim}{K}(Y,R) \xrightarrow{\underset{\sim}{K}(i,R)} \underset{\sim}{K}(X,R)$ and $\widehat{\underset{\sim}{K}}(Y,R) \xrightarrow{\widehat{\underset{\sim}{K}}(i,R)} \widehat{\underset{\sim}{K}}(X,R)$ are equivalences of spectra ($i : Y \to X$ is of course proper and eventually continuous).*

Proposition III.8 *If X and Y are metric spaces, then we define $X \times Y$ to be the metric space whose points are elements of $X \times Y$, and whose metric is given by*

$$d_{X\times Y}((x_1, y_1),(x_2, y_2)) = d_X(x_1, x_2) + d_Y(y_1, y_2)\ ,$$

suitably interpreted if $d_X(x_1, x_2)$ or $d_Y(y_1, y_2)$ is $+\infty$. Let A be a commutative ring, and suppose R_1 and R_2 are A–algebras, with A contained in the center of both R_1 and R_2. Then there is a symmetric monoidal pairing $\mu : i\mathcal{C}_X(R_1) \times i\mathcal{C}_Y(R_2) \to i\mathcal{C}_{X\times Y}(R_1 \otimes_A R_2)$, given on objects by $\mu((F_1, b_1, \varphi_1),(F_2, B_2, \varphi_2)) = (F_1 \otimes_A F_2, B_1 \times B_2, \varphi_1 \times \varphi_2)$. The pairing extends in an evident way to a symmetric monoidal pairing $\hat{\mu} : i\widehat{\mathcal{C}}_X(R_1) \times i\widehat{\mathcal{C}}_Y(R_2) \to i\widehat{\mathcal{C}}_{X\times Y}(R_1 \otimes_A R_2)$. From I.6, we obtain maps $m = m(X,Y,A)$: $\underset{\sim}{K}(X,R_1) \wedge \underset{\sim}{K}(Y,R_2) \to \underset{\sim}{K}(X \times Y, R_1 \otimes_A R_2)$ and $\hat{m} = \hat{m}(X,Y,A)$: $\widehat{\underset{\sim}{K}}(X,R_1) \wedge \widehat{\underset{\sim}{K}}(Y,R_2) \to \widehat{\underset{\sim}{K}}(X \times Y, R_1 \otimes_A R_2)$.

Pedersen and Weibel [33] continue to analyze the case of Euclidean space E^n, with standard Euclidean metric. Their result yields the following. Consider the spectrum $\underset{\sim}{K}(E^1;\mathbb{Z})$. The zeroth space of this spectrum is the group completion of the simplicial set $N.i\mathcal{C}_{E^1}(\mathbb{Z})$, so an automorphism of any object in $i\mathcal{C}_{E^1}(\mathbb{Z})$ corresponds to an element of $\pi_1(\underset{\sim}{K}(E^1;\mathbb{Z}))$. Let Z denote the object $(\mathbb{Z}[\mathbb{Z}], \mathbb{Z}, \varphi)$, where $\mathbb{Z}[\mathbb{Z}]$ denotes the free $\mathbb{Z}$–module on the set of integers $\mathbb{Z} \subseteq E^1$, $\mathbb{Z}$ denotes the basis $\mathbb{Z}$ of $\mathbb{Z}[\mathbb{Z}]$, $\varphi : \mathbb{Z} \to E^1$ is the inclusion, and let $\alpha : \mathbb{Z}[\mathbb{Z}] \to \mathbb{Z}[\mathbb{Z}]$ be the automorphism given on the basis $\mathbb{Z}$ by $\alpha(n) = n + 1$. We also let α denote the element of $\pi_1(\underset{\sim}{K}(E^1;\mathbb{Z}))$ corresponding to this α.

Theorem III.9 (Pedersen–Weibel [33], [34])

(a) *The composite*

$$\underset{\sim}{K}(X,R) \wedge S^1 \xrightarrow{\mathrm{Id}\wedge\alpha} \underset{\sim}{K}(X,R) \wedge \underset{\sim}{K}(E^1;\mathbb{Z}) \xrightarrow{\mu} \underset{\sim}{K}(X \times E^1; R)$$

induces an isomorphism on π_i for $i \geq 1$.

(b) *The natural map $\underset{\sim}{K}(X;R) \to \underset{\sim}{\widehat{K}}(X;R)$ induces an isomorphism on π_i for $i > 0$.*

The adjoint of the map of III.9 (a) gives a map $\underset{\sim}{K}(X,R) \to \Omega\underset{\sim}{K}(X \times E^1, R)$, inducing an isomorphism on π_i for $i > 0$, and we obtain a directed system of spectra whose n-th term is $\Omega^n \underset{\sim}{K}(X \times E^n; R)$. To see this, one notes that $\underset{\sim}{K}(X \times E^n; R) \cong \underset{\sim}{K}(X \times (E^1)^n; R)$, which follows from III.6. The homotopy colimit $\underset{\overrightarrow{n}}{\text{hocolim}}\, \Omega^n \underset{\sim}{K}(X \times E^n; R)$ will be denoted by $\underset{\sim}{\mathcal{K}}(X, R)$, and is in general a non-connective spectrum. One has a similar directed system $\{\Omega^N \underset{\sim}{\widehat{K}}(X \times E^n; R)\}$, and $\underset{\sim}{\widehat{\mathcal{K}}}(X; R) = \underset{\overrightarrow{n}}{\text{hocolim}}\, \Omega^n \underset{\sim}{\widehat{K}}(X \times E^n; R)$. There is a natural inclusion $\underset{\sim}{\mathcal{K}} \to \underset{\sim}{\widehat{\mathcal{K}}}$.

Proposition III.10 (Pedersen–Weibel [33])

(a) *The map $\underset{\sim}{\mathcal{K}}(X; R) \to \underset{\sim}{\widehat{\mathcal{K}}}(X; R)$ is an equivalence of spectra.*

(b) *$\pi_{-i}(\underset{\sim}{\mathcal{K}}(*, R))$ is the ith negative K-group of R, for $i > 0$, defined as in Bass [5].*

Proof: (a) is a direct consequence of III.9 (b); (b) is proved in [33]. Q.E.D.

Remark: All of the above discussion is contained implicitly or explicitly in [33] and [34].

Let $X \in \underset{\sim}{\mathcal{M}}$; we say X is *discrete* if $d(x_1, x_2) = +\infty$ whenever $x_1 \neq x_2$. The full subcategory of $\underset{\sim}{\mathcal{M}}$ on the discrete metric spaces can be identified with the category $\underline{\text{sets}}^p$ via the inclusion functor $i : \underline{\text{sets}}^p \to \underset{\sim}{\mathcal{M}}$ which assigns to any set the discrete metric space on the elements of X. Our first task is to analyze the functors $\underset{\sim}{K}(-; R) \circ i$, $\underset{\sim}{\widehat{K}}(-; R) \circ i$, and $\underset{\sim}{\mathcal{K}}(-; R) \circ i$; we will identify them with the functors $\underset{\sim}{h}^{\ell f}(-, \underset{\sim}{K}(R))$, $\underset{\sim}{h}^{\ell f}(-, \underset{\sim}{\widehat{K}}(R))$, and $\underset{\sim}{h}^{\ell f}(-, \underset{\sim}{\mathcal{K}}(R))$ respectively.

Proposition III.11 *The functors $\underset{\sim}{K}(-; R) \circ i$ and $\underset{\sim}{\widehat{K}}(-, R) \circ i$ are canonically isomorphic to the functors $X \to \prod_{x \in X} \underset{\sim}{K}(x, R)$ and $X \to \prod_{x \in X} \underset{\sim}{\widehat{K}}(x, R)$ respectively.*

Proof: From the definitions, we have isomorphisms $i\mathcal{C}_{iX}(R) \cong \prod_{x\in X} i\mathcal{C}_X(R)$ and $i\widehat{\mathcal{C}}_{iX}(R) \cong \prod_{x\in X} i\widehat{\mathcal{C}}_X(R)$ of symmetric monoidal categories; the result now follows from the definition of $\prod_{x\in X}$ and the results of [13]. Q.E.D.

Proposition III.12 *There are canonical isomorphisms of spectra*

$$\underset{\sim}{K}(X \amalg Y, R) \cong \underset{\sim}{K}(X, R) \times \underset{\sim}{K}(Y, R)$$

and

$$\underset{\sim}{\widehat{K}}(X \amalg Y, R) \cong \underset{\sim}{\widehat{K}}(X, R) \times \underset{\sim}{\widehat{K}}(Y, R),$$

and consequently induced projection maps

$$\underset{\sim}{K}(X \amalg Y, R) \to \underset{\sim}{K}(X, R),\ \underset{\sim}{K}(X \amalg Y, R) \to \underset{\sim}{K}(Y, R),$$

$$\underset{\sim}{\widehat{K}}(X \amalg Y, R) \to \underset{\sim}{\widehat{K}}(X, R),\ \underset{\sim}{\widehat{K}}(X \amalg Y, R) \to \underset{\sim}{\widehat{K}}(Y, R).$$

(Here $\times$ denotes the termwise product of spectra.)

Proof: Again, $i\mathcal{C}_{X\amalg Y}(R) \cong i\mathcal{C}_X(R) \times i\mathcal{C}_Y(R)$ and $i\widehat{\mathcal{C}}_{X\amalg Y}(R) \cong i\widehat{\mathcal{C}}_X(R) \times i\widehat{\mathcal{C}}_Y(R)$. Q.E.D.

Let $\underline{\text{sets}}^f \subseteq \underline{\text{sets}}^p$ be the full subcategory of finite sets, and let j denote the restriction of $i : \underline{\text{sets}}^p \to \underline{\mathcal{M}}$ to $\underline{\text{sets}}^f$. There is a natural transformation Θ of functors from $\underset{\sim}{K}(R) \wedge (\ )_+$ to $\underset{\sim}{K}(\ ; R) \circ j$, where $(\ )_+$ is the functor which assigns to a set X the (discrete) based space X_+, defined as follows. Let F_R denote the symmetric monoidal category of based free finitely generated left R-modules. For each $x \in X$, we have an inclusion functor $i_x : F_R \to i\mathcal{C}_X(R)$, given by $i_x((F, B)) = (F, B, \varphi_x)$, where $\varphi_x : B \to X$ is the constant function with value x. Each i_x is symmetric monoidal and hence induces a map of spectra $\vartheta_x : \underset{\sim}{K}(R) \to \underset{\sim}{K}(X; R)$, and we define $\Theta(X)$ to be $\vee_{x\in X}\vartheta_x$, where we identify $\underset{\sim}{K}(R) \wedge \{x\}_+$ with $\underset{\sim}{K}(R) \wedge S^0 = \underset{\sim}{K}(R)$.

Proposition III.13 Θ *is a natural equivalence of functors from* $\underline{\text{sets}}^f$ *to* $\underset{\sim}{S}$.

Proof: Follows directly from III.12. Q.E.D.

Corollary III.14 $\underset{\sim}{K}(-; R) \circ i$ *and* $\underset{\sim}{\widehat{K}}(-; R) \circ i$ *are naturally equivalent to* $\underset{\sim}{h}^{\ell f}(-; \underset{\sim}{K}(R))$ *and* $\underset{\sim}{h}^{\ell f}(-; \underset{\sim}{\widehat{K}}(R))$, *respectively, as functors from* $\underline{\text{sets}}^p$ *to* $\underset{\sim}{S}$.

Proof: Let X be any set, and let $\Psi_R^X : \mathcal{F}(X)^{\text{op}} \to \underset{\sim}{S}$ be defined on objects by $\Psi_R^X(U) = \underset{\sim}{K}(jU; R)$. On morphisms, $\Psi_R^X(U \subseteq V)$ is defined to be the projection $\underset{\sim}{K}(jV, R) \to \underset{\sim}{K}(jU; R)$ arising from the decomposition $\underset{\sim}{K}(jV; R) \cong$

$\underset{\sim}{K}(jU;R) \times \underset{\sim}{K}(j(V-U);R)$. The assignment $X \to (\mathcal{F}(X)^{\mathrm{op}}, \Psi_R^X)$ gives a functor η from $\underline{\mathrm{sets}}^p$ to $\underline{\mathcal{S}}^{CAT}$. To define it, one needs only to define a natural transformation $\Psi_R^X \circ f_! \to \Psi_R^Y$, if $f: X \to Y$ is a proper map. We define it to be the map $\underset{\sim}{K}(jf^{-1}U;R) \to \underset{\sim}{K}(jU;R)$ induced by $f \mid f^{-1}U$. $\underleftarrow{\mathrm{holim}} \circ \eta$ is now a functor from $\underline{\mathrm{sets}}^p$ to $\mathcal{S}$. Moreover, the natural equivalence $\Phi^X_{\underset{\sim}{K}(R)} \to \Psi_R^X$ given on U by $\Theta(U)$ gives a weak equivalence of functors $\underset{\sim}{h}^{\ell f}(-,\underset{\sim}{K}(R)) \xrightarrow{\sim} \underleftarrow{\mathrm{holim}} \circ \eta$. On the other hand, we obtain a natural transformation from $\underset{\sim}{K}(-;R) \circ i \to \underleftarrow{\mathrm{holim}} \circ \eta$ as follows. For every finite set $U \subseteq X$, we obtain a projection map $\underset{\sim}{K}(iX;R) \xrightarrow{\pi_U} \underset{\sim}{K}(jU,R)$ from the decomposition $X = (X-U) \amalg U$, using III.12. The diagrams

$$\begin{array}{ccc} \underset{\sim}{K}(iX;R) & \xrightarrow{\pi_V} & \underset{\sim}{K}(jV;R) \\ & \searrow^{\pi_U} & \downarrow \\ & & \underset{\sim}{K}(jU;R) \end{array}$$

commute for all inclusions $U \subseteq V$, where the vertical map is the projection associated to the decomposition $V = (V-U) \amalg U$, so we obtain a map

$$\underset{\sim}{K}(iX;R) \to \underset{\mathcal{F}(X)^{\mathrm{op}}}{\underleftarrow{\lim}} \Psi_R^X \to \underset{\mathcal{F}(X)^{\mathrm{op}}}{\underleftarrow{\mathrm{holim}}} \Psi_R^X$$

for each set X. This gives a natural transformation $M: \underset{\sim}{K}(-;R) \circ i \to \underleftarrow{\mathrm{holim}} \circ \eta$ since the diagrams

$$\begin{array}{ccc} \underset{\sim}{K}(iX;R) & \xrightarrow{\underset{\sim}{K}(if;R)} & \underset{\sim}{K}(iY,R) \\ \pi_{f^{-1}U} \downarrow & & \downarrow \pi_U \\ \underset{\sim}{K}(jf^{-1}U;R) & \xrightarrow{\underset{\sim}{K}(j(f \mid f^{-1}U);R)} & \underset{\sim}{K}(jU;R) \end{array}$$

all commute. That $M(X)$ is an equivalence for all sets X follows from II.4 and III.11 using the results of [13]. This gives the result for $\underset{\sim}{K}$; the proof for $\underset{\sim}{\widehat{K}}$ is identical. Q.E.D.

We also need the analogous result for $\mathcal{K}$, which is less straightforward.

Proposition III.15 *The functors $\underset{\sim}{\mathcal{K}}(-,R)\circ i$ and $\underset{\sim}{h}^{\ell f}(-;\underset{\sim}{\mathcal{K}}(R))$ are naturally equivalent.*

Proof: One readily constructs analogues to η and to the natural transformations

$$\underset{\sim}{h}^{\ell f}(-;\underset{\sim}{\mathcal{K}}(R)) \to \underset{\longleftarrow}{\operatorname{holim}} \circ \eta$$

and

$$\underset{\sim}{\mathcal{K}}(-;R)\circ i \to \underset{\longleftarrow}{\operatorname{holim}} \circ \eta,$$

and verifies that $\underset{\sim}{h}^{\ell f}(-;\underset{\sim}{\mathcal{K}}(R)) \to \underset{\longleftarrow}{\operatorname{holim}} \circ \eta$ is an equivalence. What is not immediate is that $\underset{\sim}{\mathcal{K}}(-;R)\circ i \to \underset{\longleftarrow}{\operatorname{holim}} \circ \eta$ is an equivalence. This would follow if we could show that the natural map $\underset{\sim}{\mathcal{K}}(iX;R) \to \prod_{x\in X}\underset{\sim}{\mathcal{K}}(x;R)$ is an equivalence. To see this, we verify that the map induces isomorphisms on π_i for all i. For $i \geq 0$, this follows from III.9 (a) and III.14. For $i < 0$, we proceed as follows. Recall that Karoubi [26] axiomatically defined lower K–groups $K_{-i}(\mathcal{A})$ for any additive category $\mathcal{A}$. Pedersen and Weibel [33] generalized the construction $\mathcal{C}_X(R)$ to a construction $\mathcal{C}_X(\mathcal{A})$ for any additive category, restricting to the old notion when $\mathcal{A}$ is the category R–mod of finitely generated free left R–modules. Moreover, they define a non-connective spectrum which we call $\underset{\sim}{\mathcal{L}}(X,\mathcal{A})$ (we use $\underset{\sim}{\mathcal{L}}$ only to distinguish it from our functor $\underset{\sim}{K}$), and prove that for $i > 0$, $\pi_{-i}(\underset{\sim}{\mathcal{L}}(X,\mathcal{A})) \cong K_{-i}(\mathcal{A})$. It is clear from the definitions that if $\mathcal{A} = R$–mod, then $\underset{\sim}{\mathcal{L}}(X,\mathcal{A}) \cong \underset{\sim}{\mathcal{K}}(X,R)$, that if X is any set, $\underset{\sim}{\mathcal{K}}(iX,R) \cong \underset{\sim}{\mathcal{L}}(*,\prod_{x\in X}\underline{R\text{–mod}})$, and that under these identifications, the homomorphism

$$\pi_{-i}(\underset{\sim}{\mathcal{K}}(iX,R)) \to \pi_{-i}\left(\prod_{x\in X}\underset{\sim}{\mathcal{K}}(x,R)\right), \qquad i>0,$$

corresponds to the natural map

$$K_{-i}\left(\prod_{x\in X}\underline{R\text{–mod}}\right) \to \prod_{x\in X} K_{-i}(\underline{R\text{–mod}}).$$

Thus, what is required is to show that Karoubi's groups K_{-i} commute with (perhaps infinite) products. But this result now follows from Karoubi [26] or Pedersen–Weibel [33], since a product of "flasque resolutions" of additive categories $\mathcal{A}_x$ is a flasque resolution of $\prod_{x\in X}\mathcal{A}_x$. Q.E.D.

We now wish to adapt the notion of the nerve of a covering to the situation where an object X of $\underline{\mathcal{M}}$ is covered by a family of subsets $\{X_\alpha\}_{\alpha \in A}$.

Definition III.16 *A simplicial metric space is a functor from $\underline{\Delta}^{\mathrm{op}}$ to $\underline{\mathcal{M}}$. Since $i\mathcal{C}_X(R)$ and $i\widehat{\mathcal{C}}(R)$ are functorial in X, we obtain from a simplicial metric space $X.$ simplicial spectra $\underline{K}(X., R)$, $\widehat{\underline{K}}(X., R)$, and $\underset{\sim}{\underline{K}}(X., R)$. Note that if $X.$ is a constant simplicial metric space, then all these simplicial spectra are constant.*

Recall from II.18 the definition of a disjoint union of metric spaces

Definition III.17 *Let $\mathcal{U} = \{U_\alpha\}_{\alpha \in A}$ be a covering of a metric space X by subsets. We say $\mathcal{U}$ is locally finite if the following two conditions hold.*

(a) *For any $\alpha \in A$, the set $\{\alpha' \in A \mid X_{\alpha'} \cap X_\alpha \neq \emptyset\}$ is finite*

(b) *For any set $U \subseteq X$ of finite diameter, the set $\{\alpha \in A \mid U \cap X_\alpha \neq \emptyset\}$ is finite.*

Definition III.18 *Let $\mathcal{U} = \{U_\alpha\}_{\alpha \in A}$ be any locally finite covering of a metric space X. We define a simplicial metric space $N.\mathcal{U}$ by setting*

$$N_k\mathcal{U} = \coprod_{(\alpha_0,\dots,\alpha_k)\in A^{k+1}} U_{\alpha_0} \cap \cdots \cap U_{\alpha_k},$$

and by defining face and degeneracy maps via the following formulae

(a) *$d_i|U_{\alpha_0} \cap \cdots \cap U_{\alpha_k}$ is the inclusion $U_{\alpha_0} \cap \cdots \cap U_{\alpha_k}$, to the disjoint union factor corresponding to $(\alpha_0, \dots, \widehat{\alpha}_i, \dots, \alpha_k)$*

(b) *$S_i|U_{\alpha_0} \cap \cdots \cap U_{\alpha_k}$ is the identity map on $U_{\alpha_0} \cap \cdots \cap U_{\alpha_k}$ to the factor corresponding to $(\alpha_0, \dots, \alpha_i, \alpha_i, \dots, \alpha_k)$.*

The fact that these maps are eventually continuous is clear, since they are distance nonincreasing. That they are proper follows from part (a) of the definition of local finiteness.

Definition III.19 *Let X be a metric space, and let $\mathcal{U} = \{U_\alpha\}_{\alpha \in A}$ denote a locally finite covering of X. Then we note that we have a map of simplicial metric spaces from $N.\mathcal{U}$ to $\mathcal{X}.$, the constant simplicial metric space with value X, obtained by including each $U_{\alpha_0} \cap \cdots \cap U_{\alpha_k}$ into X. This is clearly eventually continuous (it is distance nonincreasing) and it is proper due to condition (b) in the definition of local finiteness. Applying $|\ |$, we obtain maps of spectra*

$$A(\mathcal{U}) : |\underset{\sim}{K}(N.\mathcal{U}, R)| \to |\underset{\sim}{K}(\mathcal{X}., R)| \cong \underset{\sim}{K}(X, R)$$

$$\hat{A}(\mathcal{U}) : |\underset{\sim}{\hat{K}}(N.\mathcal{U}, R)| \to |\underset{\sim}{\hat{K}}(\mathcal{X}., R)| \cong \underset{\sim}{\hat{K}}(X, R)$$

$$A(\mathcal{U}) : |\underset{\sim}{K}(N.\mathcal{U}, R)| \to |\underset{\sim}{K}(\mathcal{X}., R)| \cong \underset{\sim}{K}(X, R)$$

which we call the assembly maps for these coverings.

A key point in this paper will be to show that in certain situations, the assembly map A is an equivalence. This means proving an excision result for certain coverings of metric spaces. We defer this to the next section; for now, we construct an intrinsic version of the assembly maps, independent of choice of covering.

Recall from §II the notation $\mathcal{B}_k X$ for the collection of all locally finite collections $\mathcal{A}$ of singular k–simplices in X for which there exists a number N, so that the diameter of image(σ) is less than N for all $\gamma \in \mathcal{A}$. Recall also that we defined an associated locally finite homology theory ${}^b\underset{\sim}{h}^{\ell f}(-; \underset{\sim}{A})$ to be $|\ |$ applied to the simplicial spectrum $\underline{k} \to \lim_{\mathcal{A} \in \mathcal{B}_k X} \underset{\sim}{h}^{\ell f}(\mathcal{A}; \underset{\sim}{A})$. Viewing each $\mathcal{A} \in \mathcal{B}_k X$ a a discrete metric space, we obtain an analogous simplicial spectrum $\underline{k} \to \lim_{\mathcal{A} \in \mathcal{B}_k X} \underset{\sim}{K}(\mathcal{A}; \underset{\sim}{A})$, and denote this theory by ${}^b\underset{\sim}{G}^h(X; R)$. Corollary III.14 gives a homotopy weak equivalence $\eta : {}^b\underset{\sim}{h}^{\ell f}(-; \underset{\sim}{K}(R)) \to {}^b\underset{\sim}{G}^h(X; R)$ of functors $\underline{\mathcal{M}} \to \underline{\mathcal{S}}$. We will now define a natural transformation $\alpha : {}^b\underset{\sim}{G}^h(X; R) \to \underset{\sim}{K}(X, R)$. Let $\mathcal{A}$ be any collection of singular n–simplices of X, and ζ any point of the standard n–simplex. Then we define a function $\vartheta_\zeta : \mathcal{A} \to X$ by $\vartheta_\zeta(\sigma) = \sigma(\zeta)$. ϑ_ζ is always eventually continuous, since $\mathcal{A}$ is viewed as a discrete metric space, and if $\mathcal{A}$ is locally finite ϑ_ζ is clearly symmetric monoidal, ϑ_ζ is proper. Consequently, we have the map of spectra $\underset{\sim}{K}(\vartheta_\zeta, R) : \underset{\sim}{K}(\mathcal{A}, R) \to \underset{\sim}{K}(X, R)$ given by $(F, B, \varphi) \to (F, B, \vartheta_\zeta \circ \varphi)$. Suppose further that $\mathcal{A} \in \mathcal{B}_n X$, and that N is the number required to exist by the definition of $\mathcal{B}_n X$. Then if ζ and η are both points in the standard n–simplex, we have a symmetric monoidal natural transformation $N_\zeta^\eta : \underset{\sim}{K}(\vartheta_\zeta, R) \to \underset{\sim}{K}(\vartheta_\eta, R)$, given by letting $N_\zeta^\eta(F, B, \varphi)$ be the map $(F, B, \vartheta_\zeta \circ \varphi) \mapsto (F, B, \vartheta_\eta \circ \varphi)$ which is given by the identity homomorphism from F to itself. This is a morphism in $i\mathcal{C}_X(R)$ because it and its inverse are bounded by N.

Recall that the standard n–simplex can be viewed as the nerve of the ordered set $\underline{n} = \{0, 1, \ldots, n\}$, $0 < 1 < \cdots < n$, viewed as a category. Let $\mathcal{A} \in \mathcal{B}_n X$; we define a functor $i\mathcal{C}_\mathcal{A}(R) \times \underline{n} \overset{l(\mathcal{A}, n)}{\longrightarrow} i\mathcal{C}_X(R)$ as follows. On objects, $l(\mathcal{A}, n)((F, B, \varphi), i)$ is $(F, B, \vartheta_i \circ \varphi)$, where i denotes the vertex $\Delta^n = N.\underline{n}$ corresponding to i on morphisms, it is defined by the requirement that $l(\mathcal{A}, n) \,|\, i\mathcal{C}_\mathcal{A}(R) \times j$ is the functor induced by θ_j, and that $(\mathrm{Id} \times (i \le j))$:

$((F,B,\varphi),i) \to ((F,B,\varphi),j)$ is sent to $N_i^j(F,B,\varphi)$. This is compatible with the inclusion of elements in $\mathcal{B}_n X$, so we obtain a functor $\underset{\mathcal{A}\in\mathcal{B}_n X}{\varinjlim} i\mathcal{C}_{\mathcal{A}}(R)\times\underline{n} \to i\mathcal{C}_X(R)$, and hence a map $\underset{\mathcal{A}\in\mathcal{B}_n X}{\varinjlim} N.i\mathcal{C}_{\mathcal{A}}(R) \times \Delta^n \to i\mathcal{C}_X(R)$. If $\underline{C}$ is a symmetric monoidal category, let the tth space in $\mathrm{Spt}(\underline{C})$ be denoted by $\mathrm{Spt}_t(\underline{C})$, and let $\sigma_t : S^1 \wedge \mathrm{Spt}_t(\underline{C}) \to \mathrm{Spt}_{t+1}(\underline{C})$ be the structure map for $\mathrm{Spt}(\underline{C})$. The fact that the natural transformations N_i^j are symmetric monoidal shows in addition that we obtain maps $\underset{\mathcal{A}\in\mathcal{B}_n X}{\varinjlim} \mathrm{Spt}_t(i\mathcal{C}_{\mathcal{A}}(R)) \times \Delta^n \xrightarrow{\Lambda_t} \mathrm{Spt}_t(i\mathcal{C}_X(R))$, so that the diagrams

$$\begin{array}{ccc}
\underset{\mathcal{A}\in\mathcal{B}_n X}{\varinjlim} (S^1 \wedge \mathrm{Spt}_t(i\mathcal{C}_{\mathcal{A}}(R))) \times \Delta^n & \longrightarrow & S^1 \wedge \mathrm{Spt}_t(i\mathcal{C}_X(R)) \\
\sigma_t \times \mathrm{Id} \downarrow & & \downarrow \sigma_t \\
\underset{\mathcal{A}\in\mathcal{B}_n X}{\varinjlim} \mathrm{Spt}_{t+1}(i\mathcal{C}_{\mathcal{A}}(R)) \times \Delta^n & \longrightarrow & \mathrm{Spt}_{t+1}(i\mathcal{C}_X(R))
\end{array}$$

commute. Further, it is routine that $\underset{\mathcal{A}\in\mathcal{B}_n X}{\varinjlim} \Lambda_t$ respects the equivalence defining $|\ |$, so for each t, we obtain a map from $|\underline{k} \to \underset{\mathcal{A}\in\mathcal{B}_n X}{\varinjlim} \mathrm{Spt}_t i\mathcal{C}_{\mathcal{A}}(R)|$ to $\mathrm{Spt}_t i\mathcal{C}_X(R)$, respecting the structure maps in Spt_t. Consequently, we have a map $l : {}^b\underset{\sim}{G}^h(X,R) \to \underset{\sim}{K}(X,R)$. We note that l is natural in X, with respect to morphisms in $\underline{\mathcal{M}}$. We define $\alpha : {}^b\underset{\sim}{h}^{\ell f}(-;\underset{\sim}{K}(R)) \to \underset{\sim}{K}(\ ;R)$ to the the homotopy natural transformation $l \circ \eta$, where $\eta : {}^b\underset{\sim}{h}^{\ell f}(-;\underset{\sim}{K}(R)) \to {}^b\underset{\sim}{G}^h(-;R)$ is the above defined homotopy weak equivalence of diagrams. We observe that the construction generalizes, using III.14, to homotopy natural transformations ${}^b\underset{\sim}{h}^{\ell f}(-;\widehat{\underset{\sim}{K}}(R)) \to \widehat{\underset{\sim}{K}}(-;R)$ and ${}^b\underset{\sim}{h}^{\ell f}(-;\underset{\sim}{\mathcal{K}}(R)) \to \underset{\sim}{\mathcal{K}}(-,R)$, which we also denote by α. We summarize the above discussion

Proposition III.20 *There are homotopy natural transformations*

$${}^b\underset{\sim}{h}^{\ell f}(\ ;\underset{\sim}{K}(R)) \to \underset{\sim}{K}(-;R)$$

$${}^b\underset{\sim}{h}^{\ell f}(\ ;\widehat{\underset{\sim}{K}}(R)) \to \widehat{\underset{\sim}{K}}(-;R)$$

$${}^b\underset{\sim}{h}^{\ell f}(\ ;\underset{\sim}{\mathcal{K}}(R)) \to \underset{\sim}{\mathcal{K}}(-;R)$$

defined on the category of locally compact metric spaces and uniformly continuous proper maps, which are equivalences when applied to discrete metric spaces.

IV. Excision Properties of Bounded K–theory

We are now going to consider the assembly maps $\hat{A}(\mathcal{U})$ and $\mathcal{A}(\mathcal{U})$ defined in §III for a covering $\mathcal{U} = \{Y, Z\}$ of a metric space X. In this case $|\underset{\sim}{\widehat{K}}(N.\mathcal{U}, R)|$ can be identified with the homotopy pushout of the diagram

$$\begin{array}{ccc} \underset{\sim}{\widehat{K}}(Y \cap Z; R) & \longrightarrow & \underset{\sim}{\widehat{K}}(Z; R) \\ \downarrow & & \\ \underset{\sim}{\widehat{K}}(Y; R) & & \end{array}$$

and the assembly map is the map arising from the universal mapping property of homotopy pushouts and the observation that the diagram

$$\begin{array}{ccc} \underset{\sim}{\widehat{K}}(Y \cap Z, R) & \longrightarrow & \underset{\sim}{\widehat{K}}(Z, R) \\ \downarrow & & \downarrow \\ \underset{\sim}{\widehat{K}}(Y, R) & \longrightarrow & \underset{\sim}{\widehat{K}}(X, R) \end{array}$$

commutes. In general, this map does not have good properties. For instance, consider the subset $S = [0,1] \times 0 \cup 0 \times [0, +\infty) \cup 1 \times [0, +\infty)$, viewed as a submetric space of the Euclidean plane, and consider the covering of S given by $S_0 = [0,1] \times 0 \cup 0 \times [0, +\infty)$, $S_1 = [0,1] \cup 1 \times [0, +\infty)$, $S_0 \cap S_1 = [0,1]$, and so $\underset{\sim}{\widehat{K}}(S_0 \cap S_1; R) \cong \underset{\sim}{\widehat{K}}(R)$ by III.5. The inclusion $0 \times [0, +\infty) \hookrightarrow S_0$ induces an equivalence of $\underset{\sim}{\widehat{K}}$–theory spectra, since the subset $0 \times [0, +\infty)$ satisfies hypotheses of III.7. Consequently,

$$\underset{\sim}{\widehat{K}}(S_0; R) \cong \underset{\sim}{\widehat{K}}(0 \times [0, +\infty), R) \cong \underset{\sim}{\widehat{K}}(E_+, R),$$

where E_+ denotes the non–negative real half line, and this last spectrum was shown to be contractible by Pedersen–Weibel [33] using the "Eilenberg swindle." Similarly, $\underset{\sim}{\widehat{K}}(S_1; R) \cong *$. If the map $\hat{A}(\mathcal{U})$, $\mathcal{U} = \{S_0, S_1\}$ were an equivalence, then $\pi_i(\underset{\sim}{\widehat{K}}(S; R))$ would be isomorphic to $\pi_{i-1}(\underset{\sim}{\widehat{K}}(R))$, using the Mayer–Vietoris sequence for the homotopy pushout. But, the inclusion $0 \times [0, +\infty) \hookrightarrow S$ also satisfies the conditions of III.7, so $\underset{\sim}{\widehat{K}}(S; R) \cong *$, showing that there is no good relation between the homotopy pushout associated to $\mathcal{U}$ and $\underset{\sim}{\widehat{K}}(S; R)$. To remedy this situation, we note that the associated to any covering $\mathcal{U} = \{Y, Z\}$ of a metric space and $r \in [0, +\infty)$, we have the covering $B_r\mathcal{U} = \{B_r(Y), B_r(Z)\}$, where $B_r(Y) = \{x \in X \mid \exists y \in Y,\ d(x, y) \leq r\}$, and similarly for Z. Note that for $r \leq s$, we have commutative cubes of

spectra

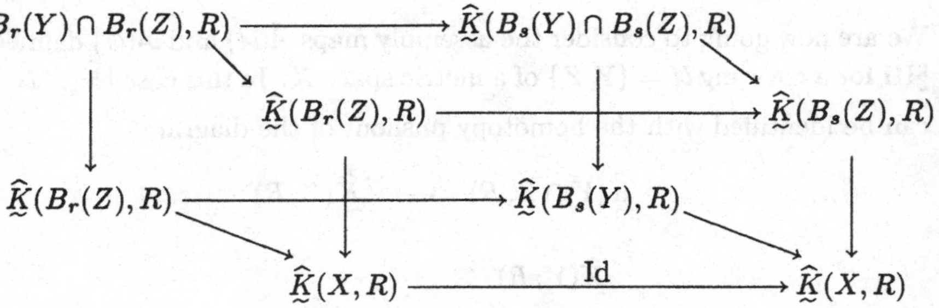

Consequently, if $\mathcal{P}_r$ denotes the homotopy pushout of the diagram

$$\begin{array}{ccc} \widehat{\underset{\sim}{K}}(B_r(Y) \cap B_r(Z); R) & \longrightarrow & \widehat{\underset{\sim}{K}}(B_r(Z), R) \\ \downarrow & & \\ \widehat{\underset{\sim}{K}}(B_r(Y); R) & & \end{array}$$

we obtain maps $\beta_{r,s} : \mathcal{P}_r \to \mathcal{P}_s$ whenever $r \leq s$, so that $\hat{A}(B_s\mathcal{U}) \circ \beta_{r,s} = \hat{A}(B_r\mathcal{U})$, and consequently a map of spectra $\varinjlim_r \mathcal{P}_r \xrightarrow{\varinjlim \hat{A}(B_r\mathcal{U})} \widehat{\underset{\sim}{K}}(X; R)$. By crossing the situation with Euclidean space R^k, and passing to the direct limit over the system defining $\underset{\sim}{K}(-, R)$, we obtain a map

$$\varinjlim_r \mathcal{N}_r \xrightarrow{\varinjlim \hat{A}(B_r\mathcal{U})} \underset{\sim}{\mathcal{K}}(X; R),$$

where $\mathcal{N}_r$ denotes the homotopy pushout of the diagram

$$\begin{array}{ccc} \underset{\sim}{\mathcal{K}}(B_r(Y) \cap B_r(Z); R) & \longrightarrow & \underset{\sim}{\mathcal{K}}(B_r(Z); R) \\ \downarrow & & \\ \underset{\sim}{\mathcal{K}}(B_r(Y); R) & & \end{array}$$

Our theorem reads as follows

Theorem IV.1 $\varinjlim_r \hat{A}(B_r\mathcal{U})$ *induces an isomorphism on* π_i *for* $i > 0$. *Consequently,* $\varinjlim_r \mathcal{A}(B_r\mathcal{U})$ *is an equivalence of spectra.*

We defer the proof of this theorem to develop some necessary technical material. Recall from §I the construction of the simplified double mapping cylinder construction associated to the diagram of symmetric monoidal categories

$$\begin{array}{ccc} i\widehat{\mathcal{C}}_{Y\cap Z}(R) & \longrightarrow & i\widehat{\mathcal{C}}_Z(R) \\ \downarrow & & \\ i\widehat{\mathcal{C}}_Y(R) & & \end{array}$$

where $\mathcal{U} = \{Y, Z\}$ is a covering of a metric space X; the results of §I concerning this construction hold since, in fact, all morphisms in all categories in the diagram are isomorphisms. Let $\mathrm{Cyl}(\mathcal{U})$ denote this simplified double mapping cylinder. Recall that in the discussion preceding I.9, we associated a symmetric monoidal functor α_X from the simplified double mapping cylinder of a diagram $\underline{B} \xleftarrow{u} \underline{A} \xrightarrow{v} \underline{C}$ of unital symmetric monoidal categories and unital symmetric monoidal functors to a unital symmetric monoidal category $\underline{X}$ with any pair of unital symmetric monoidal functors $\beta : \underline{B} \to \underline{X}$ and $\gamma : \underline{C} \to \underline{X}$ so that $\beta u = \gamma v$. We define $l : \mathrm{Cyl}(\mathcal{U}) \to i\widehat{\mathcal{C}}_X(R)$ to be the symmetric monoidal functor associated to the inclusions $j_Y : i\widehat{\mathcal{C}}_Y(R) \to i\widehat{\mathcal{C}}_X(R)$ and $j_Z : i\widehat{\mathcal{C}}_Z(R) \to i\widehat{\mathcal{C}}_X(R)$.

Proposition IV.2 *There is a homotopy commutative diagram*

$$\begin{array}{ccc} |\widehat{\underset{\sim}{K}}(N.\mathcal{U}; R)| & \searrow^{\hat{A}(\mathcal{U})} & \\ \beta \downarrow & & \widehat{\underset{\sim}{K}}(X; R) \\ \mathrm{Spt\,Cyl}\,(\mathcal{U}) & \nearrow_{\mathrm{Spt}\,(l)} & \end{array}$$

for any covering $\mathcal{U} = \{X_0, X_1\}$ of X. Further, β is an equivalence.

Proof: $\widehat{\underset{\sim}{K}}(N.\mathcal{U}; R)$ is a simplicial spectrum, i.e., a functor from $\underline{\Delta}^{\mathrm{op}}$ to $\underline{\mathcal{S}}$, and by I.2 (f), there is a natural (for Δ^{op}–diagrams in $\underline{\mathcal{S}}$) equivalence $\underset{\overrightarrow{\underline{\Delta}^{\mathrm{op}}}}{\mathrm{hocolim}}\, \widehat{\underset{\sim}{K}}(N.\mathcal{U}; R) \to |\widehat{\underset{\sim}{K}}(N.\mathcal{U}; R)|$. $\hat{A}(\mathcal{U})$ is obtained as the map induced on $|\ |$ by the simplicial map from $\widehat{\underset{\sim}{K}}(N.\mathcal{U}; R)$ to the constant simplicial spectrum with value $\widehat{\underset{\sim}{K}}(X; R)$. Consequently, $\hat{A}(\mathcal{U})$ can be identified up to a natural equivalence with the map $\underset{\overrightarrow{\underline{\Delta}^{\mathrm{op}}}}{\mathrm{hocolim}}\, \widehat{\underset{\sim}{K}}(N.\mathcal{U}; R) \to \underset{\overrightarrow{\underline{\Delta}^{\mathrm{op}}}}{\mathrm{hocolim}}\, \widehat{\underset{\sim}{K}}(X; R)$, induced by the natural transformation $\widehat{\underset{\sim}{K}}(N.\mathcal{U}; R) \to \widehat{\underset{\sim}{K}}(X; R)$. Let $n_k = \{0, 1\}^{k+1}$, and for $\tau = (\alpha_0, \ldots, \alpha_k) \in n_k$, let $\mu_k(\tau) = \cap_{i=0}^{k} X_{\alpha_i}$; we view μ_k as a functor from the discrete category n_k to $\underline{\mathcal{M}}$. Let $n.$ denote the simplicial set $\underline{k} \to n_k$, with face maps given by deletions and degeneracies given by diagonals; it may be identified with the functor $\underline{\Delta}^{\mathrm{op}} \to \underline{\mathrm{sets}}$ given on objects by $\underline{k} \mapsto F(\underline{k}, \{0, 1\})$. $n.$ can be viewed as a simplicial (discrete) category. Furthermore, for any morphism $\theta : \underline{k} \to \underline{\ell}$ in $\underline{\Delta}^{\mathrm{op}}$, we let $n.\theta$:

$n_k \to n_\ell$ denote the map of sets $n.\theta$ viewed as a functor between discrete categories. Then we have a natural transformation N_θ from μ_k to $\mu_\ell \circ (n\theta)$, given on $\tau = (\alpha_0, \dots, \alpha_k)$ by the inclusion $\cap_{i=0}^k X_{\alpha_i} \hookrightarrow \cap_{j=0}^{\ell} X_{\beta_j}$, where $n.\theta(\tau) = (\beta_0, \dots, \beta_\ell)$. One verifies directly that $N_\eta \circ N_\theta = N_{\eta\circ\theta}$, so we define a functor $\underline{\delta} : \underline{\Delta}^{\text{op}} \to \underline{\mathcal{M}}^{CAT}$ on objects by $\delta(\underline{k}) = (n_k, \mu_k)$, and on morphisms by $\underline{\delta}(\theta : \underline{k} \to \underline{S}) = (n.\theta, N_\theta)$. One can compose $\underline{\delta}$ with the functor $\underset{\sim}{\widehat{K}}(-; R)^{CAT} : \underline{\mathcal{M}}^{CAT} \to \underline{\mathcal{S}}^{CAT}$, and further with $\text{hocolim} : \underline{\mathcal{S}}^{CAT} \to \underline{\mathcal{S}}$, to obtain a functor $\delta : \underline{\Delta}^{\text{op}} \to \underline{\mathcal{S}}$, given on objects by $\underline{k} \to \underset{\overrightarrow{n_k}}{\text{hocolim}}\, \underset{\sim}{\widehat{K}}(\mu_k(-); R)$. Since n_k is finite, and since the homotopy colimit over a discrete category amounts to taking the wedge product over all the objects in the category of the value of the functor, it follows from III.12 that there is a natural equivalence of functors from δ to the functor $\underline{k} \to \underset{\sim}{\widehat{K}}(N_k\mathcal{U}; R)$, and consequently a homotopy commutative diagram

$$
\begin{array}{ccc}
\underset{\overrightarrow{\underline{\Delta}^{\text{op}}}}{\text{hocolim}}\, \underset{\sim}{\widehat{K}}(N.\mathcal{U}; R) & \searrow & \\
\uparrow & & \underset{\sim}{\widehat{K}}(X; R) \\
\underset{\overrightarrow{\underline{\Delta}^{\text{op}}}}{\text{hocolim}}\, \delta & \nearrow &
\end{array}
$$

where the vertical arrow is an equivalence and the lower diagonal arrow is induced by the various inclusions $\mu_k(\tau) \to X$. Let I be the category whose objects are the non–trivial subsets of $\{0, 1\}$, with a unique morphism from S to T if $T \subseteq S$. We define a functor $\theta : \underline{\Delta}^{\text{op}} \wr n.\mathcal{U} \to I$ as follows. A typical object of $\underline{\Delta}^{\text{op}} \wr n.\mathcal{U}$ is a pair $(k, (\alpha_0, \dots, \alpha_k))$, where k is a nonnegative integer, and $(\alpha_0, \dots, \alpha_k) \in \{0, 1\}^k$, and we define $\theta(k, \alpha_0, \dots, \alpha_k) = \{\alpha_0, \dots, \alpha_k\}$, the subset consisting of all coordinates occurring in $(\alpha_0, \dots, \alpha_k)$. Of course, there are at most two distinct coordinates. Since I is a partially ordered set, θ is determined by its behavior on orbits. As above, let $\mu. : \underline{\Delta}^{\text{op}} \wr n.\mathcal{U} \to \underline{\mathcal{M}}$ be the functor given by $\mu.(\underline{k}, (\alpha_0, \dots, \alpha_k)) = X_{\alpha_0} \cap \dots \cap X_{\alpha_k}$; it is clear that $\mu.$ factors over I, via the functor $\nu : I \to \underline{\mathcal{M}}$, given on objects by $\nu(S) = \cap_{s \in S} X_s$. We have a commutative diagram

$$
\begin{array}{ccc}
\underset{\overrightarrow{\underline{\Delta}^{\text{op}} \wr n.\mathcal{U}}}{\text{hocolim}}\, \underset{\sim}{\widehat{K}}(\mu(-); R) & \searrow & \\
\downarrow & & \underset{\sim}{\widehat{K}}(X; R) \\
\underset{\overrightarrow{I}}{\text{hocolim}}\, \underset{\sim}{\widehat{K}}(\nu(-); R) & \nearrow &
\end{array}
$$

We claim that the left hand vertical arrow is a weak equivalence; by I.2 (g), this would follow if all the categories $\{0\} \downarrow \nu$, $\{1\} \downarrow \nu$, and $\{0, 1\} \downarrow \nu$ have contractible nerves. To check that $\{0, 1\} \downarrow \nu$ has contractible nerve, we note that since there is a unique morphism from $\{0,1\}$ to any object of I, $\{0,1\} \downarrow \nu$ is isomorphic to the category $\underline{\Delta}^{op} \wr n.$. But by Thomason's homotopy colimit result, I.5 and I.2 (f), $|\underline{\Delta}^{op} \wr n.| \cong |\underset{\underline{\Delta}^{op}}{\underrightarrow{\mathrm{hocolim}}}\, n.| \cong n.$.

But $n.$ is isomorphic to the nerve of the category with objects 0 and 1, and a unique morphism between any pair of objects. This category has contractible nerve since it has an initial object. $\{0\} \downarrow \nu$ and $\{1\} \downarrow \nu$ can also be shown to have contractible nerves by similar arguments. Now I.9 shows that there is a homotopy commutative diagram

$$\begin{array}{ccc} \underset{I}{\underrightarrow{\mathrm{hocolim}}}\, \underset{\sim}{\widehat{K}}(\nu(-), R) & \searrow & \\ \downarrow & & \underset{\sim}{\widehat{K}}(X; R) \\ \mathrm{Spt}\,(\mathrm{Cyl}\,(\mathcal{U})) & \nearrow_{\mathrm{Spt}\,(l)} & \end{array}$$

The composite equivalence

$$|\underset{\sim}{\widehat{K}}(N.\mathcal{U}, R)| \xrightarrow{\sim} \underset{\underline{\Delta}^{op}}{\underrightarrow{\mathrm{hocolim}}}\, \underset{\sim}{\widehat{K}}(N.\mathcal{U}, R) \xrightarrow{\sim} \underset{\underline{\Delta}^{op}}{\underrightarrow{\mathrm{hocolim}}}\, \delta$$

$$\xrightarrow{\sim} \underset{\underline{\Delta}^{op} \wr n.}{\underrightarrow{\mathrm{hocolim}}}\, \underset{\sim}{\widehat{K}}(\mu.(-), R) \xrightarrow{\sim}$$

$$\underset{I}{\underrightarrow{\mathrm{hocolim}}}\, \underset{\sim}{\widehat{K}}(\nu(-), R) \xrightarrow{\sim} \mathrm{Spt}(\mathrm{Cyl}(\mathcal{U}))\,,$$

where the first two arrows are homotopy inverses to arrows constructed above and the middle arrow comes from I.5, is now the desired map β. Q.E.D.

To study $\hat{A}(\mathcal{U})$, it therefore suffices to study $\mathrm{Spt}(l)$.

We now prove the key lemma. Its proof is a straightforward modification of the proof of the corresponding theorem in the work of Pedersen and Weibel [33]. Let $l_r : \mathrm{Cyl}(B_r\mathcal{U}) \to i\widehat{\mathcal{C}}_X(R)$ be the functor defined above corresponding to the covering $B_r\mathcal{U}$. We have commutative diagrams

$$\begin{array}{ccc} \mathrm{Cyl}\,(B_r\mathcal{U}) & \searrow^{l_r} & \\ \downarrow & & i\widehat{\mathcal{C}}_X(R) \\ \mathrm{Cyl}\,(B_{r+s}\mathcal{U}) & \nearrow_{l_s} & \end{array}$$

of symmetric monoidal functors. For any object x of $i\widehat{\mathcal{C}}_X(R)$, we may consider the category $x \downarrow l_r$. Let $\mathrm{Fil}^r_d(x)$ denote the full subcategory of $x \downarrow l_r$ on objects $(z, z \xrightarrow{\theta} l_r z)$ so that θ and θ^{-1} both have filtration less than or equal to d. θ is an isomorphism since all morphisms in $i\widehat{\mathcal{C}}_X(R)$ are isomorphisms. Note that for any $d' \geq d$, $r' \geq r$, we have inclusion functors $\mathrm{Fil}^r_d(x) \to \mathrm{Fil}^{r'}_{d'}(x)$.

Lemma IV.3 *Let x be any object of $i\mathcal{C}_X(R) \subseteq i\widehat{\mathcal{C}}_X(R)$. Then the functor $\mathrm{Fil}^r_d(x) \to \mathrm{Fil}^{r+s}_d(x)$ induces a map on nerves which is homotopic to a constant map, whenever $s > d$.*

Proof: Let (F, B, φ) be an object of $i\widehat{\mathcal{C}}_X(R)$. Let

$$B_1 = \{b \in B \mid \varphi(b) \in B_{r+s}(Y) - B_{r+s}(Z)\},$$

$$B_2 = \{b \in B \mid \varphi(b) \in B_{r+s}(Z) - B_{r+s}(Y)\},$$

$$B_{12} = \{b \in B \mid \varphi(b) \in B_{r+s}(Y) \cap B_{r+s}(Z)\}.$$

Note that $B = B_1 \coprod B_2 \coprod B_{12}$, and we have a corresponding direct sum splitting in $i\widehat{\mathcal{C}}_X(R)$

$$(F, B, \varphi) \cong (F_1, B_1, \varphi|B_1) \oplus (F_2, B_2, \varphi|B_2) \oplus (F_{12}, B_{12}, \varphi|B_{12}),$$

where F_1, F_2, and F_{12} denote the (free) submodules of F spanned by B_1, B_2, and B_{12} respectively. Let M, M_1, M_2, and M_{12} denote the objects (F, B, φ), $(F_1, B_1, \varphi|B_1)$, $(F_2, B_2, \varphi|B_2)$, and $(F_{12}, B_{12}, \varphi|B_{12})$ respectively. M, M_2, and M_{12} can be viewed as objects in the subcategories $i\widehat{\mathcal{C}}_{B_{r+s}(Y)}(R)$, $i\widehat{\mathcal{C}}_{B_{r+s}(Z)}(R)$, and $i\widehat{\mathcal{C}}_{B_{r+s}(Y)\cap B_{r+s}(Z)}(R)$ of $i\widehat{\mathcal{C}}_X(R)$ respectively, and there is an evident bounded (by 0) isomorphism θ from M to $l_{r+s}(M_1, M_2, M_{12})$. We'll show that there is a natural transformation from the constant functor with value $((M_1, M_{12}, M_2), \theta)$ to the inclusion functor $\mathrm{Fil}^r_d \to \mathrm{Fil}^{r+s}_d(M)$. This will give the result, since natural transformations produce simplicial homotopies. To see that there is such a natural transformation, consider a typical object $((A_1, A_{12}, A_2), \eta)$ of $\mathrm{Fil}^r_d(M)$. Since η^{-1} is bounded by d, and $s > d$, η^{-1} must carry the summand A_{12} of $l_r(A_1, A_{12}, A_2) = l_{r+s}(A_1, A_{12}, A_2)$ into the summand M_{12} of M. Consequently, the idempotent e_{12} of $l_r(A_1, A_{12}, A_2)$ determined by $e_{12}(\alpha, \alpha', \alpha'') = (0, \alpha', 0)$ determines a bounded idempotent $\eta^{-1}e_{12}\eta$ of M_{12}; this exhibits A_{12} as a summand of M_{12}. Define idempotents e_1 and e_2 of $A_1 \oplus A_{12} \oplus A_2$ by $e_1(\alpha, \alpha', \alpha'') = (\alpha, 0, 0)$ and $e_2(\alpha, \alpha', \alpha'') = (0, 0, \alpha'')$. Again, since η^{-1} is bounded by d, η^{-1} must carry the summand A_1 into $M_1 \oplus M_{12}$ and the summand A_2 into $M_{12} \oplus M_2$, so $\eta^{-1}e_1\eta$ and $\eta^{-1}e_2\eta$ determine bounded idempotents of $M_1 \oplus M_{12}$ and $M_{12} \oplus M_2$, respectively. Note further that since η is bounded by d, η carries the summands M_i of M into the summands A_i of $A_1 \oplus A_{12} \oplus A_2$. Therefore, if $m \in M_i$, then $\eta^{-1}e_i\eta m = m$, so the idempotent $\eta^{-1}e_i\eta$ restricts to the identity on M_i. Let $p_1 : M_1 \oplus M_{12} \to M_{12}$

and $p_2 : M_{12} \oplus M_2 \to M_{12}$ be the evident projections, viewed as bounded idempotents of $M_1 \oplus M_{12}$ and $M_{12} \oplus M_2$ respectively. Let $\hat{p}_1 = \mathrm{Id} - p_1$ and $\hat{p}_2 = \mathrm{Id} - p_2$, viewed as idempotents of $M_1 \oplus M_{12}$ and $M_{12} \oplus M_2$ respectively. Since $\hat{p}_i$ takes values in M_i, we have $\eta^{-1} e_i \eta \hat{p}_i = \hat{p}_i$. Consider the bounded self maps $p_1 \eta^{-1} e_1 \eta$ and $p_2 \eta^{-1} e_2 \eta$ of $M_1 \oplus M_{12}$ and $M_{12} \oplus M_2$ respectively. When restricted to M_{12}, they yield idempotents of M_{12}. For

$$(p_i \eta^{-1} e_i \eta)(p_i \eta^{-1} e_i \eta) = (\mathrm{Id} - \hat{p}_i)\eta^{-1} e_i \eta (\mathrm{Id} - \hat{p}_i)\eta^{-1} e_i \eta$$

$$= (\mathrm{Id} - \hat{p}_i)\eta^{-1} e_i \eta \eta^{-1} e_i \eta - (\mathrm{Id} - \hat{p}_i)\eta^{-1} e_i \eta \hat{p}_i \eta^{-1} e_i \eta$$
$$= (\mathrm{Id} - \hat{p}_i)\eta^{-1} e_i \eta - (\mathrm{Id} - \hat{p}_i)\hat{p}_i \eta^{-1} e_i \eta = p_i \eta^{-1} e_i \eta \,.$$

Furthermore, we claim that

$$p_1 \eta^{-1} e_1 \eta + p_2 \eta^{-1} e_2 \eta + \eta^{-1} e_{12} \eta$$

is the identity idempotent of M_{12}; this follows directly from the observation that $e_1 + e_2 + e_{12}$ is the identity idempotent on $A_1 \oplus A_{12} \oplus A_2$. Let U be the object of $i\widehat{\mathcal{C}}_{B_{r+s}(Y) \cap B_{r+s}(Z)}(R)$ determined by $p_1 \eta^{-1} e_1 \eta \,|\, M_{12}$ and let V be the object determined by $p_2 \eta^{-1} e_2 \eta \,|\, M_{12}$. Then we have the bounded isomorphism $M_{12} \cong U \oplus A_{12} \oplus V$, arising from the isomorphism $A_{12} \cong \eta^{-1} e_{12} \eta M_{12}$ and from the fact that

$$p_1 \eta^{-1} e_1 \eta + p_2 \eta^{-1} e_2 \eta + \eta^{-1} e_{12} \eta = \mathrm{Id}_{M_{12}} \,.$$

Similarly, we obtain bounded isomorphisms $M_1 \oplus U \xrightarrow{\sim} A_1$ and $V \oplus M_1 \xrightarrow{\sim} A_2$. This data now produces a morphism

$$((M_1, M_{12}, M_2), \theta) \longrightarrow ((A_1, A_{12}, A_2), \eta)$$

in $\mathrm{Cyl}(B_{r+s}\mathcal{U})$ and is readily checked that it is a natural transformation. Q.E.D.

Corollary IV.4. *Let x be an object of $i\mathcal{C}_X(R) \subseteq i\widehat{\mathcal{C}}_X(R)$, and let $l : \varinjlim_r \mathrm{Cyl}(B_r\mathcal{U}) \to i\widehat{\mathcal{C}}_X(R)$ be the functor given on $\mathrm{Cyl}(B_r\mathcal{U})$ by l_r. Then $x \downarrow l$ is a contractible category.*

Proof: $x \downarrow l = \varinjlim_d \varinjlim_r \mathrm{Fil}^r_d(x)$; The result now follows immediately from IV.3. Q.E.D.

Corollary IV.5 *The functor*

$$l : \varinjlim_r \mathrm{Cyl}(B_r\mathcal{U}) \to i\widehat{\mathcal{C}}_X(R)$$

induces a map

$$\mathrm{Spt}(l) : \mathrm{Spt}(\varinjlim_r \mathrm{Cyl}(B_r\mathcal{U})) \to \widehat{\underset{\sim}{K}}(X, R),$$

which induces isomorphisms on π_i *for* $i > 0$. *Consequently* $\varinjlim_r \hat{A}(B_r\mathcal{U})$ *induces isomorphisms on* π_i *for* $i > 0$, *and* $\varinjlim_r \mathcal{A}(B_r\mathcal{U})$ *is an equivalence of spectra from* $\varinjlim_r |\underset{\sim}{\mathcal{K}}(N.B_r\mathcal{U}; R)|$ *to* $\underset{\sim}{\mathcal{K}}(X, R)$.

Proof: One observes that the image F of $\pi_0(i\mathcal{C}_X(R)) \to \pi_0(i\widehat{\mathcal{C}}(R))$ is cofinal, since every object of $i\widehat{\mathcal{C}}_X(R)$ is a summand of an object of $i\mathcal{C}_X(R)$. Let $M \subseteq \pi_0(\varinjlim_r \mathrm{Cyl}(B_r\mathcal{U}))$ be the inverse image $\pi_0(f)^{-1}(F)$; one checks easily that M is cofinal in $\pi_0(\varinjlim_r \mathrm{Cyl}(B_r\mathcal{U}))$. The first statement about $\varinjlim_r \hat{A}(B_r\mathcal{U})$ follows from IV.2. To prove the final statement, let $\mathcal{U}[k]$ denote the covering $\{Y \times E^k, Z \times E^k\}$ of $X \times E^k$, where $\mathcal{U} = \{Y, Z\}$. By the second statement applied to the covering $\mathcal{U}[k]$, we see that $\Omega^k \hat{A}(B_r\mathcal{U}[k])$ induces isomorphisms on π_i for $i > -k$. Consequently, the maps $\varinjlim_r \varinjlim_k \Omega^k \hat{A}(B_r\mathcal{U}[k])$ induces an equivalence $\varinjlim_r \widehat{\underset{\sim}{\mathcal{K}}}(N.B_r\mathcal{U}; R) \to \widehat{\underset{\sim}{\mathcal{K}}}(X, R)$, and the result follows for $\underset{\sim}{\mathcal{K}}$ by III.10 (a). Q.E.D.

Corollary IV.6 *Let* $\mathcal{U}$ *be any finite covering of a metric space* X, *and let* $B_r\mathcal{U}$ *be the covering* $\{B_rU, U \in \mathcal{U}\}$. *Then the map*

$$\varinjlim_r \mathcal{A}(B_r\mathcal{U}) : \varinjlim_r |\underset{\sim}{\mathcal{K}}(N.B_r\mathcal{U}; R)| \to \underset{\sim}{\mathcal{K}}(X, R)$$

is an equivalence of spectra.

Proof: Repeated application of IV.5. Q.E.D.

We will also need to examine the behavior of the assembly for certain infinite coverings. Specifically, we wish to study coverings $\mathcal{U} = \{U_i\}_{i \in \mathbb{Z}}$ parametrized by the integers, which satisfy the following hypothesis.

Hypothesis (A): $U_i \cap U_j = \emptyset$ *if* $|i - j| > 1$.

Thus, all triple intersections of distinct sets in the coverings are empty, and the covering is one dimensional in an appropriate sense. As in the case of finite coverings, we will need to introduce the coverings $B_r\mathcal{U} = \{B_rU_i\}_{i \in \mathbb{Z}}$.

Unfortunately, $B_r\mathcal{U}$ will no longer satisfy Hypothesis (A), and one sees that it is not reasonable to try to prove an excision theorem for $\mathcal{U}$ itself but rather for a direct limit involving coverings constructed from $\mathcal{U}$. We will have to introduce a stronger hypothesis below to ensure that even this direct limit gives an excision result. We now develop these constructions.

We consider first the notion of a map of coverings. Thus, let $\{U_\alpha\}_{\alpha\in A} = \mathcal{U}$ and $\{U'_\beta\}_{\beta\in B} = \mathcal{U}'$ be two locally finite coverings (in the sense of III.17) of a metric space X. By a map of coverings $\mathcal{U} \xrightarrow{\Theta} \mathcal{U}'$, we mean a function $\theta : A \to B$ so that $U_\alpha \subseteq U'_{\theta[\alpha]}$ for all $\alpha \in A$. A map of coverings induces a map $N.\Theta : N.\mathcal{U} \to N.\mathcal{U}'$ of simplicial metric spaces, and hence a map of simplicial spectra $\underset{\sim}{K}(N.\mathcal{U}; R) \xrightarrow{\underset{\sim}{K}(N.\Theta;R)} \underset{\sim}{K}(N.\mathcal{U}'; R)$, as well as maps $\widehat{\underset{\sim}{K}}(N.\Theta; R)$ and $\underset{\sim}{\mathcal{K}}(N.\Theta; R)$.

Consider a covering of X parametrized by the integers, say $\mathcal{U} = \{U_i\}_{i\in\mathbb{Z}}$. We associate to $\mathcal{U}$ a new covering, $\mathcal{U}'$, also parametrized by the integers, by $\mathcal{U}' = \{U'_i\}$, $U'_i = U_{2i} \cup U_{2i+1}$. We note that there is a map of coverings $\mathcal{U} \xrightarrow{\Theta} \mathcal{U}'$, given by the map $\theta : \mathbb{Z} \to \mathbb{Z}$, $\theta(2i) = \theta(2i+1) = i$. We are free to iterate this procedure; we set $\mathcal{U}(0) = \mathcal{U}$, and define $\mathcal{U}(\ell)$ inductively by $\mathcal{U}(\ell) = \mathcal{U}(\ell - 1)'$. We write $\mathcal{U}(\ell) = \{U_i(\ell)\}$, where $U_i(\ell) = U_{2i}(\ell - 1) \cup U_{2i+1}(\ell - 1)$. We have maps of coverings $\Theta_\ell : \mathcal{U}(\ell) \to \mathcal{U}(\ell + 1)$, and the diagrams

$$\begin{array}{ccc} |\underset{\sim}{K}(N.\mathcal{U}(\ell); R)| & \searrow^{A(\mathcal{U}(\ell))} & \\ |\underset{\sim}{K}(N.\Theta_\ell; R)| \downarrow & & \underset{\sim}{K}(X; R) \\ |\underset{\sim}{K}(N.\mathcal{U}(\ell+1); R)| & \nearrow_{A(\mathcal{U}(\ell+1))} & \end{array}$$

as well as their analogues for $\widehat{\underset{\sim}{K}}$ and $\underset{\sim}{\mathcal{K}}$ are easily seen to commute. Consequently, we have a map

$$\underset{\overrightarrow{\ell}}{\text{hocolim}}\, |\underset{\sim}{K}(N.\mathcal{U}(\ell) : R)| \xrightarrow{\underset{\rightarrow}{\text{hocolim}}\, A(\mathcal{U}(\ell))} \underset{\sim}{K}(X; R)$$

(and analogues for $\widehat{\underset{\sim}{K}}$ and $\underset{\sim}{\mathcal{K}}$), where $\underset{\overrightarrow{\ell}}{\text{hocolim}}$ denotes mapping telescope, i.e., homotopy colimit over the negative integers with a unique morphism from i to j whenever $i \leq j$. We wish to find conditions on $\mathcal{U}$ which will ensure that

$$\underset{\overrightarrow{\ell}}{\text{hocolim}}\, |\underset{\sim}{\mathcal{K}}(N.\mathcal{U}(\ell); R)| \xrightarrow{\underset{\rightarrow}{\text{hocolim}}\, A(\mathcal{U}(\ell))} \underset{\sim}{\mathcal{K}}(X; R)$$

is a weak equivalence. The correct condition is the following.

Hypothesis (B): *For every $r \in \mathbb{R}$, there is an ℓ so that the covering $B_r\mathcal{U}(\ell)$ satisfies Hypothesis (A).*

The rest of this section will be devoted to proving that $\underset{\overrightarrow{\ell}}{\text{hocolim}}\, \mathcal{A}(\mathcal{U}(\ell))$ is a weak equivalence when $\mathcal{U}$ is locally finite and satisfies Hypothesis (B). The idea of the proof, as in the proof of IV.1, is to reduce the proof to a simplified homotopy colimit construction involving the constructions $\text{Tor}(\underline{A} \overset{\rightarrow}{\rightarrow} \underline{B})$ studied in I.10 and mapping telescopes. We first will need to interpret $|\widehat{\underset{\sim}{K}}(N.\mathcal{U}; R)|$ as a homotopy colimit when $\mathcal{U}$ satisfies Hypothesis (A). We now develop the terminology required for this interpretation.

Let P be a partially ordered set, which we view as a category by the requirement that there be a unique morphism from X to Y when $x \geq y$; the object set is of course P itself. Suppose further that a group G operates freely on P in an order preserving way. We define the orbit category $\underline{P/G}$ as follows. The objects of $\underline{P/G}$ are the elements of P/G. Let $B \subset \overline{P \times P}$ be defined by $V = \{(x, y) \in \overline{P \times P} \mid x \geq y\}$. Then we have the induced map $\pi_1 \times \pi_2 : B/G \to P/G \times P/G$, given by $[x \times y] \mapsto ([x], [y])$, and we define $\text{Mor}_{\underline{P/G}}([x], [y]) = (\pi_1 \times \pi_2)^{-1}([x], [y])$. The composition law is obtained as follows. A typical morphism from $[x]$ to $[y]$ can thus be represented as $[g_1 x \geq g_2 y]$, for some $g_1, g_2 \in G$. By the freeness of the action, there is a unique representative of the form $x \geq gy$. If $f' : [y] \to [z]$ is a morphism, it can similarly be uniquely represented by a morphism of the form $gy \geq g'z$. The composite of the two morphisms is now $x \geq g'z$. This composition is clearly independent of the representative for $[x]$ chosen, and we do indeed have a category.

Example: Let P^a be the subset of the partially ordered subset of all non-trivial subsets of $\mathbb{Z}$ consisting of sets of the form $\{i\}$ or of the form $\{i, i+1\}$. This partially ordered set is pictured as

$$\geq \{-1\} \leq \{-1, 0\} \geq \{0\} \leq \{0, 1\} \geq \{1\} \leq \{1, 2\} \geq .$$

The group $\mathbb{Z}$ acts freely by poset isomorphism on P^a via its translation action on $\mathbb{Z}$, so $n \cdot \{i, i+1\} = \{i+n, i+n+1\}$ and $n \cdot \{i\} = \{i+n\}$. In this case, the orbit category $\underline{P^a/\mathbb{Z}}$ can be pictured as

$$[\{0,1\}] \underset{g}{\overset{f}{\rightrightarrows}} [\{0\}] .$$

Here, f is the morphism represented by $\{0, 1\} \geq \{0\}$ and g is represented by $\{0, 1\} \geq \{1\}$. Note that $\underline{P^a/\mathbb{Z}}$ is isomorphic to the category $\underline{M}$ occurring in Proposition I.10. We can also consider the category $\underline{P^a/2\mathbb{Z}}$, its picture is as follows

$$\begin{array}{ccccc} & & [\{0,1\}] & & \\ & \swarrow^{f_1} & & \searrow^{g_1} & \\ [\{0\}] & & & & [\{1\}] \\ & \nwarrow_{g_2} & & \nearrow_{f_2} & \\ & & [\{1,2\}] & & \end{array}$$

Here, the morphisms are given by

$$\begin{aligned} f_1 &= [\{0,1\} \geq \{0\}]\,, \quad f_2 = [\{1,2\} \geq \{1\}]\,, \\ g_1 &= [\{0,1\} \geq \{1\}]\,, \quad g_2 = [\{1,2\} \geq \{2\}]\,. \end{aligned}$$

We define functors $p, q : \underline{P^a/2\mathbb{Z}} \to \underline{P^a/\mathbb{Z}}$. p is defined on objects by $p[\{0,1\}] = p[\{1,2\}] = [\{0,1\}]$, $p[\{0\}] = p[\{1\}] = [\{0\}]$, and on morphisms by $p(f_i) = f$, $p(g_i) = g$. This functor is obtained by factoring out the $\mathbb{Z}/2\mathbb{Z}$–action on $\underline{P^a/2\mathbb{Z}}$. q is defined on objects by $q[\{0,1\}] = q[\{0\}] = q[\{1\}] = [\{0\}]$, $q[\{1,2\}] = [\{0,1\}]$, and on morphisms by $q(f_1) = q(g_1) = \mathrm{Id}_{[\{0\}]}$, and $qg_2 = g$, $q(f_2) = f$.

Now let X be a (perhaps infinite) simplicial complex, with free G–action. This means that G acts freely on the k–simplices of X for every k. By $\mathcal{P}(X)$, we mean the partially ordered set of simplices of X, with $\sigma \leq \tau \Leftrightarrow \sigma$ is a face of τ. Note that if X denotes the real line, viewed as a simplicial complex by $X = \cup_n [n, n+1]$, then $\mathcal{P}(X)$ is isomorphic to the poset P^a described above. We also construct from X the simplicial set $S.X$ whose k–simplices are the simplicial maps from the standard k–simplex into X. For any simplicial set $Y.$, we define its category of simplices $\underline{C}(Y.)$ as follows. The objects of $\underline{C}(Y.)$ are the simplices of $Y.$, and if $y \in Y_k$, $y' \in Y_\ell$, then

$$\mathrm{Mor}_{\underline{C}(X)}(y, y') = \{\theta : \underline{k} \to \underline{\ell} \text{ in } \underline{\Delta}^{\mathrm{op}} \text{ such that } \theta_* y = y\}\,.$$

If X is an ordered simplicial complex, then there is a functor $\underline{C}(S.X) \to \mathcal{P}(X)$ given on objects by sending a simplex x to the image of the unique non–degenerate simplex x' so that x is equal to a degeneracy operator applied to x'. Since $\mathcal{P}(X)$ is a poset, the behavior on morphisms is forced by the behavior on objects, and it is easy to check that this does define a functor. Moreover, if X is equipped with a free G–action, there is a similarly induced functor $\underline{C}((S.X)/G) \to \underline{\mathcal{P}(X)/G}$ which is characterized by the requirement that the diagram

$$\begin{array}{ccc} \underline{C} & \longrightarrow & \underline{\mathcal{P}(X)} \\ \downarrow & & \downarrow \\ \underline{C}((S.X)/G) & \longrightarrow & \underline{\mathcal{P}(X)/G} \end{array}$$

should commute. We denote the functor $\underline{C}((S.X)/G) \to \underline{\mathcal{P}(X)/G}$ by π.

Proposition IV.7 *Let $F : \underline{\mathcal{P}(X)/G} \to \underline{s \text{ sets}}$ be any functor. Then the natural map*

$$\underset{\underline{C}((S.X)/G)}{\underrightarrow{\text{hocolim}}}\ F \circ \pi \to \underset{\underline{\mathcal{P}(X)/G}}{\underrightarrow{\text{hocolim}}}\ F$$

is a weak equivalence. Consequently, the same result holds for functors from $\underline{\mathcal{P}(X)/G}$ to Spectra.

Proof: By I.2 (g), it suffices to check that for any object x of $\underline{\mathcal{P}(X)/G}$, the category $x \downarrow \pi$ has contractible nerve. But due to the freeness of the G–action, one sees that $x \downarrow \pi$ is isomorphic to the category $\tilde{x} \downarrow p$, where $\tilde{x}$ is any orbit representative in $\mathcal{P}(X)$ for x, and where $p : \underline{C}(S.X) \to \underline{\mathcal{P}(X)}$ is the functor defined above. It is further clear that if $\tilde{x}$ is a k–simplex, then $\tilde{x} \downarrow p$ is isomorphic to $\underline{C}(S.\Delta[k])$, via the inclusion $\underline{C}(S.\Delta[k]) \to \underline{C}(S.X)$ induced by the simplicial map $\Delta[k] \xrightarrow{\tilde{x}} X$. But it is a standard fact from Bousfield and Kan [11] that $N(\underline{C}(S.\Delta[k]))$ is weakly equivalent to $S.\Delta[k]$, and is therefore weakly contractible. Q.E.D.

We observe for later use that if $Y.$ is a simplicial set, then $\underline{C}(Y.)$ is isomorphic to the category $\underline{\Delta}^{\text{op}} \wr F^Y$, where $F^Y : \underline{\Delta}^{\text{op}} \to \underline{CAT}$ is the functor which assigns to $\underline{k}$ the set Y_k, viewed as a discrete category.

Now suppose that we have a locally finite covering $\mathcal{U} = \{U_i\}_{i \in \mathbb{Z}}$ of a metric space, satisfying Hypothesis (A). Let $N.\mathcal{U}$ denote the nerve of the covering, and let Z be the simplicial complex obtained by triangulating $\mathbb{R}$ with a vertex at each integer. Then we have the simplicial set $S.Z$. $S_k Z$ can be identified with the set of all sequences of integers of length $k+1$, $(n_0, \ldots, n_k)$, satisfying the condition that $\{n_0, \ldots, n_k\} \subseteq \{\ell, \ell+1\}$ for some ℓ. The sequence $(n_0, \ldots, n_k)$ corresponds to the simplicial map $\Delta[k] \to Z$ sending the j–th vertex to n_j.

$\underline{C}$ defines a functor from $\underline{s \text{ sets}}$ (and by restriction $\underline{s \text{ sets}^p}$) to $\underline{CAT}$. Via this embedding, we consider the category $\underline{\mathcal{M}}^{\underline{s \text{ sets}}}$. We define $T : \underline{\mathcal{M}}^{\underline{s \text{ sets}}} \to \underline{s.\mathcal{M}}$ on objects by $T(X., \Phi)_k = \coprod_{x \in X_k} \Phi(x)$, $T(X., \Phi)(\theta) = \Phi(x \xrightarrow{\theta} \theta_* x)$ on the disjoint union factor $\Phi(x)$. We leave the behavior of T on morphisms in $\underline{\mathcal{M}}^{\underline{s \text{ sets}^p}}$ to the reader. The local finiteness of $X.$ assures that $T(X., \Phi)$ is a simplicial object in $\underline{\mathcal{M}}$. Let $(X., \Phi)$ be any object of $\underline{\mathcal{M}}^{\underline{s \text{ sets}^p}}$, and let $f : X. \to X.'$ be a simplicial map, where $X.'$ is locally finite. We do not require that f be proper. Then we define a new object $f_!(X., \Phi)$ in $\underline{\mathcal{M}}^{\underline{s \text{ sets}^p}}$ by $f_!(X., \Phi) = (X.', f_! \Phi)$, where $(f_! \Phi)(\tilde{x}) = \coprod_{x,\, f(x) = \tilde{x}} \Phi(x)$. There is no morphism $(X., \Phi) \to f_!(X., \Phi)$ in $\underline{\mathcal{M}}^{\underline{s \text{ sets}^p}}$ if f is not proper, but there is a canonical isomorphism $T(X., \Phi) \to T(f_!(X., \Phi))$ of simplicial metric spaces. Note, though, that if $X. \xrightarrow{f} X.' \xrightarrow{g} X.''$ are simplicial maps of locally finite simplicial sets, and g is proper, then we do have a morphism

$\tilde{g} : f_!(X., \Phi) \to (g \circ f)_!(X., \Phi)$ in $\underline{\mathcal{M}}^{s\,\mathrm{sets}^p}$, so that the diagram

$$\begin{array}{ccc} & T(X., \Phi) & \\ \swarrow & & \searrow \\ T(f_!(X., \Phi)) & \xrightarrow{T(\tilde{g})} & T((g \circ f)_!(X., \Phi)) \end{array}$$

commutes, and $T(\tilde{g})$ is an isomorphism in $\underline{s.\mathcal{M}}$.

Now, let the coverings $\mathcal{U}(\ell)$ be defined as above; since $\mathcal{U}$ satisfies Hypothesis (A), so does $\mathcal{U}(\ell)$ for all $\ell \geq 0$. For each ℓ, let $\mu_\ell : \underline{\mathcal{C}}(S.Z) \to \underline{\mathcal{M}}$ be given on objects by $\mu_\ell(n_0, \ldots, n_k) = U(\ell)_{n_0} \cap \cdots \cap U(\ell)_{n_k}$, and by the evident inclusions on morphisms. Each $(S.Z, \mu_\ell)$ is an object in $\underline{\mathcal{M}}^{s\,\mathrm{sets}^p}$. Let $\pi : S.Z \to S.Z/\mathbb{Z}$ and $\bar{\pi} : S.Z \to S.Z/2\mathbb{Z}$ be the projection maps. Note that we have $\rho : S.Z/2\mathbb{Z} \to S.Z/\mathbb{Z}$, which is clearly a proper map. ρ determines a morphism $\tilde{\rho}_\ell : \bar{\pi}_!(S.Z, \mu_\ell) \to \pi_!(S.Z, \mu_\ell)$ in $\underline{\mathcal{M}}^{s\,\mathrm{sets}^p}$, where $T(\tilde{\rho}_\ell)$ is an isomorphism in $\underline{s.\mathcal{M}}$ making the diagram

(I)
$$\begin{array}{ccc} & T(S.Z, \mu_\ell) & \\ \swarrow & & \searrow \\ T(\pi_!(S.Z, \mu_\ell)) & \xrightarrow{T(\tilde{\rho}_\ell)} & T(\pi_\ell(S.Z, \mu_\ell)) \end{array}$$

commute in $\underline{s.\mathcal{M}}$. Let $\delta : Z \to Z$ be the map of simplicial complexes given on vertices by $\delta(2i) = \delta(2i+1) = i$. δ is a proper map, and we can consider the object $\delta_!(S.Z, \mu_\ell)$. Because of the properness of δ, there is a morphism $(S.Z, \mu_\ell) \xrightarrow{\tilde{\delta}} \delta_!(S.Z, \mu_\ell)$ in $\underline{\mathcal{M}}^{s\,\mathrm{sets}^p}$; $\delta_!(S.Z, \mu_\ell) = (S.Z, \delta_!\mu_\ell)$. For $(n_0, \ldots, n_k)$ a k–simplex of $S.Z$,

$$\delta_!\mu_\ell(n_0, \ldots, n_k) = U_{2n_0}(\ell) \cap \cdots \cap U_{2n_k}(\ell) \coprod U_{2n_0+1}(\ell) \cap \cdots \cap U_{2n_k+1}(\ell).$$

There is an inclusion

$$U_{2n_0}(\ell) \cap \cdots \cap U_{2n_k}(\ell) \coprod U_{2n_0+1}(\ell) \cap \cdots \cap U_{2n_k+1}(\ell)$$

$$\hookrightarrow (U_{2n_0}(\ell) \cup U_{2n_0+1}(\ell)) \cap \cdots (U_{2n_k}(\ell) \cup U_{2n_k+1}(\ell))$$

$$= U_{n_0}(\ell+1) \cap \cdots \cap U_{n_k}(\ell+1) = \mu_{\ell+1}(n_0, \ldots, n_k)$$

in $\underline{\mathcal{M}}$, which gives rise to a natural transformation $\delta_!\mu_\ell \to \mu_{\ell+1}$. We therefore obtain a morphism

$$\theta_\ell : (S.Z, \mu_\ell) \xrightarrow{\tilde{\delta}} \delta_!(S.Z, \mu_\ell) \to (S.Z, \mu_{\ell+1})$$

in $\underline{\mathcal{M}}^{s\,\mathrm{sets}^p}$. Note that the composite $S.Z \xrightarrow{\delta} S.Z \to S.Z/\mathbb{Z}$ factors uniquely through $S.Z/2\mathbb{Z}$, i.e., we have a commutative diagram

(II)
$$\begin{array}{ccc} S.Z & \xrightarrow{\delta} & S.Z \\ \downarrow & & \downarrow \\ S.Z/2\mathbb{Z} & \xrightarrow{\hat{\delta}} & S.Z/\mathbb{Z} \end{array}$$

Let $\underline{s\ \text{sets}}^f \subseteq \underline{s\ \text{sets}}^p$ denote the finite simplicial sets; $\bar{\pi}_!(S.Z, \mu_\ell)$ and $\bar{\pi}_!(S.Z, \mu_{\ell+1})$ are in the subcategory $\underline{\mathcal{M}}^{\underline{s\ \text{sets}}^f} \subseteq \underline{\mathcal{M}}^{\underline{s\ \text{sets}}^p}$. The above diagram shows that θ_ℓ induces a morphism $\bar{\theta}_\ell : \bar{\pi}_!(S.Z, \mu_\ell) \to \pi_!(S.Z, \mu_{\ell+1})$ in $\underline{\mathcal{M}}^{\underline{s\ \text{sets}}^f}$. Let $\mathbf{N}^\sigma$ denote the category obtained from $\mathbf{N}^s$ by adjoining a single new object ∞ to $\mathbf{N}^s$ declaring that for every object of $\underline{\mathbf{N}}^s$, there is a unique morphism from that object to ∞, and declaring that the only self-map of ∞ is the identity map. We now define an $\underline{\mathbf{N}}^s$ diagram λ in $\underline{\mathcal{M}}^{\underline{s\ \text{sets}}^f}$ by

$$\lambda(\{\ell\}) = \pi_!(S.Z, \mu_\ell)$$
$$\lambda(\{\ell, \ell+1\}) = \bar{\pi}_!(S.Z, \mu_\ell)$$
$$\lambda(\infty) = (*, \Phi_x)$$
$$* = \text{ one point simplicial set}$$
$$\Phi_x = \text{ constant functor with value } X.$$
$$\lambda(\{\ell, \ell+1\}) \to \{\ell\} = \tilde{\rho}_\ell$$
$$\lambda(\{\ell, \ell+1\}) \to \{\ell+1\} = \bar{\theta}_\ell$$
$$\lambda(\{\ell\} \to \infty) = (f, i)$$
$$\lambda(\{\ell, \ell+1\} \to \infty) = (g, j)$$

where $f : S.Z/\mathbb{Z} \to *$ and $g : S.Z/2\mathbb{Z} \to *$ are the constant maps, and i and j are the inclusion maps $\mu_\ell(x) \hookrightarrow X$ and $\bar{\mu}_\ell(x) \hookrightarrow X$ for $x \in S.Z/\mathbb{Z}$ and $x \in S.Z/2\mathbb{Z}$ respectively. We now define four functors from $\underline{\mathcal{M}}^{\underline{s\ \text{sets}}^f}$ to $\underline{\mathcal{S}}$, which we will eventually show are weakly equivalent.

(i) $A_0 : \underline{\mathcal{M}}^{\underline{s\ \text{sets}}^f} \to \underline{\mathcal{S}}$ is given on objects by $A_0(Y., \Phi) = |\underset{\sim}{\widehat{K}}(T(Y., \Phi), R)|$,

and the behavior on morphisms is straightforward from the functoriality of $\underset{\sim}{\widehat{K}}(-; R)$, $T(\ )$, and $|\ |$.

(ii) We view the simplicial set $Y.$ as a functor from $\underline{\Delta}^{\text{op}}$ to (discrete) categories. Thus, $\underline{C}(Y.) \cong \underline{\Delta}^{\text{op}} \wr Y.$, and we let $\Psi : \underline{\Delta}^{\text{op}} \to \underline{\mathcal{M}}^{CAT}$ be the functor so that $\Psi^v = \Phi$, where Ψ^v is defined in the discussion preceding I.5. Compose Ψ with the functor $|\underrightarrow{\text{hocolim}} \circ \underset{\sim}{\widehat{K}}(\ , R)^{CAT}|$ to define

A_1. $A_1(Y., \Phi)$ is given by $A_1(Y., \Phi) = |\underline{k} \to \underrightarrow{\text{hocolim}}_{Y_k} \underset{\sim}{\widehat{K}}(\Phi \mid Y_k; R)|$.

(iii) Let Ψ be as in (ii); then $(\underrightarrow{\text{hocolim}} \circ \underset{\sim}{\widehat{K}}(\ , R)^{CAT}) \circ \Psi$ defines a simplicial spectrum. We define $A_2(Y., \Phi)$ to be

$$\underrightarrow{\text{hocolim}}_{\underline{\Delta}^{\text{op}}}((\underrightarrow{\text{hocolim}} \circ \underset{\sim}{\widehat{K}}(\ , R)^{CAT}) \circ \Psi).$$

On objects, it is given by

$$A_2(Y., \Phi) = \underrightarrow{\text{hocolim}}_{\underline{\Delta}^{\text{op}}} |\underline{k} \to \underrightarrow{\text{hocolim}}_{Y_k} \underset{\sim}{\widehat{K}}(\Phi \mid Y_k; R)|.$$

(iv) A_3 is $\underrightarrow{\text{hocolim}} \circ \widehat{\underset{\sim}{K}}(\ ,R) \circ \Phi$; $A_3(Y.,\Phi) = \underset{\underline{C}(Y.)}{\underrightarrow{\text{hocolim}}}\, \underset{\sim}{K}(\Phi(-),R)$.

Proposition IV.8 *There are weak equivalences of functors $A_1 \to A_0$, $A_2 \to A_1$, and $A_2 \to A_3$. Consequently, there is a homotopy commutative diagram*

$$\begin{array}{ccc} \underset{\underline{\mathbb{N}}^{\sigma}}{\underrightarrow{\text{hocolim}}}\, A_0 \circ \lambda & \longrightarrow & \underset{\sim}{K}(X,R) \\ \uparrow & & \uparrow \\ \underset{\underline{\mathbb{N}}^{\sigma}}{\underrightarrow{\text{hocolim}}}\, A_3 \circ \lambda & \longrightarrow & \underset{\underline{\Delta}^{op}}{\underrightarrow{\text{hocolim}}}\, \underset{\sim}{K}(X,R) \end{array}$$

where the vertical arrows are equivalences and the horizontal arrows are induced by the morphisms to ∞ in $\underline{\mathbb{N}}^{\sigma}$.

Proof: The natural transformation $A_1 \to A_0$ is given by the equivalence $\vee_{y \in Y_k} \widehat{\underset{\sim}{K}}(y,R) \to \widehat{\underset{\sim}{K}}(Y_k;R)$ coming from III.12. $A_2 \to A_1$ is I.2 (f). $A_2 \to A_3$ is I.5. Q.E.D.

Define an $\underline{\mathbb{N}}^{\sigma}$ diagram of spectra $\hat{\lambda}$ by

$$\hat{\lambda}(\{\ell\}) = |\widehat{\underset{\sim}{K}}(N.\mathcal{U}(\ell);R)$$

$$\hat{\lambda}(\{\ell,\ell+1\}) = |\widehat{\underset{\sim}{K}}(N.\mathcal{U}(\ell),R)|$$

$$\hat{\lambda}(\infty) = \widehat{\underset{\sim}{K}}(X;R)$$

$$\hat{\lambda}(\{\ell,\ell+1\} \to \{\ell\}) = \text{Id}$$
$$\hat{\lambda}(\{\ell,\ell+1\} \to \{\ell+1\}) = |\widehat{\underset{\sim}{K}}(N.\Theta_\ell;R)|$$

$$\hat{\lambda}(\{\ell\} \to \infty) = \hat{A}(\mathcal{U}(\ell))$$
$$\hat{\lambda}(\{\ell,\ell+1\} \to \infty) = \hat{A}(\mathcal{U}(\ell)).$$

From the fact that $T(X.,\Phi) \to T(\pi_!(X,\Phi))$ and $T(X.,\Phi) \to T(\bar{\pi}_!(X,\Phi))$ are isomorphisms in $\underline{s.\mathcal{M}}$, and from diagrams (I) and (II), it follows that we have commutative diagrams (in $\underline{s.\mathcal{M}}$)

$$\begin{array}{ccccc} T(S.Z,\mu_\ell) & \overset{\text{Id}}{\longleftarrow} & T(S.Z,\mu_\ell) & \overset{T(\theta_\ell)}{\longrightarrow} & T(S.Z,\mu_{\ell+1}) \\ \downarrow \wr & & \downarrow \wr & & \downarrow \wr \\ T(\pi_!(S.Z,\mu_\ell)) & \overset{T(\tilde{\rho}_\ell)}{\longleftarrow} & T(\bar{\pi}_!(S.Z,\mu_\ell)) & \overset{T(\bar{\theta}_\ell)}{\longrightarrow} & T(\pi_!(S.Z(S.Z,\mu_{\ell+1}))) \end{array}$$

and commutative diagrams of simplicial spectra

$$\begin{array}{ccc}
\underset{\sim}{\widehat{K}}(N.\mathcal{U}(\ell),R) & \xrightarrow{\underset{\sim}{\widehat{K}}(N.\Theta_\ell;R)} & \underset{\sim}{\widehat{K}}(N.\mathcal{U}(\ell+1),R) \\
\downarrow \wr & & \downarrow \wr \\
\underset{\sim}{\widehat{K}}(T(S.Z,\mu_\ell);R) & \xrightarrow{\underset{\sim}{\widehat{K}}(T(\theta_\ell);R)} & \underset{\sim}{\widehat{K}}(T(S.Z,\mu_{\ell+1});R)
\end{array}$$

where all the vertical arrows are isomorphisms. We conclude that the $\underline{\mathbf{N}}^\sigma$ diagrams $\hat{\lambda}$ and $A_0 \circ \lambda$ are weakly equivalent diagrams of spectra. Moreover, define $\underline{\mathbf{N}}^\infty$ to be the category $\underline{\mathbf{N}}$ with an object ∞ adjoined, with a unique morphism from every object in $\underline{\mathbf{N}}$ to ∞, and so that the only self–map of ∞ is the identity. $\underline{\mathbf{N}}^\infty$ bears the same relationship toward $\underline{\mathbf{N}}$ as $\underline{\mathbf{N}}^\sigma$ does toward $\underline{\mathbf{N}}^s$, and we have the functor $\pi^\infty \mid \underline{\mathbf{N}}^s = \pi$, and $\pi^\infty(\infty) = \infty$. Define an $\underline{\mathbf{N}}^\infty$ diagram $\hat{\hat{\lambda}}$ in $\underline{S}$ as follows.

$\hat{\hat{\lambda}}(\{\ell\}) = |\underset{\sim}{\widehat{K}}(N.\mathcal{U}(\ell);R)|$

$\hat{\hat{\lambda}}(\infty) = |\underset{\sim}{\widehat{K}}(X;R)|$

$\hat{\hat{\lambda}}(\{\ell\} \to \{\ell+1\}) = |\underset{\sim}{\widehat{K}}(N.\Theta_\ell;R)|$

$\hat{\hat{\lambda}}(\{\ell\} \to \infty) = \hat{A}(\mathcal{U}(\ell))$

Then $\hat{\lambda} = \hat{\hat{\lambda}} \circ \pi^\infty$, and I.14, I.2 (g), and I.5 now show that we have a commutative diagram of spectra

$$\begin{array}{ccccc}
\underset{\overrightarrow{\underline{\mathbf{N}}^s}}{\text{hocolim}}\, \hat{\lambda} & & & & \\
\downarrow \wr & & \searrow & & \\
\underset{\overrightarrow{\underline{\mathbf{N}}}}{\text{hocolim}}\, \hat{\hat{\lambda}} & & \longrightarrow & & \underset{\sim}{\widehat{K}}(X;R) \\
\downarrow \wr & \nearrow & \underset{\overrightarrow{\ell}}{\text{hocolim}}\, \hat{A}(\mathcal{U}(\ell)) & & \\
\underset{\overrightarrow{\ell}}{\text{hocolim}}\, |\underset{\sim}{\widehat{K}}(N.\mathcal{U}(\ell);R)| & & & &
\end{array}$$

with the upper left hand arrow an equivalence and the lower one an isomorphism.

Corollary IV.9 *There is a homotopy commutative diagram of spectra*

$$\begin{array}{ccc}
\underset{\underline{\mathbb{N}}^\sigma}{\underrightarrow{\text{hocolim}}}\, A_3 \circ \lambda & \longrightarrow & \underset{\underline{\Delta}^{\text{op}}}{\underrightarrow{\text{hocolim}}}\, \underset{\sim}{K}(X,R) \\
\wr\downarrow & & \downarrow\wr \\
& \underrightarrow{\text{hocolim}}\, \hat{A}(\mathcal{U}(\ell)) & \\
\underset{\ell}{\underrightarrow{\text{hocolim}}}\, \underset{\sim}{\widehat{K}}(N.\mathcal{U}(\ell);R) & \longrightarrow & \underset{\sim}{K}(X,R)
\end{array}$$

where the vertical arrows are equivalences and the upper horizontal arrow is induced by the morphism to ∞ *in* $\underline{\mathbb{N}}^\sigma$.

Recall that we are viewing $\underline{\mathcal{M}}^{\underline{s\text{ sets}}^f}$ as a subcategory of $\underline{\mathcal{M}}^{CAT}$. We note that
$A_3 : \underline{\mathcal{M}}^{\underline{s\text{ sets}}^f} \to \underline{S}$ is actually defined on all of $\underline{\mathcal{M}}^{CAT}$, since it is the composite
$\underrightarrow{\text{hocolim}} \circ \underset{\sim}{\widehat{K}}(-;R)^{CAT}$. Recall also from the discussion above that we have

functors
$\underline{C}(S.Z/\mathbb{Z}) \to \underline{\mathcal{P}(Z)/\mathbb{Z}}$ and $\underline{C}(S.Z/2\mathbb{Z}) \to \underline{\mathcal{P}(Z)/2\mathbb{Z}}$, and one easily checks that we have factorizations

$$\begin{array}{ccc}
\underline{C}(S.Z/\mathbb{Z}) & & \\
\downarrow & \searrow^{\pi_! \mu_\ell} & \\
\underline{\mathcal{P}(Z)/\mathbb{Z}} & \xrightarrow{m_\ell} & \underline{\mathcal{M}}
\end{array}$$

and

$$\begin{array}{ccc}
\underline{C}(S.Z/2\mathbb{Z}) & & \\
\downarrow & \searrow^{\bar{\pi}_! \mu_\ell} & \\
\underline{\mathcal{P}(Z)/2\mathbb{Z}} & \xrightarrow{\bar{m}_\ell} & \underline{\mathcal{M}}
\end{array}$$

and consequently morphisms $\pi_!(S.Z,\mu_\ell) \to (\underline{\mathcal{P}(Z)/\mathbb{Z}}, m_\ell)$ and $\bar{\pi}_!(S.Z,\mu_\ell) \to (\underline{\mathcal{P}(Z)/2\mathbb{Z}}, \bar{m}_\ell)$ in $\underline{\mathcal{M}}^{CAT}$. We remark that $\mathcal{P}(Z) \cong P^a$ $\mathbb{Z}$–equivariantly, so we rewrite $(\underline{\mathcal{P}(Z)/\mathbb{Z}}, m_\ell) \cong (\underline{P^a/\mathbb{Z}}, m_\ell)$ and $(\underline{\mathcal{P}(Z)/2\mathbb{Z}}, \bar{m}_\ell) \cong (\underline{P^a/2\mathbb{Z}}, m_\ell)$. Recall the definitions of $p, q : \underline{P^a/2\mathbb{Z}} \to \underline{P^a/\mathbb{Z}}$ from above. It is now easy to verify that there are natural transformations $\bar{m}_\ell \xrightarrow{\xi_\ell} m_\ell \circ p$ and $\bar{m}_\ell \xrightarrow{\xi_\ell} m_{\ell+1} \circ q$ so that the corresponding diagrams

$$\begin{array}{ccc}
\bar{\pi}_!(S.Z,\mu_\ell) & \xrightarrow{\bar{\rho}_\ell} & \pi_!(S.Z,\mu_\ell) \\
\downarrow & & \downarrow \\
(\underline{P^a/2\mathbb{Z}}, \bar{m}_\ell) & \xrightarrow{(p,\xi_\ell)} & (\underline{P^a/\mathbb{Z}}, m_\ell)
\end{array}$$

and

$$\begin{array}{ccc}\bar{\pi}_!(S.Z,\mu_\ell) & \xrightarrow{\bar{\theta}_\ell} & \pi_!(S.Z,\mu_{\ell+1}) \\ \downarrow & & \downarrow \\ (\underline{P^a/2\mathbb{Z}},\bar{m}_\ell) & \xrightarrow{(q,\zeta_\ell)} & (P^a/\mathbb{Z},\bar{m}_{\ell+1})\end{array}$$

commute in $\underline{\mathcal{M}}^{CAT}$. We now define an $\underline{\mathbb{N}}$ diagram with values in $\underline{\mathcal{M}}^{CAT}$ by

$\lambda_1(\{\ell\}) = (P^a/\mathbb{Z}, m_\ell)$
$\lambda_1(\{\ell,\ell+1\}) = (\underline{P^a/2\mathbb{Z}}, \bar{m}_\ell)$
$\lambda_1(\infty) = (e, \Psi_X)$
$e =$ the category with one object
$\Psi_X =$ constant functor with value X.
$\lambda_1(\{\ell,\ell+1\} \to \{\ell\}) = (p, \xi_\ell)$
$\lambda_1(\{\ell,\ell+1\} \to \{\ell+1\}) = (q, \zeta_\ell)$
$\lambda_1(\{\ell\} \to \infty) = f$
$\lambda_1(\{\ell,\ell+1\} \to \infty) = g$

where f and g are morphisms corresponding to the constant functors $\underline{P^a/\mathbb{Z}} \to e$, $\underline{P^a/2\mathbb{Z}} \to e$, and the inclusions $m_\ell(x) \subseteq X$, $\bar{m}_\ell(x) \subseteq X$ respectively.

From the above discussion, it follows that we have a natural transformation $\lambda \to \lambda_1$ of $\underline{\mathbb{N}}^\sigma$ diagrams in $\underline{\mathcal{M}}^{CAT}$.

Proposition IV.10 *The natural transformation $\lambda \to \lambda_1$ is a weak equivalence of diagrams. From this it follows from IV.9 that there exists a homotopy commutative diagram*

$$\begin{array}{ccc}\underset{\underline{\mathbb{N}}^\sigma}{\underrightarrow{\text{hocolim}}}\, A_3 \circ \lambda_1 & \longrightarrow & \underset{\sim}{\widehat{K}}(X;R) \\ \downarrow \wr & \nearrow \underset{\ell}{\underrightarrow{\text{hocolim}}}\, \hat{A}(\mathcal{U}(\ell)) & \\ \underset{\ell}{\underrightarrow{\text{hocolim}}}\, |\underset{\sim}{\widehat{K}}(N.\mathcal{U}(\ell);R)| & & \end{array}$$

where the vertical arrow is an equivalence and the horizontal arrow is constructed using the morphisms to ∞ in $\underline{\mathbb{N}}^\sigma$.

We recall from I.7 that $(\underline{C},\Phi) \to \underline{C}\wr\Phi$ defines a functor $\underline{\text{SymMon}}^{CAT} \to \underline{\text{SymMon}}$, i.e., that Thomason's category theoretic colimit has the same kind of functoriality as $\underrightarrow{\text{hocolim}}$. We write $\underrightarrow{\text{scolim}}$ for this functor. We now define

$A_4 : \underline{\mathcal{M}}^{CAT} \to \underline{\mathcal{S}}$ to be the composite

$$\underline{\mathcal{M}}^{CAT} \xrightarrow{i^{CAT}} \underline{\text{SymMon}}^{CAT} \xrightarrow{\text{scolim}} \underline{\text{SymMon}} \xrightarrow{\text{Spt}} \underline{\mathcal{S}},$$

where $i : \underline{\mathcal{M}} \to \underline{\text{SymMon}}$ is given by $i(X) = i\widehat{\mathcal{C}}_X(R)$. A_4 is given on objects by $A_4(C,\Phi) = \text{Spt}\,(C\wr\Phi)$.

Proposition IV.11 *Thomason's natural equivalence* $\underset{\underline{C}}{\overrightarrow{\text{hocolim}}}(\text{Spt} \circ \Phi) \to \text{Spt}(\underline{C} \wr \Phi)$ *gives a natural equivalence of functors* $A_3 \to A_4$ *on* $\underline{\mathcal{M}}^{CAT}$. *Consequently, there is a homotopy commutative diagram*

$$\begin{array}{ccc} \underset{\underline{\mathbb{N}}^s}{\overrightarrow{\text{hocolim}}}\, A_4 \circ \lambda_1 & \longrightarrow & \widehat{K}(X;R) \\ \downarrow \wr & \nearrow \underset{\ell}{\overrightarrow{\text{hocolim}}}\, \hat{A}(\mathcal{U}(\ell)) & \\ \underset{\ell}{\overrightarrow{\text{hocolim}}}\, |\underset{\sim}{\widehat{K}}(N.\mathcal{U}(\ell);R)| & & \end{array}$$

where the vertical arrow is an equivalence and the horizontal arrow is obtained from the maps to ∞ *in* $\underline{\mathbb{N}}^\sigma$.

Let $B : \underline{\mathbb{N}}^\sigma \to \underline{\text{SymMon}}^{CAT}$ be given by

$$\begin{aligned}
&B(\{\ell\}) = \underline{P^a/\mathbb{Z}} \wr (i \circ m_\ell) \\
&B(\{\ell, \ell+1\}) = \underline{P^a/2\mathbb{Z}} \wr (i \circ \bar{m}_\ell) \\
&B(\infty) = i\widehat{\mathcal{C}}_X(R) \\
&B(\{\ell, \ell+1\} \to \{\ell\}) = (p, i(\xi_\ell)) \\
&B(\{\ell, \ell+1\} \to \{\ell+1\}) = (q, i(\zeta_\ell)) \\
&B(\{\ell\} \to \infty) = f \\
&B(\{\ell, \ell+1\} \to \infty) = g
\end{aligned}$$

where f and g are the functors induced by the constant functors $\underline{P^a/\mathbb{Z}} \to e$, $\underline{P^a/2\mathbb{Z}} \to e$, and $i(m_\ell(-) \hookrightarrow X)$, $i(\bar{m}_\ell(-) \hookrightarrow X)$, respectively.

Consider $\underline{\mathbb{N}}^s \wr B$; there is an evident functor $\varepsilon : \underline{\mathbb{N}}^s \wr B \to i\widehat{\mathcal{C}}_X(R)$, namely the composite

$$\underline{\mathbb{N}}^s \wr B \to \underline{\mathbb{N}}^s \wr (i\widehat{\mathcal{C}}_X(R)) \to i\widehat{\mathcal{C}}_X(R),$$

where $(i\widehat{\mathcal{C}}_X(R))$ is used to denote the constant functor with value $i\widehat{\mathcal{C}}_X(R)$. The following is now an immediate consequence of Thomason's homotopy colimit theorem from §I.

Corollary IV.12 *The functors* $A_r \circ \lambda$ *and* $\text{Spt} \circ B$ *are naturally equivalent spectrum valued* $\underline{\mathbb{N}}^\sigma$ *diagrams. Consequently, there is a homotopy commutative diagram*

$$\begin{array}{ccc} \text{Spt}(\underline{\mathbb{N}}^s \wr B) & \xrightarrow{\text{Spt}(\varepsilon)} & \widehat{K}(X;R) \\ \downarrow \wr & \nearrow \underset{\ell}{\overrightarrow{\text{hocolim}}}\, \hat{A}(\mathcal{U}(\ell)) & \\ \underset{\ell}{\overrightarrow{\text{hocolim}}}\, |\underset{\sim}{\widehat{K}}(N.\mathcal{U}(\ell);R)| & & \end{array}$$

where the vertical arrow is an equivalence.

This completes the translation of our problem into category theoretic language. Now consider the coverings $\mathcal{U}(\ell)$ again. For each ℓ, let $Y_\ell = \coprod_i U_i(\ell) \cap U_{i+1}(\ell)$ and $Z_\ell = \coprod_i U_i(\ell)$. We have two inclusions $f_\ell, g_\ell : Y_\ell \to Z_\ell$, where f_ℓ is the disjoint union of the inclusions $U_i(\ell) \cap U_{i+1}(\ell) \to U_i(\ell)$ and g_ℓ is the disjoint union of the inclusions $U_i(\ell) \cap U_{i+1}(\ell) \to U_{i+1}(\ell)$. The diagram $i\widehat{\mathcal{C}}_{U_\ell}(R) \underset{g_\ell}{\overset{f_\ell}{\rightrightarrows}} i\widehat{\mathcal{C}}_{Z_\ell}(R)$ is an $\underline{M}$ diagram in $\underline{\text{SymMon}}$, where $\underline{M}$ is the category of I.10, and we let $\underline{T}_\ell$ be the permutative category $\text{Tor}\,[i\widehat{\mathcal{C}}_{Y_\ell}(R) \underset{g_\ell}{\overset{f_\ell}{\rightrightarrows}} i\widehat{\mathcal{C}}_{Z_\ell}(R)]$. $\underline{T}_\ell$ is permutative since we have taken $i\widehat{\mathcal{C}}_{Y_\ell}(R)$ and $i\widehat{\mathcal{C}}_{Z_\ell}(R)$ to be permutative. An object of $\underline{T}_\ell$ is given by a pair (A, B), where A (respectively B) is an object of $i\widehat{\mathcal{C}}_{Y_\ell}(R)$ (respectively $i\widehat{\mathcal{C}}_{Z_\ell}(R)$). We define a symmetric monoidal functor $\tau_\ell : \underline{T}_\ell \to \underline{T}_{\ell+1}$ as follows. Note that every object A of $i\widehat{\mathcal{C}}_{Y_\ell}(R)$ can be canonically decomposed as $A = A^0 \oplus A^1$, where A^0 is an object of $i\widehat{\mathcal{C}}_{Y_{\ell+1}}(R)$, and where A^1 is an object of $i\widehat{\mathcal{C}}_W(R)$, with $W = \coprod_i U_{2i}(\ell) \cap U_{2i+1}(\ell) \subseteq \coprod_i U_i(\ell+1)$. Via this last inclusion, we view A' as an object in $i\widehat{\mathcal{C}}_{Z_{\ell+1}}(R)$. Define τ_ℓ on objects by $\tau_\ell(A, B) = (A^0, A^1 \oplus B)$, where B is viewed as an object of $i\widehat{\mathcal{C}}_{Z_{\ell+1}}(R)$ via the inclusion $Z_\ell \to Z_{\ell+1}$. On a morphism $(A, b) \xrightarrow{(U,V,\phi,\psi)} (\tilde{A}, \tilde{B})$, we set $\theta_\ell(U, V, \phi, \psi) = (\bar{U}, \bar{V}, \bar{\phi}, \bar{\psi})$, where $\bar{U} = U^0$, $\bar{V} = V^0$, $\bar{\phi} : A^0 \to U^0 \oplus \bar{A}^0 \oplus V^0$ is the restriction of ϕ to the factor $A^0 \subseteq A$, and $\bar{\psi}$ is the composite

$$U^0 \oplus A^1 \oplus B \oplus V^0 \xrightarrow{\text{Id} \oplus \phi \mid A^1 \oplus \text{Id} \oplus \text{Id}} U^0 \oplus U^1 \oplus \bar{A}^1 \oplus V^1 \oplus B \oplus V^0 \xrightarrow{c}$$

$$\to \bar{A}^1 \oplus U^0 \oplus U^1 \oplus B \oplus B \oplus V^1 \oplus V^0 \xrightarrow{\text{Id} \oplus \gamma_U \oplus \text{Id} \oplus \gamma_V} \bar{A}^1 \oplus U \oplus B \oplus V \xrightarrow{\text{Id} \oplus \varphi} \bar{A}^1 \oplus \bar{B}^1$$

Here, c is a commutativity isomorphism and U^0 and V^0 are viewed as objects of $i\widehat{\mathcal{C}}_{Z_{\ell+1}}(R)$ via the inclusions $f_{\ell+1}$ and $g_{\ell+1}$, respectively, A^1 is viewed as an object of $i\widehat{\mathcal{C}}_{Z_{\ell+1}}(R)$ via the inclusion $\coprod_i U_{2i}(\ell) \cap U_{2i+1}(\ell) \to \coprod_i U_i(\ell+1)$, and γ_U and γ_V are the isomorphisms $U^0 \oplus U^1 \xrightarrow{\sim} U$ and $V^1 \oplus V^0 \to V$, respectively. We define a functor $\hat{\tau}$ on $\underline{\mathbb{N}}$ with values in $\underline{\text{SymMon}}$ by $\hat{\tau}(\ell) = \underline{T}_\ell$, $\hat{\tau}(\ell \leq \ell+1) = \tau_\ell$, and consider the category $\underline{\mathbb{N}} \wr\wr \hat{\tau}$. We define a symmetric monoidal functor $\alpha : \underline{\mathbb{N}} \wr\wr \hat{\tau} \to i\widehat{\mathcal{C}}_{Z_*}(R)$. It follows from I.8 that in order to produce such an α, we must produce lax symmetric monoidal functors $\alpha_\ell : \underline{T}_\ell \to i\widehat{\mathcal{C}}_X(R)$ and symmetric monoidal natural transformation from α_ℓ to $\alpha_{\ell+1} \circ \tau_\ell$. We let α_ℓ be the symmetric monoidal functor $\nu : \underline{T}_\ell \to i\widehat{\mathcal{C}}_X(R)$ associated, as in the discussion preceding I.10, to the functor $i\widehat{\mathcal{C}}_{Z_\ell}(R) \to i\widehat{\mathcal{C}}_X(R)$ induced by the map $Z_\ell \to X$ in $\underline{M}$;

note that here we are using the local finiteness of the covering $\mathcal{U}$. Since $\mathcal{T}_{\ell+1}$ is permutative,

$$\alpha_\ell(A,B) = A \oplus B \cong A^0 \oplus A^1 \oplus B = \alpha_{\ell+1} \circ \tau_\ell(A,B)\,,$$

and this isomorphism is the required symmetric monoidal transformation from α_ℓ to $\alpha_{\ell+1} \circ \tau$. We also wish to define a symmetric monoidal functor $\underline{\mathbb{N}}^s \wr B \to \underline{\mathbb{N}} \wr \hat{\tau}$. Again by Thomason's universal mapping property I.8, we are only required to produce lax symmetric monoidal functors

$$\beta_\ell : \underline{P^a/\mathbb{Z}} \wr i \circ m_\ell \to \underline{\mathbb{N}} \wr \hat{\tau}$$
$$\beta_\ell : \underline{P^a/2\mathbb{Z}} \wr i \circ \bar{m}_\ell \to \underline{\mathbb{N}} \wr \hat{\tau}$$

together with natural transformations

$$\gamma_\ell \to \beta_\ell \circ B(\{\ell, \ell+1\} \to \{\ell\})$$
$$\gamma_\ell \to \beta_{\ell+1} \circ B(\{\ell, \ell+1\} \to \{\ell+1\})\,.$$

Recall that $\underline{P^a/\mathbb{Z}}$ is isomorphic to the category $\underline{M}$ of I.10, and further that $i \circ m_\ell$ is under this isomorphism identified with the diagram $i\widehat{\mathcal{C}}_{Y_\ell}(R) \underset{g_\ell}{\overset{f_\ell}{\rightrightarrows}} i\widehat{\mathcal{C}}_{Z_\ell}(R)$. Consequently, we have the functor $\theta : \underline{P^a/\mathbb{Z}} \wr (i \circ m_\ell) \to \mathcal{T}_\ell$ of I.10. We call this functor r_ℓ; it is symmetric monoidal. We also have the lax symmetric monoidal functor $j_\ell : \mathcal{T}_\ell \to \underline{\mathbb{N}} \wr \hat{\tau}$, $(A,B) \mapsto 1[(\ell,(A,B))]$; β_ℓ is defined to be $J_\ell \circ r_\ell$ and we set $\gamma_\ell = \beta_\ell \circ B(\{\ell,\ell+1\} \to \{\ell\})$. The natural transformation $\gamma_\ell \to \beta_\ell \circ B(\{\ell,\ell+1\} \to \{\ell\})$ is taken to be the identity natural transformation. We examine the composite $\beta_{\ell+1} \circ B(\{\ell,\ell+1\} \to \{\ell+1\})$. It is expressed as a composite $j_{\ell+1} \circ s_\ell$, where $s_\ell : \underline{P^a/2\mathbb{Z}} \wr i \circ \bar{m}_\ell \to \mathcal{T}_{\ell+1}$ is a symmetric monoidal functor, which we describe explicitly. s_ℓ can by I.8 be identified with a collection of lax symmetric monoidal functors $F_x : i \circ \bar{m}_\ell(x) \to \mathcal{T}_{\ell+1}$ for each object x of $\underline{P^a/2\mathbb{Z}}$ together with a system of natural transformations $N(f) : F_x \to F_y \circ (i \circ \bar{m}_\ell(f))$ for every morphism $x \xrightarrow{f} y$ in $\underline{P^a/2\mathbb{Z}}$ satisfying $N(f)\,N(g) = N(fg)$. Let $Y_\ell^{\text{even}} = \coprod_i U_{2i}(\ell) \cap U_{2i+1}(\ell)$, and $Y_\ell^{\text{odd}} = \coprod_i U_{2i+1}(\ell) \cap U_{2i+2}(\ell)$, $Z_\ell^{\text{even}} = \coprod_i U_{2i}(\ell)$, and $Z_\ell^{\text{odd}} = \coprod_i U_{2i+1}(\ell)$. We have inclusion functors $i\widehat{\mathcal{C}}_{Z_{\ell+1}}(R) \hookrightarrow \mathcal{T}_{\ell+1}$ and $i\widehat{\mathcal{C}}_{Y_{\ell+1}}(R) \hookrightarrow \mathcal{T}_{\ell+1}$ given on objects by $B \mapsto (0,B)$. The functor data describing s_ℓ is now given by the following table

$\underline{x}$	$\underline{F_x}$
$[\{0\}]$	$i\widehat{\mathcal{C}}_{Z_\ell^{\text{even}}}(R) \to \widehat{\mathcal{C}}_{Z_{\ell+1}}(R) \to \mathcal{T}_{\ell+1}$
$[\{1\}]$	$i\widehat{\mathcal{C}}_{Z_\ell^{\text{odd}}}(R) \to \widehat{\mathcal{C}}_{Z_{\ell+1}}(R) \to \mathcal{T}_{\ell+1}$
$[\{0,1\}]$	$i\widehat{\mathcal{C}}_{Y_\ell^{\text{even}}}(R) \to \widehat{\mathcal{C}}_{Y_{\ell+1}}(R) \to \mathcal{T}_{\ell+1}$
$[\{1,2\}]$	$i\widehat{\mathcal{C}}_{Y_\ell^{\text{odd}}}(R) \to \widehat{\mathcal{C}}_{Y_{\ell+1}}(R) \to \mathcal{T}_{\ell+1}$

The natural transformations are all identity morphisms for all objects in their domain categories except for the transformations corresponding to the two morphisms $[\{1,2\}] \xrightarrow{f_0} [\{0\}]$ and $[\{1,2\}] \xrightarrow{f_1} [\{1\}]$. $F_{[\{1,2\}]}$ is the functor $A \mapsto (A,0)$, $F_{[\{0\}]} \circ (i \circ \bar{m}_\ell(f_0))$ is the functor $A \mapsto (0, g_\ell A)$, and $F_{[\{1\}]} \circ \bar{m}_\ell(f_1)$ is the functor $A \mapsto (0, f_\ell A)$. The natural transformation $N(f_0)$ is given on an object A by the morphism $(0, A, \varphi_A, \psi_A)$, where $\varphi_A : A \to 0 \oplus A$ and $\psi_A : 0 \oplus g_\ell A \to g_\ell A$ are the canonical isomorphisms. Similarly, $N(f_1)$ is given on A by the morphism $(A, 0, \varphi^1_A, \psi^1_A)$, where $\varphi^1_A : A \to A \oplus 0$ and $\psi^1_A : f_\ell A \oplus 0 \to f_\ell A$ are the canonical isomorphisms.

The lax symmetric monoidal functor

$$\beta_\ell \circ B(\{\ell, \ell+1\} \to \{\ell\}) : \underline{P^a/2\mathbb{Z}} \wr\wr (i \circ \bar{m}_\ell) \to \underline{\mathbf{N}} \wr\wr \hat{\tau}$$

is expressed as $j_\ell \circ r_\ell \circ B(\{\ell, \ell+1\} \to \{\ell\})$, and $r_\ell \circ B(\{\ell, \ell+1\} \to \{\ell\})$ is a symmetric monoidal functor. Let G_x, N^1_f denote the functor and natural transformation data for $r_\ell \circ B(\{\ell, \ell+1\} \to \{\ell\})$ arising from Thomason's universal mapping principle I.8. The functor data is given as follows

$\underline{x}$	$\underline{G_x}$
$[\{0\}]$	$i\widehat{C}_{Z_\ell^{\text{even}}} \to i\widehat{C}_{Z_\ell} \to \widehat{\mathcal{T}}_\ell$
$[\{1\}]$	$i\widehat{C}_{Z_\ell^{\text{odd}}} \to i\widehat{C}_{Z_\ell} \to \widehat{\mathcal{T}}_\ell$
$[\{0,1\}]$	$i\widehat{C}_{Y_\ell^{\text{even}}} \to i\widehat{C}_{Y_\ell} \to \widehat{\mathcal{T}}_\ell$
$[\{1,2\}]$	$i\widehat{C}_{Y_\ell^{\text{odd}}} \to i\widehat{C}_{Y_\ell} \to \widehat{\mathcal{T}}_\ell$

In this case, all the natural transformations are identity maps. We first note that we have a symmetric monoidal natural transformation

$$\beta_\ell \circ B(\{\ell, \ell+1\} \to \{\ell\}) \to j_{\ell+1} \circ \tau_\ell \circ r_\ell \circ B(\{\ell, \ell+1\} \to \{\ell\}).$$

For any $\zeta \in \underline{P^a/2\mathbb{Z}} \wr\wr (i \circ \bar{m}_\ell)$, this is the morphism

$$1[(\ell, r_\ell \circ B(\{\ell, \ell+1\} \to \{\ell\})(\zeta))] \to 1[(\ell+1, \tau_\ell \circ r_\ell \circ B(\{\ell, \ell+1\} \to \{\ell\})(\zeta))]$$

corresponding to the morphism $\ell \mapsto \ell+1$ in $\underline{\mathbf{N}}$ and the identity map of $\tau_\ell \circ B(\{\ell, \ell+1\} \to \{\ell\})(\zeta)$. $j_{\ell+1} \circ \tau_\ell \circ r_\ell \circ B(\{\ell, \ell+1\} \to \{\ell\})$ is now a second symmetric monoidal functor $\underline{P^a/2\mathbb{Z}} \wr\wr (i \circ \bar{m}_\ell) \to \underline{\mathcal{T}}_{\ell+1}$, and one calculates directly that it is equal to $j_{\ell+1} \circ s_\ell$; indeed, $s_\ell = \tau_\ell \circ r_\ell \circ B(\{\ell, \ell+1\} \to \{\ell\})$. Moreover, the natural transformation data for s_ℓ is equal to the result of applying τ_ℓ to the natural transformation data for $r_\ell \circ B(\{\ell, \ell+1\} \to \{\ell\})$, so we have constructed the required symmetric monoidal natural transformation $\gamma_\ell \to \beta_{\ell+1} \circ B(\{\ell, \ell+1\} \to \{\ell+1\})$, and hence a symmetric monoidal functor $\underline{\mathbf{N}}^s \wr\wr B \to \underline{\mathbf{N}} \wr\wr \hat{\tau}$.

Proposition IV.13 *The diagram*

$$\begin{array}{ccc} \mathrm{Spt}\,(\underline{\mathbb{N}}^s \wr\wr B) & & \\ & \searrow^{\mathrm{Spt}\,(\varepsilon)} & \\ \downarrow & & \mathrm{Spt}\,(i\widehat{\mathcal{C}}_X(R)) = \underset{\sim}{\widehat{K}}(X,R) \\ & \nearrow^{\mathrm{Spt}\,(\alpha)} & \\ \mathrm{Spt}\,(\underline{\mathbb{N}} \wr\wr \hat{\tau}) & & \end{array}$$

is homotopy commutative, and the vertical arrow is an equivalence of spectra.

Proof: It is a direct verification that the functors ε and $\underline{\mathbb{N}}^s \wr\wr B \to \underline{\mathbb{N}} \wr\wr \hat{\tau} \xrightarrow{\alpha} i\widehat{\mathcal{C}}_X(R)$ are isomorphic as symmetric monoidal functors, the isomorphism being given by rearrangement isomorphisms, so the associated diagram of spectra commutes up to homotopy. Let $\underline{\mathbb{N}}[i]$ be the full subcategory of $\underline{\mathbb{N}}$ on the objects $\{0,1,\ldots,i\}$, and let $\underline{\mathbb{N}}^s[i]$ be the full subcategory of $\underline{\mathbb{N}}^s$ on objects $\{0\},\{1\},\ldots,\{i\},\{0,1\},\ldots,\{i-1,i\}$. By restriction, we obtain functors $\underline{\mathbb{N}}^s[i] \wr\wr B \to \underline{\mathbb{N}}[i] \wr\wr \hat{\tau}$, and $\mathrm{Spt}\,(\underline{\mathbb{N}} \wr\wr \hat{\tau}) = \underset{\overrightarrow{i}}{\mathrm{colim}}\mathrm{Spt}\,(\underline{\mathbb{N}}[i] \wr\wr \hat{\tau})$, so it will suffice to show that

$$\mathrm{Spt}\,(\underline{\mathbb{N}}^s[i] \wr\wr B) \to \mathrm{Spt}\,(\underline{\mathbb{N}}[i] \wr\wr \hat{\tau})$$

is an equivalence for all i. But let $j_i^s : \underline{P^a/\mathbb{Z}} \wr\wr (i \circ m_i) \to \underline{\mathbb{N}}(i) \wr\wr B$ and $j_i : \underline{\mathcal{T}}_i \to \underline{\mathbb{N}}[i] \wr\wr \hat{\tau}$ be the inclusion functors given on objects by $j_i^s(x) = 1[(\{i\},x)]$ and $j_i(y) = 1[(i,y)]$, where x is an object of $\underline{P^a/\mathbb{Z}} \wr\wr (i \circ m_i)$ and y is an object of $\underline{\mathcal{T}}_i$, j_i^s and j_i are both lax symmetric monoidal functors, and hence induce maps of spectra. We now have the commutative diagram

$$\begin{array}{ccc} \mathrm{Spt}\,(\underline{P^a/\mathbb{Z}} \wr\wr (i \circ m_i)) & \xrightarrow{\sim} & \mathrm{Spt}\,(\underline{\mathcal{J}}_i) \\ \mathrm{Spt}\,(j_i^s) \downarrow & & \downarrow \mathrm{Spt}\,(j_i) \\ \mathrm{Spt}\,(\underline{\mathbb{N}}^s[i] \wr\wr B) & \longrightarrow & \mathrm{Spt}\,(\underline{\mathbb{N}}[i] \wr\wr \hat{\tau}) \end{array}$$

where the upper horizontal arrow is the equivalence of spectra $\mathrm{Spt}\,(\theta)$ constructed in I.10. It is now easy to check that $\mathrm{Spt}\,(j_i)$ is an equivalence; in fact there is a symmetric monoidal functor $\underline{\mathbb{N}}[i] \wr\wr \hat{\tau} \to \underline{\mathcal{T}}_i$, and a natural transformation from the identity to the composite $\underline{\mathbb{N}}[i] \wr\wr \hat{\tau} \to \underline{\mathcal{T}}_i \xrightarrow{j_i} \underline{\mathbb{N}}[i] \wr\wr \hat{\tau}$. On the other hand, restricting the arguments of IV.8–IV.11 to the subcategory $\underline{\mathbb{N}}^s[i]$, we see that

$$\mathrm{Spt}\,(\underline{\mathbb{N}}^s[i] \wr\wr B) \cong \underset{\ell \in \underline{\mathbb{N}}[i]}{\underrightarrow{\mathrm{hocolim}}} |\underset{\sim}{\widehat{K}}(N.\mathcal{U}(\ell);R)| \cong |\underset{\sim}{\widehat{K}}(N.\mathcal{U}(i);R)|\,.$$

Of course, $\mathrm{Spt}\,\underline{P^a/\mathbb{Z}} \wr\wr (i \circ m_i) \cong |\underset{\sim}{\widehat{K}}(N.\mathcal{U}(i);R)|$, and one sees directly that $\mathrm{Spt}\,(j_i^s)$ induces the equivalence. Q.E.D.

We now begin the analysis of the functor $\alpha : \underline{\mathbb{N}} \wr\wr \hat{\tau} \to i\widehat{\mathcal{C}}_X(R)$. First, we have the following consequence of I.16.

Proposition IV.14 *The functor* $\underline{\mathbb{N}} \wr \hat{\tau} \xrightarrow{l} \underline{\mathbb{N}} \wr\wr \hat{\tau}$ *is a weak equivalence.*

Let $\varphi = \alpha \circ l$, and let φ_ℓ denote the inclusion of the full subcategory on the objects of the form (ℓ, t), where t is an object of $\underline{\mathcal{T}}_\ell$. Under the identification $t \leftrightarrow (\ell, t)$, φ_ℓ is identified with $\alpha_\ell : \underline{\mathcal{T}}_\ell \to i\widehat{\mathcal{C}}_X(R)$. For $r \geq 0$, let $B_r\mathcal{U}$ be defined as above. Let $\underline{\mathcal{T}}_{\ell,r}$ be the category

$$\underline{\mathcal{T}} \left[i\widehat{\mathcal{C}}_{Y(B_r\mathcal{U}(\ell))}(R) \underset{\underset{g_\ell}{\to}}{\overset{f_\ell}{\to}} i\widehat{\mathcal{C}}_{Z(B_r\mathcal{U}(\ell))}(R) \right].$$

The above constructions $(\tau_\ell, \hat{\tau}, \ldots)$ apply to $\underline{\mathcal{T}}_{\ell,r}$ and we let $\alpha^r_\ell : \underline{\mathcal{T}}_{\ell,r} \to i\widehat{\mathcal{C}}_X(R)$ and $\varphi^r : \underline{\mathbb{N}} \wr (\underline{\ell} \to \underline{\mathcal{T}}_{\ell,r}) \to i\widehat{\mathcal{C}}_X(R)$ be the corresponding functors. (Note that $B_r(\mathcal{U}(\ell)) = (B_r\mathcal{U})(\ell)$.) Whenever $\ell \leq \ell'$ and $r \leq r'$, we have the well defined functors $\underline{\mathcal{T}}_{\ell,r} \to \underline{\mathcal{T}}_{\ell',r'}$ induced by $\tau_{\ell'-1} \circ \tau_{\ell'-2} \circ \cdots \circ \tau_\ell$ and an evident inclusion of coverings. An object of $\underline{\mathcal{T}}_{\ell,r}$ has the form (A, B), where A is an object of $i\widehat{\mathcal{C}}_{Y(B_r\mathcal{U}(\ell))}(R)$ and B is an object of $i\widehat{\mathcal{C}}_{Z(B_r\mathcal{U}(\ell))}(R)$. For each ℓ and r and each object F of $i\widehat{\mathcal{C}}_X(R)$, we consider the category $F \downarrow \alpha_{\ell,r}$; it is the increasing union of full subcategories $\overline{\mathrm{Fil}}^r_d(\ell, R)$, where $\overline{\mathrm{Fil}}^r_d(\ell, R)$ is the full subcategory on objects $(z, \theta : F \to \alpha_{\alpha,r}(z))$ so that θ and θ^{-1} have filtration less than or equal to d. Before studying these categories, we record a preliminary lemma on idempotents.

Lemma IV.15 *Let* $Y_0, Y_1 \subseteq X$ *be subspaces of a metric space* X*; we view* $i\widehat{\mathcal{C}}_{Y_0}(R)$ *and* $i\widehat{\mathcal{C}}_{Y_1}(R)$ *as full subcategories of* $i\widehat{\mathcal{C}}_X(R)$*. Let* F_0 *and* G_1 *be objects of* $i\widehat{\mathcal{C}}_{Y_0}(R)$ *and* $i\widehat{\mathcal{C}}_{Y_1}(R)$*, respectively, and let* F *and* G *be arbitrary objects of* $i\widehat{\mathcal{C}}_X(R)$*. Suppose* $\theta : F_0 \oplus F \to G_1 \oplus G$ *is an isomorphism in* $i\widehat{\mathcal{C}}_X(R)$ *so that* θ *and* θ^{-1} *have filtration* $\leq d$*. Let* e *denote the idempotent of* $G_1 \oplus G$ *corresponding to projection on* G_1*, and suppose* $d(y_0, y_1) > d$ *for all* $y_0 \in Y_0$*,* $y_1 \in Y_1$*. Then the idempotent* $\theta^{-1}e\theta$ *of* $F_0 \oplus F$ *has the form* $0 \oplus \hat{e}$*, where* $\hat{e}$ *is an idempotent of* F*.*

Proof: This is a direct verification from the definitions of filtrations and of the morphisms in $i\widehat{\mathcal{C}}_X(R)$; we leave it to the reader. Q.E.D.

Theorem IV.16 *Suppose we are given* ℓ, r *and* d *so that* $B_{r+3d}\mathcal{U}(\ell)$ *satisfies Hypothesis (A), where* $\mathcal{U}$ *is a locally finite covering of* X*. Then the functor* $\overline{\mathrm{Fil}}^r_d(\ell, R) \to \overline{\mathrm{Fil}}^{r+3d}_d(\ell, F)$ *induces a map on nerves which is simplicially homotopic to a constant map, for any object* F *of* $i\mathcal{C}_X(R) \subseteq i\widehat{\mathcal{C}}_X(R)$*.*

Proof: The strategy is to find a natural transformation from a constant functor to the inclusion $\overline{\mathrm{Fil}}^r_d(\ell, R) \to \overline{\mathrm{Fil}}^{r+3d}_d(\ell, F)$, or rather to a functor naturally isomorphic to this inclusion. We first define the constant functor. Let $Y_i = B_{r+d}U_i(\ell) \cap B_{r+d}U_{i+1}(\ell)$, and $\hat{Z}_i = B_{r+d}U_i(\ell) - (B_{r+d}U_{i-1}(\ell) \cup B_{r+d}U_{i+1}(\ell))$. The Y_i's and Z_i's together form a partition of the metric space X, and consequently we have a decomposition

$$\text{(I)} \qquad F \cong \bigoplus_i F(i) \oplus \bigoplus_i F(i, i+1),$$

where $F(i)$ is the span of the basis elements for F whose label lies in $\hat{Z}_i$ and $F(i, i+1)$ is the span of the basis Elements of F whose labels lie in Y_i. Recall that an object of $\underline{T}_{\ell,r}$ is of the form (A, B), where $A \in i\widehat{\mathcal{C}}_{Y(B_r\mathcal{U}(\ell))}(R)$ and $B \in i\widehat{\mathcal{C}}_{Z(B_r\mathcal{U}(\ell))}(R)$. Let $W = \coprod_{\alpha \in A} W_\alpha$ be a disjoint union of metric spaces. Then any object A of $i\widehat{\mathcal{C}}_W(R)$ decomposes canonically as $A \cong \oplus_{\alpha \in A} A_\alpha$, where A_α is an object of $i\widehat{\mathcal{C}}_{W_\alpha}(R)$. Write $A_\alpha = (\bar{A}_\alpha, e_\alpha)$, where $\bar{A}_\alpha$ is an object of $i\mathcal{C}_{W_\alpha}(R) \subseteq i\widehat{\mathcal{C}}_{W_\alpha}(R)$, and e_α is a bounded idempotent of $\bar{A}_\alpha$. If the family $\{A_\alpha\}_{\alpha \in A}$ corresponds to an object A of $i\widehat{\mathcal{C}}_W(R)$, then there is a number d so that e_α has filtration $\leq d$ for all $\alpha \in A$. We say a family $\{A_\alpha\}_{\alpha \in A} = \{(\bar{A}_\alpha, e_\alpha)\}_{\alpha \in A}$, with $A_\alpha \in i\widehat{\mathcal{C}}_{W_\alpha}(R)$, is *uniformly bounded* if a d exists so that e_α has filtration $\leq d$ for all α. One readily checks that the objects of $i\widehat{\mathcal{C}}_W(R)$ are in bijective correspondence with uniformly bounded families $\{A_\alpha\}_{\alpha \in A}$, and that the morphisms correspond to families $\{f_\alpha\}_{\alpha \in A}$, where f_α is a morphism in $i\widehat{\mathcal{C}}_{W_\alpha}(R)$, and so that there is a number d so that f_α has filtration less than or equal to d for all $\alpha \in A$. Now, $Y(B_{r+3d}\mathcal{U}(\ell)) = \coprod_i Y_i(B_{r+2d}\mathcal{U}(\ell))$, where $Y_i(B_{r+3d}\mathcal{U}(\ell)) = B_{r+3d}U_i(\ell) \cap B_{r+3d}U_{i+1}(\ell)$, and $Z(B_{r+3d}\mathcal{U}(\ell)) = \coprod_i Z_i(B_{r+2d}\mathcal{U}(\ell))$, where $Z_i(B_{r+3d}\mathcal{U}(\ell)) = B_{r+3d}U_i(\ell)$. Consequently, to specify an object $(E, E') \in \underline{T}_{\ell,r+3d}$, it suffices to give uniformly bounded sequences E_i and E'_i, where $E_i \in i\widehat{\mathcal{C}}_{Y_i(B_{r+3d}\mathcal{U}(\ell))}(R)$ and $E'_i \in i\widehat{\mathcal{C}}_{Z_i(B_{r+3d}\mathcal{U}(\ell))}(R)$. We define (E, E') by

$$E_i = F(i, i+1), \qquad E'_i = F(i).$$

(In both cases, these are objects whose idempotent is the identity map, so they clearly form uniformly bounded families.) $\alpha_{\ell,r+3d}(E, E') = E \oplus E'$, so the decomposition (I) gives a filtration zero isomorphism $F \xrightarrow{\varepsilon} \alpha_{\ell,r+3d}(E, E')$. This data gives an object ξ of $\overline{\mathrm{Fil}}_d^{r+3d}(\ell, R)$, and we will show that there is a natural transformation from the constant functor with value ξ to a functor naturally isomorphic to the inclusion $\overline{\mathrm{Fil}}_d^{r}(\ell, R) \to \overline{\mathrm{Fil}}_d^{r+3d}(\ell, F)$.

Let $((A, B), F \xrightarrow{\theta} \alpha_{\ell,r}(A, B))$ be any object of $\mathrm{Fil}_d^r(\ell, F)$; we have the corresponding uniformly bounded families $\{A_i\}$ and $\{B_i\}$, where

$$A_i \in i\widehat{\mathcal{C}}_{Y_i(B_r\mathcal{U}(\ell))}(R) \text{ and } B_i \in i\widehat{\mathcal{C}}_{Z_i(B_r\mathcal{U}(\ell))}(R).$$

The object $\alpha_{\ell,r}(A, B)$ admits filtration zero idempotents ρ_i and σ_i, given by projection on A_i and B_i, respectively. Consequently, $\hat{\rho}_i = \theta^{-1}\rho_i\theta$ and

$\hat{\sigma}_i = \theta^{-1}\sigma_i\theta$ are idempotents of F, of filtration $< 2d$. We wish to describe these idempotents in terms of the decomposition (I). First, let us examine $\hat{\rho}_i$. We write $\alpha_{\ell,r}(A,B)$ as $A \oplus B = A_i \oplus A'$, where $A' = \bigoplus_{j\neq i} A_j \oplus \bigoplus_j B_j$, and

$$\text{(II)} \qquad F \cong F' \oplus F(i,i+1),$$

where $F' = \bigoplus_j F(j) \oplus \bigoplus_{j\neq i} F(j,j+1)$. $A_i \in i\widehat{\mathcal{C}}_{Y_i(B_r\mathcal{U}(\ell))}(R)$, and F' is an object of $A_i \in i\widehat{\mathcal{C}}_{X-Y_i(B_{r+d}\mathcal{U}(\ell))}(R)$. The subsets $Y_i(B_r\mathcal{U}(\ell))$ and $X - Y_i(B_{r+d}\mathcal{U}(\ell))$ and corresponding modules A_i and F' satisfy the hypotheses of IV.15, so $\hat{\rho}_i$ has the form $O \oplus \bar{\rho}_i$ in terms of the decomposition (II), where $\bar{\rho}_i$ is an idempotent of $F(i,i+1)$. Next, we turn to $\hat{\sigma}_i$. We now write $\alpha_{\ell,r}(A,B) = A \oplus B \cong B_i \oplus B'$, where $B' = \bigoplus_j A_j \oplus \bigoplus_{j\neq i} B_j$, and

$$\text{(III)} \qquad F \cong F'' \oplus (F(i-1,i) \oplus F(i) \oplus F(i,i+1)),$$

where $F'' \cong \bigoplus_{j\neq i} F(j) \oplus \bigoplus_{j\neq i-1,i} F(j,j+1)$. $B_i \in i\widehat{\mathcal{C}}_{Z_i(B_r\mathcal{U}(\ell))}(R) = i\widehat{\mathcal{C}}_{B_rU(\ell)}(R)$, and $F'' \in i\widehat{\mathcal{C}}_{X-B_{r+d}U(\ell)}(R)$. The subsets $B_rU_i(\ell)$ and $X - B_{r+d}U_i(\ell)$ and corresponding modules B_i and $F(i-1,i) \oplus F(i) \oplus F(i,i+1)$ satisfy the hypotheses of IV.15, so $\hat{\sigma}_i$ has the form $O \oplus \bar{\sigma}_i$ in terms of the decomposition (III), where $\bar{\sigma}_i$ is an idempotent of $F(i-1,i) \oplus F(i) \oplus F(i,i+1)$. We now decompose $\bar{\sigma}_i$ further in terms of the decomposition $F(i-1,i) \oplus F(i) \oplus F(i,i+1)$. Again, decompose $\alpha_{\ell,r}(A,B)$ as $\alpha_{\ell,r}(A,B) \cong B' \oplus B_i$, and write

$$\text{(IV)} \qquad F \cong F(i) \oplus F''',$$

where $F''' \cong \bigoplus_{j\neq i} F(j) \oplus \bigoplus_j F(j,j+1)$. Let $V_0 = \cup_{j\neq i} B_r U_j(\ell)$ and $V_1 = X - \cup_{j\neq i} B_{r+d} U_j(\ell)$. Then $F(i)$ is an object of $i\widehat{\mathcal{C}}_{V_1}(R)$ and B' is an object of $i\widehat{\mathcal{C}}_{V_0}(R)$. Projection on the factor B' is identified with the idempotent $\mathrm{Id} - \sigma_i$. V_0 and V_1 and the corresponding modules B' and $F(i)$ now satisfy the hypotheses of IV.15, and we conclude that in terms of the decomposition (IV), $\theta^{-1}(\mathrm{Id} - \sigma_i)\theta = \mathrm{Id} - \hat{\sigma}_i$ decomposes as $O \oplus \nu$, where ν is an idempotent of F'''; consequently, $\hat{\sigma}_i$ itself decomposes as $\mathrm{Id} \oplus \nu'$, where $\nu' = \mathrm{Id} + \nu$. Taking both decompositions into account, and writing $F(i-1,i) \oplus F(i) \oplus F(i,i+1)$, we see that $\bar{\sigma}_i$ decomposes as $\mathrm{Id}_{F(i)} \oplus \bar{\nu}$, where $\bar{\nu}$ is an idempotent of $F(i-1,i) \oplus F(i,i+1)$. Further, since $\bar{\nu}$ is the restriction of $\hat{\sigma}$ to $F(i-1,i) \oplus F(i,i+1)$, and $\hat{\sigma}_i$ has filtration $\leq 2d$, $\bar{\nu}$ has filtration $\leq 2d$. But $F(i-1,i)$ (respectively $F(i,i+1)$) is an object of $i\mathcal{C}_{Y_{i-1}(B_{r+d}\mathcal{U}(\ell))}(R)$ (respectively $F(i,i+1)$) is an object of $i\mathcal{C}_{Y_{i-1}(B_{r+d}\mathcal{U}(\ell))}(R)$ (respectively $i\mathcal{C}_{Y_i(B_{r+d}\mathcal{U}(\ell))}(R)$), and if $y_{i-1} \in Y_{i-1}(B_{r+d}\mathcal{U}(\ell))$, then $d(y_{i-1}, y_i) > 2d$, Hypothesis (A) on $B_{r+3d}\mathcal{U}(\ell)$. From these facts it follows readily that $\bar{\nu}$ decomposes as $\bar{\nu} = \nu_{i-1}^{+} \oplus \nu_i^{-}$, where ν_{i-1}^{+} is an idempotent of $F(i-1,i)$ and ν_i^{-} is an idempotent of $F(i,i+1)$. $\nu_i^{-} + \hat{\rho}_i + \nu_i^{+}$ is now an idempotent of $F(i,i+1)$, we claim $\nu_i^{-} + \hat{\rho}_i + \nu_i^{+} = \mathrm{Id}$. To check this, it suffices to show that $(\hat{\sigma}_i + \hat{\rho}_i + \hat{\sigma}_{i+1})(x) = x$ for all $x \in F(i, i+1)$, since $\nu_i^- + \hat{\rho}_i + \nu_i^+$ is the restriction of $\hat{\sigma}_i + \hat{\rho}_i + \hat{\sigma}_{i+1}$ to $F(i, i+1)$. We now write

$$\text{(V)} \qquad F \cong F(i, i+1) \oplus F',$$

where F' is defined as in the decomposition (II) above. We also write $\alpha_{\ell,r}(A, B) = C'' \oplus (B_i \oplus A_i \oplus B_{i+1})$, where $C'' \cong \bigoplus_{j \neq i} A_j \oplus \bigoplus_{j \neq i, i+1} B_j$. Now $\mathrm{Id} - (\sigma_i + \rho_i + \sigma_{i+1})$ is identified with projection on C''. Let $W_0 = \cup_{j \neq i, i+1} B_r U_j(\ell)$ and $W_1 = B_{r+d} U_i(\ell) \cap B_{r+d} U_{i+1}(\ell)$. C'' is an object of $i\widehat{\mathcal{C}}_{W_0}(R)$ and $F(i, i+1)$ is an object of $i\widehat{\mathcal{C}}_{W_1}(R)$. One now checks that W_0, W_1, and the corresponding modules C'' and $F(i, i+1)$ satisfy the hypotheses of IV.15, so

$$\theta^{-1}(\mathrm{Id} - (\sigma_i + \rho_i + \sigma_{i+1}))\theta = \mathrm{Id} - (\hat{\sigma}_i + \hat{\rho}_i + \hat{\sigma}_{i+1})$$

decomposes as $O \oplus \hat{e}$ for some idempotent of F', and hence restricts to the identity on $F(i, i+1)$.

We now define a functor $j : \overline{\mathrm{Fil}}_d^r(\ell, R) \to \overline{\mathrm{Fil}}_d^{r+3d}(\ell, F)$ by $j((A, B), \theta) = ((A', B'), \theta')$, where A', B', and θ' are defined by the following equations.

$$\begin{aligned} A_i' &= (F(i, i+1), \hat{\rho}_i) \\ B_i' &= (F(i, i-1) \oplus F(i) \oplus F(i, i+1), \hat{\sigma}_i) \end{aligned}$$

For θ', note that

$$\alpha_{\ell, r+3d}(A', B')$$

$$\cong \bigoplus_i (F(i, i+1) \oplus F(i, i+1) \oplus F(i, i+1), \nu_i^- \oplus \hat{\rho}_i \oplus \nu_i^+) \oplus \bigoplus_i F(i),$$

and we have an isomorphism l_i in $i\widehat{\mathcal{C}}_{Y_i(B_{r+3d}\mathcal{U}(\ell))}(R)$, from $(F(i, i+1), \mathrm{Id})$ to $(F(i, i+1) \oplus F(i, i+1) \oplus F(i, i+1), \nu_i^- \oplus \hat{\rho}_i \oplus \nu_i^+)$ given by $l_i(x) = (\nu_i^-(x), \hat{\rho}_i(x), \nu_i^+(x))$, for $x \in F(i, i+1)$. Let $\omega_i\, F(i, i+1) \oplus F(i, i+1) \oplus F(i, i+1) \to F(i, i+1)$ be given by $\omega_i(x_1, x_2, x_3) = x_1 + x_2 + x_3$; then the inverse to l_i is given by the composite $\omega_i \circ (\nu_i^- \oplus \hat{\rho}_i \oplus \nu_i^+)$. θ' is now the composite

$$\bigoplus_i (F(i, i+1) \oplus F(i, i+1) \oplus F(i, i+1), \nu_i^- \oplus \hat{\rho}_i \oplus \nu_i^+) \oplus \bigoplus_i F(i)$$

$$\downarrow (\textstyle\bigoplus_i \omega_i) \oplus \mathrm{Id}$$

$$\textstyle\bigoplus_i F(i, i+1) \oplus \bigoplus_i F(i)$$

$$\downarrow$$

$$F$$

where the second arrow is the degree zero isomorphism arising from the construction of the direct sum decomposition (I). One now checks that j

is a functor. Let $i : \overline{\mathrm{Fil}}^r_d(\ell, R) \to \overline{\mathrm{Fil}}^{r+3d}_d(\ell, F)$ be the inclusion functor. We will define an isomorphism of functors from i to j. θ^{-1} induces isomorphisms $A_i \to A'_i$ and $B_i \to B'_i$, whose sum gives an isomorphism

$$\alpha_{\ell,r+3d}(A, B) \to \alpha_{\ell,r+3d}(A', B')$$

compatible with θ and θ^{-1}. This gives the isomorphism from i to j. It remains to construct a natural transformation from the constant functor with value ξ to j. To define this natural transformation, we let $((A, B), \theta)$ be an object of $\overline{\mathrm{Fil}}^r_d(\ell, F)$ and let $((A', B'), \theta')$ be $j((A, B), \theta)$. Let U be the object of $i\widehat{\mathcal{C}}_{Y(B_{r+3d}\mathcal{U}(\ell))}(R)$ defined by $U_i = (F(i, i+1), \nu_i^-)$ and similarly let V be defined by $V_i = (F(i, i+1), \nu_i^+)$. We must define $\varphi : E \to U \oplus A' \oplus B$ and $\Psi : U \oplus E' \oplus V \to B'$, where U and V are viewed as objects of $i\widehat{\mathcal{C}}_{Z(B_{r+3d}\mathcal{U}(\ell))}(R)$ via the two distinct inclusions

$$Y(B_{r+3d}\mathcal{U}(\ell)) \underset{g_\ell}{\overset{f_\ell}{\rightrightarrows}} Z(B_{r+3d}\mathcal{U}(\ell)).$$

φ is given by setting φ_i equal to the isomorphism

$$\begin{array}{c} (F(i, i+1), \mathrm{Id}) \\ \downarrow l_i \\ (F(i, i+1), \nu_i^-) \oplus (F(i, i+1), \hat{\rho}_i) \oplus (F(i, i+1), \nu_i^+) \\ \| \\ (F(i, i+1), \nu_i^-) \oplus A'_i \oplus (F(i, i+1), \nu_i^+) \end{array}$$

ψ is given by setting ψ_i equal to the isomorphism

$$\begin{array}{c} (F(i-1, i), \nu_{i-1}^+) \oplus (F(i), \mathrm{Id}) \oplus (F(i, i+1), \nu_i^+) \\ \downarrow \| \\ (F(i-1, i) \oplus F(i) \oplus F(i, i+1), \nu_{i-1}^+ \oplus \mathrm{Id} \oplus \nu_i^-) \\ \| \\ (F(i-1, i) \oplus F(i) \oplus F(i, i+1), \hat{\sigma}_i) \end{array}$$

coming to the decomposition of $\hat{\sigma}_i$. These isomorphisms yield natural transformations, and this concludes the proof. Q.E.D.

As always, we suppose $\mathcal{U} = \{U_i\}_{i \in \mathbb{Z}}$ is a locally finite covering parametrized by $\mathbb{Z}$. Suppose further that each covering $B_d\mathcal{U}$ is locally finite. For each d, let $\hat{r}_d : \underline{\mathbb{N}} \to \underline{\mathrm{SymMon}}$ be the functor analogous to $\hat{r}$ associated to the covering $B_d\mathcal{U}$. We have symmetric monoidal functors $\alpha_d : \underline{\mathbb{N}} \wr\wr \hat{\tau}_d \to i\widehat{\mathcal{C}}_X(R)$ and functors $\lambda_d : \underline{\mathbb{N}} \wr \hat{\tau}_d \to \underline{\mathbb{N}} \wr\wr \hat{\tau}_d$, making the commutative diagrams

$$\begin{array}{ccccc} \underline{\mathbb{N}} \wr \hat{\tau}_d & \xrightarrow{\lambda_d} & \underline{\mathbb{N}} \wr\wr \hat{\tau}_d & & \\ & & & \searrow^{\alpha_d} & \\ \downarrow & & \downarrow & & i\widehat{\mathcal{C}}_X(R) \\ & & & \nearrow_{\alpha_e} & \\ \underline{\mathbb{N}} \wr \hat{\tau}_e & \xrightarrow{\lambda_e} & \underline{\mathbb{N}} \wr\wr \hat{\tau}_e & & \end{array}$$

when $d \le e$. For any element $\xi \in \pi_0(i\widehat{\mathcal{C}}_X(R))$, let $(\underline{\mathbb{N}} \wr \hat{\tau}_d)_\xi$ and $(\underline{\mathbb{N}} \wr\wr \hat{\tau}_d)_\xi$ denote the full subcategories of objects x so that $\alpha_d \circ \lambda_d(x)$ (respectively $\alpha_d(x)$) belongs to ξ. Let $\pi_0 = \pi_0(i\widehat{\mathcal{C}}_X(R))$. Then we have disjoint union decompositions

$$\begin{array}{ccc} \underline{\mathbb{N}} \wr \hat{\tau}_d & \cong & \coprod_{\xi\in\pi_0} (\underline{\mathbb{N}} \wr \hat{\tau}_d)_\xi \\ l_d \downarrow & & \downarrow \coprod \lambda_d \,|\, (\underline{\mathbb{N}} \wr \hat{\tau}_d)_\xi \\ \underline{\mathbb{N}} \wr\wr \hat{\tau}_d & \cong & \coprod_{\xi\in\pi_0} (\underline{\mathbb{N}} \wr\wr \hat{\tau}_d)_\xi \end{array}$$

where the disjoint union of categories has the the obvious meaning. Since a disjoint union of maps $\coprod f_\alpha : \coprod X_\alpha \to \coprod Y_\alpha$ is an equivalence if and only if each f_α is IV.14, shows that $\lambda_d \,|\, (\underline{\mathbb{N}} \wr \hat{\tau}_d)_\xi$ is a weak equivalence for each ξ. Let $i\mathcal{C}^0_X(R) \subseteq i\widehat{\mathcal{C}}(R)$ be the full subcategory on objects isomorphic to objects of $i\mathcal{C}_X(R)$; the inclusion $i\mathcal{C}_X(R) \hookrightarrow i\mathcal{C}^0_X(R)$ is an equivalence of categories. Let $\pi_f \subseteq \pi_0$ be the submonoid image $(\pi_0(i\mathcal{C}_X(R)) \to \pi_0(i\widehat{\mathcal{C}}_X(R)))$.

Proposition IV.17 *Suppose the locally finite covering $\mathcal{U}$ satisfies Hypothesis (B). Then the restriction of the functor $\varinjlim_d \alpha_d \circ \lambda_d$ to $\coprod_{\xi\in\pi_f} \varinjlim_d (\underline{\mathbb{N}} \wr \hat{\tau}_d)_\xi$ induces an equivalence on nerves*

$$N.\left(\coprod_{\xi\in\pi_f} \varinjlim_d (\underline{\mathbb{N}} \wr \hat{\tau}_d)_\xi \right) \to N.i\mathcal{C}^0_X(R)\,.$$

Consequently, the functor

$$\coprod_{\xi\in\pi_f} \varinjlim_d (\underline{\mathbb{N}} \wr\wr \hat{\tau}_d)_\xi \to i\mathcal{C}^0_X(R)$$

is a weak equivalence.

Proof: Immediate consequence of I.15 and IV.16. Q.E.D.

Let $\pi_h \subseteq \pi_0(\varinjlim_d \underline{\mathbb{N}} \wr\wr \hat{\tau}_d)$ be the inverse image $\pi_0(\varinjlim_d \alpha_d)^{-1}(\pi_f)$. Then π_h is easily seen to be a cofinal submonoid of $\pi_0(\varinjlim_d \underline{\mathbb{N}} \wr\wr \hat{\tau}_d)$. The following is now an immediate consequence of I.6 (c).

Proposition IV.18 *Suppose the locally finite covering $\mathcal{U}$ satisfies Hypothesis (B), and that each $B_d\mathcal{U}$ is locally finite. Then the map of spectra* $\mathrm{Spt}\,(\varinjlim_d \alpha_d)$ *induces isomorphisms on π_i for $i > 0$. It follows from IV.11, IV.12, and IV.13 that the map $\varinjlim_{\ell,d} \hat{A}(B_d\mathcal{U}(\ell))$ induces isomorphisms on π_i for $i > 0$.*

Theorem IV.19 *Let $\mathcal{U}$ be as in IV.18. Then the map of spectra*

$$\varinjlim_{\ell,d} \mathcal{A}(B_d(\mathcal{U}(\ell))) : \varinjlim_{\ell,d} |\mathcal{K}(N.B_d\mathcal{U}(\ell); R)| \to \underset{\sim}{\mathcal{K}}(X; R)$$

is an equivalence.

Proof: This follows directly from the fact that the coverings $\mathcal{U} \times E^k = \{U_i \times E^k\}_{i\in\mathbb{Z}}$ satisfy the hypothesis of IV.18 if $\mathcal{U}$ does, using IV.18. Q.E.D.

V. Bounded K–theory of Homogeneous Spaces

Let G be a connected Lie group, and K a maximal compact subgroup. G/K is a left G–space and can, by choosing an appropriate inner product on the tangent space $T_e(G/K)$ be given a left invariant Riemannian metric. We want to show that for the metric space G/K, the assembly map

$$^b\underset{\sim}{h}^{\ell f}(G/K, \underset{\sim}{K}(R)) \to \underset{\sim}{\mathcal{K}}(G/K; R)$$

introduced in §III is an equivalence. Our first result shows essentially that whether or not this map is an equivalence is independent of the choice of left invariant Riemannian metric.

Proposition V.1 *Let G be any group, and let X be a transitive G–space. Let d and d' be continuous metrics on X, taking only finite values (i.e., they are metrics in the standard sense, not permitted to take the value $+\infty$), and suppose that G acts by isometries with respect to both metrics. Suppose finally that all closed balls in both metrics are compact. Then the identity maps $\mathrm{Id} : (X, d) \to (X, d')$ and $\mathrm{Id} : (X, d') \to (X, d)$ are eventually continuous and, of course, proper. It follows that if (Z, d_Z) is any other metric space (in the generalized sense, i.e., with d perhaps taking the value $+\infty$), then the identity maps $(X \times Z, d \times d_Z) \to (X \times Z, d' \times d_Z)$ and $(X \times Z, d' \times d_Z) \to (X \times Z, d \times d_Z)$ are eventually continuous and proper.*

Proof: Consider X, d, d'. Fix any point $x_0 \in X$. Then for any r, there exists a number $R(r)$ so that $\overline{B_r}(x_0) \subseteq \overline{B'_{R(r)}}(x_0)$ and $\overline{B'_r}(x_0) \subseteq \overline{B_{R(r)}}(x_0)$, where $\overline{B_r}$ and $\overline{B'_r}$ denote the closed balls of radius r with respect to the metrics d and d' respectively. This follows since, e.g., the continuous function $x \mapsto d'(x, x_0)$ must be bounded above on the compact set $\overline{B_r}(x_0)$. By the transitivity and isometric nature of the G–action, we have $d(x_1, x_2) \leq r \to d'(x, x_2) \leq R(r)$ and $d'(x_1, x_2) \leq r \to d(x, x_2) \leq R(r)$. The result now follows. Q.E.D.

Corollary V.2 *Let G be a connected Lie group and K a maximal compact subgroup. Let d_1 and d_2 be two left invariant Riemannian metrics on G/K. Then, if D is any discrete set, the assembly map*

$$^b\underset{\sim}{h}^{\ell f}((D \times G/K, d_1); \underset{\sim}{K}(R)) \to \underset{\sim}{K}((D \times G/K; d_1), R)$$

is an equivalence if and only if

$$^b\underset{\sim}{h}^{\ell f}((D \times G/K, d_2); \underset{\sim}{K}(R)) \to \underset{\sim}{K}((D \times G/K; d_2), R)$$

is. (Recall that this map is really a homotopy natural transformation.)

Proof: V.1 shows that the identity map $(D \times G/K, d_1) \to (D \times G/K, d_2)$ is proper and eventually continuous. Such maps induce maps of spectra both on $^b\underset{\sim}{h}^{\ell f}(-; \underset{\sim}{K}(R))$ and on $\underset{\sim}{K}(-, R)$, compatible with the assembly map. The result is now immediate. Q.E.D.

We begin the study of G/K by considering the case of a simply connected solvable Lie group; in this case, there are no non–trivial compact subgroups, so $K = \{e\}$ and G/K is the Lie group G itself, equipped with a left invariant Riemannian metric.

Proposition V.3 *Let G be a simply connected solvable Lie group of dimension n. Then G is isomorphic to a semidirect product $T \tilde{\times} G^0$, where $G^0 \subseteq G$ is a normal simply connected solvable Lie group of dimension $n-1$, and T is isomorphic to $\mathbb{R}$ and is equipped with an action on G^0.*

Proof: The Lie algebra $\mathfrak{g}$ of G being solvable, there is an ideal $\mathfrak{g}^0 \subseteq \mathfrak{g}$ of dimension $n-1$. Let G^0 the subgroup of G belonging to $\mathfrak{g}^0$. [9, Proposition 21, p. 352] shows that G^0 is a simply connected subgroup of G, and it is solvable since the ideal $\mathfrak{g}^0$ is a solvable Lie algebra. Since $\mathfrak{g}^0$ is an ideal and G is simply connected, [9, Proposition 14, p. 316] shows that G^0 is normal and that G/G^0 is simply connected. Since G/G^0 has dimension 1, $G/G^0 \cong \mathbb{R}$. Choose any element x of $\mathfrak{g}$, $x \notin \mathfrak{g}^0$, and let T be the one parameter subgroup of G corresponding to x. Then it follows from [9, Proposition 20, p. 352] that the multiplication map $T \times G^0 \to G$ is a homeomorphism, and that T therefore surjects on G/G^0. Consequently, $G \cong T \tilde{\times} G^0$. Q.E.D.

Let $T \subseteq G$ be the subgroup of V.3. Then $\mathfrak{g}$ has a vector space splitting as $\mathfrak{g} = \mathfrak{t} \oplus \mathfrak{g}^0$, where $\mathfrak{t}$ is the Lie algebra of T. We choose a positive definite symmetric bilinear form β on $\mathfrak{g}$, so that the decomposition above is orthogonal and so that $\beta(x,x) = 1$, where x is a fixed choice of tangent vector in $\mathfrak{t}$; a choice of x gives an isomorphism $\mathbb{R} \to \mathfrak{t}$, in which the Riemannian metric $\beta \,|\, \mathfrak{t} \times \mathfrak{t}$ is identified with the standard metric on $\mathbb{R}$. Left translation by elements of G gives an orthogonal decomposition of $T_g(G)$ for every $g \in G$, $T_g(G) = T_g^h(G) \oplus T_g^v(G)$, and an inner product on $T_g(G)$ for which this decomposition is orthogonal. $T^v(G)$ is just the vertical subbundle of $T(G)$ associated to the projection $G \to T$.

Proposition V.4 *With G given the Riemannian metric associated to β, and T given the metric associated to $\beta \,|\, \mathfrak{t} \times \mathfrak{t}$, the homomorphism $G \xrightarrow{\pi} T$ is a distance non–increasing map, when G and T are viewed as metric spaces.*

Proof: Let $y \in T_g(G)$, $y = y^h + y^v$. Then the length of the projection of y into $T_{\pi(g)}(T)$ is equal to the length of y^h, and because of the orthogonal nature of the decomposition $T^v \oplus T^h$, the length of y is equal to the sum of the lengths of y^h and y^v. This shows that the Jacobian of π is length non–increasing, which gives the result. Q.E.D.

Let $s : T \to G$ denote the inclusion. From the choice of metric on T, we see that s is also length non–increasing, so by V.5 s is an isometry onto its image. Identify $\mathfrak{t}$ with $\mathbb{R}$ via the choice of $x \in \mathfrak{t}$. We can then consider the subspace $\pi^{-1}([a,b]) \subseteq G$, for $a < b$, and let $z \in [a,b]$. We define a deformation $\rho_t(a,b,z) : \pi^{-1}([a,b]) \to \pi^{-1}(z)$ by the formula $\rho_t(a,b,z)(g) = g(t(z - \pi(g)))$.

Proposition V.5

(a) *Let d denote the distance function in G, and suppose $0 \le t \le 1$. Then for all $g \in \pi^{-1}([a,b])$, $d(g, \rho_t(a,b,z)(g)) \le b - a$.*

(b) *For all a, b, and z as above, $\rho_1(a,b,z)$ is a proper map and $\rho_t(a,b,z)$ is a proper homotopy from the identity map on $\pi^{-1}([a,b])$ to the composite*

$$\pi^{-1}([a,b]) \xrightarrow{\rho_1(a,b,z)} \pi^{-1}(z) \hookrightarrow \pi^{-1}([a,b]) .$$

Consequently, the inclusion $\pi^{-1}(z) \to \pi^{-1}([a,b])$ is a proper homotopy equivalence.

(c) $B_r\, \pi^{-1}([a,b]) = \pi^{-1}([a-r, b+r])$.

Proof: (a) We have

$$d(g, \rho_t(a,b,z)(g)) = d(g, gs(t(z - \pi(g)))) = d(e, s(t(z - \pi(g)))) .$$

But since s is an isometry onto its image,

$$d(e, s(t(z - \pi(g)))) = |t(z - \pi(g))| = |t|\,|z - \pi(g)| \leq |t|\,(b - a)\,,$$

which is the result. Q.E.D.

(b) Any compact set C in $\pi^{-1}(z)$ is contained in a closed ball of radius r with center $g \in G$. (a) now shows that

$$\{(t, g) \mid \rho_t(a, b, z)(g) \in C\} \subseteq [0, 1] \times B_{r+2(b-a)}(g)\,,$$

which is a compact set. Consequently, $\rho_t(a, b, z)$ is a proper homotopy and $\rho_1(a, b, z)$ is a proper map. Q.E.D.

(c) $B_r\,\pi^{-1}([a, b]) \subseteq \pi^{-1}([a - r, b + r])$ since π is distance non–increasing. Suppose $g \in \pi^{-1}([a - r, b + r])$, so that there is a path φ in T of length less than or equal to r with $\varphi(0) = \pi(g)$ and $\varphi(1) \in [a, b]$. Let $\hat{\varphi}$ be the unique path in G with $\pi \circ \hat{\varphi} = \varphi$ so that the tangent vector to $\hat{\varphi}$ at $\varphi(t)$ lies in $T^h_{\varphi(t)}(G)$. From the definition of the Riemannian metric on G, $\text{length}\,(\hat{\varphi}) = \text{length}\,(\varphi) \leq r$, and $\hat{\varphi}(1) \in \pi^{-1}([a, b])$. Q.E.D.

Corollary V.6 *Let a, b, c, and G be as above, and let Z be any metric space.*

(a) $\underset{\sim}{h}^{\ell f}(Z \times \pi^{-1}(z); \underset{\sim}{A}) \to \underset{\sim}{h}^{\ell f}(Z \times \pi^{-1}([a, b]); \underset{\sim}{A})$ *is a weak equivalence of spectra.*

(b) *The inclusion $Z \times \pi^{-1}(z) \to Z \times \pi^{-1}([a, b])$ induces equivalences on $\underset{\sim}{K}(-; R)$, $\widehat{\underset{\sim}{K}}(-; R)$, and $\underset{\sim}{\mathcal{K}}(-; R)$.*

Proof: (a) follows from V.5 (b) and II.14, since a product of proper homotopy equivalences is a proper homotopy equivalence. (b) is a direct consequence of V.5 (a) and III.7. Q.E.D.

We can now prove the main result of this section for simply connected solvable Lie groups.

Theorem V.7 *Let G be a simply connected solvable Lie group, equipped with a left invariant Riemannian metric. Then if D is any countable discrete metric space, the assembly map*

$$^{b}\underset{\sim}{h}^{\ell f}(G \times D; \underset{\sim}{\mathcal{K}}(R)) \to \underset{\sim}{\mathcal{K}}(G \times D; R)$$

is an equivalence of spectra.

Proof: The proof proceeds by induction on $\dim(G)$. For the case $\dim(G) = 0$, $G = \{e\}$, this follows from III.15. Now suppose the inductive hypothesis is known for groups of dimension less than $\dim(G)$. Let T and G^0 be as in V.3, and let μ be the Riemannian metric associated to the inner product β introduced in the discussion preceding V.4. By V.2, it suffices to prove the result for the metric μ. We identify T with $\mathbb{R}$ as before, and define a covering $\mathcal{U} = \{U_i\}_{i\in\mathbb{Z}}$ by $U_i = \pi^{-1}([i-1/3, i+4/3])$. Let $\mathcal{U}(\ell)$ be the coverings constructed from $\mathcal{U}$ as in § IV, and let

$$B_r\mathcal{U}(\ell) = \{B_r U_i\}_{i\in\mathbb{Z}} = (B_r\mathcal{U})(\ell).$$

Also, let $U_i' \subseteq T$, $U_i' = [i-1/3, i+4/3]$, and let $\mathcal{U}', \mathcal{U}'(\ell), B_r\mathcal{U}'$, etc., be defined analogously in the space T. We note that each covering $B_r\mathcal{U}(\ell)$ is excisive. This follows from the fact that $\mathcal{U}^0 = \{U_i^0\}_{i\in\mathbb{Z}}$ is an open covering of X, and that hence for each r and ℓ, $B_r\mathcal{U}^0(\ell)$ is an open covering of X. Each covering $B_r\mathcal{U}(\ell)$ is clearly locally finite in the sense of III.17. We also verify that $\mathcal{U}$ satisfies Hypothesis (B) of § IV. But from V.5 (c), it is clear that it suffices to check this for the covering $\mathcal{U}'$. For a given number r, one checks that if $\ell > \log_2(d+1/3)+1$, then

$$B_rU_i'(\ell) \cap B_rU_j'(\ell) = \emptyset \text{ if } |i-j| > 1.$$

Let $\mathcal{U}\times D$, $\mathcal{U}(\ell)\times D$, and $B_r\mathcal{U}(\ell)\times D$ denote the coverings $\{U_i\times D\}_{i\in\mathbb{Z}}$, $\{U_i(\ell)\times D\}_{i\in\mathbb{Z}}$, and $\{B_rU_i\times D\}_{i\in\mathbb{Z}}$ of $G\times D$. These coverings are also excisive, locally finite, and the covering $\mathcal{U}\times D$ also satisfies Hypothesis (B); this follows from the corresponding results for $\mathcal{U}$. Note that

$$(B_r\mathcal{U}(\ell))\times D = B_r(\mathcal{U}(\ell)\times D) = B_r((\mathcal{U}\times D)(\ell)).$$

We now consider the commutative diagram

(*)
$$\begin{array}{ccc}
\underset{\ell,r}{\underrightarrow{\operatorname{colim}}}\,|{}^b\underset{\sim}{h}^{\ell f}(N.B_r(\mathcal{U}\times D(\ell)), R)| & \longrightarrow & \underset{\ell,r}{\underrightarrow{\operatorname{colim}}}\,|\underset{\sim}{\mathcal{K}}(N.B_r(\mathcal{U}\times D(\ell)), \underset{\sim}{\mathcal{K}}(R))| \\
\downarrow & & \downarrow \underset{\ell,r}{\underrightarrow{\operatorname{colim}}}\mathcal{A}(B_r(\mathcal{U}\times D)(\ell)) \\
{}^b\underset{\sim}{h}^{\ell f}(G\times D, \underset{\sim}{\mathcal{K}}(R)) & \longrightarrow & \underset{\sim}{\mathcal{K}}(G\times D, R)
\end{array}$$

where the horizontal maps are the assembly maps ${}^b\underset{\sim}{h}^{\ell f} \to \underset{\sim}{\mathcal{K}}$. Since the covering $\mathcal{U}\times D$ satisfies Hypothesis (B), it follows from IV.19 that the right hand vertical arrow is an equivalence. We claim that the left hand vertical arrow is also an equivalence. To see this, consider the diagram

(**)
$$\begin{array}{ccc}
\underset{\ell,r}{\underrightarrow{\operatorname{colim}}}\,{}^b\underset{\sim}{h}^{\ell f}(N.B_r(\mathcal{U}\times D(\ell)), \underset{\sim}{\mathcal{K}}(R)) & \longrightarrow & \underset{\ell,r}{\underrightarrow{\operatorname{colim}}}\,\underset{\sim}{h}^{\ell f}(N.B_r(\mathcal{U}\times D(\ell)), \underset{\sim}{\mathcal{K}}(R)) \\
\downarrow & & \downarrow \\
{}^b\underset{\sim}{h}^{\ell f}(G\times D; \underset{\sim}{\mathcal{K}}(R)) & \longrightarrow & \underset{\sim}{h}^{\ell f}(G\times D; \underset{\sim}{\mathcal{K}}(R))
\end{array}$$

The horizontal arrows are equivalences by an application of II.21; one notes that $N_k B_r(\mathcal{U} \times D(\ell))$ is a countable disjoint union of metric spaces of the form $\pi^{-1}([a,b])$, for some a and b. To prove that the right hand vertical arrow is an equivalence, it will suffice to prove that

$$\underset{\sim}{h}^{\ell f}(N.B_r(\mathcal{U} \times D(\ell)), \underset{\sim}{K}(R)) \to \underset{\sim}{h}^{\ell f}(G \times D, \underset{\sim}{K}(R))$$

is an equivalence for a cofinal set of values of ℓ and r. Since $\mathcal{U} \times D$ satisfies Hypothesis (B) we are free to verify this only for pairs (ℓ, r) for which $B_r U_i(\ell) \cap B_r U_j(\ell) = \emptyset$ when $|i-j| > 1$. Consider such a choice of (ℓ, r). Let $\mathcal{H}_{\text{even}} = \coprod_i B_r U_{2i}(\ell)$ and $\mathcal{H}_{\text{odd}} = \coprod_i B_r U_{2i+1}(\ell)$. In view of the hypotheses on (ℓ, r), the evident maps $\mathcal{H}_{\text{even}} \to G$ and $\mathcal{H}_{\text{odd}} \to G$ are inclusions. Moreover, one easily checks that if we let $\mathcal{H} = \{\mathcal{H}_{\text{even}}, \mathcal{H}_{\text{odd}}\}$ denote the two element covering, and let $\mathcal{H} \times D = \{\mathcal{H}_{\text{even}} \times D, \mathcal{H}_{\text{odd}} \times D\}$, then there is an isomorphism of simplicial metric spaces $N.B_r(\mathcal{U}(\ell) \times D) \to N.(\mathcal{H} \times D)$, making the diagram

$$\begin{array}{ccc} N.B_r(\mathcal{U}(\ell) \times D) & \longrightarrow & N.(\mathcal{U} \times D) \\ & \searrow \quad \swarrow & \\ & G \times D & \end{array}$$

commute, where $G \times D$ denotes the constant simplicial metric space with value $G \times D$. It follows from the fact that $B_r(\mathcal{U}(\ell) \times D)$ is excisive that $\mathcal{H} \times D$ is excisive. Consequently, we have the commutative diagram

$$\begin{array}{ccc} \underset{\sim}{h}^{\ell f}(N.B_r(\mathcal{U}(\ell) \times D); \underset{\sim}{K}(R)) & \xrightarrow{\approx} & \underset{\sim}{h}^{\ell f}(N.(\mathcal{U} \times D), \underset{\sim}{K}(R)) \\ & \searrow & \downarrow \wr \\ & & \underset{\sim}{h}^{\ell f}(G \times D, \underset{\sim}{K}(R)) \end{array}$$

where the right hand vertical arrow is an equivalence by II.15, and the horizontal arrow is induced by an isomorphism of simplicial metric spaces. Therefore, the right hand vertical arrow in $(**)$ is an equivalence, hence so is the left hand vertical arrow. Thus, to show that the lower horizontal arrow in $(*)$ is an equivalence, it will suffice to show that the upper one is. The upper horizontal arrow is induced by a map of simplicial spectra, and it will suffice to show that it is an equivalence in each level. For this, we must show that for a cofinal set of ℓ and r, and each k, that the map

$${}^b\underset{\sim}{h}^{\ell f}(N_k B_r(\mathcal{U}(\ell) \times D), \underset{\sim}{K}) \to \underset{\sim}{K}(N_k B_r(\mathcal{U}(\ell) \times D), R)$$

is an equivalence of spectra. We take our cofinal set of (ℓ, r)'s to be the set of all ℓ and r so that $B_r\mathcal{U}(\ell)$ satisfies Hypothesis (A). In this case, by V.5 (c), $N_k(B_r(\mathcal{U}(\ell) \times D))$ is a finite disjoint union of metric spaces, each of which is of the form

(a) $$\coprod_i \pi^{-1}([2^\ell i - 1/3 - r, 2^\ell(i+1) + 1/3 + r]) \times D$$

or

(b) $$\coprod_i \pi^{-1}([2^\ell i - 1/3 - r, 2^\ell i + 1/3 + r]) \times D\,,$$

so it suffices to show that ${}^b\underset{\sim}{h}^{\ell f}(-,\underset{\sim}{K}(r)) \to \underset{\sim}{K}(-;R)$ induces an isomorphism on metric spaces of this form. Consider (a); we have the inclusion

(i) $$\coprod_i \pi^{-1}(2^\ell(i+1/2))\times D \hookrightarrow \coprod_i \pi^{-1}([2^\ell i-1/3-r, 2^\ell(i+1)+1/3+r])\times D\,.$$

But, $\coprod_i \pi^{-1}(2^\ell(i+1/2))$ is isometric to $\mathbb{Z}\times\pi^{-1}(2^{\ell-1})$, where $\mathbb{Z}$ is viewed as a discrete metric space, and similarly $\coprod_i \pi^{-1}([2^\ell i-1/3-r, 2^\ell(i+1)+1/3+r])$ is isometric to $\mathbb{Z}\times\pi^{-1}([-1/3-r, 2^\ell+1/3+r])$, and under these identifications, the inclusion (i) corresponds to the inclusion $\mathrm{Id}_{\mathbb{Z}}\times(\pi^{-1}(2^{\ell-1})\hookrightarrow \pi^{-1}([-1/3-r, 2^\ell+1/3+r]))$. Let $A^0 = \coprod_i \pi^{-1}(2^\ell(i+1/2))\times D$ and $A = \coprod_i \pi^{-1}([2^\ell i - 1/3 - r, 2^\ell i + 1/3 + r])$; we have the inclusion $A^0 \hookrightarrow A$. We have the commutative diagrams

$$\text{(I)}\qquad \begin{array}{ccc} {}^b\underset{\sim}{h}^{\ell f}(A^0,\underset{\sim}{K}(R)) & \longrightarrow & \underset{\sim}{K}(A^0;R) \\ \downarrow & & \downarrow \\ {}^b\underset{\sim}{h}^{\ell f}(A,\underset{\sim}{K}(R)) & \longrightarrow & \underset{\sim}{K}(A,R) \end{array}$$

and

$$\text{(II)}\qquad \begin{array}{ccc} {}^b\underset{\sim}{h}^{\ell f}(A^0,\underset{\sim}{K}(R)) & \longrightarrow & \underset{\sim}{h}^{\ell f}(A^0,\underset{\sim}{K}(R)) \\ \downarrow & & \downarrow \\ {}^b\underset{\sim}{h}^{\ell f}(A,\underset{\sim}{K}(R)) & \longrightarrow & \underset{\sim}{h}^{\ell f}(A,\underset{\sim}{K}(R)) \end{array}$$

In each case, the right hand vertical arrow is an equivalence; this follows by V.6 (b) in case (I) and V.6 (a) in case (II), using the identifications $A^0 \cong \mathbb{Z}\times\pi^{-1}(2^{\ell-1})$ and $A \cong \mathbb{Z}\times\pi^{-1}([-1/3-r, 2^\ell+1/3+r])$. But the horizontal arrows in (II) are also equivalences, since both A and A^0 satisfy the hypotheses of II.21. Therefore, the left hand vertical arrow in (I) is also an equivalence. We are required to show that the lower horizontal arrow is an equivalence; it now suffices to show that the upper one is. But, $A^0 \cong \pi^{-1}(2^{\ell-1})\times\mathbb{Z}\times D$, and $\mathbb{Z}\times D$ is countable. Moreover, $\pi^{-1}(2^{\ell-1})$ is isometric to G^0, with some G^0 invariant metric (not necessarily coming from a Riemannian metric on G^0) in which all closed balls are compact. V.1 together with the inductive result shows that ${}^b\underset{\sim}{h}^{\ell f}(A^0,\underset{\sim}{K}(R)) \to \underset{\sim}{K}(A^0,R)$ is an equivalence. A similar argument works in case (b). This gives the result. Q.E.D.

We now carry out the analysis of G/K, where G is connected and K is a maximal compact subgroup. Let $\mathfrak{n}$ and $\mathfrak{r}$ denote the maximal nilpotent ideal and radical of $\mathfrak{g}$, respectively. See [9, Ch. 1, § 4.4 and § 5.2] for definitions of these ideals. Let N and R be the corresponding subgroups of G. By [9, Ch. III, § 9.7, Proposition 23] R and N are closed normal subgroups of G. Recall also from [9, Ch. I, § 5.3] the definition of the nilpotent radical $\mathfrak{s} \subseteq \mathfrak{n} \subseteq \mathfrak{g}$. We say a Lie algebra $\mathfrak{g}$ is *reductive* if its nilpotent radical consists only of 0. This implies that $\mathfrak{g} \cong \mathfrak{l} \oplus \mathfrak{a}$, where $\mathfrak{l}$ is semisimple and $\mathfrak{a}$ is Abelian. It follows from [9, Ch. I, § 6.4, Corollary to Proposition 6] that $\mathfrak{g}/\mathfrak{s}$ is a reductive Lie algebra, and hence by part (c) of the Corollary to Proposition 5 of Ch. 1, § 6.4 in [9], that $\mathfrak{g}/\mathfrak{n}$ or any quotient of $\mathfrak{g}/\mathfrak{n}$ is a reductive if its Lie algebra is; G/N is therefore a reductive group. The following Proposition is Exercise 22 for Ch. III, § 9 of [9].

Proposition V.9 *Let G be a connected reductive Lie group. Then $G \cong (S \times V)/D$, where V is a finite dimensional real vector space, S is a simply connected semisimple Lie group, and D is a discrete subgroup of the center of G.*

Definition V.10 *Let G be a reductive Lie group, written as $G \cong (S\times V)/D$, with S simply connected and semisimple and V a vector group. We say G is T–compact if $V/V \cap D$ is compact. We say G is Z–compact if the entire center of G is compact. For G any connected Lie group, we say G is quasi–linear if there is a simply solvable normal subgroup $G^0 \subseteq G$ so that G/G^0 is a Z–compact reductive group.*

Now consider a connected Lie group G, with N the subgroup corresponding to the the maximal nilpotent ideal in $\mathfrak{g}$, the Lie algebra of G. The following follows from the discussion in § I of [38].

Proposition V.11 *N contains a unique maximal compact subgroup N_K, which is invariant under all automorphisms of N. Consequently, K_N is normal in G, and if $K \subseteq G$ is any maximal compact subgroup, then $K_N \subseteq K$, and we have a diffeomorphism of left G–spaces*

$$G/K \to G/K_N \Big/ N/K_N \ ,$$

where the action on $G/K \to G/K_N\, /N/K_N$ is obtained by pulling the G/K_N action back along the projection $G \to G/K_N$. Consequently, in order to prove that

$$^{b}\underset{\sim}{h}^{\ell f}(G/K \times D; \underset{\sim}{K}(R)) \to \underset{\sim}{K}(G/K \times D, R)$$

is an equivalence for all G, all invariant Riemannian metrics on G/K, and all countable discrete metric spaces D, it suffices to restrict attention to *groups G for which the subgroup corresponding to the maximal nilpotent ideal of the Lie algebra of G is simply connected. We refer to such groups as nilsimple.*

Now consider a nilsimple group G, with $N \subseteq G$ the subgroup corresponding to the maximal nilpotent ideal of the Lie algebra of G. G/N is reductive, and we write $G \cong S \times V/D$ as in V.9. $V/V \cap D$ is a connected Abelian Lie group, hence it can be written as $T \times E$, where T is a compact torus, and E is a vector group. If $\pi : G \to G/N$ is the projection, then $\pi^{-1}(E)$ is a simply connected solvable normal subgroup of G, and the quotient $G/\pi^{-1}(E)$ is a T compact reductive group. Thus, for any nilsimple group G, there is a normal solvable simply connected subgroup $G^0 \subseteq G$ so that G/G^0 is T compact reductive. We can any such subgroup G^0 *r–characteristic.*

Proposition V.12 *Let G be any nilsimple group. Then there is a local isomorphism $G \to \bar{G}$ of connected Lie groups, where $\bar{G}$ is nilsimple and contains an r–characteristic subgroup $\bar{G}^0$ so that $\bar{G}/\bar{G}^0$ is Z compact. If G^0 is the Lie subgroup of G having the same Lie algebra as $\bar{G}^0$, then G^0 is r–characteristic for G, and $G/G^0 \to \bar{G}/\bar{G}^0$ is a local isomorphism of reductive Lie groups.*

Proof: We consider the universal covering group $\tilde{G}$. Let $G^0 \subseteq G$ be any r–characteristic subgroup; because of the simple connectivity of G^0, $\tilde{G}$ can be viewed as the pullback of the diagram

$$\begin{array}{ccc} & & G \\ & & \downarrow \\ \widetilde{G/G^0} & \longrightarrow & G/G^0 \end{array}$$

where $\widetilde{G/G^0}$ denotes the universal cover of G/G^0. Consequently we have an exact sequence

$$\{e\} \to G^0 \to \tilde{G} \to \widetilde{G/G^0}$$

of connected Lie groups, where $\widetilde{G/G^0}$ is simply connected reductive, hence isomorphic to $S \times V$, with S semisimple and V a vector group. The local isomorphisms from $\tilde{G}$ to connected Lie groups are parametrized by discrete subgroups of the center of $\tilde{G}$; similarly, discrete subgroups of the center of $\widetilde{G/G^0}$ parametrize the local isomorphism from G/G^0 to connected Lie groups. It follows from the definitions that if D is the discrete subgroup of the center of $\widetilde{G/G^0}$ defining the reductive group G/G^0, then there is a corresponding discrete subgroup D' of the center of $\tilde{G}$ so that D' maps isomorphically to D under the projection $\tilde{G} \to \widetilde{G/G^0}$. To prove the result, it will clearly suffice to produce a discrete cocompact subgroup D_m of the center of $\widetilde{G/G^0}$ containing D and a discrete subgroup D'_m of the center of $\tilde{G}$ containing D' which maps isomorphically to D_m under the projection $\tilde{G} \to$

$\widetilde{G/G^0}$. To find D'_m, we consider the composite $\tilde{G} \to \widetilde{G/G^0} \cong S \times V \to S$, with S simply connected and semisimple. From [9, Ch. III, § 6.6, Corollary 1 and Ch. I, § 6, Theorem 5, and Corollary 1] it follows that $\tilde{G}$ is the semidirect product of S with $\pi^{-1}(V)$, where $\pi : \tilde{G} \to \widetilde{G/G^0}$ is the projection, along some homomorphism $\varphi : S \to \mathrm{Aut}\,(\pi^{-1}(V))$. But from [9, Ch. III, § 10.2, Theorem I (iv)] it follows that if we let $\mathfrak{h}$ denote the Lie algebra of $\pi^{-1}(V)$, then the homomorphism $\mathrm{Aut}\,(\pi^{-1}(V)) \to \mathrm{Aut}\,(\mathfrak{h})$ is injective, since $\pi^{-1}(V)$ is connected. The composite

$$S \xrightarrow{\varphi} \mathrm{Aut}\,(\pi^{-1}(V)) \to \mathrm{Aut}\,(\mathfrak{h})$$

can be viewed as a linear representation of S, and from [9, Ch. III, § 9, Exercise 7(a)] it follows that the composite (and hence φ) factors through a quotient S/L, where L is a subgroup of finite index in the center of S. Note that for any $\ell \in L$, the element $(\ell, e) \subseteq S\tilde{\times}_\varphi \pi^{-1}(V) \cong \tilde{G}$ is an element of the center of $\tilde{G}$, and in fact $\ell \to (\ell, e)$ defines a homomorphism σ from L into the center of $\tilde{G}$ so that the composite

$$L \xrightarrow{\sigma} \mathrm{center}\,(\tilde{G}) \to \mathrm{center}\,(S)$$

is the inclusion of L into center$\,(S)$. Now consider the subgroup $p(D) \subseteq$ center$\,(S)$, where $p : S \times V \to S$ is the projection. $L \cap p(D)$ is a subgroup L, and we can choose a free Abelian subgroup $L^0 \subseteq L$ so that $L^0 \cap p(D) = \{0\}$ and so that $L^0 + (p(D) \cap L)$ has finite index in L. We now let $D_m = D + L^0$ and $D'_m = D' + \sigma(L^0)$. These subgroups clearly have the required properties. Q.E.D.

Consider now any nilsimple connected Lie group G, and let G^0 be r–characteristic. The Lie algebra of the T compact reductive group G/G^0 decomposes as $\mathfrak{s} \oplus \mathfrak{a}$, where $G/G^0 \cong S \times V$, and $\mathfrak{s}$ and $\mathfrak{a}$ are the Lie algebras of S and V, respectively. S is semisimple and V is Abelian, as usual. Consider any Cartan decomposition $\mathfrak{s} = \mathfrak{t} \oplus \mathfrak{p}$ for the semisimple Lie algebra $\mathfrak{s}$; see [25, Ch. III, § 7] for a discussion of this.

Proposition V.13 *Suppose G/G^0 is Z–compact. Then the Lie subgroup of G/G^0 corresponding to the subalgebra $\mathfrak{t} \oplus \mathfrak{a}$ is a maximal compact subgroup of G/G^0.*

Proof: With G/G^0 expressed as $S \times V/D$, we see that the image of V is a central compact subgroup of G/G^0, with Lie algebra $\mathfrak{a}$, and the quotient group $G/G^0 \,/\, \mathrm{image}\,(V) \cong S/p(D)$, where $p : S \times V \to S$ is the projection. It follows that the maximal compact subgroups of G/G^0 are in bijective correspondence with the maximal compact subgroups of $S/p(D)$ via projection. The center of $S/p(D)$ is discrete and compact, hence finite, so the subgroup of $S/p(D)$ corresponding to $\mathfrak{t}$ is maximal compact in $S/p(D)$ by [25, Ch. VI, Theorem 1.1 (i)]. The result now follows. Q.E.D.

Corollary V.14 *Let G be nilsimple, with G^0 an r-characteristic subgroup.. Let $G/G^0 \cong S \times V/D$, with $\mathfrak{s}$ and $\mathfrak{a}$ the Lie algebras of S and V, respectively, and let $\mathfrak{t} \oplus \mathfrak{p}$ be a Cartan decomposition of $\mathfrak{s}$. Let $K \subseteq G/G^0$ be the Lie subgroup corresponding to $\mathfrak{t}$. Then there is a section $s : K \to G$, so that the composite $K \xrightarrow{s} G \to G/G^0$ is the inclusion G/G^0. Moreover, $\mathrm{Ad}_G(s(K))$ is compact.*

Proof: Consider first the case where G/G^0 is Z compact. Then K is compact, and G^0 is simply connected solvable, so by [9, Ch. III, § 9, Exercise 19] we have a section $s : K \to G$ as required. The compactness of $\mathrm{Ad}_G(s(K))$ follows from the compactness of K. For a general nilsimple group, we apply V.12. Thus, consider the diagram

$$\begin{array}{ccc} G & \longrightarrow & \hat{G} \\ \downarrow & & \downarrow \\ G/G^0 & \longrightarrow & \hat{G}/G^0 \end{array}$$

where the horizontal maps are local isomorphisms and $\hat{G}/G^0$ is Z compact. Let $K \subseteq G/G^0$ be the subgroup corresponding to t, and $\hat{K}$ the subgroup of $\hat{G}/G^0$ corresponding to $\mathfrak{t}$. We have a section $s : \hat{K} \to \hat{G}$ by the above discussion, and consider the composite $l : K \to \hat{K} \xrightarrow{s} \hat{G}$. Both vertical maps induce isomorphisms on π_1, and consequently it follows that

$$\mathrm{image}\,(\pi_1(l)) \subseteq \mathrm{image}\,(\pi_1(G) \to \pi_1(\hat{G})).$$

By covering space theory, we see that l lifts to a unique group homomorphism $K \to G$, which is the required map. $\mathrm{Ad}_G(K)$ is isomorphic to $\mathrm{Ad}_{\hat{G}}(\hat{K})$, since $G \to \hat{G}$ is a local isomorphism, so the proof is complete. Q.E.D.

We now recall the Iwasawa decomposition for a semisimple Lie group. See [25, Ch. VI, § 3,4,5] for details. Let G be a semisimple Lie group with Lie algebra $\mathfrak{g}$, and let $\mathfrak{t} \oplus \mathfrak{p}$ be a Cartan decomposition for $\mathfrak{g}$. Then there are subalgebras $\mathfrak{h}$ and $\mathfrak{n}$ of $\mathfrak{g}$, with $\mathfrak{h}$ Abelian, $\mathfrak{n}$ nilpotent, and $\mathfrak{h} + \mathfrak{n}$ a subalgebra with $\mathfrak{n}$ an ideal in $\mathfrak{h} + \mathfrak{n}$, so that $\mathfrak{g}$ is the vector space direct sum $\mathfrak{t} \oplus \mathfrak{h} \oplus \mathfrak{n}$.

Theorem V.15 (See [25, Ch. VI, Theorems 1.1 and 5.1]) *Let K be the subgroup of G corresponding to $\mathfrak{t}$, and let B be the subgroup corresponding to $\mathfrak{h} + \mathfrak{n}$. Then K contains the center of G, the multiplication map $B \times K \to G$ is a diffeomorphism and is left B equivariant and right K equivariant. Moreover, B is solvable and simply connected.*

We now establish a similar result for T compact reductive groups. Let G be T compact reductive, with $G \cong S \times V/D$, with D discrete subgroup of the center of $S \times V$. $\mathfrak{g}$, the Lie algebra of G, breaks up as $\mathfrak{s} \oplus \mathfrak{a}$, with $\mathfrak{s}$ the Lie algebra of S and $\mathfrak{a}$ the Lie algebra of V. Let $\mathfrak{s} = \mathfrak{t} \oplus \mathfrak{h} \oplus \mathfrak{n}$ be an Iwasawa decomposition for $\mathfrak{s}$, and let K denote the subgroup corresponding to $\mathfrak{t}+\mathfrak{a}$ and let B the subgroup corresponding to $\mathfrak{h}\oplus\mathfrak{n}$.

Proposition V.16 *The multiplication map $B\times K \to G$ is a diffeomorphism. It is left B equivariant and right K equivariant, and B is solvable and simply connected.*

Proof: Consider the projection $G \xrightarrow{p} S/p(D)$. Then the images of B and K give an Iwasawa decomposition for $S/p(D)$. This means that for any element $g \in G$, there are elements $b \in B$ and $k \in K$ so that $bgk \in V/V\cap D$. But $V/V\cap D \subseteq K$, so by modifying k, we can assure that $bgk = e$. This shows that the multiplication map is surjective. Suppose $b\cdot k = e$, with $b \in B$, $k \in K$. From V.15, we conclude that $k \in V/V\cap D$, so $b \in B\cap(V/V\cap D)$. But, let $\tilde{B}$ be the subgroup of $S\times V$ corresponding to $\mathfrak{h}+\mathfrak{n}$. Then $\tilde{B} \subseteq S$; consequently, if $B\cap(V/V\cap D)$ is non trivial, then $p(\tilde{B})$ has a non trivial intersection with $p(D)$. But $p(D)$ is contained in the center of S, and the center of S is contained in K, again by V.15, so $p(\tilde{B})\cap p(D) = \{e\}$. Consequently, the multiplication map is injective. The equivariance statements are clear. Let $B^0 \subseteq S/p(D)$ be the subgroup corresponding to $\mathfrak{h}+\mathfrak{n}$. Then $p: G \to S/p(D)$ restricts to a local isomorphism from B to B^0, and B^0 is solvable and simply connected by V.15. Consequently, B is solvable and simply connected. Q.E.D.

Now, let G be a nilsimple connected Lie group, and let G^0 be r–characteristic. Let B and K be the subgroups of G/G^0 as in V.16. Let $\bar{B}$ denote the inverse image of B under the projection; $\bar{B}$ is solvable and simply connected. Let $\bar{K}$ denote the image of a section $K \xrightarrow{s} G$, whose existence is guaranteed by V.14

Proposition V.17 *The multiplication map $\bar{B}\times\bar{K} \to G$ is a diffeomorphism, and is left $\bar{B}$ equivariant and right $\bar{K}$ equivariant. $\mathrm{Ad}_G(\bar{K})$ is compact.*

Proof: The first statement follows directly from V.16. The equivariance is clear, and the compactness of $\mathrm{Ad}_G(\bar{K})$ follows from V.14. Q.E.D.

Proposition V.18 *$\bar{K}$ contains a maximal compact subgroup K^0 of G, which is normal in $\bar{K}$.*

Proof: According to [25, Ch. VI, Theorem 2.2] the subgroup K' of $S/p(D)$ corresponding to the subalgebra $\mathfrak{t}$ contains a maximal compact subgroup K'' of $S/p(D)$, and further decomposes as a product of K'' with a vector group. Consequently, K'' is normal in K'. Because of the compactness of $V/V\cap D$, and the fact that it is central, it is clear that K, which corresponds to $\mathfrak{t}+\mathfrak{a}$, contains a unique maximal compact subgroup projecting to K''. Let K^0 be the image of this compact subgroup under the section s; it is a compact normal subgroup of $\bar{K} = s(K)$. We claim that K^0 is a maximal compact subgroup of G. But, its projection in G/G^0 is maximal compact. Consequently, if $\tilde{K}$ is any compact subgroup of G containing K^0, its image in G/G^0 must be equal to the image of K^0, and hence $\tilde{K}$ would contain non trivial elements of G^0. Therefore, $\tilde{K} \cap G^0$ would be a nontrivial compact subgroup of G^0, of which there are one, since G^0 is solvable and simply connected. Thus, K^0 is maximal compact and normal in $\bar{K}$. Q.E.D.

Consider the coset space G/K^0. Since $K^0 \triangleleft \bar{K}$, $\bar{K}/K^0$ acts on the right and G acts on the left, giving a $G \times \bar{K}/K^0$ action on G/K^0. We wish to produce a $G \times \bar{K}/K^0$ invariant Riemannian metric on G/K^0.

Proposition V.19 *Suppose $L^0 \triangleleft L^1 \subseteq G$ and Lie subgroups of a connected Lie group, and suppose $\mathrm{Ad}_G(L^1)$ is compact. G/L^0 is then a $G \times (L^1/L^0)$ space, and there is a $G \times (L^1/L^0)$ invariant Riemannian metric on G/L^0.*

Proof: The tangent space to G/L^0 at e is $\mathfrak{g}/\mathfrak{l}$, where $\mathfrak{l}$ is the Lie algebra corresponding to L^0, any G invariant metric on G/L^0 is determined by its value at e. L^1 acts by the adjoint representation on $\mathfrak{g}/\mathfrak{l}$, since L^0 is normal in L^1, and any $\mathrm{Ad}_G(L^1)$ invariant positive definite inner product on $\mathfrak{g}/\mathfrak{l}$ yields a $G \times L^1/L^0$ invariant Riemannian metric on G/L^0. Such an inner product exists by the compactness of $\mathrm{Ad}_G(L^1)$. Q.E.D.

Proposition V.20 *Let G be any nilsimple connected Lie group, and let $\bar{B}$ and $\bar{K}$ be as in V.17 and let K^0 be as in V.18, so K^0 is maximal compact in G. Then the coset space G/K^0 is a $G \times \bar{K}/K^0$ space, and admits a $G \times \bar{K}/K^0$ invariant Riemannian metric.*

Proof: Direct application of V.17 and V.19. Q.E.D.

Theorem V.21 *Let G be any connected Lie group, $K \subseteq G$ a maximal compact subgroup, and let G/K be equipped with a left invariant Riemannian metric. Then the assembly map*

$$\underset{\sim}{h}^{\ell f}(G/K \times D, \underset{\sim}{K}(R)) \to \underset{\sim}{K}(G/K \times D; R)$$

is an equivalence for any countable discrete metric space D.

Proof: By V.1, it suffices to prove the result for any particular metric. We choose to prove it for a $G \times \bar{K}/K^0$ invariant metric, whose existence is guaranteed by V.20. But using the notation of V.17, it follows from V.17 that G/K^0 is a simply transitive $B \times \bar{K}/K^0$ space, and $B \times \bar{K}/K^0$ is a simply connected solvable Lie group. The result now follows from V.7. Q.E.D.

VI. Equivariant bounded K–theory

Definition VI.1 *By an action of a group Γ on a metric space (X, d), we mean a left action of Γ on the set X so that $d(\gamma x, \gamma y) = d(x, y)$, i.e., an* *action by isometries. A weak action of* Γ *on* (X, d) *is a left* Γ *action on* X *so that for every* $r \in [0, +\infty)$ *and* $\gamma \in \Gamma$, *there is a number* $\hat{r} \in [0, +\infty)$ *so that* $d(x_1, x_2) \leq r$ *implies* $d(\gamma x_1, \gamma x_2) \leq \hat{r}$. *We will require this weaker notion in the sequel to this paper.*

Given a weak action of Γ on a metric space X, one sees easily that Γ acts by symmetric monoidal isomorphisms on the categories $i\mathcal{C}_X(R)$ and $i\widehat{\mathcal{C}}_X(R)$. The action is given on objects of $i\mathcal{C}_X(R)$ by $\gamma\cdot(G, B, \varphi) = (F, B, \gamma \circ \varphi)$ and on morphisms by $\gamma(L) = L$, and similarly on $i\widehat{\mathcal{C}}_X(R)$. For us, it will be necessary to have a version of bounded K–theory which behaves *equivariantly* like $\underset{\sim}{h}^{\ell f}(-; \underset{\sim}{K}(R))$. We have seen that if Γ acts freely on a locally finite simplicial complex X, then

$$\underset{\sim}{h}^{\ell f}(X; \underset{\sim}{K}(R))^{\Gamma} \cong \underset{\sim}{h}^{\ell f}(\Gamma\backslash X; \underset{\sim}{K}(R)).$$

On the other hand, it is easy to see that if γ acts freely on the metric space X, then the only fixed object of $i\mathcal{C}_X(R)$ is the zero object. Consequently, we must produce an equivariant bounded K–theory, similar in spirit to equivariant complex K–theory [3] and equivariant stable homotopy theory [40]. Thomason [44] has given a general construction which applies in this situation.

Let $\underline{C}$ be any category with left Γ action. Let $\underline{E\Gamma}$ denote the category with one object for every element of Γ, and a unique morphism between any two objects of $\underline{E\Gamma}$. Γ acts on the left (and on the right) of $\underline{E\Gamma}$ via the action given by left (or right) multiplication on the objects; the description on objects forces the behavior on morphisms. We let $\underset{\sim}{\underline{C}}$ denote the category whose objects are functors from $\underline{E\Gamma}$ to $\underline{C}$, and whose morphisms are the natural transformations between functors. Γ acts on $\underset{\sim}{\underline{C}}$ by $\gamma(F)(x) = \gamma F(\gamma^{-1}x)$ and $\gamma(F)(f) = \gamma F(\gamma^{-1}f)$, where x and f denote an object and a morphism in $\underline{E\Gamma}$, respectively and, if $N : F \to G$ is a natural transformation, $\gamma N : \gamma F \to \gamma G$ is the natural transformation given on an object $x \in \underline{E\Gamma}$ by $\gamma(N(\gamma^{-1}x)) : \gamma F(\gamma^{-1}x) \to \gamma G(\gamma^{-1}x)$. The fixed point subcategory $\underset{\sim}{\underline{C}}^{\Gamma}$ now is the category whose objects are the equivariant functors from $\underline{E\Gamma}$ to $\underline{C}$ and whose morphisms are the equivariant natural transformations. There is an equivariant functor $\underline{C} \to \underset{\sim}{\underline{C}}$ given by assigning to an object $x \in \underline{C}$ the constant functor with value x. There is a non equivariant functor $\underset{\sim}{\underline{C}} \to \underline{C}$ given by restriction to the subcategory consisting only of the identity element of Γ. The composite $\underline{C} \to \underset{\sim}{\underline{C}} \to \underline{C}$ is equal to the identity functor. There is a natural isomorphism from the identity functor on $\underset{\sim}{\underline{C}}$ to the composite $\underset{\sim}{\underline{C}} \to \underline{C} \to \underset{\sim}{\underline{C}}$, arising from the evident isomorphism of functors from $\underline{E\Gamma} \to * \to \underline{E\Gamma}$ to the identity on $\underline{E\Gamma}$, where $* \to \underline{E\Gamma}$ is the inclusion on the subcategory on the identity element of Γ. Therefore, the equivariant functor $\underline{C} \to \underset{\sim}{\underline{C}}$ is (non equivariantly) an equivalence of categories.

Remark: $\widetilde{\underline{C}}^{\Gamma}$ should be thought of as a category theoretic first approximation to the homotopy fixed point set $(N.\underline{C})^{h\Gamma}$, since the nerve of $\underline{E\Gamma}$ is a contractible free left Γ space.

We remark that if $\underline{C}$ is symmetric monoidal and Γ acts on $\underline{C}$ by symmetric monoidal isomorphisms, then $\widetilde{\underline{C}}$ becomes a symmetric monoidal category by pointwise sum of functors, and the Γ action on $\widetilde{\underline{C}}$ is symmetric monoidal. Spt $(\widetilde{\underline{C}})$ is thus a spectrum with Γ action, and the functor $\widetilde{\underline{C}} \to \underline{C}$ induces an equivalence Spt $(\widetilde{\underline{C}}) \to$ Spt $(\underline{C})$. This equivalence is of course not equivariant.

We might now proceed to define equivariant bounded K–theory in terms of $i\widetilde{\mathcal{C}}_X(R)$ and $i\widetilde{\widehat{\mathcal{C}}}_X(R)$, when Γ acts on X. The theory one obtains this way will be useful in the sequel to this paper, but for our immediate purposes, we will need a slightly more refined theory. Suppose a group Γ acts on a metric space X. Then let $i\widetilde{\mathcal{C}}^0_X(X)(R) \subseteq i\widetilde{\mathcal{C}}_X(R)$ be the full subcategory on functors $F : \underline{E\Gamma} \to i\mathcal{C}_X(R)$ for which $F(f)$ and $F(f)^{-1}$ has filtration 0 for all morphisms f in $\underline{E\Gamma}$. $i\widetilde{\mathcal{C}}^0_X(R)$ is clearly closed under the Γ action on $i\widetilde{\mathcal{C}}_X(R)$. We also have the forgetful functor $i\widetilde{\mathcal{C}}^0_X(R) \to i\widetilde{\mathcal{C}}_X(R) \to i\mathcal{C}_X(R)$, which restricts a functor F to the object $C \in \underline{E\Gamma}$, and this forgetful functor is an equivalence of categories since the "diagonal functor" $i\mathcal{C}_X(R) \to i\widetilde{\mathcal{C}}_X(R)$ defined above has image in $i\widetilde{\mathcal{C}}^0_X(R)$. We similarly define $i\widetilde{\widehat{\mathcal{C}}}^0_X(R)$, and define equivariant K–theory spectra $\underset{\sim}{K}^{\Gamma}(X;R)$ and $\underset{\sim}{\widehat{K}}^{\Gamma}(X;R)$ by applying Spt to $i\widetilde{\mathcal{C}}^0_X(R)$ and $\widetilde{\widehat{\mathcal{C}}}^0_X(R)$, respectively. As in the non equivariant case, we produce a non connective theory $\underset{\sim}{\mathcal{K}}^{\Gamma}(X;R)$ by stabilizing over $X \times E^k$, where E^k denotes Euclidean k space with trivial Γ action. We summarize the evident properties of these functors as follows. Let $\underline{\mathcal{M}}^{\Gamma}$ denote the category of metric spaces with Γ action and equivariant, eventually continuous, proper maps, and let $\underline{\mathcal{S}}^{\Gamma}$ denote the category of spectra with Γ action and equivariant maps of spectra. We have forgetful functors $\underline{\mathcal{M}}^{\Gamma} \to \underline{\mathcal{M}}$ and $\underline{\mathcal{S}}^{\Gamma} \to \underline{\mathcal{S}}$, which just forgets the Γ action, and we refer to these both as ρ.

Proposition VI.2 *$\underset{\sim}{K}^{\Gamma}(-;R)$, $\underset{\sim}{\widehat{K}}(-;R)$, and $\underset{\sim}{\mathcal{K}}^{\Gamma}(-;R)$ all define functors $\underline{\mathcal{M}}^{\Gamma} \to \underline{\mathcal{S}}^{\Gamma}$, and we have natural equivalences of functors $\rho \circ \underset{\sim}{K}^{\Gamma}(-;R) \to \underset{\sim}{K}(-;R) \circ \rho$, $\rho \circ \underset{\sim}{\widehat{K}}^{\Gamma}(-;R) \to \underset{\sim}{\widehat{K}}(-;R) \circ \rho$, and $\rho \circ \underset{\sim}{\mathcal{K}}^{\Gamma}(-;R) \to \underset{\sim}{\mathcal{K}}(-;R) \circ \rho$.*

The next step will be to evaluate the fixed spectra $\underset{\sim}{K}^{\Gamma}(-;R)^{\Gamma}$, $\underset{\sim}{\widehat{K}}^{\Gamma}(-;R)^{\Gamma}$, and $\underset{\sim}{\mathcal{K}}^{\Gamma}(-;R)^{\Gamma}$. In order to do this, we must introduce a generalization of our bounded K–theory construction.

Definition VI.3 *Let X be any set, and r a ring. By a filtration sheaf on X*

with coefficients in R, we mean a family $\mathcal{F} = \{\mathcal{F}_d\}$, $d \in [0, \infty)$, which assign to every pair of elements of X an additive subgroup of R, subject to the compatibility constraint $\mathcal{F}_d(x, y) \cdot \mathcal{F}_e(y, z) \subseteq \mathcal{F}_{d+e}(x, z)$, where $\cdot$ denotes the multiplicative product in R and so that $1 \in \mathcal{F}_0(x, x)$ $\forall x \in X$. We define a symmetric monoidal category $\mathcal{C}_X(R; \mathcal{F})$ as follows. The objects are triples (F, B, φ), where F is a free left R–module, B is a basis for F, and $\varphi : B \to X$ is a labeling function so that for any fixed $\beta \in B$ and $d \in [0, +\infty)$, the set $\{\tilde{\beta} \in B \mid \mathcal{F}_d(\varphi\beta, \varphi\tilde{\beta}) \neq \{0\}\}$ is finite. If (F, B, φ) and (F', B', φ') are objects of $\mathcal{C}_X(R; \mathcal{F})$, we say an R–module homomorphism $F \xrightarrow{f} F'$ is bounded by d if for all $\beta \in B$

$$f(\beta) = \sum_{\beta' \in B'} r_{\beta'\beta}\beta',$$

with $r_{\beta'\beta} \in \mathcal{F}_d(\beta, \beta')$. The compatibility conditions show that if f is bounded by d and g is bounded by e, then $g \circ f$ is bounded by $d + e$. One defines $\widehat{\mathcal{C}}_X(R; \mathcal{F})$ to be the idempotent completion of $\mathcal{C}_X(R; \mathcal{F})$, and defines the (symmetric monoidal) categories of isomorphisms $i\mathcal{C}_X(R, \mathcal{F})$ and $i\widehat{\mathcal{C}}_X(R, \mathcal{F})$. $\underset{\sim}{K}(X; R; \mathcal{F})$ and $\widehat{\underset{\sim}{K}}(X, R, \mathcal{F})$ are now defined to be Spt $(i\mathcal{C}_X(R; \mathcal{F}))$ *and* Spt *$(i\widehat{\mathcal{C}}_X(R; \mathcal{F}))$; note that in this generality X is not required to be a metric space However, if X is a metric space and $\mathcal{F}$ is a filtration sheaf on the set X with coefficients in R, we say $\mathcal{F}$ is subordinate to the metric on X if $\mathcal{F}_d(x, y) = \{0\}$ whenever $d(x, y) > d$.*

Examples:

(i) Let X be a metric space, and R a ring. Then we define a filtration sheaf $\mathcal{F}_X$ on X by $(\mathcal{F}_X)_d(x, y) = \{0\}$ if $d(x, y) > d$, and $(\mathcal{F}_X)_d(x, y) = R$ if $d(x, y) \leq d$.

One verifies that the categories $i\mathcal{C}_X(R, \mathcal{F}_X)$ and $i\mathcal{C}_X(R)$ are isomorphic symmetric monoidal categories (here $i\mathcal{C}_X(R)$ denotes the usual metric space category constructed in § III), and hence we have equivalences of spectra $\underset{\sim}{K}(X; R) \cong \underset{\sim}{K}(X; R; \mathcal{F}_X)$.

(ii) Let $\underline{A}$ be a filtered ring, i.e., a ring A equipped with a family of additive subgroups A_i so that $A_i \cdot A_j \subseteq A_{i+j}$. Then for any set X, we define a filtration sheaf $\mathcal{F}^{\underline{A}}$ on X by $\mathcal{F}^{\underline{A}}_d(x, y) = A_d$.

(iii) Let X be a metric space, and let $\underline{A}$ be a filtered ring. Then we define a filtration sheaf on X $\mathcal{F}^{\underline{A}}_X$, which is a mixed version of (i) and (ii), by $(\mathcal{F}^{\underline{A}}_X)_d(x, y) = A_d$ if $d(x, y) \leq d$.

Now suppose Γ acts freely and properly discontinuously on a metric space X, i.e., that for fixed x and y, the infimum over $\gamma \in \Gamma$ of $d(\gamma x, y)$ is attained. We then define the orbit space metric Δ on $\Gamma\backslash X$ by $\Delta([x], [y]) =$

$\inf_{\gamma\in\Gamma} d(\gamma x, y)$. Let $s : \Gamma\backslash X \to X$ be any section, i.e., any choice of orbit representatives. s is not for the moment required to have any metric properties. Let A be a ring. We define a filtration sheaf $\mathcal{G}(X, s, A)$ on $\Gamma\backslash X$ with coefficients in the group ring $A[\Gamma]$ by letting $\mathcal{G}(X, s, A)_d(z_1, z_2)$ be the A linear span of the set $\{\gamma \in \Gamma \mid d(sz_1, \gamma sz_2) \leq d\}$. One checks that this gives a filtration sheaf on $\Gamma\backslash X$, subordinate to the orbit space metric.

Proposition VI.4 *There are equivalences of spectra*

$$\underset{\sim}{K}^{\Gamma}(X, A)^{\Gamma} \cong \underset{\sim}{K}(\Gamma\backslash X, R[\Gamma], \mathcal{G}(X, s, A))$$

and

$$\underset{\sim}{\widehat{K}}^{\Gamma}(X, A)^{\Gamma} \cong \underset{\sim}{\widehat{K}}(\Gamma\backslash X, R[\Gamma], \mathcal{G}(X, s, A)) \ .$$

The equivalences are natural with respect to maps in $\underline{\mathcal{M}}^{\Gamma}$ respecting choices of section.

Proof: We only carry out the case of $\underset{\sim}{K}^{\Gamma}$; $\underset{\sim}{\widehat{K}}^{\Gamma}$ is similar. Consider the category $i\widehat{\mathcal{C}}^0_X(R)^{\Gamma}$. In [44], it is observed that if $\underline{C}$ is a category with trivial Γ action, then $\widehat{\underline{C}}$ is just the category of representations of Γ in $\underline{C}$. $i\mathcal{C}_X(R)$ has objects of the form (F, B, φ), and the Γ action is given by $\gamma \cdot (F, B, \varphi) = (F, B, \gamma \circ \varphi)$. Thus, the action is trivial on the "modules" and "bases", affecting only the labeling functions φ, in the sense that the forgetful functor from the category $i\mathcal{C}_X(R)$ to the category of based free R–modules is equivariant when we equip the latter category with the trivial Γ action. Consequently, from [44], we see that there is a functor from $i\widetilde{\mathcal{C}}_X(R)^{\Gamma}$ to the category of all representations of Γ in the category of based free left R–modules. Under the assignment, if $\Phi : \underline{E\Gamma} \to i\mathcal{C}_X(R)$ is an equivariant functor, then $\Phi(x) = \Phi(y)$ for all $x, y \in \underline{E\Gamma}$, and $\Phi(g \to h)$ is left multiplication by $h^{-1}g$. In our situation, the functor Φ is given by $\Phi(g) = (G, B, g \circ \varphi)$, where (F, B) is a fixed based module. Let the action of Γ be given by ρ, so each $\rho(g)$ is an automorphism of (F, B). The requirement that $\Phi \in i\widetilde{\mathcal{C}}^0_X(R)$ amounts to the condition that $\rho(h^{-1}g)$, viewed as an isomorphism in $i\mathcal{C}_X(R)$ from $(F, B, g \circ \varphi)$ to $(F, B, h \circ \varphi)$ has filtration zero. This means that if $\beta \in B$, and $\rho(h^{-1}g)(\beta) = \sum_{i=1}^{n} r_i\beta_i$, $r_i \neq 0$, then

$$h \circ \varphi(\beta_i) = g \circ \varphi(\beta), \quad \text{or} \quad \varphi(\beta_i) = h^{-1}g\varphi(\beta).$$

Thus, the requirement can be reduced to the condition that the representation ρ of Γ admits a basis B with labeling function $\varphi : B \to X$ so that for any $\gamma \in \Gamma$, and $\beta \in B$, $\rho(\gamma)(\beta) \in \text{span}\,(\varphi^{-1}(\gamma\varphi(\beta)))$. If we now decompose B as $\coprod_{x\in X} \varphi^{-1}(x)$, and write the corresponding sum splitting of F as $F \cong \oplus_{x\in X} F_x$, $F_x = \text{span}\,(\varphi^{-1}(x))$, then for each γ, $\rho(\gamma)$ restricts to an isomorphism $F_x \to F_{\gamma x}$ for all x. The section s now gives F the structure of a free left $R\Gamma$–module on the generators $\varphi^{-1}(s(\Gamma\backslash X)) \subseteq B$. Define a labeling function $\psi : \varphi^{-1}(s(\Gamma\backslash X)) \to \Gamma\backslash X$ to be the projection; $(F, \varphi^{-1}(s(\Gamma\backslash X)), \psi)$ is now readily seen to be an object of $i\mathcal{C}_{\Gamma\backslash X}(R\Gamma; \mathcal{G}(X,s,R))$. A morphism $\Phi \xrightarrow{f} \Phi'$ in $i\widetilde{\mathcal{C}}^0_X(R)^\Gamma$ gives rise to an isomorphism $F \xrightarrow{\bar{f}} F'$ of the corresponding free left R–modules, which is compatible with the operators $\rho(\gamma)$ and $\rho'(\gamma)$. This is just an $R\Gamma$–linear isomorphism from F to F', viewed as free left $R\Gamma$–modules, satisfying certain boundedness hypotheses, which we now analyze. Suppose $\Phi(e) = (F, B, \varphi)$ and $\Phi'(e) = (F', B', \varphi')$. Let $\beta \in \varphi^{-1}(s(\Gamma\backslash X))$ and $\beta' \in (\varphi')^{-1}(s(\Gamma\backslash X))$, and let $\bar{f}^{\beta'}_{\beta}$ be the $\beta'\beta$ matrix element of the matrix of $\bar{f}$ relative to the bases $\varphi^{-1}(s(\Gamma\backslash X))$, $(\varphi')^{-1}(s(\Gamma\backslash X))$. Write

$$\bar{f}^{\beta'}_{\beta} = \sum_{i=1}^{n} r_i\gamma_i, \qquad r_i \neq 0, \qquad \gamma_i \in \Gamma.$$

Then in order for the original morphism f to have had filtration d, it is necessary that

$$d(\varphi(\beta), \gamma_i\varphi(\beta')) \leq d \text{ for all } i,$$

or that x is in the span of the set $\{\gamma\beta' \mid d(\beta, \gamma\beta') \leq d\}$, or that $\bar{f}^{\beta'}_{\beta} \in \mathfrak{g}(X,s,R)_d$. One checks that the original f is bounded by d if and only if $\bar{f}^{\beta'}_{\beta} \in \mathcal{G}(X,s,R)_d$ for all β, β'. We conclude that we have an equivalence of categories $i\widetilde{\mathcal{C}}^0_X(R)^\Gamma \cong i\mathcal{C}_{\Gamma\backslash X}(R, \mathcal{G}(X,s,R))$, which gives the result. Q.E.D.

In certain cases, it is possible to identify the K–theory spectra associated with this filtration sheaf with the K–theory spectra defined in § III, with labels in $\Gamma\backslash X$.

Definition VI.5 *Let $\mathcal{F}, \mathcal{G}$ be two filtration sheaves on a set X with coefficients in a ring R. We say $\mathcal{F}$ and $\mathcal{G}$ are comparable if for each d, $0 \leq d < +\infty$, there is an e, $0 \leq e < +\infty$, so that $\mathcal{F}_d(x,y) \subseteq \mathcal{G}_e(x,y)$ and so that $\mathcal{G}_d(x,y) \subseteq \mathcal{F}_e(x,y)$, independent of x and y. Comparable filtration sheaves clearly have weakly equivalent K–theory spectra.*

Definition VI.6 *Let $\underset{\sim}{R} = \{R_i\}_{i\geq 0}$ denote a filtered ring, and let $R = \cup_{i=0}^{\infty} R_i$. If $\mathcal{F}$ is a filtration sheaf on a metric space X with coefficients in R, then we say $\mathcal{F}$ is $\underset{\sim}{R}$ constant if it is comparable to the filtration sheaf $\mathcal{F}^{\underset{\sim}{R}}_X$ introduced in Example (iii) following Definition VI.3.*

Let Γ be any finitely generated group, and let $\Omega = \{g_1, \ldots, g_n\}$ be any finite generating set. Define $\ell : \Gamma \to \{0, 1, \ldots, n, \ldots\}$ by

$$\ell(\gamma) = \min\{n \mid \exists i_1, \ldots, i_n, \text{ with } g_{i_1} \cdots g_{i_n} = \gamma\}.$$

Defining $A[\Gamma]_i = A[\{\gamma \mid \ell(\gamma) \leq i\}]$, the group ring $A[\Gamma]$ becomes a filtered ring, which we denote by $\underset{\sim}{A[\Gamma]}$. If Ω' is any other finite generating set, and $A[\Gamma]'_i$ denote the associated filtration, then these two filtrations are equivalent in the sense that for all d, $0 \leq d < +\infty$, there exists e, $0 \leq e < +\infty$, so that $A[\Gamma]_d \subseteq A[\Gamma]'_e$ and $A[\Gamma]'_d \subseteq A[\Gamma]_e$. It follows easily that the filtration sheaves $\mathcal{F}_X^{\underset{\sim}{A[\Gamma]}}$ and $\mathcal{F}_X^{\underset{\sim}{A[\Gamma]}'}$ are comparable, and hence that a given filtration sheaf on X with coefficients in $A[\Gamma]$ is $\underset{\sim}{A[\Gamma]}$ constant if and only if it is $\underset{\sim}{A[\Gamma]}'$ constant.

Proposition VI.7 *Let Γ be a finitely generated group acting freely and properly discontinuously on a metric space X. Let Ω be any finite generating set, and let $\Gamma_d \subseteq \Gamma$, $\Gamma_d = \{\gamma \mid \ell(\gamma) \leq d\}$. $\underset{\sim}{A[\Gamma]}$ will denote the associated filtered ring. Suppose further that there is a section $s : \Gamma\backslash X \to X$ so that the following three conditions hold.*

(A) *s is a morphism in $\underset{\sim}{\mathcal{M}}$, i.e., s is eventually continuous.*

(B) *For all $d \in [0, +\infty)$ there exists N such that for every $z \in \Gamma\backslash X$ and $\gamma \in \Gamma_d$, $d(sz, \gamma sz) \leq N$.*

(C) *For all $N \in [0, +\infty)$ there exists a d so that for all $z \in \Gamma\backslash X$, $B_N(sz) \cap \Gamma \cdot sz \subseteq \Gamma_d \cdot sz$.*

Then $\mathcal{G}(X, s, A)$ is $\underset{\sim}{A[\Gamma]}$ constant.

Remark: These conditions clearly hold when X is a Riemannian manifold with free, properly discontinuous Γ action and $\Gamma\backslash X$ compact. s is obtained by choosing a bounded fundamental domain for the action. More generally, the condition holds for the Γ manifold $X \times Z$, where X is as above, the Γ action on $X \times Z$ is given by $\gamma(x, z) = (\gamma x, z)$ and $X \times Z$ is given the product metric with respect to some Riemannian metric on Z.

Proof: Let $r = \sum_{i=1}^n \alpha_i \gamma_i \in A[\Gamma]$, where the α_i's are all non zero. Then $r \in (\mathcal{F}_{\Gamma\backslash X}^{\underset{\sim}{A[\Gamma]}})_k(z_1, z_2)$ if there are numbers $d, e \geq 0$ so that $\gamma_i \in \Gamma_d$ for all i, $\Delta(z_1, z_2) \leq e$, and $d + e \leq k$. Here Δ denotes the orbit space metric. Choose N so large that $d(sz, \gamma sz) \leq N$ for all $\gamma \in \Gamma_k$, which is possible by Hypothesis (B), and so that $d(sz_1, sz_2) \leq N$ for all $z_1, z_2 \in \Gamma\backslash X$ such that $\Delta(z_1, z_2) \leq k$, which is possible by Hypothesis (A). If $r \in (\mathcal{F}_{\Gamma\backslash X}^{\underset{\sim}{A[\Gamma]}})_k(z_1, z_2)$, we find that

$$d(sz_1, \gamma_i sz_2) \leq d(sz_1, sz_2) + d(sz_2, \gamma_i sz_2) \leq 2N,$$

so

$$(\mathcal{F}_{\Gamma\backslash X})_k^{\widetilde{A[\Gamma]}}(z_1, z_2) \subseteq \mathfrak{g}(X, s, A)_{2N}(z_1, z_2),$$

independent of z_1 and z_2. Suppose now that $r \in \mathcal{G}(X, s, A)_N(z_1, z_2)$. This is equivalent to the statement that $d(sz_1, \gamma_i sz_2) \leq N$ for all i. Consequently, $\Delta(z_1, z_2) \leq N$; choose N' so large that $d(sz_1, sz_2) \leq N'$ whenever $\Delta(z_1, z_2) \leq N$, possible by Hypothesis (A). Now,

$$d(sz_2, \gamma_i sz_2) \leq d(sz_2, sz_1) + d(sz_1, \gamma_i sz_2) \leq N' + N.$$

By Hypothesis (C), it is possible to choose d so large that $d(sz, \gamma sz) \leq N' + N$ implies $\gamma \in \Gamma_d$. It is now clear that

$$\mathcal{G}(X, s, A)_N(z_1, z_2) \subseteq (\mathcal{F}_{\Gamma\backslash X})_{d+N}^{\widetilde{A[\Gamma]}}(z_1, z_2).$$

This is the result. Q.E.D.

We note that for any metric space X and filtration sheaf $\mathcal{F}$ on X with coefficients in R, subordinate to the metric on X, there are functors $i\mathcal{C}_X(R, \mathcal{F}) \to i\mathcal{C}_X(R)$ and $i\widehat{\mathcal{C}}_X(R, \mathcal{F}) \to i\widehat{\mathcal{C}}_X(R)$ induced by the inclusion of the filtration sheaf $\mathcal{F}$ into the filtration sheaf $\mathcal{G}$ defined by $\mathcal{G}_d(x, y) = \{0\}$ if $d(x, y) > d$ and $\mathcal{G}_d(x, y) = R)$ if $d(x, y) \leq d$. This gives maps of spectra $\underset{\sim}{K}(X, R, \mathcal{F}) \to \underset{\sim}{K}(X, R)$ and $\widehat{\underset{\sim}{K}}(X, R, \mathcal{F}) \to \widehat{\underset{\sim}{K}}(X, R)$. In particular, VI.4 shows that we have natural maps $\underset{\sim}{K}^\Gamma(X, A)^\Gamma \to \underset{\sim}{K}(\Gamma\backslash X, A[\Gamma])$ and $\widehat{\underset{\sim}{K}}^\Gamma(X, A)^\Gamma \to \widehat{\underset{\sim}{K}}(\Gamma\backslash X, A[\Gamma])$, when Γ acts freely and properly discontinuously on X.

Corollary VI.8 *Let X be a connected Riemannian manifold with a smooth, isometric, free, properly discontinuous Γ action, and suppose $\Gamma\backslash X$ is compact. Let A be a ring. Then we have equivalences*

$$\underset{\sim}{K}^\Gamma(X, A)^\Gamma \cong \underset{\sim}{K}(\Gamma\backslash X, A[\Gamma]) \cong \underset{\sim}{K}(A[\Gamma])$$

and

$$\widehat{\underset{\sim}{K}}^\Gamma(X, A)^\Gamma \cong \widehat{\underset{\sim}{K}}(\Gamma\backslash X, A[\Gamma]) \cong \widehat{\underset{\sim}{K}}(A[\Gamma]).$$

Proof: One checks first that the functor

$$i\mathcal{C}_{\Gamma\backslash X}(A[\Gamma], \mathcal{F}_{\Gamma\backslash X}^{\widetilde{A[\Gamma]}}) \to i\mathcal{C}_{\Gamma\backslash X}(A[\Gamma])$$

is an equivalence of categories, since the compactness of $\Gamma\backslash X$ implies that the objects in $i\mathcal{C}_{\Gamma\backslash X}(A[\Gamma])$ are actually finitely generated. This shows that the map

$$\underset{\sim}{K}(\Gamma\backslash X, A[\Gamma], \mathcal{F}_{\Gamma\backslash X}^{\widetilde{A[\Gamma]}}) \to \underset{\sim}{K}(\Gamma\backslash X, A[\Gamma])$$

is an equivalence. Secondly, $\underset{\sim}{K}(\Gamma\backslash X, A[\Gamma])$ is equivalent to $\underset{\sim}{K}(A[\Gamma])$ by III.5. The result for $\underset{\sim}{\widehat{K}}$ follows similarly. Q.E.D.

We wish to prove the analogous result for $\underset{\sim}{\mathcal{K}}$; this requires a preliminary lemma. Let $\underset{\sim}{A}$ denote a filtered ring, and let X denote any metric space. Let $\mathcal{F}_{\Gamma\backslash X}^{\underset{\sim}{A}}$ be the filtration sheaf defined in Example (iii) following Definition VI.3.

Lemma VI.9 *Let $X = E^k$, the k–dimensional Euclidean space, and suppose $A = \cup_{i=0}^{\infty} A_i$. Then the maps*

$$\underset{\sim}{K}(X, A, \mathcal{F}_X^{\underset{\sim}{A}}) \to \underset{\sim}{K}(X, A)$$

and

$$\underset{\sim}{\widehat{K}}(X, A, \mathcal{F}_X^{\underset{\sim}{A}}) \to \underset{\sim}{\widehat{K}}(X, A)$$

are equivalences of spectra.

Proof: One verifies that the maps induce isomorphisms on π_i. For $i \geq k$, this follows from [33, Theorem 3.4] applied to the filtered additive category $\mathcal{A}$ of finitely generated based A–modules, where a morphism has filtration d if the elements in the matrix representing it all lie in A_d. For $i < k$, this follows from [33, § 5] and the observation that, in the notation of [33], the category $\mathcal{C}_+(\mathcal{A})$ is flasque and $\mathcal{A}$ filtered. Q.E.D.

Corollary VI.10 *Let Γ and X be as in VI.8. Then there is weak equivalence of spectra*

$$\underset{\sim}{\mathcal{K}}^{\Gamma}(X, A)^{\Gamma} \cong \underset{\sim}{\mathcal{K}}(\Gamma\backslash X; A[\Gamma]) \cong \underset{\sim}{\mathcal{K}}(A[\Gamma])\,.$$

Proof: It follows from VI.7 that

$$\underset{\sim}{\widehat{K}}^{\Gamma}(X \times E^k, A)^{\Gamma} \cong \underset{\sim}{\widehat{K}}((\Gamma\backslash X) \times E^k, A[\Gamma], \mathcal{F}_{\Gamma\backslash X}^{\widetilde{A[\Gamma]}}),$$

and VI.9 shows that this latter spectrum is equivalent to $\underset{\sim}{\widehat{K}}((\Gamma\backslash X)\times E^k, A[\Gamma])$. The result now follows from the definition of $\underset{\sim}{\mathcal{K}}$. Q.E.D.

We will now examine the case of discrete metric spaces.

Proposition VI.11 *Let Y be any metric space. Then the two functors $X \to \underset{\sim}{K}^{\Gamma}(X \times Y, A)^{\Gamma}$ and $X \to \underset{\sim}{K}((\Gamma\backslash X) \times Y, A)$ are naturally equivalent as functors from the category of free Γ sets and proper, equivariant maps (viewed as discrete metric spaces) to $\underline{S}$. Here, $X \times Y$ is given the Γ action $\gamma(x,y) = (\gamma x, y)$. Similarly for $\underset{\sim}{\widehat{K}}^{\Gamma}$.*

Proof: For any object $(F, B, \varphi) \in i\mathcal{C}_{X\times Y}(R)$ and $x \in X$, let F_x denote the span of the basis elements β so that $\varphi(\beta) = (x,y)$ for some $y \in Y_j$. Of course F is the direct sum $\oplus_{x\in X} F_x$. An object of $i\widetilde{\mathcal{C}}^{0}_{X\times Y}(R)^{\Gamma}$ is an equivariant functor $\underline{E\Gamma} \xrightarrow{\Psi} i\mathcal{C}_{X\times Y}(R)$ with values in the filtration zero isomorphisms in $i\mathcal{C}_{X\times Y}(R)$. Write $\Psi(e) = (F(\Psi), B(\Psi), \varphi(\Psi))$. Since Ψ is a functor from $\underline{E\Gamma}$, we see that it determines, for each $\gamma \in \Gamma$, isomorphisms $\Psi_\gamma : F(\Psi)_e \to F(\Psi)_{\gamma^{-1}}$. Let $i\widetilde{\mathcal{C}}'_{X\times Y}(R)^{\Gamma}$ denote the full subcategory of $i\widetilde{\mathcal{C}}^{0}_{X\times Y}(R)^{\Gamma}$ on the functors Ψ so that Ψ_γ is an identity map for each γ. It is easy to see that the inclusion

$$i : i\widetilde{\mathcal{C}}'_{X\times Y}(R)^{\Gamma} \hookrightarrow i\widetilde{\mathcal{C}}^{0}_{X\times Y}(R)^{\Gamma}$$

is an equivalence of categories, i.e., that there is a functor

$$i\widetilde{\mathcal{C}}^{0}_{X\times Y}(R)^{\Gamma} \xrightarrow{j} i\widetilde{\mathcal{C}}'_{X\times Y}(R)^{\Gamma}$$

so that $j \circ i$ is the identity and so that the functor $i \circ j$ is isomorphic to the identity functor on $i\widetilde{\mathcal{C}}^{0}_{X\times Y}(R)$. Thus $X \to \mathrm{Spt}\,(i\widetilde{\mathcal{C}}'_{X\times Y}(R)^{\Gamma})$ is a functor from the category of free Γ sets and proper equivariant maps to $\underline{S}$, and there is a natural equivalence from $[X \to \mathrm{Spt}\,(i\widetilde{\mathcal{C}}'_{X\times Y}(R)^{\Gamma})]$ to $[X \to \mathrm{Spt}\,(i\widetilde{\mathcal{C}}^{0}_{X\times Y}(R)^{\Gamma})]$. Let $\bar{X} \subseteq X$ denote any set of orbit representatives for X; we have the projection $\pi : \bar{X} \to \Gamma\backslash X$, which is a bijection. We now define a functor

$$\rho : i\widetilde{\mathcal{C}}'_{X\times Y}(R)^{\Gamma} \to i\mathcal{C}_{\bar{X}\times Y}(R)$$

as follows. For $\Psi \in i\widetilde{\mathcal{C}}'_{X\times Y}(R)^{\Gamma}$, we let $\overline{B(\Psi)} = \varphi(\Psi)^{-1}(\bar{X} \times Y)$, and let $\overline{F(\Psi)}$ be the R linear span on $\overline{B(\Psi)}$. Now define ρ on objects by

$$\rho(\Psi) = (\overline{F(\Psi)}, \overline{B(\Psi)}, \varphi(\Psi) \,|\, \overline{B(\Psi)})\,.$$

Let $p : i\mathcal{C}_{\bar{X}\times Y}(R) \to i\mathcal{C}_{(\Gamma\backslash X)\times Y}(R)$ the functor induced by $\pi \times \mathrm{Id}$; we now have $p \circ \rho : i\widetilde{\mathcal{C}}'_{X\times Y}(R)^{\Gamma} \to i\mathcal{C}_{(\Gamma\backslash X)\times Y}(R)$. It follows from the definition of

$i\widetilde{\mathcal{C}}'_{X\times Y}(R)^\Gamma$ that the functor $p\circ\rho$ is independent of the choice of $\bar{X}$, and that it is a natural transformation of functors from the category of free Γ sets and proper equivariant maps. It also follows directly from IV.4 that $p\circ\rho$ is an equivalence of categories. The point is that in this case, the filtration sheaf $\mathcal{G}(X\times Y,s,R)$ is given by

$$\begin{cases} \mathcal{G}(X\times Y,s,R)_d((z_1,y_1),(z_2,y_2)) = \{0\} & \text{if } z_1 \neq z_2, \\ \mathcal{G}(X\times Y,s,R)_d((z,y_1),(z,y_2)) = \{0\} & \text{if } d(y_1,y_2) > d, \\ \mathcal{G}(X\times Y,s,R)_d((z,y_1),(z,y_2)) = R \subseteq R[\Gamma] & \text{if } d(y_1,y_2) \leq d. \end{cases}$$

Note that $\bigcup_d \mathcal{G}(X\times Y,s,R)_d \neq R[\Gamma]$. We now have a diagram

$$\mathrm{Spt}\,(i\widetilde{\mathcal{C}}^0_{X\times Y}(R)^\Gamma) \leftarrow \mathrm{Spt}\,(i\widetilde{\mathcal{C}}'_{X\times Y}(R)^\Gamma) \rightarrow \mathrm{Spt}\,(i\mathcal{C}_{\Gamma\backslash X\times Y}(R))$$

of equivalences of functors in X, and hence a homotopy natural equivalence

$$\underset{\sim}{K}^\Gamma(X\times Y,R)^\Gamma \rightarrow \underset{\sim}{K}((\Gamma\backslash X)\times Y,R)\,.$$

Q.E.D.

Corollary VI.12 *The functors $\underset{\sim}{\mathcal{K}}^\Gamma(-,R)^\Gamma$ and $\underset{\sim}{\mathcal{K}}(\Gamma\backslash(-),R)$ are naturally equivalent as functors on the category of free Γ sets and proper equivariant maps.*

Proof: Apply VI.11 with $Y = E^k$ and pass to limits over k. Q.E.D.

Proposition VI.13 *For any free Γ set X, the natural maps $\underset{\sim}{K}^\Gamma(X,R)^\Gamma \rightarrow \underset{\sim}{K}^\Gamma(X,R)^{h\Gamma}$, $\underset{\sim}{\widehat{K}}^\Gamma(X,R)^\Gamma \rightarrow \underset{\sim}{\widehat{K}}^\Gamma(X,R)^{h\Gamma}$, and $\underset{\sim}{\mathcal{K}}^\Gamma(X,R)^\Gamma \rightarrow \underset{\sim}{\mathcal{K}}^\Gamma(X,R)^{h\Gamma}$ are equivalences of spectra.*

Proof: Consider the case of $\underset{\sim}{K}^\Gamma$; the other cases are identical. It follows from II.4 and III.14 that as a Γ–module

$$\pi_i(\underset{\sim}{K}^\Gamma(X;R)) \cong \pi_i(\widehat{\underset{\sim}{K}(R)})[X] \cong \mathrm{Hom}_{\mathbb{Z}}(\mathbb{Z}[X],\pi_i(\underset{\sim}{K}(R)))\,,$$

where the action on the last group is obtained by using the multiplication action on $\mathbb{Z}[X]$. Thus, $\pi_i(\underset{\sim}{K}^\Gamma(X;R))$ is "coinduced", and it follows that

$$H^p(\Gamma;\pi_i(\underset{\sim}{K}^\Gamma(X;R))) = 0 \ \text{ if } \ p > 0.$$

Consequently, the spectral sequence of I.3 converging to $\pi_*(\underset{\sim}{K}^\Gamma(X;R)^{h\Gamma})$ collapses at the E_2 level, with $E_2^{p,q} = 0$ if $p \neq 0$, and

$$E_2^{p,q} = \pi_q(\underset{\sim}{K}^\Gamma(X;R))^\Gamma \cong \pi_q(\widehat{\underset{\sim}{K}(R)})[\Gamma\backslash X]\,.$$

On the other hand, it also follows from II.4, III.14, and VI.11 that

$$\pi_q(\underset{\sim}{K}^\Gamma(X;R))^\Gamma \cong \pi_q(\widehat{\underset{\sim}{K}(R)})[\Gamma\backslash X],$$

and one readily verifies that the natural map $\underset{\sim}{K}^\Gamma(X,R)^\Gamma \to \underset{\sim}{K}^\Gamma(X,R)^{h\Gamma}$ induces the identification of homotopy groups. In the case of $\underset{\sim}{\mathcal{K}}^\Gamma$, one must of course use III.15 and VI.12 in place of III.14 and VI.11. Q.E.D.

We now wish to construct an equivariant version of the assembly map constructed in III.20. For a given group Γ, we will compare the functors

$${}^b\underset{\sim}{h}^{\ell f}(-;\underset{\sim}{K}(R)) : \underline{\mathcal{M}}_f^\Gamma \to \underline{\mathcal{S}}^\Gamma$$

and

$$\underset{\sim}{K}^\Gamma(-;R)^\Gamma : \underline{\mathcal{M}}_f^\Gamma \to \underline{\mathcal{S}}^\Gamma,$$

where $\underline{\mathcal{M}}_f^\Gamma$ will denote the category of metric spaces with free, properly discontinuous, isometric Γ action. We first consider a general construction. Let $\underline{E}$ and $\underline{C}$ be categories, with $\underline{C}$ symmetric monoidal, and suppose a group Γ acts on $\underline{E}$ and $\underline{C}$, with Γ acting on $\underline{C}$ by symmetric monoidal automorphisms. Let $\underline{\text{Fun}}\,(\underline{E},\underline{C})$ denote the category whose objects are the functors from $\underline{E}$ to $\underline{C}$, and whose morphisms are the natural transformations. Under pointwise sum $\underline{\text{Fun}}\,(\underline{E},\underline{C})$ becomes a symmetric monoidal category, and Γ acts on $\underline{\text{Fun}}\,(\underline{E},\underline{C})$ by symmetric monoidal automorphisms via $(\gamma\circ F)(x) = \gamma F(\gamma^{-1}x)$. If $X.$ is any space and $\underset{\sim}{Z}$ is a spectrum, let $F(X., \underset{\sim}{Z})$ denote the function spectrum which can be defined by $F(X., \underset{\sim}{Z})_k = F(X., Z_k)$ if $\underset{\sim}{Z}$ is a Kan Ω spectrum.

Proposition VI.14 *There is an equivariant natural (with respect to equivariant symmetric monoidal functors $\underline{C} \to \underline{C}'$ between symmetric monoidal categories with symmetric monoidal Γ actions) transformation*

$$\text{Spt}\,(\underline{\text{Fun}}\,(\underline{E},\underline{C})) \to F(N.\underline{E};\text{Spt}\,(\underline{C})),$$

which is an equivalence if $\underline{E}$ contains an initial or terminal object. Note we do not claim the transformation is an equivariant equivalence.

Proof: Suppose first that $\underline{E}$ and $\underline{C}$ are arbitrary categories. There is an evaluation functor $\underline{E} \times \underline{\text{Fun}}\,(\underline{E},\underline{C}) \to \underline{C}$, and an induced map $N.\underline{E} \times N.\text{Fun}\,(\underline{E},\underline{C})$, and hence an adjoint map

$$N.\underline{\text{Fun}}(\underline{E},\underline{C}) \to F(N.\underline{E}, N.\underline{C}).$$

The spectrum level statement follows from the observation that the adjoint is symmetric monoidal and the equivariance from the fact that the evaluation is equivariant. The equivalence follows since $e \in \underline{E}$ is either initial or terminal, the restriction functor

$$\underline{\text{Fun}}\,(\underline{E}, \underline{C}) \to \underline{\text{Fun}}\,(e, \underline{C}) \cong \underline{C}$$

is an equivalence of categories, and $N.\underline{E}$ is weakly contractible. Q.E.D.

We let K_0, K_1, and $K_2 : \underline{\mathcal{M}}^\Gamma \to \underline{\mathcal{S}}^\Gamma$ denote the bounded K–theory spectra $\underset{\sim}{K}(-, R)$, $\widehat{\underset{\sim}{K}}(-, R)$, and $\underset{\sim}{\mathcal{K}}(-, R)$, respectively, equipped with "naive" Γ action given by $\gamma(F, B, \varphi) = (F, B, \gamma \circ \varphi)$. Define functors

$$\Theta_i^{(i)} : \underline{\mathcal{M}}^\Gamma \to \underline{\mathcal{S}}^\Gamma$$

by

$$\Theta_i^{(i)}(X) = F(N.\underline{E\Gamma}, K_i(X)),$$

where Γ acts by conjugation of maps. Let K_0^Γ, K_1^Γ, and K_2^Γ denote the functors $\underset{\sim}{K}^\Gamma(\,;R)$, $\widehat{\underset{\sim}{K}}^\Gamma(\,;R)$, and $\underset{\sim}{\mathcal{K}}^\Gamma(-, R)$, respectively. A map $f : \underset{\sim}{X} \to \underset{\sim}{Y}$ of spectra with Γ action is said to be a weak Γ equivalence if each map $f^\Gamma : \underset{\sim}{X}^\Gamma \to \underset{\sim}{Y}^\Gamma$ is a weak equivalence; a natural transformation $N : F \to G$ of functors from a category $\underline{C}$ to $\underline{\mathcal{S}}^\Gamma$ is a weak Γ equivalence if $N(x)$ is a weak Γ equivalence for all $x \in \underline{C}$. Let $F, G : \underline{C} \to \underline{\mathcal{S}}^\Gamma$ be functors; a *homotopy natural transformation* is a sequence of functors $H_i : \underline{C} \to \underline{\mathcal{S}}^\Gamma$, $i = 0, \ldots, n$ with $H_0 = F$, $H_n = G$, and natural transformations $\rho_i : H_i \to H_{i-1}$, $i = 1, \ldots, n$, and $q_i : H_i \to H_{i+1}$, $i = 0, \ldots, n-1$, so that each ρ_i is weak Γ equivalence. If each q_i is also a weak Γ equivalence, we say the homotopy natural homotopy transformation is a *homotopy weak equivalence* from F to G. A homotopy natural transformation is just a morphism in the homotopy category of $\underline{\mathcal{S}}^\Gamma$ valued diagrams on $\underline{C}$ obtained by inverting the weak Γ equivalences of functors. Note that we have taken the strongest notion of weak Γ equivalence; weak Γ equivalences become equivariant homotopy equivalences after geometric realization. We note that we can compose homotopy natural transformations just as in the non equivariant case introduced in § I.

Let

$$\ell_0(X) = \text{Spt}\,(\underline{\text{Fun}}\,(\underline{E\Gamma}, i\mathcal{C}_X(R))),$$
$$\ell_1(X) = \text{Spt}\,(\underline{\text{Fun}}\,(\underline{E\Gamma}, i\widehat{\mathcal{C}}_X(R))),$$
$$\ell_2(X) = \varinjlim_k \Omega^k \ell_1(X \times E^k),$$

where E^k denotes Euclidean space with trivial Γ action. Note that ℓ_i is a functor from $\underline{\mathcal{M}}^\Gamma$ to $\underline{\mathcal{S}}^\Gamma$. The natural transformation of VI.14 provides an equivariant map $\ell_i(X) \to F(N.\underline{E\Gamma}, K_i(X))$; moreover, K_0 and K_1 are defined as the spectra associated to full, Γ invariant, symmetric monoidal subcategories of $i\mathcal{C}_X(R)$ and $i\widehat{\mathcal{C}}_X(R)$, respectively, so we obtain equivariant maps $K_i^\Gamma \to \ell_i$. These maps are all natural in X, so we obtain natural transformations $\eta_i^{(i)} : K_i^\Gamma \to \Theta_i^{(i)}$ of functors from $\underline{\mathcal{M}}^\gamma$ to $\underline{\mathcal{S}}^\Gamma$. Let $\underline{\Gamma\text{-sets}}^p \subseteq \underline{\mathcal{M}}^\Gamma$ denote the full subcategory of discrete metric spaces with free Γ action.

Proposition VI.15 $\eta_i^{(i)}$ *restricted to* $\underline{\Gamma\text{-sets}}^p$ *is a weak* Γ *equivalence.*

Proof: The natural maps $(K_i^\Gamma(X))^\Gamma \to (K_i^\Gamma(X))^{h\Gamma}$ are weak equivalences by VI.13. For any Γ spectrum $\underset{\sim}{X}$ whatsoever, $F(N.\underline{E\Gamma}; \underset{\sim}{X})^\Gamma \to F(N.\underline{E\Gamma}; \underset{\sim}{X})^{h\Gamma}$ can be identified with the map $F(N.\underline{E\Gamma}; \underset{\sim}{X})^\Gamma \to F(N.\underline{E\Gamma} \times N.\underline{E\Gamma}; \underset{\sim}{X})^\Gamma$ induced by the projection map $N.\underline{E\Gamma} \times N.\underline{E\Gamma} \to N.\underline{E\Gamma}$ on the first factor, which is an equivariant homotopy equivalence, so $F(N.\underline{E\Gamma}, \underset{\sim}{X})^\Gamma \to F(N.\underline{E\Gamma}, \underset{\sim}{X})^{h\Gamma}$ is an equivalence. Consequently, $\Theta_i^{(i)}(X)^\Gamma \to \Theta_i^{(i)}(X)^{h\Gamma}$ is a weak equivalence. By I.2 (c), to check that $\eta_i^{(i)}$ is a weak Γ equivalence, it now suffices to check that $\eta_i^{(i)}$ is an equivalence non equivariantly. This now follows directly from VI.2. Q.E.D.

Recall that in the proofs of III.14 and III.15, we constructed functors $\mu_i : \underline{\Gamma\text{-sets}}^p \to \underline{\mathcal{S}}$, $i = 0, 1, 2$, and natural equivalences $\underset{\sim}{h}^{\ell f}(-, \mu_i(\text{point})) \to \mu_i(-)$ and $K_i(-) \to \mu_i(-)$. For instance, μ_0 was defined on an object X to be $\underset{U \in \mathcal{F}(X)^{op}}{\underleftarrow{\text{holim}}} \underset{\sim}{K}(U; R)$. Note $\mu_0(\text{point}) \cong \underset{\sim}{K}(R)$, $\mu_1(\text{point}) \cong \underset{\sim}{\widehat{K}}(R)$, and $\mu_2(\text{point}) \cong \underset{\sim}{\mathcal{K}}(R)$. The functoriality of the construction shows that all the functors can be viewed as functors from $\underline{\Gamma\text{-sets}}^p$ to $\underline{\mathcal{S}}^\Gamma$, and that the natural transformations are functorial. Define $\Theta_i^{(ii)}$ and $\Theta_i^{(iii)}$, $i = 0, 1, 2$, from $\underline{\Gamma\text{-sets}}^p$ to $\underline{\mathcal{S}}^\Gamma$ by

$$\Theta_i^{(ii)}(X) = F(N.\underline{E\Gamma}, \mu_i(X))$$

and

$$\Theta_i^{(iii)}(X) = F(N.\underline{E\Gamma}, \underset{\sim}{h}^{\ell f}(X, \mu_i(\text{point}))).$$

We define $\eta_i^{(ii)}$ and $\eta_i^{(iii)}$ to be the natural transformations

$$F(N.\underline{E\Gamma}, K_i(-)) \to F(N.\underline{E\Gamma}, \mu_i(-))$$

and

$$F(N.\underline{E\Gamma}, \underset{\sim}{h}^{\ell f}(-, \mu_i(\text{point}))) \to F(N.\underline{E\Gamma}, \mu_i(-))$$

induced from the transformations above via the functor $F(N.\underline{E\Gamma}, -)$. Thus $\eta_i^{(ii)} : \Theta_i^{(i)} \to \Theta_i^{(ii)}$ and $\eta_i^{(iii)} : \Theta_i^{(iii)} \to \Theta_i^{(ii)}$.

Proposition VI.16 *$\eta_i^{(ii)}$ and $\eta_i^{(iii)}$ are both weak Γ equivalences.*

Proof: The maps $(\Theta_i^{(i)})^\Gamma \to (\Theta_i^{(i)})^{h\Gamma}$, $(\Theta_i^{(ii)})^\Gamma \to (\Theta_i^{(iii)})^{h\Gamma}$, and $(\Theta_i^{(iii)})^\Gamma \to (\Theta_i^{(iii)})^{h\Gamma}$ are all weak equivalences; this was shown for $\Theta_i^{(i)}$ in VI.15, and the other two cases follow by the identical argument. Consequently, it suffices to show that $\eta_i^{(ii)}$ and $\eta_i^{(iii)}$ are equivalences, non equivariantly. But this follows from the proofs of III.14 and III.15. Q.E.D.

Finally, let $\Theta_i^{(iv)}$ be the functor $\underset{\sim}{h}^{\ell f}(-, \mu_i(\text{point}))$. Then the map $N.\underline{E\Gamma} \to$ point induces a natural transformation

$$\underset{\sim}{h}^{\ell f}(-, \mu_i(\text{point})) \to F(N.\underline{E\Gamma}, \underset{\sim}{h}^{\ell f}(-, \mu_i(\text{point}))),$$

i.e., a natural transformation $\Theta_i^{(iv)} \xrightarrow{\eta_i^{(iv)}} \theta_i^{(iii)}$.

Proposition VI.17. *If Γ is torsion free, then $\eta_i^{(iv)}$ is a weak Γ equivalence.*

Proof: Follows from II.22 by the argument used in proving the similar result for $\Theta_i^{(i)}$ in VI.15. Q.E.D.

The diagram

$$\underset{\sim}{h}^{\ell f}(-, \mu_i(\text{point})) \overset{\eta_i^{(iii)} \circ \eta_i^{(iv)}}{\longrightarrow} \Theta_i^{(ii)} \overset{\eta_i^{(ii)} \circ \eta_i^{(i)}}{\longleftarrow} K_i^\Gamma$$

defines a weak Γ equivalence of functors

$$\underset{\sim}{h}^{\ell f}(-, \mu_i(\text{point})) = \underset{\sim}{h}^{\ell f}(-, K_i(\text{point})) \to K_i^\Gamma$$

from $\underline{\Gamma\text{-sets}}^p$ to $\underline{S}^\Gamma$.

Now let X be a metric space with isometric left Γ action, and let $\mathcal{U} = \{U_\alpha\}_{\alpha \in A}$ be a Γ invariant locally finite covering. Then as in § III, we obtain a simplicial object $N.\mathcal{U}$ in the category $\underline{\mathcal{M}}^\Gamma$. Applying the functors $\underset{\sim}{\mathcal{K}}^\Gamma$, we obtain an equivariant map

$$\mathcal{A}^\Gamma(\mathcal{U}) : |\underset{\sim}{\mathcal{K}}^\Gamma(N.\mathcal{U}; R)| \to \underset{\sim}{\mathcal{K}}^\Gamma(X; R).$$

Suppose that we have a smooth compact closed $K(\Gamma, 1)$ manifold X; equip it with a Riemannian metric and a triangulation τ. Note that because of the compactness of X, the simplices in the triangulation have uniformly bounded diameter, hence the same is true when we consider the induced Riemannian metric and triangulation $\hat{\tau}$ on the universal cover $\tilde{X}$. Let $\mathcal{U}$ denote the covering of $\tilde{X}$ by simplices of $\hat{\tau}$; then $\mathcal{U}$ is Γ invariant and locally finite. We examine the fixed point set

$$|\underset{\sim}{K}^{\Gamma}(N.\mathcal{U};R)|^{\Gamma} = |\underset{\sim}{K}^{\Gamma}(N.\mathcal{U};R)^{\Gamma}|.$$

For each k, the Γ metric space $N_k\mathcal{U}$ is a disjoint union of metric spaces of uniformly bounded diameter, by the above remark. For a metric space X, let $\pi_0 X$ be defined as in II.18. Because of the uniformly bounded diameter of the components of $N_k\mathcal{U}$, the map $N_k\mathcal{U} \xrightarrow{P} \pi_0 N_k\mathcal{U}$ is a morphism in $\underline{\mathcal{M}}^{\Gamma}$. By choosing an element in $P^{-1}(x)$ for each element in a set of orbit representatives for the (free) Γ action on $\pi_0(N_k\mathcal{U})$, we obtain a Γ equivariant map $\pi_0(N_k\mathcal{U}) \to N_k(\mathcal{U})$ in $\underline{\mathcal{M}}^{\Gamma}$, and it follows easily from III.7, VI.4, and VI.12 that $\underset{\sim}{\mathcal{K}}(P,R)$ is a Γ equivariant equivalence. Consequently, the map

$$|\underset{\sim}{\mathcal{K}}^{\Gamma}(N.\mathcal{U},R)| \to |\underset{\sim}{\mathcal{K}}^{\Gamma}(\pi_0 N.\mathcal{U},R)|$$

is a Γ equivariant equivalence of spectra, so

$$|\underset{\sim}{\mathcal{K}}^{\Gamma}(N.\mathcal{U},R)|^{\Gamma} \to |\underset{\sim}{\mathcal{K}}^{\Gamma}(\pi_0 N.\mathcal{U},R)|^{\Gamma}$$

is a weak equivalence of spectra. But,

$$|\underset{\sim}{\mathcal{K}}^{\Gamma}(\pi_0 N.\mathcal{U},R)|^{\Gamma} \xrightarrow{\sim} |\underset{\sim}{\mathcal{K}}^{\Gamma}(\pi_0 N.\mathcal{U},R)^{\Gamma}|,$$

and by VI.12, we have an equivalence

$$|\underset{\sim}{\mathcal{K}}^{\Gamma}(\pi_0 N.\mathcal{U},R)^{\Gamma}| \xrightarrow{\sim} |\underset{\sim}{\mathcal{K}}(\Gamma\backslash\pi_0 N.\mathcal{U},R)|.$$

But one readily checks that $\pi_0 N.\mathcal{U}$ is the simplicial set associated with the barycentric subdivision of the triangulation $\hat{\tau}$, and hence that $\Gamma\backslash\pi_0 N.\mathcal{U}$ is the simplicial set associated with the barycentric subdivision of τ. Consequently, by III.15, we have an equivalence

$$|\underset{\sim}{\mathcal{K}}(\Gamma\backslash\pi_0 N.\mathcal{U},R)| \xrightarrow{\sim} |\underset{\sim}{h}^{\ell f}(\Gamma\backslash\pi_0 N.\mathcal{U},\underset{\sim}{\mathcal{K}}(R))|,$$

and II.17, $|\underset{\sim}{h}^{\ell f}(\Gamma\backslash\pi_0 N.\mathcal{U},\underset{\sim}{\mathcal{K}}(R))|$ is naturally equivalent to $\underset{\sim}{h}^{\ell f}(\Gamma\backslash X,\underset{\sim}{\mathcal{K}}(R))$.

We summarize:

Proposition VI.18 *There is a well defined (up to homotopy) equivalence*

$$\underset{\sim}{h}^{\ell f}(\Gamma\backslash X,\underset{\sim}{\mathcal{K}}(R)) \to |\underset{\sim}{\mathcal{K}}^{\Gamma}(N.\mathcal{U},R)|^{\Gamma}.$$

On the other hand, it follows from VI.10 that $\underset{\sim}{K}^{\Gamma}(X,R)^{\Gamma} \cong \underset{\sim}{K}(R[\Gamma])$. *Thus, the map induced on fixed point sets by* $\mathcal{A}^{\Gamma}(\mathcal{U})$ *may be identified with a homotopy class of maps from*

$$\underset{\sim}{h}^{\ell f}(\Gamma\backslash X, \underset{\sim}{K}(R)) \to \underset{\sim}{K}(R[\Gamma])\,.$$

Since $\Gamma\backslash X$ *is compact,*

$$\underset{\sim}{h}^{\ell f}(\Gamma\backslash X, \underset{\sim}{K}(R)) \cong (\Gamma\backslash X)_+ \wedge \underset{\sim}{K}(R)\,,$$

and the classifying map for the universal cover defines an equivalence $\Gamma\backslash X_+ \to B\Gamma_+$. *Therefore, the induced map on fixed point sets of* $\mathcal{A}^{\Gamma}(\mathcal{U})$ *determines a homotopy class of maps from* $B\Gamma_+ \wedge \underset{\sim}{K}(R)$ *to* $\underset{\sim}{K}(R[\Gamma])$.

Proposition VI.19 *The above mentioned homotopy class of maps is independent of the manifold* X, *of the metric chosen, and of the triangulation chosen. We say a map* $B\Gamma_+ \wedge \underset{\sim}{K}(R)$ *to* $\underset{\sim}{K}(R[\Gamma])$ *is an assembly map if it belongs to this homotopy class.*

Proof: Elementary, we leave the proof to the reader. Q.E.D.

Remark: The assembly map can, of course, be defined without reference to X; or to the metric or triangulation; see [27]. With enough work, one can even construct an assembly map in our context (i.e., realized as a fixed point map of a Γ equivariant map of bounded K theory spectra) with reference only to Γ. We do not carry this out here, in the interest of brevity, but return to this point in a later paper in this series.

Definition VI.20 *We say a map* $f : \underset{\sim}{W} \to \underset{\sim}{Z}$ *of spectra is split injective if there is a map* $g : \underset{\sim}{Z} \to \underset{\sim}{W}$ *so that* $g \circ f$ *is homotopic to the identity map of* $\underset{\sim}{W}$. *Note that this means* $\pi_i(f)$ *is the inclusion on a direct summand.*

Proposition VI.21 *Suppose* $\underset{\sim}{W}$ *and* $\underset{\sim}{Z}$ *are spectra with* Γ *action, and let* $f : \underset{\sim}{W} \to \underset{\sim}{Z}$ *be an equivariant map of spectra. Suppose further that*

(A) *The map* $\underset{\sim}{W}^{\Gamma} \to \underset{\sim}{W}^{h\Gamma}$ *is an equivalence*

(B) f *is an equivalence, non equivariantly.*

Then the induced map on fixed point sets, $f^{\Gamma} : \underset{\sim}{W}^{\Gamma} \to \underset{\sim}{Z}^{\Gamma}$, *is split injective.*

Proof: Consider the following square.

$$\text{(I)} \qquad \begin{array}{ccc} \underset{\sim}{W}^{\Gamma} & \xrightarrow{f^{\Gamma}} & \underset{\sim}{Z}^{\Gamma} \\ \downarrow & & \downarrow \\ \underset{\sim}{W}^{h\Gamma} & \xrightarrow{f^{h\Gamma}} & \underset{\sim}{Z}^{h\Gamma} \end{array}$$

The left hand vertical arrow is an equivalence by Hypothesis (A). The lower horizontal arrow is an equivalence by I.2 (c), using Hypothesis (B). Therefore, the composite $\underset{\sim}{W}^{\Gamma} \xrightarrow{f^{\Gamma}} \underset{\sim}{Z}^{\Gamma} \to \underset{\sim}{Z}^{h\Gamma}$ is an equivalence, and composing it on the left with any homotopy inverse gives the result. Q.E.D.

Lemma VI.22 *Let Γ be a group, X a closed compact smooth $K(\Gamma,1)$ manifold, equipped with a Riemannian metric and a triangulation τ. Let $\mathcal{U}$ be the covering of $\tilde{X}$ by the simplices of $\hat{\tau}$, the associated triangulation of $\tilde{X}$. Then the natural map*

$$|\underset{\sim}{\mathcal{K}}^{\Gamma}(N.\mathcal{U},R)|^{\Gamma} \to |\underset{\sim}{\mathcal{K}}^{\Gamma}(N.\mathcal{U},R)|^{h\Gamma}$$

is an equivalence.

Proof: In the discussion preceding VI.18 above, we saw that the projection map

$$\underset{\sim}{\mathcal{K}}^{\Gamma}(P,R) : |\underset{\sim}{\mathcal{K}}^{\Gamma}(N.\mathcal{U},R)| \to |\underset{\sim}{\mathcal{K}}^{\Gamma}(\pi_0 N.\mathcal{U},R)|$$

is an equivariant equivalence. On the other hand, VI.15, VI.16, and VI.17 show that $|\underset{\sim}{\mathcal{K}}^{\Gamma}(\pi_0 N.\mathcal{U},R)|$ is equivariantly equivalent to $|\underset{\sim}{h}^{\ell f}(\pi_0 N.\mathcal{U},\underset{\sim}{\mathcal{K}}(R))|$. The result now follows by II.23. Now Γ satisfies the hypothesis of II.23 since $\tilde{X}$ is the required finite dimensional contractible Γ complex. Q.E.D.

Lemma VI.23 *Let $\Gamma, X, \mathcal{U}$ be as in VI.22. Suppose further that the natural map*

$${}^{b}\underset{\sim}{h}^{\ell f}(\tilde{X},\underset{\sim}{\mathcal{K}}(R)) \to \underset{\sim}{\mathcal{K}}(\tilde{X},R)$$

of III.20 is an equivalence. Then $\mathcal{A}^{\Gamma}(\mathcal{U})$ is an equivalence, non equivariantly.

Proof: In view of VI.2, it suffices to show that $\mathcal{A}(\mathcal{U})$ is an equivalence. Consider the following commutative square

$$\begin{array}{ccc} |{}^{b}\underset{\sim}{h}^{\ell f}(N.\mathcal{U},\underset{\sim}{\mathcal{K}}(R))| & \longrightarrow & {}^{b}\underset{\sim}{h}^{\ell f}(\tilde{X},\underset{\sim}{\mathcal{K}}(R)) \\ \downarrow & & \downarrow \\ |\underset{\sim}{\mathcal{K}}(N.\mathcal{U},R)| & \xrightarrow{\mathcal{A}(\mathcal{U})} & \underset{\sim}{\mathcal{K}}(\tilde{X},R) \end{array}$$

The right hand vertical arrow is an equivalence by hypothesis. To study the left hand vertical arrow, we consider the projection $P : N.\mathcal{U} \to \pi_0 N.\mathcal{U}$, and the commutative square

$$\text{(II)}\qquad \begin{array}{ccc} |{}^b\underset{\sim}{h}^{\ell f}(N.\mathcal{U}, \underset{\sim}{\mathcal{K}}(R))| & \longrightarrow & |{}^b\underset{\sim}{h}^{\ell f}(\pi_0 N.\mathcal{U}, \underset{\sim}{\mathcal{K}}(R))| \\ \downarrow & & \downarrow \\ |\underset{\sim}{\mathcal{K}}(N.\mathcal{U}, R)| & \longrightarrow & |\underset{\sim}{\mathcal{K}}(\pi_0 N.\mathcal{U}, R)|. \end{array}$$

The lower horizontal arrow is an equivalence by an application of III.7, using the uniform boundedness of the simplices in the triangulation. The upper horizontal arrow is an equivalence since

$${}^b\underset{\sim}{h}^{\ell f}(N.\mathcal{U}, \underset{\sim}{\mathcal{K}}(R)) \to \underset{\sim}{h}^{\ell f}(N.\mathcal{U}, \underset{\sim}{\mathcal{K}}(R))$$

and

$${}^b\underset{\sim}{h}^{\ell f}(\pi_0 N.\mathcal{U}, \underset{\sim}{\mathcal{K}}(R)) \to \underset{\sim}{h}^{\ell f}(\pi_0 N.\mathcal{U}, \underset{\sim}{\mathcal{K}}(R))$$

are both equivalences (using again the uniformly bounded diameter of the components $N.\mathcal{U}$) and since $P : N.\mathcal{U} \to \pi_0 N.\mathcal{U}$ is a proper homotopy equivalence (use II.14). The right hand vertical arrow is an equivalence by III.15 and the fact that ${}^b\underset{\sim}{h}^{\ell f}(-, \underset{\sim}{\mathcal{K}}(R))$ and $\underset{\sim}{h}^{\ell f}(-, \underset{\sim}{\mathcal{K}}(R))$ coincide on discrete metric spaces. Consequently the left hand vertical arrow in (II) is an equivalence, hence so is the one in (I). Therefore, to prove that $\mathcal{A}(\mathcal{U})$ is an equivalence, it will suffice to show that the upper horizontal arrow is an equivalence. To see this, we first observe that X and all the metric spaces $N_k\mathcal{U}$ satisfy the hypotheses of II.21, so that it suffices to check that

$$|\underset{\sim}{h}^{\ell f}(N.\mathcal{U}, \underset{\sim}{\mathcal{K}}(R))| \to \underset{\sim}{h}^{\ell f}(\tilde{X}, \underset{\sim}{\mathcal{K}}(R))$$

is an equivalence. For $\varepsilon > 0$, let $B_\varepsilon\mathcal{U}$ denote the covering consisting of the ε neighborhoods of the simplices of the triangulation $\hat{\tau}$. For sufficiently small $\varepsilon > 0$, the map $N.\mathcal{U} \to N.B_\varepsilon\mathcal{U}$ is a proper homotopy equivalence, so it will suffice to show that

$$|\underset{\sim}{h}^{\ell f}(N.B_\varepsilon\mathcal{U}, \underset{\sim}{\mathcal{K}}(R))| \to \underset{\sim}{h}^{\ell f}(\tilde{X}, \underset{\sim}{\mathcal{K}}(R))$$

is an equivalence. From the choice of ε, for any set $B_\varepsilon U$, $U \in \mathcal{U}$, $\Gamma B_\varepsilon \cong \coprod_{\gamma \in \Gamma} \gamma B_\varepsilon U$; let $\widetilde{\mathcal{U}}$ denote the covering of $\tilde{X}$ by the sets of the form $\Gamma \cdot B_\varepsilon U$. Note that $\widetilde{\mathcal{U}}$ is a finite covering. The fact that $\Gamma B_\varepsilon U$ is a disjoint union of the spaces $\gamma B_\varepsilon U$ allows one to show that the map

$$|\underset{\sim}{h}^{\ell f}(N.B_\varepsilon\mathcal{U}, \underset{\sim}{\mathcal{K}}(R))| \to |\underset{\sim}{h}^{\ell f}(N.\widetilde{\mathcal{U}}, \underset{\sim}{\mathcal{K}}(R))|$$

induced by the map of coverings $B_\varepsilon \to \gamma B_\varepsilon U$ is an isomorphism of spectra, and the map

$$\mathcal{A}(\widetilde{\mathcal{U}}) : |\underset{\sim}{h}^{\ell f}(N.\widetilde{\mathcal{U}}, \underset{\sim}{\mathcal{K}}(R))| \to \underset{\sim}{h}^{\ell f}(\tilde{X}, \underset{\sim}{\mathcal{K}}(R))$$

is an equivalence by iterated application of II.15. Q.E.D.

Theorem VI.24 *Suppose* Γ *is a group, and that* X *is a* $K(\Gamma, 1)$ *manifold. If* $\tilde{X}$ *admits a* Γ *invariant Riemannian metric so that the transformation*

$${}^b\underset{\sim}{h}^{\ell f}(\tilde{X}, \underset{\sim}{\mathcal{K}}(R)) \to \underset{\sim}{\mathcal{K}}(\tilde{X}, R)$$

of III.20 is an equivalence, then any assembly map $B\Gamma_+ \wedge \underset{\sim}{\mathcal{K}}(R) \to \underset{\sim}{\mathcal{K}}(R[\Gamma])$ *is a split injection.*

Proof: VI.22 and VI.23 show that

$$\mathcal{A}^\Gamma(\mathcal{U}) : |\underset{\sim}{\mathcal{K}}^\Gamma(N.\mathcal{U}, R)| \to \underset{\sim}{\mathcal{K}}^\Gamma(X, R)$$

satisfies the hypotheses of VI.21, where $\mathcal{U}$ is the covering associated to any triangulation of X. Q.E.D.

Theorem VI.25 *let* Γ *be a discrete, cocompact, torsion free subgroup of a connected Lie group* G. *Then any assembly map* $B\Gamma_+ \wedge \underset{\sim}{\mathcal{K}}(R) \to \underset{\sim}{\mathcal{K}}(R[\Gamma])$ *is a split injection.*

Proof: Let $K \subseteq G$ be any maximal compact subgroup, and let $\tilde{X} = G/K$ be equipped with a left invariant Riemannian metric. Then the metric is also Γ invariant, and when $\tilde{X}$ is equipped with this metric ${}^b\underset{\sim}{h}^{\ell f}(\tilde{X}; \underset{\sim}{K}(R)) \to \underset{\sim}{K}(\tilde{X}; R)$ is an equivalence by V.21. Q.E.D.

References

[1] J.F. Adams, *Stable Homotopy and Generalized Homology*, Chicago Lectures in Mathematics, University of Chicago Press 1974.

[2] M.F. Atiyah, *Characters and cohomology of finite groups*, Inst. Hautes Études Sci. Publ. Math. **9** (1961), 23–64.

[3] M.F. Atiyah and G.B. Segal, *Equivariant K–theory and completion*, Journal of Differential Geometry **3** (1969), 1–18.

[4] D. Barden, *Structure of Manifolds,* Thesis, Cambridge University 1963.

[5] H. Bass, *Algebraic K–theory,* Mathematics Lecture Notes Series, W.A. Benjamin 1968.

[6] M.G. Bökstedt, W.C. Hsiang, and I. Madsen, *The cyclotomic trace and the K–theoretic analogue of Novikov's conjecture,* Proc. Nat. Acad. Sci. U.S.A. **86** (1989), 8607–8609.

[7] A. Borel and J.C. Moore, *Homology theory for locally compact spaces,* Mich. Math. J. **7** (1960), 137–159.

[8] A. Borel and J.P. Serre, *Cohomologie d'immeubles et de groupes S–arithmétiques,* Topology **15** (1976), 211–232.

[9] N. Bourbaki, *Elements of Mathematics: Lie Groups and Lie Algebras,* Chapters 1–III, Springer Verlag 1989.

[10] A.K. Bousfield and E.M. Friedlander, *Homotopy theory of Γ–spaces, spectra, and bisimplicial sets,* in "Geometric Applications of Homotopy Theory II," Lecture Notes in Mathematics **658**, Springer Verlag 1978, 80–130.

[11] A.K. Bousfield and D.M. Kan, *Homotopy, Limits, Completions, and Localizations,* Lecture Notes in Mathematics **304**, Springer Verlag 1972.

[12] G. Carlsson, *Segal's Burnside ring conjecture and the homotopy limit problem,* in "Homotopy Theory," E. Rees and J.D.S. Jones editors, London Math. Soc. Lect. Notes **117**, Cambridge University Press 1987, 6–34.

[13] G. Carlsson, *On the algebraic K-theory of infinite product categories,* submitted to K-theory.

[14] W.G. Dwyer and D.M. Kan, *A classification theorem for diagrams of simplicial sets,* Topology **23** (1984), 139–155.

[15] S. Eilenberg and G.M. Kelly, *Closed categories,* in "Proceedings of the Conference on Categorical Algebra, La Jolla," Springer Verlag 1965, 421–562.

[16] F. T. Farrell and W.C. Hsiang, *On Novikov's conjecture for non-positively curved manifolds I,* Annals of Math. **113** (1981), 109–209.

[17] F. T. Farrell and W.C. Hsiang, *The Whitehead group of poly–(finite or cyclic) groups,* J. Lond. Math. Soc. **24** (1981), 308–324.

[18] F. T. Farrell and L. Jones, *K–theory and dynamics I*, Annals of Math. **124** (1986), 531–569.

[19] F. T. Farrell and L. Jones, *K–theory and dynamics II*, Annals of Math. **126** (1987), 451–493.

[20] P. Freyd, *Abelian Categories*, Harper and Row 1964.

[21] E.M. Friedlander, *Etale Homotopy of Simplicial Schemes*, Annals of Math. Studies, Princeton University Press 1982.

[22] S.M. Gersten, *On the spectrum of algebraic K–theory*, Bull. Amer. Math. Soc. **78** (1972), 216–219.

[23] M.J. Greenberg and J.R. Harper, *Algebraic Topology: a First Course*, Benjamin Cummings 1981.

[24] A.E. Hatcher and J.B. Wagoner, *Pseudoisotopies of compact manifolds*, Astérisque **6** 1973.

[25] S. Helgason, *Differential Geometry, Lie Groups, and Symmetric Spaces*, Academic Press 1978.

[26] M. Karoubi, *Foncteurs derivés et K–théorie*, Lecture Notes in Mathematics **136**, Springer Verlag 1970.

[27] J.L. Loday, *K–théorie algebrique et representations de groupes*, Ann. Sci. Norm. Sup. **9** (1976).

[28] J.P. May, *Simplicial Objects in Algebraic Topology*, Van Nostrand 1967.

[29] J.P. May, *Pairings of categories and spectra*, J. Pure and Applied Algebra **19** (1980), 299–346.

[30] J.P. May, *E_∞ ring spaces and E_∞ ring spectra*, Lecture Notes in Math. **577**, Springer Verlag 1977.

[31] B. Mazur, *Differential topology from the point of view of simple homotopy theory*, Publ. Math. Inst. Hautes Études Sci. **15**, 1963.

[32] P.F. Mostad, *Bounded K–theory of the Bruhat–Tits building for the special linear group over the p–adics with applications to the assembly map*, Thesis, Princeton University 1991.

[33] E.K. Pedersen and C. Weibel, *A non–connective delooping of algebraic K–theory*, Lecture Notes in Math. **1126**, Springer Verlag 1985, 166–181.

[34] E.K. Pedersen and C. Weibel, *K–theory homology of spaces,* in "Algebraic Topology," Lecture Notes in Math. **1370**, Springer Verlag 1989, 346–361.

[35] D.G. Quillen, *Higher algebraic K–theory I,* Lecture Notes in Math. **341**, Springer Verlag 1973.

[36] D.G. Quillen, *On the cohomology and K–theory of the general linear group over a finite field,* Annals of Math. **28** (1972), 552–568.

[37] F.S. Quinn, *Applications of topology with control,* Proceedings of the International Congress of Math., Berkeley 1986, 598–605.

[38] M.S. Ragunathan, *Discrete Subgroups of Lie Groups,* Springer Verlag 1972.

[39] A.A. Ranicki, *Algebraic L–theory and topological manifolds,* Cambridge Tracts in Mathematics 102, Cambridge University Press 1992.

[40] G.B. Segal, *Equivariant stable homotopy theory,* Actes Congrès Int. des. Math., Tome 2, 1970, 59–63.

[41] A.A. Suslin, *On the K–theory of algebraically closed fields,* Inv. Math. **73** (1983), 241–245.

[42] R.W. Thomason, *Homotopy colimits in the category of small categories,* Math. Proc. Camb. Phil. Soc. **85** (1979), 91–109.

[43] R.W. Thomason, *First quadrant spectral sequences in algebraic K–theory via homotopy colimits,* Comm. in Algebra **10** (1982), 1589–1668.

[44] R.W. Thomason, *The homotopy limit problem,* Proc. of the Northwestern Homotopy Theory Conf., Contemp. Math. **19** (1983), 407–420.

[45] R.W. Thomason, *Algebraic K–theory and étale cohomology,* Ann. Sci. Ec. Norm. Sup. **18** (1985), 437–552.

[46] J. B. Wagoner, *Delooping classifying spaces in algebraic K–theory,* Topology **11** (1972), 349–370.

[47] F. Waldhausen, *Algebraic K–theory of generalized free products,* Annals of Math. **108** (1978), 135–256.

[48] F. Waldhausen, *Algebraic K–theory of spaces,* in "Algebraic and Geometric Topology," Lecture Notes in Mathematics **1126**, Springer Verlag 1985, 318–419.

[49] C.T.C. Wall, *Finiteness obstructions for CW–complexes I,* Annals of Math. **81** (1965), 56–59.

[50] C.T.C. Wall, *Surgery on Compact Manifolds,* Academic Press 1970.

DEPARTMENT OF MATHEMATICS, STANFORD UNIVERSITY, STANFORD CA94305, USA

email: `gunnar@gauss.stanford.edu`

Rigidity of the index on open manifolds

Jürgen Eichhorn

1 Introduction

Let (M^n, g) be open complete, $(S, (.,.), \nabla) \longrightarrow M$ a Clifford bundle, D the generalized Dirac operator. D depends on ∇ and g, so we write $D = D^{\nabla, g}$. Let $\mathcal{I}_D$ be the corresponding index form, $\mathbf{m}$ a functional on $\mathcal{I}_D$, $\mathcal{B}$ an operator algebra, ζ a cyclic cohomology class, $Ind\, D \in K_i\mathcal{B}$, $ind_t D = \langle \mathcal{I}_D, \mathbf{m} \rangle$, $ind_a D = \langle Ind\, D, \zeta \rangle$ the topological and analytical index respectively. Rigidity amounts to the description of the admissible variations of ∇, g such that $[\mathcal{I}_D]$, $Ind\, D$, $ind_t D$, $ind_a D$ remain unchanged. We consider spaces $\mathcal{C}$ of Clifford connections. A special case is given by g of bounded geometry up to order $k > \frac{n}{2} + 1$ and the space $\mathcal{C}(B_k)$ of Clifford connections of bounded geometry up to order k, $\nabla \in \mathcal{C}(B_k)$. The key approach is to introduce suitable Banach or Sobolev topologies by means of uniform structures on $\mathcal{C}(B_k)$, thus obtaining spaces $\mathcal{C}^r(B_k)$. The choice of the uniform structure has to be adapted to the choice of i, $\mathcal{B}$, $\mathbf{m}$, ζ. Then we obtain many rigidity theorems, depending on the choice of $\mathcal{B}$, e.g. the following :

Theorem *Assume (M^n, g) open, complete, of bounded geometry up to order k, $k \geq r > n/2 + 1$, $\nabla, \nabla_1 \in \mathcal{C}(B_k)$, $\nabla_1 \in$ component of $\nabla \in \mathcal{C}^r(B_k)$, $D = D^{\nabla}$, $D_1 = D^{\nabla_1}$, S and D graded, $D : \Omega^{0,1} \longrightarrow \Omega^{0,0} = L_2(S)$ Fredholm. Then D_1 is Fredholm too and*

$$Ind\, D^+ = Ind\, D_1^+.$$

Remark For ∇, $\nabla_1 \in \mathcal{C}(B_k)$ arbitrary or other topologies this is in general definitely wrong. On compact manifolds this is always trivially true. For more general spaces of Clifford connections one has to describe what is a continuous family of deformations D^{∇_t}. As a matter of fact, for $\nabla_t = t\nabla + (1-t)\nabla_1$ it is in general not. Moreover, in general index theories on open manifolds the index is a real number. It is is by no means clear whether the real number $ind\, D^{\nabla_t}$ should continuously vary or jump or be constant.

2 The general scheme of index theory

Given a Riemannian manifold (M^n, g), hermitian vector bundles $E, F \longrightarrow M$ and an elliptic operator $D : C_0^\infty(E) \longrightarrow C_0^\infty(F)$, the general scheme of index theory consists of the following. One chooses a suitable extension of D, an operator algebra $\mathcal{B}$, a functor $K_i(\mathcal{B})$, constructs an element $Ind\,D \in K_i(\mathcal{B})$, defines an element $\mathcal{I}_D$ of cohomological nature, mainly a differential form, defines pairings $\langle \mathcal{I}_D, \mathbf{m} \rangle$, $\langle Ind\,D, \zeta \rangle$ and sets $ind_t D = \langle \mathcal{I}_D, \mathbf{m} \rangle$, $ind_a D = \langle Ind\,D, \zeta \rangle$. In most cases $\langle Ind\,D, \zeta \rangle$ is a pairing with a cyclic cohomology class. This gives a diagram

$$\begin{array}{ccc} D & \longrightarrow & Ind\,D \in K_i(\mathcal{B}) \\ \downarrow & & \downarrow \\ \mathcal{I}_D & \to \langle \mathcal{I}_D, \mathbf{m} \rangle = ind_t D \overset{?}{=} & ind_a D = \langle Ind\,D, \zeta \rangle. \end{array}$$

The commutative closure of the diagram by an equality $ind_t D = ind_a D$ means the establishing of an index theorem.

Working out this program, one has to make concrete choices of the class of the D's, $\mathcal{B}$, i, $\zeta \in HC^*(\mathcal{B})$, and the functional $\mathbf{m}$, and one has to construct $\mathcal{I}_D$ and $IndD$. We start with very general definitions from [7], [8]. Let $S \longrightarrow M$ be a graded Clifford bundle, η the grading, $H = H_+ \oplus H_- = L_2(S)$ and $A \in L(H)$.

Definition 1 *A has* bounded propagation *if there is a number $R > 0$ such that for any $s \in H$*

$$supp(As) \cup supp(A^*s) \subseteq Pen(supp(s), R),$$

where $Pen(supp(s), R) = \{x \in M \mid d(x, supp(s)) \leq R\}$.

Example. If D is the generalized Dirac operator belonging to (S, ∇) then e^{itD} has bounded propagation.

Let $\mathcal{A} = \mathcal{A}_H$ be the set of all operators having bounded propagation. $\mathcal{A}$ is a unital *-algebra.

A positive operator $A \in L(H)$ is said to be *locally traceable*, if for all compactly supported continuous functions f on M, the operator fAf is of trace class. A general operator is locally traceable if it is a finite linear combination of positive locally traceable operators.

Lemma 2.1 *The set $\mathcal{B} = \mathcal{B}_H$ of all locally traceable operators with bounded propagation is a $*$-ideal in $\mathcal{A}_H$.* □

Definition 2 *Let A be a unital super algebra, η be a grading, B a super ideal in A. An element $F \in A$ will be called a generalized Fredholm operator if F is odd relative to the super structure, i.e., $F\eta + \eta F = 0$ and $F^2 - 1 \in B$.*

By well known constructions of algebraic K-theory there is an index $Ind\,F\in K_0(B)$ defined as follows. Namely, there exists an element $G\in A$ such that $G^2=1$ and $G-F\in B$. Then

$$Ind\,F=\left[G(\frac{1+\eta}{2})G\right]-\left[\frac{1-\eta}{2}\right]\in K_0(B).$$

Definition 3 *Let D be an unbounded self adjoint operator on H. D is called* generalized elliptic *if there is a constant $c>0$ such that for all $t\in\mathbf{R}$ the unitary operator e^{itD} belongs to $\mathcal{A}_H$, has propagation bound $\leq c\mid t\mid$, and if there is an $n>0$ such that $(1+D^2)^{-n}$ is locally traceable. D will be called an* even *elliptic operator if $D\eta+\eta D=0$.*

Example The generalized Dirac operator of S is generalized elliptic. Let W the algebra of functions on $\mathbf{R}$ having compactly supported Fourier transform. A chopping function is a smooth function $\psi:\mathbf{R}\to\mathbf{R}$ such that $\psi(x)\to\pm1$ for $x\to\pm\infty$ and ψ' belongs to W.

Lemma 2.2 *Chopping functions exist. If ψ is a chopping function then $\psi^2-1\in W$. If ψ_0 and ψ_1 are two chopping functions then $\psi_0-\psi_1\in W$.* □

Lemma 2.3 *Let D be a generalized elliptic operator. If $\phi\in W$ then $\phi(D)\in\mathcal{B}_H$. If ψ is a chopping function then $\psi(D)\in\mathcal{A}_H$.* □

Let D be even elliptic, ψ an odd chopping function, $F=\psi(D)$. Then $F\in\mathcal{A}_H$, $F^2-1\in\mathcal{B}_H$, $F\eta+\eta F=0$ and $Ind\,F\in K_0(\mathcal{B}_H)$ is well defined. According to [7], $Ind\,F$ is independent of the choice of ψ and we define $Ind\,D:=Ind\,\psi(D)$.

Definition 4 *Let $(D_t)_t$ be a family of elliptic operators parametrized by $t\in[a,b]$. $(D_t)_t$ is called a* continuous family *if it satisfies the following conditions.*

1. *The operators D_t have a common dense domain $\mathcal{D}$.*
2. *The graph norms on $\mathcal{D}$ induced by the operators D_t are equivalent.*
3. *The map $t\to D_t$ is continuous from $[a,b]$ to $L(\mathcal{D},H)$, where $\mathcal{D}$ is equipped with the graph norm.*

Proposition 2.1 *Let $(D_t)_t$ be a continuous family of even elliptic operators on H. Then the image of $Ind\,D_t$ in $K_0(\bar{\mathcal{B}}_H)$ is independent of t, where $\bar{\mathcal{B}}_H$ denotes the C^*-algebra obtained as the norm closure of $\bar{\mathcal{B}}_H$ in $L(H)$.*

See [7] for a proof. □

Lemma 2.4 *The C^*-algebra $\bar{\mathcal{B}}_H$ contains the compact operators.*

This is Lemma 4.12 in [8]. □

3 Sobolev topologies in spaces of Clifford connections

Rigidity theory for the index is strongly adapted to the index theory under consideration, i.e., to the choice of the algebra $\mathcal{A}$, the ideal $\mathcal{B}$, the corresponding K-functors and so on. We here consider three cases, the case of section 2, i.e., $\mathcal{A}_H$, $\mathcal{B}_H$, $\bar{\mathcal{B}}_H$, the case of uniform operators considered in [9], and finally the Fredholm case, i.e., the case $\mathcal{B}$ =compact operators. We permit "continuous" deformations of the Clifford connection and the metric on M and have therefore to describe what are "continuous" deformations, i.e., we have to define intrinsic and natural topologies in the space of Clifford connections and metrics. If M is compact this is a very trivial matter. In the noncompact case this task seems to us to be rather nontrivial. We start our approach shortly recalling some very simple definitions from general topology.

Let X be a set, $\mathcal{U} \subset \mathcal{P}(X \times X)$ = set of all subsets of $X \times X$. $\mathcal{U}$ is called a *uniform structure* if it satisfies the following conditions.
(F_1) $V \in \mathcal{U}, V_1 \supseteq V$ implies $V_1 \in \mathcal{U}$
(F_2) $V_1, ..., V_n \in \mathcal{U}$ implies $V_1 \cap \cdots \cap V_n \in \mathcal{U}$.
(U_1) Every $V \in \mathcal{U}$ contains the diagonal $\Delta \subset X \times X$.
(U_2) $V \in \mathcal{U}$ implies $V^{-1} \in \mathcal{U}$.
(U_3) If $V \in \mathcal{U}$ then there exists $W \in \mathcal{U}$ such that $W \circ W \subseteq V$.
The sets of $\mathcal{U}$ are called neighborhoods of the uniform structure, $(X, \mathcal{U})$ a *uniform space*.

$\mathcal{B} \subseteq \mathcal{U}$ is called a *fundamental system* or *basis* for $\mathcal{U}$ if each neighborhood of $\mathcal{U}$ contains an element of $\mathcal{B}$. $\mathcal{B} \subset \mathcal{P}(X \times X)$ is a fundamental system for a uniquely determined uniform structure if and only if it satisfies the following conditions.
(B_1) If $V_1, V_2 \in \mathcal{B}$ then $V_1 \cap V_2$ contains an element of $\mathcal{B}$.
(U_1') Each $V \in \mathcal{B}$ contains the diagonal $\Delta \subset X \times X$.
(U_2') For each $V \in \mathcal{B}$ there exists $V' \in \mathcal{B}$ such that $V' \subseteq V^{-1}$.
(U_3') For each $V \in \mathcal{B}$ there exists $W \in \mathcal{B}$ such that $W \circ W \subseteq V$.

Every uniform structure induces a topology on X. Let $(X, \mathcal{U})$ be a uniform space. Then for every $x \in X$ $\mathcal{U}(x) = \{V(x)\}_{V \in \mathcal{U}}$ is a neighborhood filter for a uniquely determined topology on X. This topology is called the uniform topology generated by the uniform structure $\mathcal{U}$.

The uniform space $(X, \mathcal{U})$ is called *Hausdorff* if $\bigcap_{V \in \mathcal{U}} V = \Delta$. A uniform space is metrizable if and only if $(X, \mathcal{U})$ is Hausdorff and has a countable basis $\mathcal{B}$. Let $(X, \mathcal{U})$ be a uniform space. Then there exists a complete uniform space $(\tilde{X}, \tilde{\mathcal{U}})$ such that $\tilde{X}$ is isomorphic to a dense subset of X. If $(X, \mathcal{U})$ is additionally Hausdorff then $(\tilde{X}, \tilde{\mathcal{U}})$ is determined up to isomorphism. $(\tilde{X}, \tilde{\mathcal{U}})$ is called the *completion* and is built up by Cauchy filters. We refer to [10], pp. 126–127 for the proof. Let $(Y, \mathcal{U}_Y)$ be a Hausdorff uniform space, $X \subset Y$ a dense subspace. If X is metrizable by a metric ϱ then ϱ is extendable to a metric on Y which metrizes the uniform space $(Y, \mathcal{U}_Y)$. In conclusion, if $(X, \mathcal{U})$ is a metrizable uniform space, $(\tilde{X}, \tilde{\mathcal{U}})$ and $(\tilde{X}^\varrho, \tilde{\mathcal{U}}^\varrho)$ are its uniform and metric completions, respectively, then

$$\tilde{X} = \tilde{X}^\varrho$$

as metrizable topological spaces. We want to endow the space of Clifford connections in a canonical manner with an intrinsic uniform structure.

Let (M^n, g) be open, complete, $S \longrightarrow M$ a Clifford bundle without fixed connection. Set

$$\mathcal{C} = \mathcal{C}_S = \{\nabla \mid \nabla \textit{ is a Clifford connection in } S\}.$$

Clifford connection means here $\nabla(X \cdot s) = \nabla X \cdot s + X \cdot \nabla s$. Consider $\Omega^1(\mathbf{g}_S^{cl})$ =space of all smooth 1-forms η with values in the skew symmetric endomorphisms of S satisfying $\eta(X \cdot s) = X \cdot \eta(s)$. Given a Clifford connection ∇^S in S, ∇^S induces a connection in $\mathbf{g}_S^{cl}$ by

$$\nabla\varphi = [\nabla^S, \varphi], \; i.e. \; (\nabla\varphi)(s) = \nabla^S\varphi(s) - \varphi\nabla^S s.$$

We denote in the sequel $\mathbf{g}_S^{cl} \equiv \mathbf{g}$. Set now

$$V_\delta = \Big\{(\nabla, \nabla_1) \in \mathcal{C} \times \mathcal{C} \mid \nabla - \nabla_1 \in C_0^\infty(T^*M \otimes \mathbf{g}) \text{ and } {}^{b,m}|\nabla - \nabla_1|_\nabla =$$

$$= \sum_{i=0}^{m} \sup_{x \in M} |\nabla^i(\nabla - \nabla_1)|_x < \delta\Big\}.$$

Proposition 3.1 $\mathcal{B} = \{V_\delta\}_{\delta > 0}$ *is a basis for a metrizable uniform structure.*

Proof. (B_1) and (U_1') are trivial. (U_2') really needs serious work. It would be proved if we could show

$$ {}^{b,m}|\nabla - \nabla'|_{\nabla'} \leq P({}^b|\nabla^i(\nabla - \nabla')|), \tag{3.1}$$

where P is a polynomial without constant term in ${}^b|\nabla^i(\nabla - \nabla')|, i = 0, \ldots, m$. If (3.1) would be established then for given $\delta > 0$ there exists $\delta' > 0$ such that ${}^{b,m}|\nabla - \nabla'|_{\nabla'} < \delta'$ implies $P({}^b|\nabla^i(\nabla - \nabla')|) < \delta$. Assume now ${}^{b,m}|\nabla - \nabla'|_{\nabla'} < \delta'$. Then ${}^{b,m}|\nabla - \nabla'|_\nabla \leq P({}^b|\nabla'^i(\nabla' - \nabla)|) < \delta$, i.e. $(\nabla', \nabla) \in V_{\delta'}$ implies $(\nabla, \nabla') \in V_\delta, V_{\delta'} \subseteq V_\delta^{-1}$. Therefore we have to establish (3.1). Set $\eta = \nabla' - \nabla$. For the zero-th derivatives there is nothing to show. Consider the first derivative. $|\;|$ shall denote the pointwise norm. Then

$$|\nabla'\eta| \leq |(\nabla' - \nabla)\eta| + |\nabla\eta| \leq C_1|\eta|^2 + |\nabla\eta| \leq C_2(|\eta|^2 + |\nabla\eta|),$$

$$|\eta| + |\nabla'\eta| \leq C_3(|\eta| + |\eta|^2 + |\nabla\eta|),$$

$${}^{b,1}|\eta|_{\nabla'} \leq C_3({}^b|\eta| + {}^b|\eta|^2 + |\nabla\eta|).$$

We conclude similarly

$$\begin{aligned}|\nabla'^2\eta| &\le |(\nabla'-\nabla)\nabla'\eta| + |\nabla\nabla'\eta| \le |(\nabla'-\nabla)(\nabla'-\nabla)\eta| + \\ &+ |(\nabla'-\nabla)\nabla\eta| + |\nabla(\nabla'-\nabla)\eta| + |\nabla^2\eta| \le \\ &\le C_4(|\eta|^3 + |\eta||\nabla\eta| + |\eta||\nabla\eta| + |\nabla^2\eta|),\end{aligned}$$

$$^{b,2}|\eta|_{\nabla'} \le C_5(^b|\eta| +^b |\eta|^2 +^b |\eta|^3 +^b |\nabla\eta| +^b |\eta| \cdot^b |\nabla\eta| +^b |\nabla^2\eta|).$$

Now we turn to the general case. Assume we are done for $\nabla'\eta, \ldots, \nabla'^{r-1}\eta$. Then

$$\nabla'^r\eta = \sum_{i=1}^{r} \nabla^{i-1}(\nabla'-\nabla)\nabla'^{r-i}\eta + \nabla^r\eta.$$

It remains to consider the terms

$$\nabla^{i-1}(\nabla'-\nabla)\nabla'^{r-i}\eta.$$

Again iterating the procedure, i.e. applying it to ∇'^{r-i} and so on, we have to estimate expressions of the kind

$$\nabla^{i_1}(\nabla'-\nabla)^{i_2}\nabla^{i_3}(\nabla'-\nabla)^{i_4}\cdots\nabla^{i_{r-2}}(\nabla'-\nabla)^{i_{r-1}}\nabla^{i_r}\eta \tag{3.2}$$

with $i_1 + \cdots + i_r = r, i_r < r$. We derive that (3.2) splits into a sum of terms, each of which can be estimated by

$$C_{n_1\cdots n_s}|\nabla^{n_1}\eta|\cdots|\nabla^{n_s}\eta|, \quad n_1+1+\cdots+n_s+1 = r+1,$$

i.e.

$$^b|\nabla'^r\eta| \le \sum_{n_1+1+\cdots+n_s+1\le r+1} {}^b|\nabla^{n_1}\eta| \cdots^b |\nabla^{n_s}\eta| +^b |\nabla^r\eta|,$$

which implies (U_2'). Next we turn to the condition (U_3'). Given $\delta > 0$, we have to show that there exists $\delta' > 0$ such that $V_{\delta'} \circ V_{\delta'} \subset V_\delta$, i.e. if

$$(\nabla_1, \nabla_2) \in V_{\delta'} \circ V_{\delta'} = \{(\nabla_1, \nabla_2) \in \mathcal{C} \times \mathcal{C} \mid \textit{There exists } \nabla \\ \textit{with } (\nabla_1, \nabla) \in V_{\delta'} \textit{ and } (\nabla, \nabla_2) \in V_{\delta'}\}$$

then

$$^{b,m}|\nabla_1 - \nabla_2|_{\nabla_1} < \delta.$$

(U_3') would be proved if we could show

$$^{b,m}|\nabla_1 - \nabla_2|_{\nabla_1} \le P(^{b,i}|\nabla_1 - \nabla|_{\nabla_1}, ^{b,j}|\nabla - \nabla_2|_\nabla),$$

where P is a polynomial in $^{b,i}|\nabla_1 - \nabla|_{\nabla_1}$, $^{b,j}|\nabla - \nabla_2|_\nabla$, $i, j = 0, \ldots, m$, without constant term.

We start with

$$|\nabla_1 - \nabla_2| \le |\nabla_1 - \nabla| + |\nabla - \nabla_2|,$$
$${}^b|\nabla_1 - \nabla_2| \le^b |\nabla_1 - \nabla| +^b |\nabla - \nabla_2| = P_0({}^b|\nabla_1 - \nabla|,^b |\nabla - \nabla_2|).$$

Next we consider

$$|\nabla_1(\nabla_1 - \nabla_2)| \le |\nabla_1(\nabla_1 - \nabla)| + |\nabla_1(\nabla - \nabla_2)|. \tag{3.3}$$

The critical point in (3.3) is the estimation of $|\nabla_1(\nabla - \nabla_2)|$. But

$$|\nabla_1(\nabla - \nabla_2)| \le |(\nabla_1 - \nabla)(\nabla - \nabla_2)| + |\nabla(\nabla - \nabla_2)|,$$
$${}^b|\nabla_1(\nabla - \nabla_2| \le C_1({}^b|\nabla_1 - \nabla| \cdot^b |\nabla - \nabla_2| +^b |\nabla(\nabla - \nabla_2)|),$$
$${}^b|\nabla_1(\nabla_1 - \nabla_2)| \le C_2({}^b|\nabla_1(\nabla_1 - \nabla)| +^b |\nabla_1 - \nabla| \cdot^b |\nabla - \nabla_2| +$$
$$+^b|\nabla(\nabla - \nabla_2)|) =$$
$$= P_1({}^b|\nabla_1 - \nabla|,^b |\nabla_1(\nabla_1 - \nabla)|,^b |\nabla - \nabla_2|,^b |\nabla(\nabla - \nabla_2)|).$$

We conclude analogously for $\nabla_1^2(\nabla_1 - \nabla_2)$,

$$|\nabla_1^2(\nabla_1 - \nabla_2)| \le |\nabla_1^2(\nabla_1 - \nabla)| + |\nabla_1^2(\nabla - \nabla_2)|,$$
$$|\nabla_1^2(\nabla - \nabla_2)| \le |(\nabla_1 - \nabla)(\nabla_1 - \nabla)(\nabla - \nabla_2)| + |(\nabla_1 - \nabla)\nabla(\nabla - \nabla_2)| +$$
$$+|\nabla(\nabla_1 - \nabla)(\nabla - \nabla_2)| + |\nabla^2(\nabla - \nabla_2|),$$
$${}^b|\nabla_1^2(\nabla - \nabla_2)| \le C_3({}^b|\nabla_1 - \nabla|^2 \cdot^b |\nabla - \nabla_2| +^b |\nabla_1 - \nabla| \cdot^b |\nabla(\nabla - \nabla_2)| +$$
$$+^b|\nabla(\nabla_1 - \nabla)| \cdot^b |\nabla - \nabla_2| +^b |\nabla_1 - \nabla| \cdot^b |\nabla(\nabla - \nabla_2)| +^b |\nabla^2(\nabla - \nabla_2)|).$$

Using

$$|\nabla(\nabla_1 - \nabla)| \le |(\nabla - \nabla_1)(\nabla - \nabla_1)| + |\nabla_1(\nabla_1 - \nabla)|,$$

we obtain finally

$${}^b|\nabla_1^2(\nabla_1 - \nabla_2)| \le C_4({}^b|\nabla_1^2(\nabla_1 - \nabla)| +^b |\nabla_1 - \nabla|^2 \cdot^b |\nabla - \nabla_2| +$$
$$+^b|\nabla_1 - \nabla| \cdot^b |\nabla(\nabla - \nabla_2)| + ({}^b|\nabla - \nabla_1|^2 +^b |\nabla_1(\nabla_1 - \nabla)|) \cdot^b |\nabla - \nabla_2| +$$
$$+^b|\nabla^2(\nabla - \nabla_2)|) =$$
$$= P_2({}^b|\nabla_1^i(\nabla_1 - \nabla)|,^b |\nabla^j(\nabla - \nabla_2)|).$$

In the general case

$$|\nabla_1^r(\nabla_1 - \nabla_2)| \le |\nabla_1^r(\nabla_1 - \nabla)| + |\nabla_1^r(\nabla - \nabla_2)|, \tag{3.4}$$

and we have to estimate

$$\nabla_1^r(\nabla - \nabla_2) \equiv \nabla_1^r \eta.$$

We start as above

$$\nabla_1^r\eta = (\nabla_1 - \nabla)\nabla_1^{r-1}\eta + \nabla\nabla_1^{r-1}\eta,$$
$$\nabla_1^r\eta = \sum_{i=1}^r \nabla^{i-1}(\nabla_1 - \nabla)\nabla_1^{r-i}\eta + \nabla^r\eta.$$

Iterating the procedure, we have to estimate expressions of the kind

$$\nabla^{i_1}(\nabla_1 - \nabla)^{i_2}\cdots\nabla^{i_{r-2}}(\nabla_1 - \nabla)^{i_{r-1}}\nabla^{i_r}\eta$$

with $i_1 + \cdots + i_r = r$, $i_r < r$. Each expression splits into a sum of terms each of which can be estimated by

$$C_{n_1\cdots n_s}|\nabla^{n_1}(\nabla_1 - \nabla)|\cdots|\nabla^{n_{s-1}}(\nabla_1 - \nabla)|\cdot|\nabla^{n_s}\eta|, \tag{3.5}$$

$n_1 + 1 + \cdots + n_s + 1 = r + 1$. According to our proof of (U_2'),

$${}^b|\nabla^{n_i}(\nabla_1 - \nabla)| \leq Q_{n_i}({}^b|\nabla_1^j(\nabla_1 - \nabla)|). \tag{3.6}$$

Here Q_{n_i} is a polynomial in the indicated variables without constant term. (3.4) – (3.6) yield

$${}^b|\nabla_1^r(\nabla_1 - \nabla_2)| \leq P_r({}^b|\nabla_1^i(\nabla_1 - \nabla)|, {}^b|\nabla^j(\nabla - \nabla_2)|),$$
$${}^{b,m}|\nabla_1 - \nabla_2| \leq \sum_{r=0}^m P_r =$$
$$= P({}^b|\nabla_1^i(\nabla_1 - \nabla)|, {}^b|\nabla^j(\nabla - \nabla_2)|).$$

Therefore we have established (U_3'). Denote by ${}^{b,m}\mathcal{U}(\mathcal{C})$ the corresponding uniform structure. It is metrizable since it is trivially Hausdorff and $\{V_{1/n}\}_{n\geq n_0}$ is a countable basis. □

Let ${}^b_m\mathcal{C}$ denote $\mathcal{C}$ endowed with the corresponding topology and denote by ${}^{b,m}\tilde{\mathcal{C}}$ the completion. For any $\nabla \in {}^{b,m}\tilde{\mathcal{C}}$

$$\{{}^{b,m}U_\varepsilon(\nabla)\}_{\varepsilon>0} = \{\{\nabla' \in {}^{b,m}\tilde{\mathcal{C}} \mid {}^{b,m}|\nabla - \nabla'|_\nabla < \varepsilon\}\}_{\varepsilon>0}$$

is a neighborhood basis in this topology.

Proposition 3.2 *${}^{b,m}\tilde{\mathcal{C}}$ is locally contractible.*

Proof. Let $\nabla' \in {}^{b,m} U_\varepsilon(\nabla)$. Then $\nabla^t = t\nabla' + (1-t)\nabla = \nabla + t(\nabla' - \nabla) \in {}^{b,m}U_\varepsilon(\nabla)$ since $|\nabla^i(t\nabla' + (1-t)\nabla - \nabla)| = |\nabla^i t(\nabla' - \nabla)| = t|\nabla^i(\nabla' - \nabla)|$ and since a Cauchy sequence for ∇, ∇' gives such a sequence for ∇^t. □

Corollary 3.1 *In ${}^{b,m}\tilde{\mathcal{C}}$, the components and arc components coincide.*

Proof. ${}^{b,m}\tilde{\mathcal{C}}$ is locally arcwise connected since it is locally contractible. □

Remarks. 1. The elements of ${}^{b,m}\tilde{\mathcal{C}}$ are of differentiability class C^m.

2. Let ${}^{b,m}\tilde{\Omega}^1(\mathbf{g}, \nabla)$ be the completion of $C_0^\infty(T^*M\otimes\mathbf{g})$ with respect to ${}^{b,m}|\ |$. Then by construction of our completion

$${}^{b,m}U_\varepsilon(\nabla) \subseteq \nabla + {}^{b,m}\tilde{\Omega}^1(\mathbf{g}, \nabla).$$

Before we calculate the components of ${}^{b,m}\tilde{\mathcal{C}}$ we need the following

Proposition 3.3 . *Let ∇ and ∇_1 be C^{m-1}-connections and $\nabla - \nabla_1 \in {}^{b,m-1}\tilde{\Omega}^1(\mathbf{g}, \nabla)$. Then*

$$^{b,m}\tilde{\Omega}^1(\mathbf{g}, \nabla) = {}^{b,m}\tilde{\Omega}^1(\mathbf{g}, \nabla_1) \tag{3.7}$$

as equivalent Banach spaces.

Proof. By assumption, ∇, ∇_1 are connections of class C^{m-1}, i.e. their connection coefficients are of class C^{m-1}. Then (3.7) makes sense since $\nabla, \ldots, \nabla^m, \nabla_1, \ldots, \nabla_1^m$ are well defined. We prove

$$^{b,m}|\ |_\nabla \sim {}^{b,m}|\ |_{\nabla_1}.$$

We obtain from

$$\begin{aligned} |\nabla_1\varphi| &\leq |(\nabla_1 - \nabla)\varphi| + |\nabla\varphi| \\ |\varphi| + |\nabla_1\varphi| &\leq C'(|\nabla_1 - \nabla| \cdot |\varphi| + |\nabla\varphi| + |\varphi| \leq \\ &\leq C_1(\nabla, \nabla_1)(|\varphi| + |\nabla\varphi|). \end{aligned}$$

Quite analogously

$$\begin{gathered} |\varphi| + |\nabla\varphi| \leq D_1(\nabla, \nabla_1)(|\varphi| + |\nabla_1\varphi|, \\ ^{b,1}|\ | \sim {}^{b,1}|\ |, \quad {}^{b,1}\tilde{\Omega}^1(\mathbf{g}, \nabla) = {}^{b,1}\tilde{\Omega}^1(\mathbf{g}, \nabla_1), \end{gathered}$$

in particular $\nabla - \nabla_1 \in {}^{b,1}\tilde{\Omega}^1(\mathbf{g}, \nabla_1)$.

Assume now the assertion for $r - 1 \leq m - 1$,

$$\begin{gathered} ^{b,r-1}|\ | \sim {}^{b,r-1}|\ |, \\ ^{b,r-1}\tilde{\Omega}^1(\mathbf{g}, \nabla) = {}^{b,r-1}\tilde{\Omega}^1(\mathbf{g}, \nabla_1), \end{gathered}$$

in particular

$$\nabla - \nabla_1 \in {}^{b,r-1}\tilde{\Omega}^1(\mathbf{g}, \nabla_1).$$

Then

$$\nabla_1^r\varphi = \sum_{i=1}^{r} \nabla^{i-1}(\nabla_1 - \nabla)\nabla_1^{r-i}\varphi + \nabla^r\varphi. \tag{3.8}$$

For $i = 1$

$$|(\nabla_1 - \nabla)\nabla_1^{r-1}\varphi| \leq C'(\nabla, \nabla_1)|\nabla_1^{r-1}\varphi|, \tag{3.9}$$

by induction assumption

$$^{b}|(\nabla_1 - \nabla)\nabla_1^{r-1}\varphi| \leq C_1(\nabla, \nabla_1)\, {}^{b,r-1}|\varphi|_\nabla. \tag{3.10}$$

Consider $\nabla^{i-1}(\nabla_1 - \nabla)\nabla_1^{r-i}\varphi$, $i \geq 2$. Iterating the procedure, i.e. applying it to ∇_1^{r-i} and so on, we have to estimate expressions of the kind

$$\nabla^{i_1}(\nabla_1 - \nabla)^{i_2} \cdots \nabla^{i_{r-2}}(\nabla_1 - \nabla)^{i_{r-1}}\nabla^{i_r}\varphi \tag{3.11}$$

with $i_1 + \cdots + i_r = r$, $i_1 + i_3 + \cdots + i_{r-2} \leq r-1$, $i_r \leq r-1$. If we give ∇, $\nabla_1 - \nabla$, φ each degree one, then each term of (3.11) has degree $i_1 + \cdots + i_r + 1 = r+1$. (3.11) splits into a sum of terms each of which can be estimated by

$$C_{n_1 \cdots n_s}|\nabla^{n_1}(\nabla_1 - \nabla)| \cdots |\nabla^{n_{s-1}}(\nabla_1 - \nabla)| \cdot |\nabla^{n_s}\varphi|, \tag{3.12}$$

$n_1 + 1 + \cdots + n_s + 1 = r+1, n_1 + n_2 + \cdots + n_{s-1} \leq r-1$. By assumption ${}^b|\nabla^{n_1}(\nabla_1 - \nabla)| < \infty, {}^b|\nabla^{n_2}(\nabla_1 - \nabla)| < \infty, \cdots$, and we obtain

$$\begin{aligned} C_{n_1 \cdots n_s}\, {}^b|\nabla^{n_1}(\nabla_1 - \nabla)| \cdots {}^b|\nabla^{n_{s-1}}(\nabla_1 - \nabla)| \cdot {}^b |\nabla^{n_s}\varphi| \leq \\ \leq C_2(\nabla, \nabla_1)\, {}^{b,n_s}|\varphi|_\nabla. \end{aligned} \tag{3.13}$$

The induction assumption and (3.8) – (3.13) yield

$${}^{b,r}|\varphi|_{\nabla_1} \leq C_4(\nabla, \nabla_1)\, {}^{b,r}|\varphi|_\nabla,$$

i.e. $\nabla - \nabla_1 \in {}^{b,r-1}\tilde{\Omega}^1(\mathbf{g}, \nabla)$ implies

$${}^{b,r}|\ |_\nabla \sim {}^{b,r}|\ |_{\nabla_1},$$

which finishes the induction. □

Proposition 3.4 . *Let comp(∇) be the component of ∇ in ${}^{b,m}\tilde{\mathcal{C}}$. Then*

$$comp(\nabla) = \nabla + {}^{b,m}\tilde{\Omega}^1(\mathbf{g}, \nabla).$$

Proof. We show at first $comp(\nabla) \subseteq \nabla + {}^{b,m}\tilde{\Omega}^1(\mathbf{g}, \nabla)$. Let $\nabla' \in comp(\nabla)$ and let ∇^t be an arc between $\nabla = \nabla^0$ and $\nabla' = \nabla^1$. For every $\varepsilon > 0$ this arc can be covered by a finite number of ε-neighborhoods ${}^{b,m}U(\nabla^{t_\sigma})$ in ${}^{b,m}\tilde{\mathcal{C}}$, $\sigma = 0, \ldots, s$, $t_0 = 0$, $t_s = t$, such that additionally ${}^{b,m}|\nabla^{t_\sigma} - \nabla^{t_{\sigma+1}}|_{\nabla^{t_{\sigma+1}}} < \varepsilon$, $\sigma = 0, \ldots, s-1$. We set $\nabla^{t_\sigma} \equiv \nabla_\sigma$. Then $\nabla - \nabla_1 \in {}^{b,m}\tilde{\Omega}^1(\mathbf{g}, \nabla_1)$. According to Proposition 3.3, $\nabla - \nabla_1 \in {}^{b,m}\tilde{\Omega}^1(\mathbf{g}, \nabla)$. $\nabla_1 - \nabla_2 \in {}^{b,m}\tilde{\Omega}^1(\mathbf{g}, \nabla_2)$, which implies $\nabla_1 - \nabla_2 \in {}^{b,m}\tilde{\Omega}^1(\mathbf{g}, \nabla_1)$, $\nabla_1 - \nabla_2 \in {}^{b,m}\tilde{\Omega}^1(\mathbf{g}, \nabla)$,

$$\nabla - \nabla_1 + \nabla_1 - \nabla_2 = \nabla - \nabla_2 \in {}^{b,m}\tilde{\Omega}^1(\mathbf{g}, \nabla).$$

We conclude from a trivial induction $\nabla - \nabla' \in {}^{b,m}\tilde{\Omega}^1(\mathbf{g}, \nabla)$, $\nabla' = \nabla + \nabla' - \nabla \in \nabla + {}^{b,m}\tilde{\Omega}^1(\mathbf{g}, \nabla)$. On the other hand, if $\nabla' \in \nabla + {}^{b,m}\tilde{\Omega}^1(\mathbf{g}, \nabla)$, then $\nabla^t = \nabla + t(\nabla' - \nabla) = t\nabla + (1-t)\nabla$ is an arc in ${}^{b,m}\tilde{\mathcal{C}}$ connecting ∇ and ∇', as we have already seen. Therefore

$$comp(\nabla) = \nabla + {}^{b,m}\tilde{\Omega}^1(\mathbf{g}, \nabla). \qquad \square$$

For many purposes and other operator algebras or index theories, respectively, it is necessary to consider additional geometric assumptions and other topologies in the space of Clifford connections.

Let as above (M^n, g) be open complete, $(S, \nabla) \to M$ a Clifford bundle with with metric Clifford connection ∇. Consider the following conditions:

$$\begin{array}{ll} (I) & r_{inj}(M) = \inf_{x\in m} r_{inj}(x) > 0. \\ (B_k(M)) & |\nabla^i R| \le C_i,\ 0 \le i \le k. \\ (B_k(S,\nabla)) & |(\nabla^S)^i R^S| \le D_i,\ 0 \le i \le k. \end{array}$$

Here R or R^S denotes the curvature tensor of M or S, respectively. If M and S satisfy these conditions, we say they have *bounded geometry* up to order k. Let $D = D^\nabla$ be the generalized Dirac operator belonging to ∇, $(Ds)(x) = \sum_{i=1}^n e_i \cdot \nabla_{e_i} s(x)$, where $\cdot$ is the Clifford multiplication and $e_1, \ldots, e_n \in T_x M$ is an orthonormal basis. Consider

$$\Omega_r(S,\nabla) \equiv \Omega_r^0(S,\nabla) =$$
$$= \left\{ s \in C^\infty(S) \;\middle|\; |s|_{\nabla,r} = \left(\int \textstyle\sum_{i=0}^r |\nabla^i s|_x^2 \, d\,vol_x(g)\right)^{1/2} < \infty \right\}$$

and $\bar\Omega^{0,r}(S,\nabla) \equiv \bar\Omega^r(S,\nabla) \overset{def}{=}$ completion of $\Omega_r(S,\nabla)$ with respect to $|\ |_{\nabla,r}$, $\tilde\Omega^r(S,\nabla) \overset{def}{=}$ completion of $C_0^\infty(S)$ with respect to $|\ |_{\nabla,r}$ and $\Omega^r(S,\nabla) \overset{def}{=}$ space of all distributional sections s such that $|s|_{\nabla,r} < \infty$. Similarly, define

$$\Omega_r(S,D) = \left\{ s \in \Omega(S) \;\middle|\; |s|_{D,r} = \left(\int \sum_{i=0}^r |D^i s|_x^2 \, d\,vol_x(g) \right)^{1/2} < \infty \right\}$$

and $\bar\Omega^r(S,D), \tilde\Omega^r(S,D), \Omega^r(S,D)$. Then

$$\tilde\Omega^r(S,\nabla) \subseteq \bar\Omega^r(S,\nabla) \subseteq \Omega^r(S,\nabla),$$
$$\tilde\Omega^r(S,D) \subseteq \bar\Omega^r(S,D) \subseteq \Omega^r(S,D),$$
$$\tilde\Omega^r(S,\nabla) \subseteq \tilde\Omega^r(S,D),\ \bar\Omega^r(S,\nabla) \subseteq \bar\Omega^r(S,D),\ \Omega^r(S,\nabla) \subseteq \Omega^r(S,D)$$

as continuous embeddings, and all spaces are Hilbert spaces.

Proposition 3.5 *Assume* (I), $(B_k(M))$ *and* $r \le k+2$. *Then*

$$\tilde\Omega^r(S,\nabla) = \bar\Omega^r(S,\nabla) = \Omega^r(S,\nabla),$$
$$\tilde\Omega^r(S,D) = \bar\Omega^r(S,D) = \Omega^r(S,D).$$

We refer to [3] for the proof. □

Proposition 3.6 *Assume* $(I), (B_k(M))$ *and* $(B_k(S)), k \ge 0$. *Then* $\Omega^r(S,\nabla)$ *and* $\Omega^r(S,D)$ *are for* $r \le k$ *equivalent.*

The proof is carried out in [2]. □

Proposition 3.7 *Assume (I) and* $(B_0(M))$, $r > n/2 + m$. *Then there exists a continuous embedding*

$$\bar{\Omega}^r(S,\nabla) \hookrightarrow^{b,m} \Omega(S,\nabla).$$

The proof is contained in [3]. □

Corollary 3.2 *Assume (I),* $(B_k(M))$, $(B_k(S,\nabla))$, $k > n/2$. *Then there exists a continuous embedding*

$$\Omega^r(S,D) \hookrightarrow^{b,0} \Omega(S,\nabla) =^b \Omega(S). \quad \square$$

Define for $r \in \mathbf{Z}_+$

$$\Omega^1_r(\mathbf{g},\nabla) \stackrel{\text{def}}{=}$$
$$\left\{ \varphi \in C^\infty(T^*M \otimes \mathbf{g}) \,\middle|\, |\varphi|_{\nabla,r} = \left(\int \sum_{i=0}^{r} |\nabla^i \varphi|_x^2 \, dvol_x(g) \right)^{1/2} < \infty \right\}$$

and define in similar manner as above

$$\tilde{\Omega}^{1,r}(\mathbf{g},\nabla), \bar{\Omega}^{1,r}(\mathbf{g},\nabla), \Omega^{1,r}(\mathbf{g},\nabla)$$

by completion in the first two cases or taking distributional sections in the third case, respectively.

Let (M_n, g) be open, complete with $(B_k(M))$, $k \geq 1$, $S \to M$ a Clifford bundle without fixed connection. Set

$$\mathcal{C}(B_k) = \{\nabla \,|\, \nabla \text{ Clifford connection satisfying } (B_k(S,\nabla))\}$$

and for $r \leq k, r > n/2+1, \delta > 0$

$$V_\delta = \{(\nabla,\nabla') \in \mathcal{C}(B_k) \times \mathcal{C}(B_k) \,|\, |\nabla - \nabla'|_{\nabla,r} < \delta\}.$$

Proposition 3.8 *Assume (I),* $(B_k(M)), r > n/2 + 1$. *Then* $\mathcal{B} = \{V_\delta\}_{\delta>0}$ *is a fundamental system for a metrizable uniform structure* $\mathcal{U}^r(\mathcal{C}(B_k))$.

The rather long and nontrivial proof is contained in [5]. □

Denote $\mathcal{C}_r(B_k)$ as $\mathcal{C}(B_k)$ endowed with the corresponding topology and by $\mathcal{C}^r(B_k)$ the completion. Let $comp(\nabla)$ be the component of ∇ in $\mathcal{C}^r(B_k)$.

Proposition 3.9 *Assume* k, r *as above,* $\nabla, \nabla' \in \mathcal{C}(B_k)$, $\nabla' \in comp(\nabla) \subseteq \mathcal{C}^{r-1}(B_k)$. *Then*

$$\Omega^{1,i}(\mathbf{g},\nabla) = \Omega^{1,i}(\mathbf{g},\nabla')$$

and

$$\Omega^{0,i}(S,\nabla) = \Omega^{0,i}(S,\nabla'), 0 \leq i \leq r.$$

The rather nontrivial proof is carried out in [4]. □

Proposition 3.10 *Assume k, r as above, $\nabla \in \mathcal{C}(B_k)$. Then*

$$comp(\nabla) = \nabla + \bar{\Omega}^{1,r}(\mathbf{g}, \nabla).$$

For the proof we refer to [5]. □

Corollary 3.3 *Assume $(I), (B_k(M)), k \geq r > n/2 + 1$. Then*

$$comp(\nabla) = \nabla + \tilde{\Omega}^{1,r}(\mathbf{g}, \nabla).$$

This follows immediately from the corresponding version of Proposition 3.5 which holds for any vector bundle as proved in [4], [5]. □

If $\nabla \in \mathcal{C}(B_k)$, $\eta \in C_0^\infty(T^*M \otimes \mathbf{g})$, then $\nabla + \eta \in \mathcal{C}(B_k)$. The latter is no longer true if η has bounded (b,k)-norm. It remains true if we assume ${}^{b,i}|\eta| < \infty$, $i \geq k+1$. Therefore we still define another basis in $\mathcal{C}(B_k)$ by $\mathcal{B} = \{V_\delta\}_{\delta>0}$,

$$V_\delta = \{(\nabla, \nabla') \in \mathcal{C}(B_k) \times \mathcal{C}(B_k) \,\Big|\, {}^{b,i}|\nabla - \nabla'|_\nabla < \delta\},$$

thus getting completed spaces ${}^{b,i}\mathcal{C}(B_k)$, $i \geq k+1$ and components $comp(\nabla) = \nabla + {}^{b,i}\Omega(\mathbf{g}, \nabla)$.

Finally we have to consider completed spaces of Clifford connections without curvature restrictions as in Proposition 3.1 but now without the assumption that $\nabla - \nabla'$ should have compact support. Let $m \geq 0$. Set for $\delta > 0$

$$V_\delta = \{(\nabla, \nabla') \in \mathcal{C} \times \mathcal{C} \,\Big|\, {}^{b,m}|\nabla - \nabla'|_\nabla < \delta\}.$$

Proposition 3.11 *$\mathcal{B} = \{V_\delta\}_{\delta>0}$ is a basis for a metrizable uniform structure.*

The proof is completely analogous to that of Proposition 3.1. □

Denote ${}^b_m\mathcal{C}$ for $\mathcal{C}$ endowed with the corresponding topology and ${}^{b,m}\mathcal{C}$ for its completion. Once again we are interested in the components.

Set

$${}^b_m\Omega^1(\mathbf{g}, \nabla) = \{\eta \in \Omega^1(\mathbf{g}) \mid {}^{b,m}|\eta|_\nabla < \infty\}$$

and ${}^{b,m}\Omega(\mathbf{g}, \nabla)$ =completion of ${}^b_m\Omega^1(\mathbf{g}, \nabla)$ with respect to ${}^{b,m}|\ |$.

Proposition 3.12 *The component of $\nabla \in {}^{b,m}\mathcal{C}$ is given by*

$$comp(\nabla) = \nabla + {}^{b,m}\Omega(\mathbf{g}, \nabla).$$

The proof is quite similar to that of Proposition 3.4. □

Proposition 3.13 *Assume* $\nabla, \nabla' \in {}^{b,0}\mathcal{C} = {}^{b}\mathcal{C}$, $\nabla - \nabla' \in {}^{b,0}\Omega(\mathbf{g}) = {}^{b}\Omega(\mathbf{g})$, *i.e.* $\nabla' \in comp(\nabla)$, $D = D^{\nabla}$, $D' = D^{\nabla'}$. *Then*

$$\bar{\Omega}^{0,1}(S, D) = \bar{\Omega}^{0,1}(S, D').$$

The same holds for $\nabla, \nabla' \in {}^{b,0}\tilde{\mathcal{C}}$, $\nabla' \in comp(\nabla)$, *since the component of* ∇ *in* ${}^{b,0}\tilde{\mathcal{C}}$ *is contained in the corresponding component in* ${}^{b,0}\mathcal{C}$. □

We finish this section with a version of the Sobolev embedding theorem which we need in Theorem 4.4.

Proposition 3.14 *Assume* (M^n, g) *open, complete with* (I) *and* $(B_0(M))$, $r > n/2 + m$. *Then there exists a continuous embedding*

$$\bar{\Omega}^{1,r}(\mathbf{g}, \nabla) \hookrightarrow {}^{b,m}\, \Omega(\mathbf{g}, \nabla).$$

For the proof see Theorem 1.13 of [3]. □

4 The rigidity of the index

Now we are ready to establish part of our theorems concerning the rigidity of the index. There are several definitions of the analytical index depending on the choice of the ideal $\mathcal{B}$. As mentioned above, we consider here three cases, $\mathcal{B}$ =compact operators, i.e. the Fredholm case, $\mathcal{B} = \mathcal{B}_H$, the locally traceable operators with bounded propagation speed and $\mathcal{B}$ =the uniform operators as discussed in [9]. These index theories are quite different. The Fredholm case always leads to a zero index in the $\bar{\mathcal{B}}_H$-case since the Fredholm property implies a gap in the spectrum (around zero) and then the index in $K_0(\bar{\mathcal{B}}_H)$ always vanishes as indicated in [7], [8]. Nevertheless, the Fredholm theory makes sense since there are many examples of manifolds and elliptic differential operators on open manifolds of bounded geometry having the Fredholm property. We consider components in the uncompleted and completed spaces of Clifford connections. The uncompleted case is particularly trivial. We restrict ourselves to the graded (even) case and discuss the rigidity if the operator in question is Fredholm.

Theorem 4.1 *Let* (M^n, g) *be open, complete, satisfying* (I) *and* (B_k), $k \geq r > n/2+1$, $\nabla, \nabla' \in \mathcal{C}(B_k)$, $\nabla' \in comp(\nabla)$ *in* $\mathcal{C}^r(B_k)$, $D = D^{\nabla}$, $D' = D^{\nabla'}$, S *and* D *graded and let* $D : \Omega^{0,1}(S, D) \to \Omega^{0,0}(S) = L_2(S)$ *be Fredholm. Then* D' *is also Fredholm and*

$$Ind\, D^{+} = Ind\, D'^{+}.$$

Proof. According to Proposition 3.6 and 3.9,

$$\Omega^{0,1}(S,D) = \Omega^{0,1}(S,\nabla) = \Omega^{0,1}(S,\nabla') = \Omega^{0,1}(S,D').$$

According to corollary 3.3, $\nabla' = \nabla + \eta$, where $\eta \in \tilde{\Omega}^{1,r}(\mathbf{g},\nabla)$. Now η is the $|\ |_{\nabla,r}$-limit of a sequence $(\eta_i)_i$ with compact support, and according to the Sobolev embedding theorem (Proposition 3.7) also the ${}^{b,1}|\ |$-limit of this sequence. In particular, ${}^{b,1}|\eta| < \infty$. Let $\varphi \in C_0^\infty(T^*M \otimes \mathbf{g})$. Then φ^{cl}, $\varphi^{cl}(s)(x) = \sum_{i=1}^n e_i \cdot \varphi_{e_i}(s(x))$ defines a compact operator $\Omega^{0,1}(S,\nabla) \to \Omega^{0,0}(S) = L_2(S)$. The pointwise operator norm $|\varphi^{cl}|_{op,x}$ coincides with $|\varphi|_x$ since $e_i \cdot \varphi_{e_i}(s(x)) = \varphi_{e_i}(e_i \cdot s(x))$ and $e_i \cdot$ is an isometry. Moreover,

$$\Big(\int |\sum e_i \cdot \varphi_{e_i}(s)|_x^2 \, d\,vol_x(g)\Big)^{1/2} \le C \cdot \Big(\int |\varphi|_x^2 \, |s(x)|^2 \, d\,vol\Big)^{1/2} \le$$

$$\le C \cdot {}^b|\varphi| \cdot \Big(\int |s(x)|^2 \, d\,vol\Big)^{1/2} = C \cdot {}^b|\varphi| \cdot |s|_{\nabla,0} \le C \cdot {}^b|\varphi| \cdot |s|_{\nabla,1}$$

implies $|\varphi^{cl}|_{op} \le C \cdot {}^b|\varphi|$. Therefore $\eta_i \to \eta$ with respect to $|\ |_{op}$ and φ^{cl} is compact. But $D'^{+} = D^{+} + \varphi^{cl,+}$. □

Theorem 4.2 *Let (M^n, g) be open, complete, satisfying $\nabla' \in comp(\nabla) \subset {}^{b,0}\tilde{\mathcal{C}}$, $D = D^\nabla : \bar{\Omega}^{0,1}(S,\nabla) \to \Omega^{0,0}(S) = L_2(S)$ Fredholm, S and D graded. Then $D' = D^{\nabla'}$ is also Fredholm and*

$$Ind\, D^{+} = Ind\, D'^{+}.$$

Proof. We obtain from Proposition 3.13 the equivalence

$$\Omega^{0,1}(S,D) = \Omega^{0,1}(S,D').$$

From our assumptions it follows immediately that $\nabla' = \nabla + \eta$, $\eta = \lim_{i\to\infty} \eta_i$ with respect to ${}^b|\ |$, $\eta_i \in C_0^\infty(T^* \otimes \mathbf{g})$. Once again we have $D'^{+} = D^{+} + \eta^{cl,+}$, $\eta^{cl,+}$ a compact operator. □

Remark. If M^n is compact then ${}^{b,m}\tilde{\mathcal{C}} = {}^{b,m}\mathcal{C}$ and $\mathcal{C}^r(B_k) = \mathcal{C}^r$ consist of one component and we obtain once again the well known rigidity of the index. If M^n is noncompact then the connection spaces above have uncountably many components and the simple intuition that $t\nabla' + (1-t)\nabla$ is an arc is totally wrong and the index can jump. This is a special feature of noncompactness.

Now we turn to the quite opposite case, $\mathcal{B} = \mathcal{B}_H$. D^+ is Fredholm if and only if the spectrum of D has a gap around zero and $\dim(ker\ D) < \infty$. Considering $\mathcal{B} = \mathcal{B}_H$, the index in $K_0(\bar{\mathcal{B}}_H)$ is automatically zero if the spectrum of D has a gap.

Denote by $\overline{Ind\, D}$ the image of $Ind\, D$ in $K_0(\bar{\mathcal{B}}_H)$.

Theorem 4.3 *Let (M^n, g) be open, complete, $(S,\nabla) \to M$ a graded Clifford bundle, $D = D^\nabla$ the graded generalized Dirac operator. Assume $\nabla' \in {}^b\mathcal{C}, D' = D^{\nabla'}$. Then*

$$\overline{Ind\, D} = \overline{Ind\, D'}.$$

Proof. Set $\nabla_t = t \cdot \nabla' + (1-t) \cdot \nabla$, $D_t = D^{\nabla_t}$. According to Proposition 3.13, $(D_t)_t$ is a continuous family. Apply Proposition 2.1. □

Remark. In ${}^b\mathcal{C}$ we have very "large" components and therefore a weaker rigidity.

Theorem 4.4 *Assume* (M_n, g) *open, complete with* $(I), (B_k(M)), k \geq r > n/2+1$, $(S, \nabla) \to M$ *a graded Clifford bundle,* $\nabla \in \mathcal{C}(B_k)$, $\nabla' \in comp(\nabla) \subseteq \mathcal{C}^r(B_k)$, $D = D^{\nabla}$, $D' = D^{\nabla'}$ *generalized graded Dirac operators. Then*

$$\overline{Ind\, D} = \overline{Ind\, D'}.$$

Proof. Set $\nabla_t = t \cdot \nabla' + (1-t)\nabla$, $D_t = D^{\nabla_t}$. Apply the Sobolev embedding theorem, Proposition 3.14 and Theorem 4.3. □

As the very simple proofs indicate, the main problem in studying rigidity is to define the right topology in $\mathcal{C}$. We solved this problem quite naturally an generally by construction of canonical uniform structures.

J. Roe defined in [9] the algebra $\mathcal{U}$ of uniform operators. We will not recall this definitions and constructions but refer to [9]. He assumed (M^n, g) with (I) and $(B_\infty(M))$, considered $\mathcal{C}(B_\infty)$ and assumed the existence of a so-called regular exhaustion. By means of such an exhaustion he defines a trace τ and a functional $\mathbf{m}$ on the bounded cohomology ${}^bH^n(M)$. Finally one starts with (M^n, g), S, D, constructs $Ind\, D \in K_0(\mathcal{U})$, $\tau \in HC^0(\mathcal{U})$, a form $\mathcal{I}_D \in^b H^n(M, g)$, a functional $\mathbf{m}$ on ${}^bH^n(M)$ and defines

$$ind_a\, D^+ = \langle Ind\, D^+, \tau \rangle, \; ind_t\, D^+ = \langle \mathcal{I}_D, \mathbf{m} \rangle.$$

Then J. Roe proved the following index theorem, assuming (I), $(B_\infty(M))$, $(B_\infty(S, \nabla))$.

Theorem 4.5

$$ind_a\, D^+ = ind_t\, D^+.$$

We have shown that (I), $(B_k(M))$, $(B_k(S, \nabla))$, $k > n/2$, are sufficient.

Assuming the index theorem above, we can prove the rigidity of the analytical index in this situation by proving the rigidity of the topological index. This shall be done in the next section.

5 The rigidity of the topological index

We recall some simple facts from bounded Chern–Weil theory ([6]). Let $(E, \nabla) \to M$ be a hermitian vector bundle with metric connection $\nabla = \nabla^E$ over (M^n, g) of rank N, U a bundle chart, $s_1, \ldots, s_N : U \to E|_U$ a local basis, $\Omega = \Omega^\nabla \equiv R^\nabla$ the curvature of E, $\Omega s_i = \sum_j \Omega_{ij} \otimes s_j$, where (Ω_{ij})

is an $N \times N$ matrix of 2-forms on U, and let M_N be the ring of $N \times N$ matrices. An invariant polynomial $P : M_N \to \mathbf{C}$ defines in a well known manner a closed graded differential form $P = P(\Omega) = P_0 + P_1 + \cdots$, where P_u is homogeneous, $P_u(\Omega) = 0$ for $2u > n$. The determinant is an example of a possibility for P. Let $\sigma_u(\Omega)$ be the $2u$-homogeneous part (in the sense of forms) of $\det(1 + \Omega_{ij})$.

Lemma 5.1 *Each invariant polynomial is a polynomial in* $\sigma_1, \ldots, \sigma_N$. □

Lemma 5.2 *If* $\nabla \in \mathcal{C}(B_0)$ *and* $u \geq 1$ *then* $\sigma_u \in {}^b\Omega^{2u}(M)$.

Proof. Let $|\ |_x$ be the pointwise norm. We have

$$|\Omega|_x^2 = \frac{1}{2} \sum_{i,j} \sum_{k<l} |\Omega_{ij,kl}|_x^2,$$

where $\Omega_{ij,kl} = \Omega_{ij}(e_k, e_l)$ and $e_1, \ldots, e_n$ is an orthonormal basis of T_xM. According to our assumption, $|\Omega|_x \leq a$ for all $x \in M$. The proof would be done if we could estimate $|\sigma_u(\Omega)|_x$ from above by $|\Omega|_x$. By definition

$$\sigma_u(\Omega) = \frac{1}{u!} \sum \varepsilon^{i_1 \cdots i_u}_{j_1 \cdots j_u} \Omega_{i_1 j_1} \wedge \cdots \wedge \Omega_{i_u j_u}, \tag{5.1}$$

where the summation runs over all $1 \leq i_1 < \cdots < i_u \leq N$ and all permutations $(i_1, \ldots, i_u) \to (j_1, \ldots, j_u)$. ε denotes the sign of the permutation. We perform induction. For $u = 1$ follows $\sigma_1(\Omega) = \sum \Omega_{ii}$. The inequality

$$|\Omega_{ij}|_x^2 \leq \sum_{s,t} |\Omega_{s,t}|_x^2 = 2|\Omega|_x^2 \tag{5.2}$$

implies in particular $|\sigma_1(\Omega)|_x^2 \leq b|\Omega|_x^2$. For arbitrary forms φ, ψ one has

$$|\varphi \wedge \psi|_x \leq |\varphi|_x \cdot |\psi|_x. \tag{5.3}$$

(5.1)–(5.3) and an easy induction argument therefore yield

$$|\sigma_u(\Omega)|_x^2 \leq c|\Omega|_x^{2u},$$

together with $|\Omega|_x^2 \leq a^2$. Finally $|\sigma_u(\Omega)|_x \leq d$. □

Corollary 5.1 *Let* P *be an invariant polynomial,* $\nabla \in \mathcal{C}(B_0), u \geq 1$. *Then each form* $[P_u(\Omega)]$ *is an element of* ${}^bH^{2u}(M)$. □

Corollary 5.2 *Under the assumptions of corollary 5.1,* P *and* ∇ *define well defined classes* $[P_u(\Omega^\nabla)] \in {}^bH^{2u}(M)$, $u = 1, 2, \ldots$ □

Now the natural question arises: how does $[P_u(\Omega^\nabla)]$ depend on ∇?

We denote $I = [0,1], i_t : M \to I \times M$ the embedding $i_t(x) = (t,x)$ and endow $I \times M$ with the product metric $\begin{pmatrix} 1 & 0 \\ 0 & g \end{pmatrix}$.

Let ${}^b\Omega^{q,d}$ be the space of all C^1 q–forms φ on M such that φ and $d\varphi$ are bounded and define the bounded cohomology ${}^bH^*(M)$ as the cohomology of the complex

$$\cdots \xrightarrow{d} {}^b\Omega^{q+1,d} \xrightarrow{d} {}^b\Omega^{q,d} \xrightarrow{d} {}^b\Omega^{q-1,d} \xrightarrow{d} \cdots,$$

${}^bH^q = {}^bZ^q / {}^bB^q$.

Lemma 5.3 *For every $q \geq 0$ there exists a linear bounded mapping K : ${}^b\Omega^{q+1,d}(I \times M) \to {}^b\Omega^{q,d}(M)$ such that $dK + Kd = i_1^* - i_0^*$.*

Proof. Since $g_{I\times M} = \begin{pmatrix} 1 & 0 \\ 0 & g \end{pmatrix}$, $i_t : M \to I \times M$ is an isometric embedding and i_t^* is bounded. i_t^* maps into ${}^b\Omega^{q,d}(M)$ because $|di^*\varphi|_x = |i^*d\varphi|_x \leq c|d\varphi|_x$. Denote $X_0 = \frac{\partial}{\partial t}$ and for $\varphi \in {}^b\Omega^{q+1,d}(I \times M)$ $\varphi_0(X_1, \ldots, X_n) := \varphi(X_0, X_1, \ldots, X_q)$. Then $\varphi_0 \in {}^b\Omega^{q+1,d}(I \times M), |\varphi_0|_{t,x} \leq |\varphi|_{t,x}$, and we define

$$(K\varphi)(X_1, \ldots, X_q) := \int_0^1 i_t^*\varphi_0(X_1, \ldots, X_q)dt.$$

Therefore K is bounded too. The equation $dK + Kd = i_1^* - i_0^*$ is a well known fact. □

Lemma 5.4 *Let $f, g : M \to N$ be smooth mappings between Riemannian manifolds, $F : I \times M \to N$ a smooth homotopy, $f^*, g^* : {}^b\Omega^{q,d}(N) \to {}^b\Omega^{q,d}(M)$, $F^* : {}^b\Omega^{q,d}(N) \to {}^b\Omega^{q,d}(I \times M)$ bounded and $\varphi \in {}^b\Omega^{q,d}(N)$ closed. Then $(g^* - f^*)\varphi \in {}^bB^q(M)$.*

Proof. According to our assumption, $KF^*\varphi \in {}^b\Omega^{q-1,d}(M)$ and $(g^* - f^*)\varphi = ((F \circ i_1)^* - (F \circ i_0)^*)\varphi = (i_1^*F^* - i_0^*F^*)\varphi = (dK + Kd)F^*\varphi = dKF^*\varphi$. □

Now we are able to prove our main proposition.

Proposition 5.1 *Let $P : M_N \to \mathbf{C}$ be an invariant polynomial, $u \geq 1$. Then each component U of ${}^{b,2}\mathcal{C}(B_1)$ determines a uniquely determined cohomology class $[P_u(\Omega^U)] \in {}^bH^{2u}(M)$.*

Proof. All elements of U are at least of class C^2. Assume ∇_0, $n\nabla_1 \in U$; set $\eta := \nabla_1 - \nabla_0$ and $\nabla_t = \nabla_0 + t\eta$, $t \in [0, 1]$. We have to show $[P_u(\Omega^{\nabla_0})] =$

$[P_u(\Omega^{\nabla_1})]$. Consider $\Omega_t := \Omega^{\nabla_t}$,

$$\Omega_t = \Omega_0 + td^{\nabla_0}\eta + \frac{1}{2}t^2[\eta, \eta].$$

$|\Omega_t|_x$ is bounded on M. If $\bar{p} : [0,1] \times M \to M$ denotes the projection $(t, x) \mapsto x$, $' = p^*E$ the lifting of the bundle spaces, then $p^*\nabla_0, p^*\nabla_1$ are connections for the lifted bundles. $tp^*\nabla_1 + (1-t)p^*\nabla_0 = p^*\nabla_0 + tp^*\eta$ is again a connection ∇'.

$$\Omega^{\nabla'} = p^*\Omega^{\nabla_0} + td^{p^*\nabla_0}p^*\eta + \frac{1}{2}t^2[p^*\eta, p^*\eta]$$

is bounded again. ∇', $\Omega^{\nabla'}$ define a bounded cocycle on $[0, 1] \times M$. Let i_t again be the mapping $x \mapsto (t, x)$. Then $i_0^*(E', \nabla')$ resp. $i_1^*(E', \nabla')$ can be identified with (E, ∇_0) resp. (E, ∇_1). i_t, $0 \le t \le 1$, is a smooth bounded homotopy between i_0 and i_1. According to 5.6, $i_0^*P_u(\Omega^{\nabla'})$ and $i_1^*P_u(\Omega^{\nabla'})$ are cohomologous in ${}^bH^{2u}(M)$, i.e. $P_u(\Omega^{\nabla_0})$ and $P_u(\Omega^{\nabla_1})$ are cohomologous in ${}^bH^{2u}(M)$. □

Definition 5 *We define for a component U of ${}^{b,2}\mathcal{C}(B_1)$ the u-th Chern class $c_u(E, U)$ by*

$$c_u(E, U) := \frac{1}{2\pi\imath}u\left[\sigma_u(\Omega^U)\right].$$

Then $c_u \in {}^bH^{2u}(M)$.
Let (E, ∇) be a real Riemannian vector bundle, (E^C, ∇^C) its complexification. If ∇ satisfies (B_1) then the same holds for ∇^C. There exists an inclusion of the components $U \subset {}^{b,2}\mathcal{C}_E(B_1)$ into the components $U^C \subset {}^{b,2}\mathcal{C}_{E^C}(B_1)$. Then we define the u-th Pontrjagin class $p_u(E, U)$ by

$$p_u(E, U) := (-1)^u c_{2u}(E^C, U^C).$$

Theorem 5.1 *Assume (M^n, g) open complete with (I) and (B_k), $(S, \nabla) \to M$ a Clifford bundle with (B_k), $k > n/2 + 1$, η a grading, D the generalized Dirac operator, $D\eta + \eta D = 0$. Then the bounded cohomology class of the index form $\mathcal{I}_D$ is an invariant of comp(∇) in ${}^{b,k+1}\mathcal{C}(B_k)$.*

Proof. According to [1], the most general Clifford module is of the form $\Delta \otimes V$, where Δ is the spin representation and V a vector space. Translating this to bundles, the index form of the corresponding generalized Dirac operator is given by

$$\mathcal{I}_{D_V} = \hat{A}(TM) \cdot ch(V),$$

where $ch(V)$ denotes the Chern character. Now apply Proposition 5.1 to $ch(V)$. □

Corollary 5.3 *Assume the assumptions of Theorem 5.1 and a regular exhaustion for M. Then the analytical index* $ind_a D^+ = \langle Ind\, D^+, \tau \rangle$ *is an invariant of* $comp(\nabla)$ *in* ${}^{b,k+1}\mathcal{C}(B_k)$.

Proof. The assertion follows from Theorem 5.1 and the index theorem 4.5. □

Remark 1 . We also introduced canonical intrinsic topologies in several spaces of Riemannian metrics. Similar results as in sections 3, 4, 5 can be obtained if the metric g of (M^n, g) varies in its component. For reasons of technical length we will not present them here.

References

[1] M.Atiyah, R.Bott, V.Patodi. *On the heat equation and the index theorem.* Inventiones mathematicæ **19** (1973), 279–330

[2] U.Bunke. *Spektraltheorie von Diracoperatoren auf offenen Mannigfaltigkeiten.* Thesis Greifswald 1991

[3] J.Eichhorn. *Elliptic differential operators on noncompact manifolds of bounded geometry.* Teubner–Texte zur Mathematik 106 (1988), 4–169

[4] J.Eichhorn. *The Manifold Structure of Maps Between Open Manifolds.* Ann. of Global Analysis and Geometry **3** (1993), 253–300

[5] J.Eichhorn. *Gauge Theory on open Manifolds of bounded Geometry.* International J. of Modern Physics A **7**, No. 17 (1992), 3927–3977

[6] J.Eichhorn. *Characteristic classes and numbers for fibre bundles over noncompact manifolds.* Proc. Conf. Diff. Geom. and Appl., Reidel Publ. (1987), 13–29

[7] J.Roe. *Exotic cohomology and index theory for complete Riemannian manifolds.* Preprint Oxford 1990

[8] J.Roe. *Coarse cohomology and index theory on Riemannian manifolds,* Mem. Amer. Math. Soc., No. 497 (1993).

[9] J.Roe. *An index theorem on open manifolds.* J. Diff. Geom. 27 (1988), 87–113

[10] H.Schubert. *Topologie.* Stuttgart 1964

FACHBEREICH MATHEMATIK, UNIVERSITÄT GREIFSWALD, JAHNSTRASSE 15A, 17487 GREIFSWALD, GERMANY

email: `eichhorn@math-inf.uni-greifswald.dbp.de`

Remarks on Steenrod homology

Steven C. Ferry

1. Generalities and motivation

The purpose of this note is to introduce readers to Steenrod homology theory and to explain why this theory arises in studies of the Novikov Conjecture. In a nutshell, Steenrod homology is the homology theory which is appropriate for the study of compact metric spaces which have bad local properties. Studies of the integral Novikov Conjecture and Borel Conjecture for the fundamental group of an aspherical polyhedron K frequently make use of compactifications of the universal cover $\tilde{K}$. The space at infinity in such a compactification cannot be assumed to have good local properties,[1] so homological studies of the space at infinity require something like Steenrod homology. To see that singular homology is insufficient for the study of spaces with bad local properties, we begin by recalling the statement of Alexander Duality:

Theorem (Alexander Duality). *If A is a compact subset of an oriented n-manifold M, then*

$$\check{H}^q(A) \cong H_{n-q}(M, M-A)$$

where $\check{H}^*(\ )$ denotes Čech cohomology. In case $M = \mathbb{R}^n$, we have

$$\check{H}^q(A) \cong \bar{H}_{n-q-1}(\mathbb{R}^n - A),$$

where $\bar{H}_*(\ )$ denotes reduced singular homology theory. □

If the space A has bad local properties, the use of Čech cohomology rather than singular cohomology on the left side of this isomorphism is necessary.

The author was partially supported by NSF grant number DMS-9305758.

[1] The universal cover of the figure eight is compactified by a Cantor set. Davis and Januszkiewicz [DJ] have constructed aspherical manifolds whose universal covers are negatively curved polyhedra which are compactified by homology manifolds which are not locally simply connected.

If A is the topologist's sine curve

$A =$ 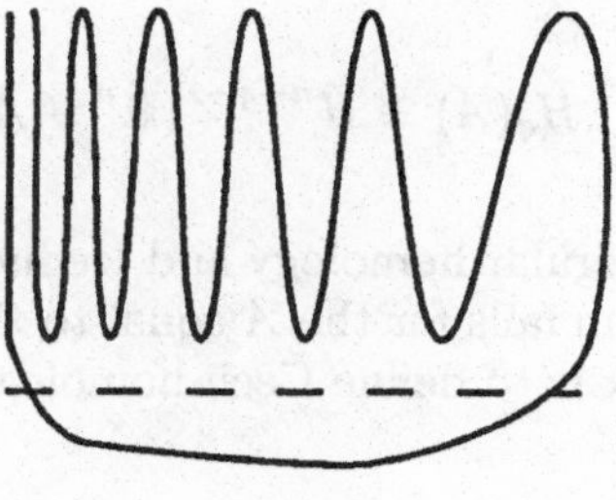

the singular complex of A is chain contractible, so the singular cohomology of A is isomorphic to that of a point. To see this, cut A into two pieces above and below the dotted horizontal line pictured. The upper half has two contractible path components, so its singular homology is that of two points. The lower half is an interval and the Mayer-Vietoris sequence for the union shows that A has the singular homology of a point. The singular chain complex of A is therefore chain contractible and the singular cohomology is also trivial. On the other hand, A separates $\mathbb{R}^2$, so duality "predicts" $H^1(A) \cong \mathbb{Z}$, which fails in singular cohomology.

Definition. *If A is compact metric, let $\mathcal{U}_i$ be a sequence of finite open covers of A such that $\mathcal{U}_{i+1}$ refines $\mathcal{U}_i$ and so that*

$$\lim_{i\to\infty} \sup\{\operatorname{diam}(U) \mid U \in \mathcal{U}_i\} = 0.$$

Let N_i be the nerve of $\mathcal{U}_i$, i. e., the simplicial complex with vertices $\{\langle U\rangle \mid U \in \mathcal{U}_i\}$ such that $\langle U_0, \ldots, U_k\rangle \in N_i$ if and only if $U_0 \cap \cdots \cap U_k \neq \emptyset$. Then we define Čech cohomology by the formula

$$\check{H}^q(A) = \varinjlim\{H^q(N_i), s_{i*}\}$$

where $H^q(N_i)$ is singular cohomology and the maps $s_{i+1} : N_{i+1} \to N_i$ are induced by refinement.[2]

Using the fact that every open cover of A is refined by the $\mathcal{U}_i$'s for i greater than some i_0, one shows that any two such sequences of $\mathcal{U}_i$'s are cofinal. This implies that the definition is independent of the choice of $\mathcal{U}_i$'s.

---

[2] If N_{i+1} refines N_i, then for each $U \in N_{i+1}$ there is a $V_U \in N_i$ with $U \subset V_U$. If $\{\psi_U\}$ is a partition of unity subordinate to N_{i+1}, the map s_{i+1} is given by the formula $s_{i+1}(x) = \sum \psi_U(x)\langle V_U\rangle$.

If A is a finite polyhedron, we have a similar duality theorem from the homology of A to the cohomology of the complement:

$$H_q(A) \cong \bar{H}^{n-q-1}(\mathbb{R}^n - A)$$

where we have used singular homology and (reduced) singular cohomology. Again, this isomorphism fails for the A equal to the topologist's sine curve. A first attempt at a fix is to define Čech homology by

$$\check{H}_q(A) = \varprojlim H_q(N_i)$$

where the N_i's are as before, and use Čech homology on the left side of the isomorphism. This restores the isomorphism in the case of the topologist's sine curve, but fails for the dyadic solenoid.

Example. Let $d_2 : S^1 \to S^1$ be the degree two map $z \to z^2$. The *dyadic solenoid* is

$$\Sigma = \varprojlim\{S^1, d_2\} = \left\{ (z_1, z_2, \dots) \in \prod_{i=1}^{\infty} S^1 \;\middle|\; d_2(z_{i+1}) = z_i \right\}.$$

The dyadic solenoid can be embedded in $\mathbb{R}^3$ as the intersection of solid tori T_i, where a generator of $\pi_1 T_{i+1}$ represents twice a generator of $\pi_1 T_i$.

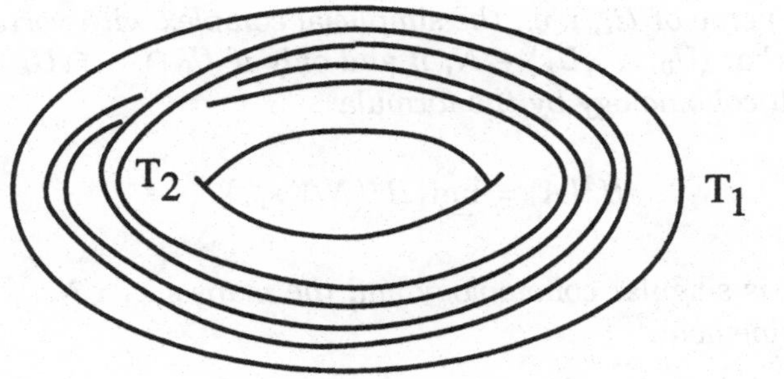

$\check{H}_i(\Sigma)$ is equal to $\mathbb{Z}$ for $i = 0$ and 0 for all $i > 0$. $S^3 - \Sigma$ is a union of tori $\bar{T}_i = S^3 - \mathring{T}_i$ where the inclusion of the core of T_i into T_{i+1} represents twice the generator of π_1. The cohomology of $S^3 - \Sigma$ may be computed using Milnor's $\varprojlim^1$ sequence in cohomology.

Definition. *If $\{A_i, h_i\}$ is an inverse system of abelian groups A_i and homomorphisms $h_i : A_i \to A_{i-1}$, define a shift map $\Delta : \prod A_i \to \prod A_i$ by $\Delta(a_1, a_2, \dots) = (a_1 - h_2(a_2), a_2 - h_3(a_3), \dots)$. Then $\varprojlim\{A_i, h_i\} = ker(\Delta)$ and $\varprojlim^1\{A_i, h_i\} = coker(\Delta)$.*

The functor $\varprojlim^1$ is the first derived functor of the inverse limit functor. For countable systems, the higher derived functors vanish, so an exact sequence of inverse systems gives rise to a six term sequence of $\varprojlim$'s and $\varprojlim^1$'s. The group $\varprojlim^1\{A_i, h_i\}$ is clearly trivial for sequences of surjections. This leads to easy calculations of $\varprojlim^1$ in some cases. If $A = \{A_i, h_i\}$ where each A_i is isomorphic to $\mathbb{Z}$ and h_i is multiplication by 2, $B = \{B_i, g_i\}$ where each B_i is isomorphic to $\mathbb{Z}$ and g_i is the identity, and $C = \{C_i, k_i\}$ with C_i cyclic of order 2^i and k_i the quotient surjection, then there is a short exact sequence $0 \to A \to B \to C \to 0$ where the map $A_i \to B_i$ is multiplication by 2^i and the map $B_i \to C_i$ is the quotient surjection. The six-term sequence in this case gives us

$$0 \to \mathbb{Z} \to \varprojlim\{\mathbb{Z}/2^i\mathbb{Z}, k_i\} \to \varprojlim{}^1\{\mathbb{Z}, \times 2\} \to 0$$

which shows that $\varprojlim^1\{\mathbb{Z}, \times 2\}$ is uncountable.

Theorem (Milnor's cohomology $\varprojlim^1$ sequence [Mi2]**).** *If $X = \bigcup U_i$ where $U_i \subset U_{i+1} \subset \dots$ are open sets, then there is a short exact sequence*

$$0 \to \varprojlim{}^1 H^{q-1}(U_i) \to H^q(X) \to \varprojlim H^q(U_i) \to 0. \quad \square$$

Applied to the complement of the solenoid, this gives a short exact sequence

$$0 \to \varprojlim{}^1\{\mathbb{Z}, \times 2\} \to H^2(S^3 - \Sigma) \to 0 \to 0.$$

This shows that $H^2(S^3 - \Sigma) \cong \varprojlim^1\{\mathbb{Z}, \times 2\}$.

This computation shows that Čech homology theory is not strong enough to give duality in the case of the solenoid, since duality predicts

$$\mathbb{Z} \cong \check{H}_0(\Sigma) \cong H^2(\mathbb{R}^3 - \Sigma) \cong \varprojlim{}^1\{\mathbb{Z}, \times 2\} \oplus \mathbb{Z},$$

which is wildly false.

Remark. The astute reader will have noticed that the singular homology of the dyadic solenoid is the same as its Steenrod homology. One might hope from this example that replacing the inverse limit by a homotopy inverse limit and *then* taking singular homology would lead to the Steenrod theory. This process fails for the object pictured below, which is the one-point compactification of an infinite-holed surface.

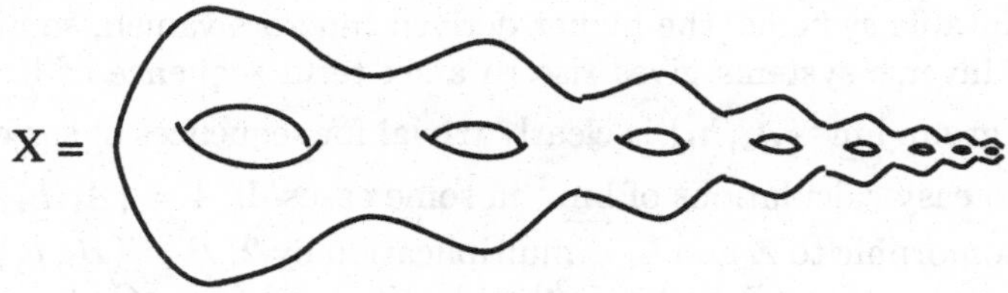

The space X is the inverse limit of surfaces of increasing genus. Since X separates $\mathbb{R}^3$ into two components, duality requires that $H_2(X)$ be equal to $\mathbb{Z}$. Since increasing numbers of singular simplices are required to represent the top classes of these surfaces, passing to a homotopy inverse limit and taking singular homology fails for this example.

Even though the solenoid Σ is connected, the duality calculation above shows that its 0th homology group in the "correct" homology theory must be uncountably generated. This may seem counterintuitive, but it turns out that $\pi_1(\mathbb{R}^3/\Sigma)$ is uncountably generated, which is consistent with the H_0 calculation. One way to show this is to use the Steenrod homology theory $H_*^{st}(\ )$ which will be defined in the next section. The long exact sequence of the pair (S^3, Σ) gives an isomorphism $H_1^{st}(S^3, \Sigma) \cong \bar{H}_0^{st}(\Sigma)$ and invariance under relative homeomorphisms gives $H_1^{st}(S^3, \Sigma) \cong H_1^{st}(S^3/\Sigma)$. An appropriate Hurewicz theorem then shows that $\pi_1(\mathbb{R}^3/\Sigma)$ is uncountable.

2. The construction of Steenrod homology theories

In [St], Steenrod constructed a homology theory $H_*^{st}(\ )$ for compact pairs of metric spaces which enabled him to prove the following duality theorem.

Theorem (Steenrod Duality). *For all compact* $A \subset \mathbb{R}^n$,

$$H_q^{st}(A) \cong \bar{H}^{n-q-1}(\mathbb{R}^n - A).$$

To see what the definition of $H_*^{st}(\ )$ must entail, it is instructive to examine the maps in Milnor's $\varprojlim^1$ sequence. Thinking of $H^n(X)$ as $[X, K(\mathbb{Z}, n)]$, restriction gives maps $H^n(X) \to H^n(U_i)$ and a map $H^n(X) \to \varprojlim H^n(U_i)$. If $\alpha : X \to K(\mathbb{Z}, n)$ is a map such that $\nu_i : \alpha|U_i \sim *$ is a nullhomotopy for each i, the ν_i's do not necessarily piece together to give a nullhomotopy of d_2. In fact, for each i, ν_i and $\nu_{i+1}|U_i$ piece together to give a map $\Sigma U_i \to K(\mathbb{Z}, n)$, which yields an element of $H^n(\Sigma U_i) \cong H^{n-1}(U_i)$. This gives an element of $\varprojlim^1 H^{n-1}(U_i)$ which is trivial if and only if the ν_i's may be rechosen so that they piece together to give a nullhomotopy of α.

These cohomological considerations also suggest the correct definition of the appropriate homology theory — the missing ingredient in the naive Čech definition above is *coherence* of the cycles on the various nerves of covers of X. An n-dimensional Steenrod *regular cycle* α on X should be something like a collection of n-cycles α_i on the N_i's together with an explicit choice of $(n+1)$-chains β_i on each N_i such that $\partial\beta_i = \alpha_i - s_{i+1\#}(\alpha_{i+1})$. Here are Steenrod's words on the subject from p. 833 of [St]:
"In this paper, we introduce a new type of cycle. In essence, it is a single infinite cycle (in the compact metric space X) with the regularity requirement that the diameters of successive simplices shall converge to zero. These cycles lead to homology groups $H^q(X)$ which are new topological invariants. The Vietoris[3] homology group of one less dimension $V^{q-1}(X)$ proves to be a homomorphic image of $H^q(X)$. Explicitly, to each Vietoris cycle there corresponds a regular cycle of one higher dimension and conversely. However, to certain bounding Vietoris cycles there correspond non-bounding regular cycles. In this way we succeed in strengthening the requirements for bounding ... "[4]

From p. 834 of the same paper, we have:

"A (locally finite) q-chain of a simplicial complex K is a function defined over the q-simplices of K with values in the (coefficient) group."

Definition. A *regular map of a complex K in X* is a function f defined over the vertices of K with values in X such that, for any $\epsilon > 0$, all but a finite number of simplices have their vertices imaging onto sets of diameter $< \epsilon$.

---

[3] Steenrod's formal definition, which follows, is self-contained and does not require a prior knowledge of Vietoris homology.

[4] Evidently, the convention of writing homology with lower indices had not yet taken hold in 1940. We will denote Steenrod's groups by H_q^{st} and shift the dimension by one to be consistent with current usage. A modern treatment of Steenrod's theory from a somewhat different point of view is given in [Ma].

Definition. A *regular q-chain of X* is a set of three objects: a complex A, a regular map f of A in X, and a (locally finite) q-chain C^q of A. If C^q is a q-cycle, (A, f, C^q) is called a *regular q-cycle*.

Definition. Two regular q-cycles (A_1, f_1, C_1^q) and (A_2, f_2, C_1^q) of X are homologous if there exists a $(q+1)$-chain (A, f, C^{q+1}) such that A_1 and A_2 are closed (not necessarily disjoint) subcomplexes of A, f agrees with f_1 on A_1 and f_2 on A_2, and $\partial C^{q+1} = C_1^q - C_2^q$.

In modern terminology, the reduced Steenrod homology group $\bar{H}_q^{st}(X)$ is the abelian group of homology classes of regular $(q+1)$-cycles in X. Note the dimension shift. Steenrod's definition yields reduced homology. Our H_q^{st} will denote *unreduced* Steenrod homology.

Recall that the *locally finite singular homology* of a locally compact metric space X is defined to be the homology theory based on infinite singular chains with the requirement that if $K \subset X$ is compact, then K meets the images of only finitely many simplices from any chain. The boundary of a locally finite chain is a locally finite chain, so we can form the locally finite singular chain complex of X and take homology in the usual manner. We will denote locally finite singular homology by the symbol $H_q^{lf}(X)$. If X is compact, locally finite chains are finite and H_q^{lf} is ordinary singular homology.

We will now argue that the homology theory constructed by Steenrod for a compact finite-dimensional metric space X is isomorphic to the locally finite homology of the complement of X in a disk D^n if X is embedded as a subset of the boundary.

If X is a compact metric space with finite covering dimension, X can be embedded in $\partial D^n \subset D^n$ for $n \geq 2 \cdot \dim(X) + 2$.

Proposition. *For compact* $X \subset \partial D^n \subset D^n$, $\bar{H}_q^{st}(X)$ *is isomorphic to* $H_{q+1}^{\ell f}(D^n - X)$.

Remark. At first glance, this proposition may look like a restatement of Steenrod's duality theorem. In fact, its character is quite different. Instead of relating homology to cohomology in dual dimensions, this result relates homology to locally finite homology with a dimension shift of one. What we are really showing here is that the Steenrod homology of X is isomorphic to the "homology of the end"[5] of $D^n - X$ with a dimension shift. The modern

[5] For a locally finite CW complex Z, there is an inclusion $\mathcal{S}_*(Z) \to \mathcal{S}_*^{lf}(Z)$ of the sin-

reader who is more comfortable with locally finite homology on noncompact manifolds than with Vietoris cycles on compact metric spaces can *define* the Steenrod homology of X to be the locally finite homology of $D^n - X$ with no appreciable loss of understanding or integrity. Later in this section, we will use the *fundamental complex* to generalize this definition to arbitrary compact metric spaces. Note that this theorem would be false if we used S^n in place of D^n, since when X is a polyhedron, the homology of the end of $S^n - X$ is the homology of the boundary of a regular neighborhood of X rather than the homology of X itself. Geometrically, the key point is that cycles in the complement can't link cycles in X when X is in the boundary of a disk. It turns out that when X is embedded in the boundary of a disk D, the proper homotopy type of $D - X$ depends only on X. Later, we will prove the analogous fact that the proper homotopy type of a fundamental complex of X depends only on X.

Note that if X is a finite polyhedron in ∂D, then D collapses to the cone (from the center of D) on X rel the cone on X. This shows that $D - X$ is proper homotopy equivalent to the open cone on X. □

Proof of Proposition. For the proof, we consider D^n to be $D^{n-1} \times [0,1]$ and assume that $X \subset D^{n-1} \times [0,1]$. If (A, f, C^{q+1}) is a Steenrod cycle, the map $f : A^{(0)} \to X$ extends to a map

$$\hat{f} : A \to D^{n-1} \times \{1\} \subset D^{n-1} \times [0,1]$$

by convexity. Let $\rho : \hat{A} \to [0,1]$ be a function from the 1-point compactification $\hat{A} = A \cup \{\infty\}$ of A to $[0,1]$ so that $\rho^{-1}(1) = \infty$. Now let $\bar{f} : A \to D^{n-1} \times [0,1]$ be given by

$$\bar{f}(a) = (\mathrm{proj}_{D^{n-1}} \circ \hat{f}(a), \rho(a)).$$

$\bar{f}(C^{q+1})$ is a locally finite cycle on $D^{n-1} \times [0,1] - X$. A relative version of this construction shows that homologous Steenrod cycles give homologous locally finite cycles. Conversely, triangulating $D^{n-1} \times [0,1] - X$ by a triangulation in which simplices get smaller and smaller near X shows that every locally finite cycle on $D^{n-1} \times [0,1] - X$ comes from a Steenrod cycle.

---

gular chains into the locally finite singular chains. The quotient complex $\mathcal{S}_*^{lf}(Z)/\mathcal{S}_*(Z) = \mathcal{S}_{*-1}^e(Z)$ is often called the chain complex of the end and its homology is called the homology of the end. The long exact homology sequence associated to $0 \to \mathcal{S}_*(Z) \to \mathcal{S}_*^{lf}(Z) \to \mathcal{S}_*^e(Z) \to 0$ shows that the locally finite homology of Z is isomorphic to the homology of the end when Z is contractible.

The required map on vertices simply sends a vertex v of this triangulation which is within ϵ of X to a point of $X \cap B_\epsilon(v)$. □

To see how this relates to Čech theory, let X be compact metric with a cofinal sequence of covers $\mathcal{U}_i$ with nerves N_i and refinement maps $s_{i+1} : N_{i+1} \to N_i$ as in the definition of Čech cohomology from the first section. Let $\mathcal{U}_0 = \{X\}$ and $N_0 = pt$. Form the mapping telescope of the $\{N_i, s_i\}$'s, starting with $N_0 = *$.

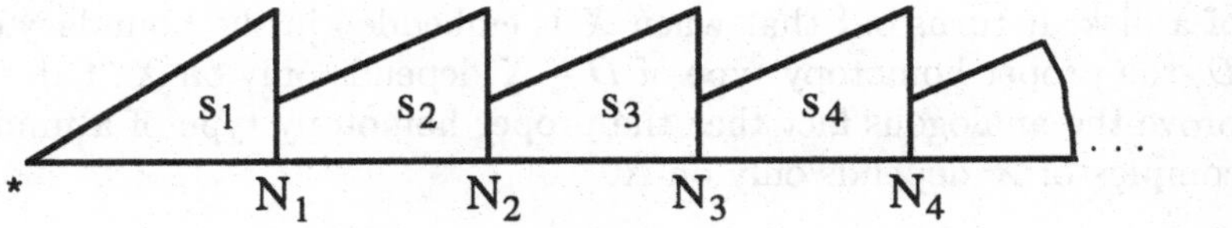

This is a Lefschetz *fundamental complex* of X. We will denote it by $OFC(X)$, for *open fundamental complex*. In the proof of Theorem 2 of [Mil], Milnor shows that if the sequence of covers $\{N_i\}$ and maps s_i are chosen carefully,[6] then the telescope can be compactified by adding a copy of X at infinity. This compactification can be obtained as the inverse limit of the sequence $\{M(s_1) \cup \cdots \cup M(s_i), p_i\}$ where

$$p_i : M(s_1) \cup \cdots \cup M(s_i) \to M(s_1) \cup \cdots \cup M(s_{i-1})$$

is the mapping cylinder projection on the last mapping cylinder. We will refer to this as $CFC(X)$. Lefschetz has shown (see [Bor]) that a compact metric space is an ANR if and only if it is ϵ-homotopy dominated by some finite polyhedron for each $\epsilon > 0$. The stages in this sequence dominate the inverse limit, so $CFC(X)$ is a contractible ANR. We can repeat the argument of the previous Proposition using the mapping cylinder rays in place of the collar structure to show that the Steenrod homology of X is isomorphic to the locally finite homology of the fundamental complex. Here is the relation to embedding X in a disk: if we were to replace each K_i by

[6] To get this result, Milnor has to be *much* more careful with the maps. The ones we've described are simplicial, so the inverse limit space obtained this way is guaranteed to contain topological simplices, which do not abound in generic compact metric spaces. The interested reader should check [Mil1] for the proof. Nevertheless, the open fundamental complex obtained as I describe here is proper homotopy equivalent to the one Milnor constructs and therefore suffices for a definition of Steenrod homology. Technically, $\varprojlim\{N_i, s_i\}$ is a compactum X' which is (strong) shape equivalent to X and which therefore has Steenrod homology isomorphic to that of X. For X a finite polyhedron, the open cone on X is a good model for an open fundamental complex.

$K_i \times Q$, where X is the Hilbert cube, and each s_i by a tiny generic perturbation of $(s_i \circ proj_{K_i}) \times 0$, then the resulting fundamental complex would be homeomorphic to the Hilbert cube with the inverse limit X embedded in a "boundary." The proof of this is not difficult, but would take us too far afield for a short survey. The fundamental nature of the fundamental complex is illustrated by the following, which is a version of Chapman's Complement Theorem [Si].

Proposition. *The simple homotopy type of the fundamental complex of $\{K_i, \alpha_i\}$ is invariant under homotopies of the bonding maps α_i and passage to subsequences. That is, the simple homotopy type of the fundamental complex of $\{K_i, \alpha_i\}$ is a pro-homotopy invariant of $\{K_i, \alpha_i\}$.*

Proof. M. Cohen's book [Co] on simple homotopy theory contains two mapping cylinder lemmas.

Lemma. *If $f, g : K \to L$ are homotopic maps between finite CW complexes, then $M(f)$ is simple-homotopy equivalent to $M(g)$ rel $K \amalg L$.*

Lemma. *If $f : K \to L$ and $g : L \to P$ are maps between finite CW complexes, then $M(g \circ f)$ is simple-homotopy equivalent to $M(f) \cup_L M(g)$ rel $K \amalg P$.*

Invariance under homotopies of the bonding maps is an immediate consequence of the first, while invariance under passage to subsequences is a similarly immediate consequence of the second. □

Corollary. *The Steenrod homology of X is the locally finite homology of the fundamental complex of $\{N_i, s_i\}$, where the N_i's are the nerves of any cofinal sequence of covers of X.*

It is not difficult[7] to prove Axioms 1–9 for this theory. A map between compact metric spaces induces a map between cofinal systems of nerves of covers and a proper homotopy class of maps between fundamental complexes. Let $\mathcal{M}$ be the category of compact metric spaces and continuous maps between them. Let $\mathcal{A}$ be the category of abelian groups. We can strengthen the above corollary to obtain:

[7] But we will refrain. Aside from functoriality, which we describe here, Axioms 8 and 9 (see below) are the most interesting. Axiom 8 follows from an excision and standard facts about cell-like maps, while the proof of Axiom 9 uses a direct construction to produce a locally finite cycle in $OFC(\bigvee X_i)$ representing the product of locally finite cycles on the $OFC(X_i)$'s. When the X_i,s are finite polyhedra, this construction is straightforward if we use the fundamental complex obtained by writing $\bigvee X_i$ as the inverse limit of $\bigvee_{i=1}^n X_i$'s.

Proposition. *There is a functor $S : \mathcal{M} \to \mathcal{A}$ which takes a compact metric space X to $H_*^{\ell f}(OFC(X))$. This functor is naturally equivalent to the reduced Steenrod homology functor $\bar{H}_{*-1}^{st}$.*

Proof. The equivalence is given by the isomorphisms

$$H_*^{\ell f}(OFC(X)) \cong \bar{H}_*^{st}(CFC(X)/X) \cong H_*^{st}(CFC(X), X) \cong \bar{H}_{*-1}^{st}(X)$$

where the first isomorphism is given by the isomorphism between the locally finite homology of a space and the reduced Steenrod homology of its 1-point compactification and the last isomorphism comes from the long exact sequence of the pair $(CFC(M), M)$ and the fact that $CFC(M)$ is contractible. □

If X and Y are compact metric spaces, a proper homotopy class of maps between fundamental complexes $OFC(X)$ and $OFC(Y)$ does not induce a homotopy class of maps $X \to Y$. Since a proper map $OFC(X) \to OFC(Y)$ does induce a homomorphism $H_*^{\ell f}(OFC(X)) \to H_*^{\ell f}(OFC(Y))$ and therefore a homomorphism $\bar{H}_*^{st}(X) \to \bar{H}_*^{st}(Y)$, it is desirable for many purposes to expand the notion of map between compact metric spaces so that proper maps between open fundamental complexes $OFC(X)$ and $OFC(Y)$ are in 1-1 correspondence with "maps" between X and Y. This is essentially the definition of a *strong shape morphism* from X to Y [DS], [EH].

It is worthwhile to interpret these strong shape morphisms in terms of nerves of covers. After passing to suitable subsequences, a proper map $\alpha : OFC(X) \to OFC(Y)$ gives rise to a sequence of maps $\alpha_i : N(\mathcal{U}_i) \to N(\mathcal{V}_i)$ together with choices of coherence homotopies making the diagrams

$$\begin{array}{ccc} N(\mathcal{U}_i) & \xrightarrow{\alpha_i} & N(\mathcal{V}_i) \\ \uparrow & & \uparrow \\ N(\mathcal{U}_{i+1}) & \xrightarrow{\alpha_{i+1}} & N(\mathcal{V}_{i+1}) \end{array}$$

homotopy commute. Two such systems are equivalent if, after passing to subsequences, there are homotopies of the corresponding α_i's and homotopies of the coherences. This last is what makes the shape theory "strong." If we think of X and Y as being embedded in euclidean space, this means that a strong shape morphism from X to Y is represented by a sequence of continuous maps from smaller and smaller neighborhoods of X to smaller and smaller neighborhoods of Y together with homotopies connecting maps at one stage to maps at the next. The similarity to the various notions of "approximate map" appearing in the analytic approaches to the Novikov Conjecture is striking.

Steenrod shows that the homology theory resulting from any of these equivalent definitions satisfies the Eilenberg-Steenrod axioms for all compact pairs. In [Mil], Milnor proves that Steenrod homology satisfies two extra axioms — invariance under relative homeomorphism and a cluster axiom, which says that the homology of an infinite compact wedge is the product of the homologies. One of the main theorems of [Mil] says that these axioms characterize Steenrod homology theory for compact metric spaces in the same way that the usual Eilenberg-Steenrod axioms characterize ordinary homology theory on finite complexes. It is the extension of this result to extraordinary homology theories which comes into play in the study of the Novikov Conjecture. Here is the statement of Milnor's theorem:

Theorem (Characterization of Steenrod homology). *There exists one and only one homology theory*[8] $H_*(\ )$ *defined for pairs of compact metric spaces which satisfies the following two Axioms as well as the seven Eilenberg-Steenrod Axioms and which satisfies* $H_0(pt) = G$.

The new axioms are:

Axiom 8. *Invariance under relative homeomorphism.*

Axiom 9. (Cluster Axiom). *If* X *is the union of countably many compact subsets* $X_1, X_2, \ldots$ *which intersect pointwise at a single point* b, *and which have diameters tending to zero, then* $H_q(X, b)$ *is naturally isomorphic to the direct product of the groups* $H_q(X, b)$.

Axiom 9 is closely related to the following:

Theorem (Milnor $\varprojlim^1$ sequence in homology). *If* X *is the inverse limit of a sequence of maps of compact metric spaces* $X_{i+1} \to X_i$ *there is defined a short exact sequence*

$$0 \to \varprojlim{}^1 H^{st}_{q+1}(X_i) \to H^{st}_q(X) \to \varprojlim H^{st}_q(X_i) \to 0. \quad \square$$

The most useful case of this is the case in which a compact metric space X is written as an inverse limit of finite polyhedra, $X = \varprojlim\{K_i, s_i\}$. In this case, the sequence allows us to compute the Steenrod homology of X in terms of the singular homology groups of the K_i's. For instance, if Σ is the solenoid of the first section, we have

$$0 \to \varprojlim{}^1 H_1(S^1) \to H^{st}_0(\Sigma) \to \varprojlim H_0(S^1) \to 0$$

[8] This homology theory is, of course, $H^{st}_*(\ )$.

which shows that $H_0^{st}(\Sigma)$ is isomorphic to $H^2(\mathbb{R}^3 - \Sigma)$, as computed in the first section. The proof, which we take directly from [Mil] is simple:[9]

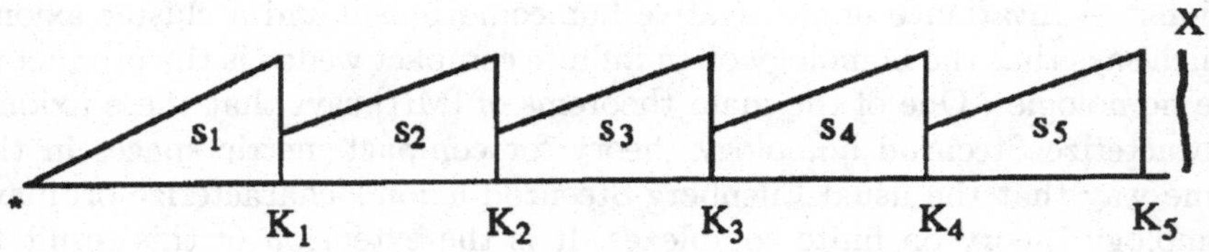

Let K_0 be a point and form the mapping telescope compactified by a copy of X at "infinity" as in the construction of the closed fundamental complex. The compactified telescope T is a contractible ANR containing X. Choose a point x_0 in X and let R be a base ray running down the mapping cylinders from x_0 to $*$. The boundary map for the pair (T, X) and the relative homeomorphism axiom comparing (T, X) with $(T/X, [X])$ give an isomorphism $\bar{H}_{k+1}^{st}(T/X) \cong \bar{H}_k^{st}(X)$. Let $A \subset T/X$ consist of the union of R with all of the K_i's. Collapsing R shows that A is homotopy equivalent to a cluster as in Axiom 9. On the other hand, $(T/X)/A$ is also a cluster, this time of the suspensions of the K_i's. The long exact sequence of (T, A) and Axiom 8 give:

$$\bar{H}_{k+2}((T/X)/A) \xrightarrow{\partial} H_{k+1}(A) \to H_{k+1}(T/X) \to \bar{H}_{k+1}((T/X)/A) \xrightarrow{\partial} H_k(A).$$

Applying Axiom 9 and the isomorphism $\bar{H}_{k+1}^{st}(T/X) \cong \bar{H}_k^{st}(X)$ gives

$$\prod \bar{H}_{k+1}(K_i) \to \prod \bar{H}_{k+1}(K_i) \to \bar{H}_{k+1}(T/X) \to \prod \bar{H}_k(K_i) \to \prod \bar{H}_k(K_i).$$

Examination of the boundary map ∂ shows that it is the shift map Δ used in the definition of $\varprojlim$ and $\varprojlim^1$.

There is a considerable literature extending these results to larger classes of spaces. To quote from [In]: "Different constructions of Steenrod (exact) homology have been given by many authors (Steenrod, Milnor, Kolmogorov, Chogoshvili, Sitnikov, Borel, Moore, Sklyarenko, Massey, Kuzminov, Shvedov, Berikashvili, Inasaridze and others)." References to some of the appropriate papers are included in the bibliography of this paper.

[9] We repeat it here because of the argument's beauty and because the included picture may help readers to understand the argument.

Steenrod homology theories with coefficients in any spectrum were developed by Edwards-Hastings [EH] and Kahn-Kaminker-Schochet [KKS]. In modern algebraic topology, it is usual to think of the homology of a finite polyhedron K with coefficients in a spectrum $\mathcal{S}$ as being obtained by taking homotopy groups of the homology spectrum $\mathcal{H}(K)$, where $\mathcal{H}(K)$ is the smash product of K with $\mathcal{S}$. A definition of Steenrod homology which works for any spectrum $\mathcal{S}$ is to start with a compact metric space X, form the sequence of nerves N_i as above, form the homology spectra $\mathcal{H}(N_i;\mathcal{S}) = N_i \wedge \mathcal{S}$, and then define

$$H_k(X;\mathcal{S}) = \pi_k \operatorname{holim}\{\mathcal{H}(N_i;\mathcal{S})\}.$$

Thus, the Čech homology construction works if we "spacify" before taking the limit and then take homotopy groups! There is a lemma, due to Milnor ([BK] IX.3.1) , which says that if $Z = \varprojlim\{Z_i, \alpha_i\}$ and the maps $\alpha_i : Z_i \to Z_{i-1}$ are fibrations, then there is a short exact sequence

$$0 \to \varprojlim{}^1 \pi_{k+1} Z_i \to \pi_k Z \to \varprojlim \pi_k Z_i \to 0.$$

If $\{N_i\}$ is a sequence of nerves of covers of a compact metric space X as above, this translates into a Milnor exact sequence in Steenrod $\mathcal{S}$-homology

$$0 \to \varprojlim{}^1 H_{k+1}(N_i;\mathcal{S}) \to H_k(X;\mathcal{S}) \to \varprojlim H_k(N_i;\mathcal{S}) \to 0.$$

The correspondence between Steenrod homology theories on compacta and locally finite homology theories on locally compact ANR's carries over to homology with coefficients in a spectrum. Thus, one can use a good notion of "locally finite cycle" to define the Steenrod homology of a compactum to be the locally finite homology of its fundamental complex or, conversely, one can use a definition of Steenrod homology with coefficients in a spectrum to define the locally finite homology of a locally compact metric space X as the reduced Steenrod homology of its 1-point compactification.

3. Steenrod homology and the Novikov Conjecture

Here is how Steenrod homology arises in one geometric approach to the L-theory integral Novikov conjecture. In order to show that the assembly map is a monomorphism, it suffices to show that the map preceding it in the surgery exact sequence is zero. If M and N are aspherical manifolds and $f : M \to N$ is a homotopy equivalence which is a homeomorphism on the boundary, this amounts to showing that f is tangential (see [Mi3]) rel boundary. We can pass to universal covers and form the diagram:

$$\begin{array}{ccc} \widetilde{N} \times_\Gamma \widetilde{N} & \xrightarrow{\tilde{f} \times_\Gamma \tilde{f}} & \widetilde{M} \times_\Gamma \widetilde{M} \\ \downarrow{\scriptstyle proj_1} & & \downarrow{\scriptstyle proj_1} \\ N & \xrightarrow{f} & M. \end{array}$$

Here $\Gamma = \pi_1 M = \pi_1 N$ acts diagonally on $M \times M$ and $N \times N$ and the maps $proj_1$ are induced by projection onto the first factor. It turns out that $\widetilde{N} \times_\Gamma \widetilde{N}$ and $\widetilde{M} \times_\Gamma \widetilde{M}$ are bundles with fiber $\widetilde{N}$ and $\widetilde{M}$ over N and M which are micro-equivalent to the topological tangent bundles of N and M. To show that f is tangential, it suffices to show that $\tilde{f} \times_\Gamma \tilde{f}$ is proper[10] homotopic to a fiber-preserving map which restricts to a homeomorphism on each fiber. This approach goes back to Farrell-Hsiang [FH].

One approach to this problem is via bounded surgery theory [FP]. The map $\tilde{f} \times_\Gamma \tilde{f}$ restricts to a copy of $\tilde{f}$ on each fiber. Thus, the problem of homotoping these maps to homeomorphisms can be viewed as a parameterized bounded surgery problem. We can proceed by induction on skeleta in N to boundedly homotop maps over each skeleton to homeomorphisms. Assuming that we have succeeded over $\partial\Delta^k$, the obstruction to succeeding over the interior lies in $\mathcal{S}^{bdd}\begin{pmatrix} \widetilde{M} \times \Delta^k & \\ \downarrow & \text{rel } \partial(\widetilde{M} \times \Delta^k) \\ \widetilde{M} & \end{pmatrix}$. The bounded surgery sequence which computes this is:

$$\cdots \to \mathcal{S}^{bdd}\begin{pmatrix} \widetilde{M} \times \Delta^k & \\ \downarrow & \text{rel } \partial(\widetilde{M} \times \Delta^k) \\ \widetilde{M} & \end{pmatrix} \to H^{\ell f}_{n+k}(\widetilde{M}; G/TOP) \\ \to L^{bdd}_{n+k,\widetilde{M}}(e) \to \cdots$$

If these structure sets vanish, we conclude that the bounded L-groups are isomorphic to locally finite homology with coefficients in the L-spectrum. Thus, it is natural to study the functorial properties of the bounded L-groups with an eye to showing that they are, in fact, locally finite homology groups with coefficients in L-theory. See [PW] and [Ran].

Variations on this procedure are possible. In particular, if $\widetilde{M}$ admits a nice compactification, we could use continuously controlled surgery theory

[10] I'm cutting some corners here. The *proper* condition is not necessary to show tangentiality, but it is necessary to show vanishing in the integral surgery sequence. Without the word proper, we would be proving rational Novikov.

[AnCFK], [CP] in place of bounded surgery theory. So far, success in this program has come mainly in such cases [FW1], [CP]. This is where Steenrod homology enters the picture, since the space at infinity need not be an ANR. If $M \cup X = \bar{M}$ is a contractible compactification of $\widetilde{M}$, we have $H_*^{\ell f}(\widetilde{M}) = H_*^{st}(\bar{M}, X) \cong H_{*-1}^{st}(X)$. Thus, in case $\widetilde{M}$ admits a nice compactification, we can also proceed by trying to show that the bounded L-theory of $\widetilde{M}$ (or the continuously controlled L-theory of $\bar{M}$ at X) is the Steenrod homology of the space X at infinity.

Carlsson [Car2] applied locally finite homology to the K-theory Novikov Conjecture in the context of equivariant homotopy theory. On the analytical side, the identification of elliptic operators with K-homology cycles allows one to trace related constructions back much farther through [GL], [Kas], and even to [Lus]. The relation between locally finite homology with coefficients in K- and L- theory and controlled topology was studied by Quinn [Q1], [Q2]. Using Quinn's obstruction theory, the geometric approach to the L-theory Novikov Conjecture in [FW2] can be interpreted in terms of locally finite L-homology. Similar considerations also arise in the Higson-Roe approach [HigR], [R] to the Novikov Conjecture via coarse geometry. In the Higson-Roe approach, the goal is to prove that an assembly map

$$K_*^{\ell f}(\widetilde{M}) \to K_*(B_H)$$

is an isomorphism. The first term is locally finite K-theory of the universal cover of M and the second term is the K-theory of the bounded propagation operators on $\widetilde{M}$. Again, one approach to the goal is to show that the right-hand side is a locally finite homology theory or, in favorable cases, that it is the Steenrod homology of a "space at infinity."

One of the fundamental problems of controlled topology is expressed by the question "how big is epsilon?" Quinn [Q1], [Q2], Chapman [Ch] and the author, [F], [ChF] have proven stability theorems in various categories showing that "epsilon control for epsilon sufficiently small implies arbitrary control." When $\widetilde{M}$ is a nonpositively curved Riemannian manifold, these theorems show that bounded control suffices, that is, that a boundedly controlled geometric or algebraic object can be deformed to one with arbitrarily fine control. For aspherical complexes K where bounded control in $\tilde{K}$ implies epsilon control for any epsilon, the argument at the beginning of this section will prove the L-theory Novikov Conjecture for $\pi_1(K)$: In the obstruction theory of [FP], an element of the bounded L-group is represented by a "bounded quadratic chain complex." The question, then, is whether such objects and the relations between them can always be subdivided to obtain ϵ-controlled quadratic chain complexes for any ϵ. These last are the cycles in Quinn's theory, which is known to be locally finite homology theory with coefficients in the connective L-spectrum. In the presence of such subdivisions, the assembly map

$$H^{\ell f}_{n+k}(\widetilde{M}; G/TOP) \to L^{bdd}_{n+k,\widetilde{M}}(e)$$

is an isomorphism, the structure set $\mathcal{S}^{bdd}\left(\begin{array}{c}\widetilde{M}\times\Delta^k\\ \downarrow\\ \widetilde{M}\end{array}\quad \mathrm{rel}\ \partial(\widetilde{M}\times\Delta^k)\right)$ is trivial, and the homotopy equivalence f is tangential. Unfortunately, the author currently knows of no general procedure for performing such subdivisions. The proofs[11] that the author is familiar with all rely on some sort of geometry in the universal cover which allows one to squeeze a bounded cycle to an epsilon cycle. Even the argument of [CP] has a trace of this flavor – the vanishing of a controlled surgery obstruction near infinity gives homeomorphisms near infinity on the tangent bundles and an Alexander trick as in [FH] does the squeezing. In this sense, a nice compactification is a vestigial nonpositive curvature condition. The situation in the absence of geometry is far from clear.

References

[AnCFK] D. R. Anderson, F. X. Connolly, S. C. Ferry, and E. K. Pedersen, *Algebraic K-theory with continuous control at infinity*, J. Pure Appl. Alg. **94** (1994), 25–48.

[BD] P. Baum and R. G. Douglas, *Relative K-homology and C*-algebras*, K-Theory **5** (1991), 1–46.

[Ber1] N. A. Berikashvili, *Axiomatics of the Steenrod-Sitnikov homology theory on the category of compact Hausdorff spaces*, Trudy Mat. Inst. Steklov. **154** (1983), 24–37.

[Ber2] ———, *The Steenrod-Sitnikov homology theory on the category of compact spaces*, Dokl. Akad. Nauk SSSR **254** (1980), 1289–1291.

[Bor] K. Borsuk, *Theory of Retracts*, PWN-Polish Scientific Publishers, Warsaw, 1966.

[BotK] B. I. Botvinnik and V. I. Kuzminov, *A condition for the equivalence of homology theories on categories of bicompacta*, Sibirsk. Mat. Zh. **20** (1979), 1233–1240, 1407; English translation, Siberian Math. J. **20** (1979), 873–879.

[BK] A. K. Bousfield and D. Kan, *Homotopy limits, completions and localizations*, Lecture Notes in Math., vol. 304, Springer-Verlag, Berlin, Heidelberg, New York, 1972.

[11] Proofs of subdivision, that is. The Novikov proofs which rely on Cappell's splitting theorem [Cap1], [Cap2] seem not to involve this sort of geometry.

[Cap1] S. E. Cappell, *A splitting theorem for manifolds*, Invent. Math. **33** (1976), 69–170.

[Cap2] ———, *On homotopy invariance of higher signatures*, Invent. Math. **33** (1976), 171–179.

[Car1] G. Carlsson, *Bounded K-theory and the assembly map in algebraic K-theory, I*, these proceedings.

[Car2] ———, *Homotopy fixed points in the algebraic K-theory of certain infinite discrete groups*, Advances in Homotopy Theory (S. Salamon, B. Steer and W. Sutherland, eds.), London Math. Soc. Lecture Notes, vol. 139, 1989, pp. 5–10.

[CarP] G. Carlsson and E. K. Pedersen, *Controlled algebra and the Novikov conjectures for K- and L-theory*, Topology (to appear).

[Ch] T. A. Chapman, *Approximation results in topological manifolds*, Mem. Amer. Math. Soc. no. 251 (1981).

[ChF] T. A. Chapman and S.C. Ferry, *Approximating homotopy equivalences by homeomorphisms*, Amer. J. Math. **101** (1979), 583–607.

[Co] M. M. Cohen, *A course in Simple-homotopy theory*, Springer-Verlag, Berlin, Heidelberg, New York, 1973.

[DJ] M. Davis and T. Januszkiewicz, *Hyperbolization of polyhedra*, J. Diff. Geometry **34** (1991), 347–388.

[DS] J. Dydak and J. Segal, *Shape theory: an introduction*, Lecture Notes in Math., vol. 688, Springer-Verlag, Berlin, Heidelberg, New York.

[EH] D.A. Edwards and H.M. Hastings, *Čech and Steenrod homotopy theories with applications to geometric topology*, Lecture Notes in Math., vol. 542, Springer-Verlag, Berlin, Heidelberg, New York, 1976.

[FH] F. T. Farrell and W.-C. Hsiang, *On Novikov's conjecture for nonpositively curved manifolds, I*, Ann. of Math. **113** (1981), 199–209.

[F] S. C. Ferry, *The homeomorphism group of a compact Q-manifold is an ANR*, Ann. of Math. **106** (1977), 101–120.

[FP] S. Ferry and E. K. Pedersen, *Epsilon surgery theory*, these proceedings.

[FW1] S. Ferry and S. Weinberger, *A coarse approach to the Novikov Conjecture*, these proceedings.

[FW2] ———, *Curvature, tangentiality, and controlled topology*, Invent. Math. **105** (1991), 401–414.

[GL] M. Gromov and H. B. Lawson, Jr., *Positive scalar curvature in the presence of a fundamental group*, Ann. of Math. **111** (1980), 209–230.

[HigR] N. Higson and J. Roe, *On the coarse Baum-Connes conjecture*, these proceedings.

[In] Hvedri Inassaridze, *On the Steenrod homology theory of compact spaces*, Michigan Math. J. **38** (1991), 323–338.

[KKS] D. S. Kahn, J. Kaminker and C. Schochet, *Generalized homology theories on compact metric spaces*, Michigan Math. J. **24** (1977), 203–224.

[KS] J. Kaminker and C. Schochet, *Analytic equivariant K-homology*, Geometric Applications of Homotopy Theory, I, Lecture Notes in Math., vol. 657, Springer-Verlag, 1978, pp. 292–302.

[Kas] G. G. Kasparov, *Equivariant KK-theory and the Novikov conjecture*, Invent.

Math. **91** (1988), 147–201.

[LM] Ju. T. Lisica and S. Mardesic, *Steenrod-Sitnikov homology for arbitrary spaces*, Bull. Amer. Math. Soc. **9** (1983), 207–210.

[Lus] G. Lusztig, *Novikov's higher signature and families of elliptic operators*, J. Diff. Geom. **7** (1972), 229–256.

[Ma] W. Massey, *Homology and Cohomology Theory*, Marcel Dekker, Inc., New York and Basel, 1978.

[Mi1] J. Milnor, *On the Steenrod homology theory*, these proceedings.

[Mi2] ———, *On axiomatic homology theory*, Pacific J. Math. **12** (1962), 337–341.

[Mi3] ———, *Microbundles, Part I*, Topology **3** (1964), 53–80.

[Mit] W. J. R. Mitchell, *Local homology and cohomology*, Math. Proc. Cambridge Philos. Soc. **89** (1981), 309–324.

[PW] E. K. Pedersen and C. Weibel, *K-theory homology of spaces*, Algebraic Topology, Proceedings Arcata 1986 (G. Carlsson, R. L. Cohen, H. R. Miller and D. C. Ravenel eds.), Lecture Notes in Math., vol. 1370, Springer-Verlag, Berlin, Heidelberg, New York, 1989, pp. 346–361.

[Q1] F. Quinn, *Ends of maps, II*, Invent. Math. **68** (1982), 353–424.

[Q2] ———, *An Obstruction to the Resolution of Homology Manifolds*, Michigan Math. J. **301** (1987), 285–292.

[Ran] A. Ranicki, *Lower K- and L-Theory*, London Math. Soc. Lecture Notes, vol. 178, Cambridge Univ. Press, Cambridge, 1992.

[R] J. Roe, *Coarse cohomology and index theory on complete Riemannian manifolds*, Mem. Amer. Math. Soc. no. 497 (1993).

[Sa1] S. A. Saneblidze, *The extraordinary Steenrod homology theory*, Soobshch. Akad. Nauk Gruzin. SSR **104** (1981), 281–284.

[Sa2] ———, *On a uniqueness theorem for the Steenrod-Sitnikov homology theory on the category of compact Hausdorff spaces*, Soobshch. Akad. Nauk Gruzin. SSR **105** (1982), 33–36.

[Si] L. C. Siebenmann, *Chapman's classification of shapes: A proof using collapsing*, Manuscripta Math. **16** (1975), 373–384.

[Sk1] E. G. Sklyarenko, *Homology theory and the exactness axiom*, Uspekhi Mat. Nauk **24** no. 5 (1969), 87–140; English translation, Russian Math. Surveys **24** no. 5 (1969), 91–142.

[Sk2] ———, *On the homology theory associated with Aleksandrov-Čech cohomology*, Uspekhi Mat. Nauk **34** no. 6 (1979), 90–110; English translation, Russian Math. Surveys **34** no. 6 (1979), 103–137.

[St] N. Steenrod, *Regular cycles of compact metric spaces*, Annals of Math. **41** (1940), 833–851.

Department of Mathematical Sciences, SUNY at Binghamton, Binghamton, NY 13901, U.S.A.

email: `steve@math.binghamton.edu`

Epsilon Surgery Theory

Steven C. Ferry and Erik K. Pedersen

Contents

1. Introduction

In classical surgery theory one begins with a Poincaré duality space X and a normal map

$$\begin{array}{ccc} \nu_M & \xrightarrow{\tilde{f}} & \xi \\ \downarrow & & \downarrow \\ M & \xrightarrow{f} & X \end{array}$$

The problem is to vary (M, f) by a normal cobordism to obtain a homotopy equivalence $f' : M' \to X$.

It is desirable to have an epsilon or controlled version of surgery theory. Thus, X comes equipped with a reference map to a metric space K, and the aim is to produce a homotopy equivalence $f' : M' \to X$, which is small measured in K. The existence of such an f' implies that X is a small Poincaré duality space in the sense that cells are close to their dual cells, measured in K. That X be a small Poincaré duality space must therefore be part of the original data.

For many applications the most interesting question is whether such a map f' exists with arbitrarily small control in K. In this case, X would have to be an ϵ-Poincaré duality space for all $\epsilon > 0$. Unfortunately, there are technical difficulties in defining and dealing with such ϵ-Poincaré duality spaces. Our approach is to work instead with a bounded surgery theory. Our bounded surgery theory generalizes epsilon surgery theory in the same sense that simple proper surgery theory (as developed in [39, 41, 18, 26]) generalizes classical compact surgery theory. The fact that bounded categories are categories avoids many technical difficulties.

Consider a classical surgery problem as above. Cross with the real line and look for an infinite simple homotopy equivalence $f' : M' \to X \times \mathbb{R}$. Such a manifold M' has the form $N' \times \mathbb{R}$ for some closed manifold N', so $f' : N' \to X$ solves the classical surgery problem. This means that the two-ended simple surgery theory is as good for applications as the compact theory. The two-ended theory is more general, though, since it applies to any two-ended manifold with fundamental group equal to the fundamental group of each end.

This is our approach to epsilon surgery theory and its generalization. We consider surgery problems parameterized over $K \times \mathbb{R}$, where $K \times \mathbb{R}$ is given a metric so that $K \times \{t\}$ becomes t times as big as $K \times \{1\}$ for $t > 1$. Call this space $O(K_+)$. (This description is not quite accurate. See §2 for precise definitions and details).[1]

---

[1] It is easy to see that if Z is a Poincaré duality space with a map $Z \to K$ such that Z has ϵ-Poincaré duality for all $\epsilon > 0$ when measured in K (after subdivision), e.g. a homology manifold, then $Z \times \mathbb{R}$ is an $O(K_+)$-bounded Poincaré complex. The converse

We let X be a complex with bounded Poincaré duality over this control space. Given a proper normal map

$$\begin{array}{ccccc} \nu_M & \xrightarrow{\bar{f}} & \xi & & \\ \downarrow & & \downarrow & & \\ M & \xrightarrow{f} & X & \longrightarrow & O(K_+) \end{array}$$

we study the problem of producing a proper normal cobordism to $f' : M \to X$ such that f' is a bounded simple homotopy equivalence measured in $O(K_+)$. The obstruction groups obtained are the desired obstruction groups for epsilon surgery. This is our codification of the idea that an ϵ-Poincaré space (for all $\epsilon > 0$) is a sequence of smaller and smaller Poincaré duality spaces joined by smaller and smaller Poincaré cobordisms.

Thus, in case X is a homology manifold (homology manifolds are naturally epsilon Poincaré for all epsilon) with a reference map $\varphi : X \to K$, we replace an epsilon surgery problem

$$\begin{array}{ccccc} \nu_M & \xrightarrow{\bar{f}} & \xi & & \\ \downarrow & & \downarrow & & \\ M & \xrightarrow{f} & X & \longrightarrow & K \end{array}$$

by the bounded surgery problem

$$\begin{array}{ccccc} \nu_M \times 1 & \xrightarrow{\bar{f}} & \xi \times 1 & & \\ \downarrow & & \downarrow & & \\ M \times \mathbb{R} & \xrightarrow{f} & X \times \mathbb{R} & \longrightarrow & O(K_+) \end{array}$$

If $f \times id$ is properly normally cobordant to a bounded simple equivalence $f' : M \to X \times \mathbb{R}$, we split M' near the end to obtain a sequence of more and more controlled solutions to the original problem. Our approach generalizes this sort of epsilon surgery in case other data happen to be available. For an application in which a parameterization over an open cone appears naturally, see Theorem 18.1, which is *not* a bounded translation of an epsilon problem. The other applications in §18 do, however, illustrate how one passes between the bounded and epsilon worlds.

Naturally, we require hypothesis in addition to the general situation described above. In the first part of the paper, our main hypotheses are that the control map $X \to K$ have constant coefficients in the sense that it "looks like" a product on π_1, and that K be a finite complex. This restricts

---

(while true) will not concern us here.

applicability in, say, the case of group actions, to semifree group actions. In §14 we show how to extend the theory to treat a more general equivariant case, but for readability, we have chosen to give most details in the special case of constant coefficients.

Some surprising phenomena come up. The method allows studying many more objects than exist in the compact world. An n-manifold parameterized by $\mathbb{R}^{n+k}$ is an object of dimension $-k$, so we have objects that in some sense correspond to negative dimensional manifolds. This leads to nonconnective surgery spectra. The necessary algebra for our theory has been developed in [23, 27, 9].

Needless to say, we have benefited from discussions with many colleagues. We should like particularly to mention Anderson, Hambleton, Hughes, Lück, Munkholm, Quinn, Ranicki, Taylor, Weibel, Weinberger, and Williams. In fact, related theories have been developed by Hughes, Taylor and Williams, [14], Madsen and Rothenberg [17], and Weinberger [44].

Finally we want to acknowledge that the very stimulating atmosphere of the SFB at Göttingen has had a major effect on the developments of this paper. The first author would also like to thank Odense University for its support in the fall of 1987.

2. Algebraic preliminaries

Let M be a metric space, and let R be a ring with anti-involution. For definiteness, the reader should keep in mind the model case in which M is the infinite open cone $O(K)$ on a complex $K \subset S^n \subset \mathbb{R}^{n+1}$ and $R = \mathbb{Z}\pi$, with π a finitely presented group. The category $\mathcal{C}_M(R)$ is defined as follows:

Definition 2.1. An object A of $\mathcal{C}_M(R)$ is a collection of finitely generated free right R-modules A_x, one for each $x \in M$, such that for each ball $C \subset M$ of finite radius, only finitely many A_x, $x \in C$, are nonzero. A morphism $\varphi : A \to B$ is a collection of morphisms $\varphi^x_y : A_x \to B_y$ such that there exists $k = k(\varphi)$ such that $\varphi^x_y = 0$ for $d(x, y) > k$.

The composition of $\varphi : A \to B$ and $\psi : B \to C$ is given by $(\psi \circ \varphi)^x_y = \sum_{z \in M} \psi^z_y \varphi^x_z$. The composition $(\psi \circ \varphi)$ satisfies the local finiteness and boundedness conditions whenever ψ and φ do.

Definition 2.2. The dual of an object A of $\mathcal{C}_M(R)$ is the object A^* with $(A^*)_x = A^*_x = \mathrm{Hom}_R(A_x, R)$ for each $x \in M$. A^*_x is naturally a left R-module, which we convert to a right R-module by means of the anti-involution. If $\phi : A \to B$ is a morphism, then $\phi^* : B^* \to A^*$ and $(\phi^*)^x_y(h) = h \circ \phi^y_x$, where $h : B_x \to R$ and $\phi^y_x : A_y \to B_x$. ϕ^* is bounded whenever ϕ is. Again, ϕ^* is naturally a left module homomorphism which induces a homomorphism of right modules $B^* \to A^*$ via the anti-involution.

Definition 2.3. There are functors $\oplus$ and Π from $\mathcal{C}_M(R)$ to Mod(R), the category of free modules over R. $\oplus A = \bigoplus_{x\in M} A_x$ and $\Pi A = \prod_{x\in M} A_x$. Notice that $\Pi A^* = (\oplus A)^*$.

Definition 2.4. Consider a map $p : X \to M$

(i) The map $p : X \to M$ is *eventually continuous* if there exist k and a covering $\{U_\alpha\}$ of X, such that the diameter of $p(U_\alpha)$ is less than k.
(ii) A *bounded CW complex* over M is a pair (X, p) consisting of a CW complex X and an eventually continuous map $p : X \to M$ such that there exists k such that $\mathrm{diam}(p(C)) < k$ for each cell C of X. (X, p) is called *proper* if the closure of $p^{-1}(D)$ is compact for each compact $D \subset M$. We consider (X, p_1) and (X, p_2) to be the same, if there exists k so that $d(p_1(x), p_2(x)) < k$ for all x.

Remark 2.5. We do not require the control map p to be continuous in the above definition. It is, however, often the case that p may be chosen to be continuous. This is the case if the metric space is "boundedly highly connected" in an appropriate sense. See Definition 5.2.

Definition 2.6. Consider a bounded CW complex (X, p)

(i) The bounded CW complex (X, p) is (-1)*-connected* if there is $k \in \mathbb{R}_+$ so that for each point $m \in M$, there is a point $x \in X$ such that $d(p(x), m) < k$.
(ii) (X, p) is 0*-connected* if for every $d > 0$ there exist $k = k(d)$ so that if $x, y \in X$ and $d(p(x), p(y)) \leq d$, then x and y may be joined by a path in X whose image in M has diameter $< k(d)$. Notice that we have set up our definitions so that 0-connected does not imply -1-connected.

Definition 2.7. Let $p : X \to M$ be 0-connected, but not necessarily (-1)-connected.

(i) (X, p) has *trivial bounded fundamental group* if for each $d > 0$, there exist $k = k(d)$ so that for every loop $\alpha : S^1 \to X$ with $\mathit{diam}(p \circ \alpha(S^1)) < d$, there is a map $\bar{\alpha} : D^2 \to X$ so that the diameter of $p \circ \bar{\alpha}(D^2)$ is smaller than k.
(ii) (X, p) has *bounded fundamental group* π if there is a π-cover $\widetilde{X}$ so that $\widetilde{X} \to M$ has trivial bounded fundamental group.

If X is a CW complex, we will denote the cellular chains of $\widetilde{X}$ by $C_\#(X)$, considered as a chain complex of free right $\mathbb{Z}\pi$-modules. When $p : X \to M$ is a proper bounded CW complex with bounded fundamental group, we can consider $C_\#(X)$ to be a chain complex in $\mathcal{C}_M(\mathbb{Z}\pi)$ as follows: For each cell $C \in X$, choose a point $c \in C$ and let $D_\#(X)_y$ be the free submodule of $C_\#(X)$ generated by cells for which $p(c) = y$. The boundary map is bounded, since cells have a fixed maximal size. We will denote the cellular chains of $\widetilde{X}$ by $D_{\#}(X)$ when we consider them as a chain complex in $\mathcal{C}_M(\mathbb{Z}\pi)$ and by $C_{\#}(X)$ when we consider them as an ordinary chain complex of $\mathbb{Z}\pi$ modules. We will denote $D_{\#}(X)^*$ by $D^{\#}(X)$. If $(X, \partial X)$ is a bounded CW pair, $D_{\#}(X, \partial X)$ denotes the relative cellular chain complex regarded as a chain complex in $\mathcal{C}_M(\mathbb{Z}\pi)$.

Lemma 2.8. *When $p : X \to M$ is a proper bounded CW complex with bounded fundamental group, we have the following formulas*

(i) $\oplus D_{\#}(X) = C_{\#}(X)$
(ii) $\oplus D^{\#}(X) = C^{\#}_{cs}(X)$ *(≡ cochains with compact support)*
(iii) $\Pi D_{\#}(X) = C^{lf}_{\#}(X)$ *(≡ locally finite chains)*
(iv) $\Pi D^{\#}(X) = C^{\#}(X)$.

Proof. Statement (i) is clear, and (iv) follows from the formula $\Pi(A^*) = (\oplus A)^*$. Statements (ii) and (iii) follow easily from the fact that p is proper. In case p is not proper this suggests an extension of the concepts of homology with locally finite chains and cohomology with compact support to concepts requiring compactness or locally finiteness only in a designated direction. □

We recall the *open cone construction* from [27]. If K is a subset of $S^n \subset \mathbb{R}^{n+1}$, we define $O(K)$ to be the metric space $O(K) = \{t \cdot x | 0 \leq t, x \in K\} \subset \mathbb{R}^{n+1}$. Here, $O(K)$ inherits a metric from $\mathbb{R}^{n+1}$. We think of this as a 1-parameter family of metrics on K, in which distance grows larger with increasing t. We state the main result of [27]:

Theorem 2.9. *Let K be of the homotopy type of a finite complex. The K-theory of the categories $\mathcal{C}_{O(K)}(R)$ is given by*

$$K_*(\mathcal{C}_{O(K)}(R)) = KR_{*-1}(K)$$

where KR is the nonconnective homology theory associated to the algebraic K-theory of the ring R.

We refer the reader to [27] for further facts about the O construction and the K-theory of $\mathcal{C}_{O(K)}(R)$.

Definition 2.10. Let R be a ring with unit. We denote $\mathrm{Coker}(K_*(\mathcal{C}_M(\mathbb{Z})) \to K_*(\mathcal{C}_M(R)))$ by $\widetilde{K}_*(\mathcal{C}_M(R))$. If R is a group ring, $R = \mathbb{Z}\pi$, we denote $\widetilde{K}_1(\mathcal{C}_M(R))/\pm\pi$ by $Wh_M(\pi)$.

Definition 2.11. We define

(i) a metric space M is *allowable* if there exist a bounded finite-dimensional simplicial complex K and a map $p : K \to M$ which is -1, 0- and 1-connected.

(ii) a map $f : X \to Y$ between metric spaces is *eventually Lipschitz* if the inverse image of every bounded set is bounded and if there are numbers $k > 0$ and $d > 0$ so that $d(f(x), f(x')) \leq max(d, kd(x, x'))$ for all $x, x' \in X$. We say that X and Y are *eventually Lipschitz equivalent* if there exist a number $M > 0$ and eventual Lipschitz maps $f : X \to Y$, $g : Y \to X$ so that $d(f \circ g, id) < M$ and $d(g \circ f, id) < M$.

In connection with our study of the resolution problem, we will want to apply this theory to open cones on compact finite-dimensional ANR's, so we prove that such spaces are allowable.

Proposition 2.12. *If $X \subset S^n$ is a compact* ANR, *then $O(X)$ is an allowable metric space.*

Proof. Let $r_t : U \to X$ be a homotopy from a neighborhood U of X in S^m to X such that $r_0 = id$, $r_1 : U \to X$ is a retraction, and $r_t|X = id$ for all t. Let $X = \bigcap_{i=1}^{\infty} P_i$, where $P_1 \supset P_2 \supset \ldots$ are finite polyhedra and $r_t(P_{i+1}) \subset P_i$ for all $t \in [0, 1]$. We may assume that $P_n \subset N_{\frac{1}{n}}(X)$. Form the telescope $K = \bigcup_{n=1}^{\infty} O(P_n) \cap B(n, 0)$, where $B(n, 0)$ is the closed ball of radius n around $0 \in \mathbb{R}^{m+1}$. The map r_1 induces a map $\bar{r} : K \to O(X)$ which is $-1, 0$ and 1-connected. □

Remark 2.13. Notice the following

(i) If $M = O(X)$ is the open cone on a compact ANR and $p : Z \to M$ is a finite-dimensional bounded CW complex, then p may be chosen to be continuous. The argument is an induction over the skeleta of Z starting with p restricted to the 0-skeleton.

(ii) Assume that K is a connected PL complex. Then we have

$$\widetilde{K}_1(\mathcal{C}_{O(K)}(R)) = K_1(\mathcal{C}_{O(K)}(R)) \text{ and } Wh_{O(K)}(\pi) = \widetilde{K}_1(\mathcal{C}_{O(K)}(\mathbb{Z}\pi)),$$

except when K is empty, in which case $O(K) = pt$. The argument is an Eilenberg swindle.

Definition 2.14. Let M be a metric space and let $p : X \to M$ and $q : Y \to M$ be maps. A map $f : X \to Y$ is said to be *bounded over M* (or simply *bounded*) if there is a number $k > 0$ so that $d(q \circ f(x), p(x)) < k$ for all $x \in X$. We say that f is a *bounded homotopy equivalence* if there exist $g : Y \to X$ and homotopies $h : f \circ g \sim id$, $l : g \circ f \sim id$ so that f, g, h and l are bounded.

Theorem 2.15. *Let M be an allowable metric space and let $p_X : X \to M$, $p_Y : Y \to M$ be proper bounded finite-dimensional CW complexes, both -1 and 0-connected with bounded fundamental group π. If $f : X \to Y$ is a cellular map such that:*

(i) $f \circ p_Y = p_X$.
(ii) *f induces a* π_1*-isomorphism.*
(iii) *f induces a (bounded) homotopy equivalence of chain complexes in* $\mathcal{C}_M(\mathbb{Z}\pi)$, $f_\# : D_\#(X) \to D_\#(Y)$.

Then f is a bounded homotopy equivalence.

Proof. This is proved in [2] for the case of a continuous control map. An alternative approach that works in our generality is to replace X and Y by proper regular neighborhoods $N(X) \subset N(Y)$ in some high-dimensional euclidean space and then apply the bounded h-cobordism theorem 2.17. The problem with torsion is solved by crossing with S^1, thus killing the torsion, thus getting a bounded homotopy equivalence of $X \times S^1 \to Y \times S^1$, and hence $X \to Y$ since X is bounded homotopy to $X \times \mathbb{R}$ which is the cyclic cover of $X \times S^1$. □

Remark 2.16. The theorem above plays a rôle in our theory which is analogous to the Whitehead theorem's rôle in standard surgery theory.

Finally we note that we have the bounded analogue of the s-cobordism theorem.

Theorem 2.17. (Bounded s-cobordism theorem). *Assume* $W \to M$ *is a* -1 *and* 0*-connected manifold with bounded fundamental group* π *such that the boundary of* W *has 2 components* $\partial_0 W \subset W$ *and* $\partial_1 W \subset W$*, such that the inclusions are bounded homotopy equivalences. The torsion* τ *of* W *is defined by the torsion of the contractible chain complex* $D_\#(W, \partial_0 W) \in Wh_M(\pi)$ *see* [37]. *For* $\dim(W) > 5$ *we have that* W *is isomorphic to* $\partial_0 W \times I$ *if and only if* $\tau = 0$.

Proof. The s-cobordism theorem in [25] is only stated for the parameter space $\mathbb{R}^n$, and in that context a bounded *and* thin h-cobordism theorem is proved. As far as the bounded s-cobordism theorem is concerned, the arguments only use that the h-cobordism is -1 and 0-connected with bounded fundamental group. □

3. Bounded Poincaré complexes

Given a bounded CW complex $p : X \to M$ with bounded fundamental group π, an element $[y] \in H_n^{lf}(X, \mathbb{Z})$ induces a cap-product $y \cap - : D^\#(X) \to D_{n-\#}(X)$. Here, $y \cap -$ is a homomorphism of chain complexes in $\mathcal{C}_M(\mathbb{Z})$ and is well-defined up to chain homotopy. The formula defining $y \cap -$ is the usual one using the Alexander-Whitney diagonal approximation. The size estimate on $y \cap -$ follows from the fact that the diagonal approximation takes the generator $c \in (D_n(X))_m$ to a sum $\sum c_i \otimes c_i'$ where $c_i \in (D_\#(X))_{m_i}$,

$c'_i \in (D_\#(X))_{m'_i}$, and $d(p(m_i), p(m'_i)) \leq 2k$, where k is the bound on the diameter of the cells of X as measured in M.

Our notational conventions in the following definition are based on [43, pp. 21–22]

Definition 3.1. Let $p : X \to M$ be a proper bounded CW complex with bounded fundamental group π, and let $\widetilde{X} \to X$ be an orientation double covering. Then X is a *bounded n-dimensional Poincaré duality space* if there is an element $[X] \in H_n^{lf}(X; \mathbb{Z})$, such that $[X] \cap - : D^\#(X) \to D_{n-\#}(X)$ is a bounded homotopy equivalence of chain complexes. Here, $\mathbb{Z}$ is made into a left $\mathbb{Z}\pi$ module using the antiinvolution on $\mathbb{Z}\pi$. X is a *simple* bounded Poincaré duality space if the torsion of $[X] \cap -$ is trivial in $Wh_M(\pi)$. If $p : X \to M$ is a disjoint union of spaces satisfying this condition, we shall also call X a Poincaré space. Notice that X may have infinitely many components, but the properness of $p : X \to M$ ensures that locally there are only finitely many components.

Definition 3.2. Let $p : (X, \partial X) \to M$ be a proper bounded CW pair so that X has bounded fundamental group π. The pair $(X, \partial X)$ is *an n-dimensional bounded Poincaré duality pair* if ∂X is an $(n-1)$-dimensional bounded Poincaré complex with orientation double covering the pullback of the orientation double covering on X and if there is an element $[X] \in H_n^{lf}(X, \partial X; \mathbb{Z})$, such that $[X] \cap - : D^\#(X) \to D_{n-\#}(X, \partial X)$ is a bounded homotopy equivalence of chain complexes. $(X, \partial X)$ is a *simple* bounded Poincaré duality space, if the torsion of $[X] \cap -$ is trivial in $Wh_M(\pi)$.

Remark 3.3. We note that $[X] \cap -$ is independent up to chain homotopy of the choice of representative chain for $[X]$. This is true since any other choice is of the form $X + \delta z$ and $(X+\delta z) \cap y - x \cap y = \delta z \cap y = \delta(z \cap y) - z \cap \delta y$, so $z \cap -$ is a chain homotopy between the two maps $X\cap$ and $(X + \delta z)\cap$.

Example 3.4. If X happens to be a manifold with a bounded handle decomposition, the usual proof of Poincaré duality produces a bounded Poincaré structure on X.

Definition 3.5. Let $\phi : (W, \partial W) \to (X, \partial X)$ be a map of bounded Poincaré duality pairs such that $\phi_*([W]) = [X]$. We define $K^\#(W, \partial W)$ to be the algebraic mapping cone of $\phi^* : D^\#(X, \partial X) \to D^\#(W, \partial W)$. We define $K_\#(W, \partial W)$ to be the dual of $K^\#(W, \partial W)$. We have short exact sequences of chain complexes

$$0 \longrightarrow D^\#(X, \partial X) \xrightarrow{\phi^*} D^\#(W, \partial W) \longrightarrow K^\#(W, \partial W) \longrightarrow 0$$

$$0 \longrightarrow K_\#(W, \partial W) \longrightarrow D_\#(W, \partial W) \xrightarrow{\phi^*} D_\#(X, \partial X) \longrightarrow 0.$$

As in classical surgery theory we have the following:

Lemma 3.6. *Let $\phi : (W, \partial W) \to (X, \partial X)$ be a map of bounded Poincaré duality pairs such that $\phi_*([W]) = [X]$. Then cap product with $[W]$ and $[X]$ induces a bounded chain homotopy equivalence from $K_\#(W, \partial W)$ to $K^{n-\#}(W)$.*

Remark 3.7. We have

(i) This definition of $K_\#(W, \partial W)$ gives $K_\#(W, \partial W)$ the same indexing as the kernel complex in [4] and [43]. Except for a shift in the index and changes in signs, $K_\#(W, \partial W)$ is just the algebraic mapping cone of $D_\#(W, \partial W) \to D_\#(X, \partial X)$.

(ii) Parameterizing an open manifold by the identity, our constructions give a simple proof of Poincaré duality on open manifolds from homology with locally finite coefficients to standard cohomology, or from cohomology with compact supports to standard homology. One applies Π and $\oplus$ respectively to the chain homotopy equivalence $[X] \cap -$.

4. Spivak normal fibre space

In this section we construct the Spivak normal fibre space of a bounded Poincaré duality space. Since bounded Poincaré complexes are certainly open Poincaré complexes in the sense of Taylor [41], we could simply refer to [41], but for the readers' convenience we give the existence proof:

Construct a proper embedding of $X \subset \mathbb{R}^n$, $n - \dim X \geq 3$. Let W a regular neighborhood of X and $r : W \to X$ a retraction. $W \to M$ has a bounded fundamental group, and we can triangulate sufficiently finely to get a bounded CW structure on W. Let F be the homotopy fibre of the map $\partial W \to X$.

Lemma 4.1. *The fibre F is homotopy equivalent to a sphere of dimension $n - \dim X - 1$.*

Proof. By the codimension 3 condition, F is simply connected. It is clear that F is also the fibre of the pullback to the universal cover of X, so consider the relative fibration $(*, F) \to (\widetilde{W}, \widetilde{\partial W}) \to \widetilde{X}$. We have at $\mathbb{Z}\pi$-module chain level homotopy equivalence $D_\#(W, \partial W) \cong D^{n-\#}(W) \cong D^{n-\#}(X) \cong D_{\dim X - n + \#}(X)$, so applying $\oplus$ we get $C_\#(W, \partial W) \cong C_{\# + \dim X - n}(X)$.

Therefore $H_*(\widetilde{W}, \partial\widetilde{W}) = H_{*+\dim X-n}(\widetilde{X})$. The usual spectral sequence argument shows that

$$H_i(*, F) = \begin{cases} 0 & i < n - \dim X \\ \mathbb{Z} & i = n - \dim X \\ 0 & i > n - \dim X \end{cases}$$

and F is thus a sphere, since $\pi_1 F = 0$. Considering $W \subset \mathbb{R}^n \subset \mathbb{R}^n_+ = S^n$ and collapsing everything outside W produces a spherical Thom class. □

The proof of uniqueness of the Spivak normal fibre space is also standard, and is left to the reader.

5. Surgery below the middle dimension

Our definition of "bounded surgery problem" is a straightforward translation of Wall's "surgery problem" [43, p. 9] into bounded topology.

Definition 5.1. Let X^n be a bounded Poincaré duality space over a metric space M and let ν be a (TOP, PL or O) bundle over X. A *bounded surgery problem* is a triple (W^n, ϕ, F) where $\phi : W \to X$ is a proper map from an n-manifold W to X such that $\phi_*([W]) = [X]$ and F is a stable trivialization of $\tau_W \oplus \phi^*\nu$. Two problems (W, ϕ, F) and $(\overline{W}, \bar{\phi}, \bar{F})$ are *equivalent* if there exist an $(n+1)$-dimensional manifold P with $\partial P = W \coprod \overline{W}$, a proper map $\Phi : P \to X$ extending ϕ and $\bar{\phi}$, and a stable trivialization of $\tau_P \oplus \Phi^*\nu$ extending F and $\bar{F}$. See [43, p. 9] for further details.

We will use the notation $\begin{array}{c} W \xrightarrow{\phi} X \\ \downarrow \\ M \end{array}$ to denote a bounded surgery problem. When M is understood, we will shorten the notation to $\phi : W \to X$ or even to ϕ. In all cases, the bundle information is included as part of the data. Our theorem on surgery below the middle dimension and its proof are parallel to Theorem 1.2 on p. 11 of [43]. In order to state the theorem, we need a definition.

Definition 5.2. If $p : X \to M$ is a control map, we will say that $f : Y \to X$ is *boundedly k-connected over M* if for every $c > 0$ there is a number $d > 0$ so that for each $-1 \le l < k$ and map $\alpha : S^l \to Y$ with extension $\beta : D^{l+1} \to X$ of $f \circ \alpha$ with $\operatorname{diam}(p \circ \beta(D^{l+1})) \le c$, there exist a map $\gamma : D^{l+1} \to Y$ and a homotopy $h : D^{l+1} \to X$ with $h_0 = f \circ \gamma$, $h_1 = \beta$, and $\operatorname{diam}(p \circ h(D^{l+1} \times I)) \le d$.

Note that if X is a bounded CW complex over M, then $X^{(k)} \to X$ is boundedly k-connected. The notion of boundedly k-connected *over M* differs from the notion of bounded connectivity of the *control map* discussed in §2. In particular, there is a dimension shift which is analogous to the dimension shift one normally encounters in discussing the connectivity of the *space* X as compared to the connectivity of the *map* $X \to *$.

Here is our theorem on surgery below the middle dimension.

Theorem 5.3. *Let $(X^n, \partial X)$ be a bounded Poincaré duality space over M, $n \geq 6$, or $n \geq 5$ if ∂X is empty. Consider a bounded surgery problem $\phi : (W, \partial W) \to (X^n, \partial X)$. Then $\phi : (W, \partial W) \to (X^n, \partial X)$ is equivalent to a bounded surgery problem $\bar{\phi} : (\overline{W}, \partial\overline{W}) \to (X^n, \partial X)$ such that $\bar{\phi}$ is boundedly $[\frac{n}{2}]$-connected over M and $\bar{\phi}| : \partial\overline{W} \to \partial X$ is boundedly $[\frac{n-1}{2}]$-connected.*

Proof. We start by considering the case in which $\partial X = \emptyset$. Triangulate W so that the diameters $p \circ \phi(\sigma)$, σ a simplex of W, are bounded. Replacing X by the mapping cylinder of ϕ, we can assume that $W \subset X$.

We inductively define a bordism $U^{(i)}$, $-1 \leq i \leq [\frac{n+1}{2}]$ and maps $\Phi^{(i)} : U^{(i)} \to W \cup X^{(i)}$, so that $\partial U^{(i)} = W \coprod \overline{W}^{(i)}$ and so that $\Phi^{(i)}$ is a bounded homotopy equivalence. We begin by setting $U^{(-1)} = W \times I$, and letting $\Phi^{(i)} \to X$ be $\phi \circ \mathrm{proj}$. Let $U^{(0)}$ be obtained from $U^{(-1)}$ by adding a disjoint $(n+1)$-ball corresponding to each 0-cell of $X - W$. The map $\Phi^{(0)}$ is constructed by collapsing each new ball to a point and sending the point to the corresponding 0-cell of $X - W$.

Assume that $\Phi^{(i)} : U^{(i)} \to X$ has been constructed in such a way that $U^{(i)}$ is an abstract regular neighborhood of a complex consisting of W together with cells in dimensions $\leq i$ corresponding to the cells of $X - W$ in those dimensions. Assume further that $\Phi^{(i)}$ is the composition of the regular neighborhood collapse with a map which takes cells to corresponding cells. Each $(i+1)$-cell of $X - W$ induces an attaching map $S^i \to U^{(i)}$. If $2i+1 \leq n$, general position allows us to move this map off of the underlying complex and approximate the attaching map by an embedding $S^i \to \overline{W}^{(i)}$. The bundle information tells us how to thicken this embedding to an embedding of of $S^i \times D^{n-i}$ and attach $(i+1)$-handles to $U^{(i)}$, forming $U^{(i+1)}$. We extend $\Phi^{(i)}$ to $\Phi^{(i+1)}$ in the obvious manner. This process terminates with the construction of $U^{[\frac{n+1}{2}]}$. Turning $U^{[\frac{n+1}{2}]}$ upside down, we see that $U^{[\frac{n+1}{2}]}$ is obtained from $\overline{W}^{[\frac{n+1}{2}]}$ by attaching handles of index $> [\frac{n+1}{2}]$. Thus, the composite map $\overline{W}^{[\frac{n+1}{2}]} \to X$ is boundedly $[\frac{n}{2}]$-connected over M.

In case $\partial X \neq \emptyset$, the argument is similar. We first construct U over the boundary (and, therefore, over a collar neighborhood of the boundary) and then construct U over the interior. □

Remark 5.4. Notice the following

(i) The direct manipulation of cells and handles has replaced the usual appeals to homotopy theory and the Hurewicz-Namioka Theorem.

This is a general technique for adapting arguments from ordinary algebraic topology to the bounded category.

(ii) The construction in the proof yields somewhat more – we wind up with $(\overline{W}, \partial\overline{W}) \subset (X, \partial X)$. When $n = 2k+1$, $\overline{W}$ and X are equal through the k-skeleton. When $n = 2k$, $\partial\overline{W}$ is equal to ∂X through the $(k-1)$-skeleton and $\overline{W}$ contains every k-cell of $X - \partial X$. Since $\overline{W} \to X$ is k-connected, every k-cell in ∂X is homotopic rel boundary to a map into W. By attaching a $k+2$-cell to this homotopy along a face, we can guarantee that for every k-cell in ∂X there is a $k+1$-cell in X so that half of the boundary of the $k+1$-cell maps homeomorphically onto the k-cell and the other half maps into W.

6. Controlled cell-trading

In this section we prove bounded versions of Whitehead's cell-trading lemma. There are algebraic and geometric versions of this lemma. We will need to use both in this paper. These operations apply equally well to cells in a bounded CW complex and to handles in a bounded handle decomposition. We will use cell terminology throughout, except for the term "handle addition."

We need some notation. If we have a sequence

$$A \xrightarrow{(f,g)} B \oplus C \xrightarrow{k+l} D$$

of objects and morphisms, we can represent it pictorially as:

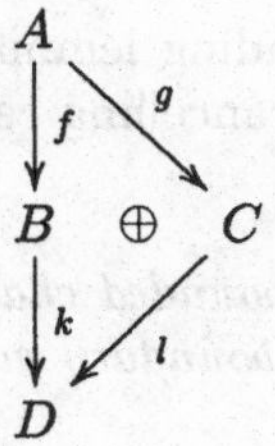

Performing the elementary operation – sliding handles corresponding to basis elements in B over handles in C – corresponding to a bounded homomorphism $m : B \to C$ results in the diagram

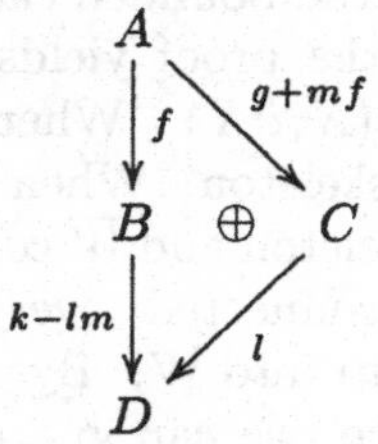

We will write this operation schematically as

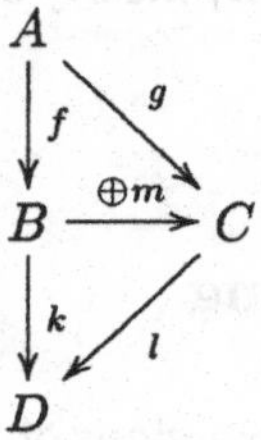

and call it *adding the B-cells to C via m*. When the sequence $A \to B \oplus C \to D$ is a part of a cellular $\mathbb{Z}\pi$-chain complex, this operation is realized geometrically by handle-addition by taking each generator x in B and sliding it across $m(x)$. Changing the attaching maps of the cells this way clearly has the effect described above on the cellular chains.

The next construction is *cancellation of cells.* If a portion of a chain complex looks like $\ldots A \to B \oplus C \to D \oplus C' \ldots$ and the composite $C \to B \oplus C \to D \oplus C' \to C'$ is an isomorphism sending generators to generators, then the chain complex is bounded chain homotopy equivalent to $\ldots A \to B \to D \ldots$. This has a geometric counterpart in cancellation of n- and $(n+1)$-cells. Note that it is not sufficient that the map $C \to C'$ be an isomorphism. It must send generators to generators. The complementary process of changing $\ldots A_n \xrightarrow{\partial} A_{n-1} \to \ldots$ to $\ldots A_n \oplus D \xrightarrow{\partial \oplus 1} A_{n-1} \oplus D$ is called *introducing cancelling* $n-1$ *and* n *cells.*

Here is our algebraic cell-trading lemma. This process involves introducing cells, adding cells, and cancelling cells, and results in n-cells being "traded for" $(n+2)$-cells.

Lemma 6.1. *Suppose given a bounded chain complex decomposed as modules as* $B_\# \oplus A_\#$ *for which the boundary map has the form*

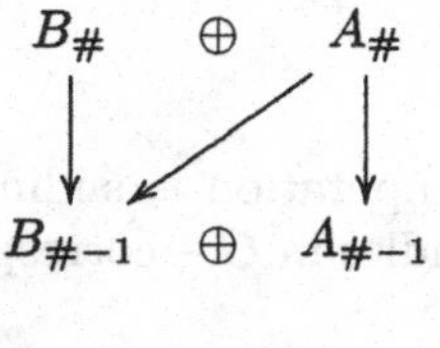

If there is a bounded chain homotopy s, with $(s|B_{\#}) = 0$, from the identity to a morphism which is 0 on $A_{\#}$ for $\# < k$, then $B_{\#} \oplus A_{\#}$ is boundedly chain-homotopy equivalent to $B_{\#} \oplus A'_{\#}$ where $A'_{\#} = 0$ for $\# < k$ and $A'_{\#} = A_{\#}$ for $\# \geq k+2$.

Proof. First introduce cancelling 1- and 2-cells corresponding to A_0 to obtain

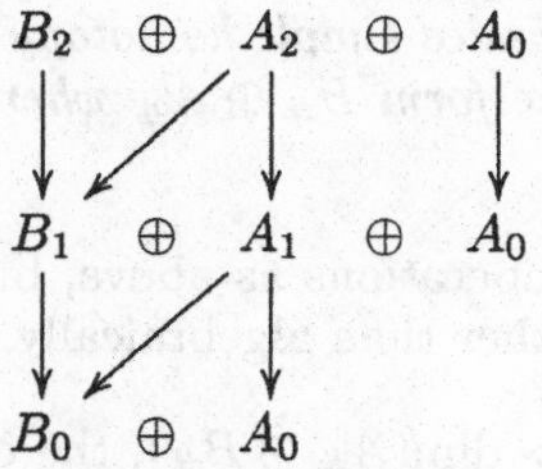

Now add the new A_0-cells in dimension 1 to A_1 via s to obtain

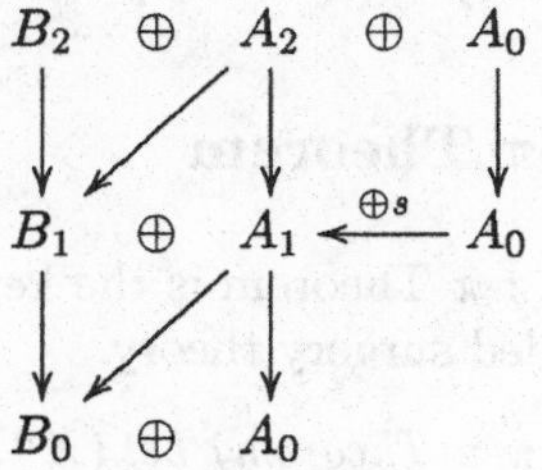

The lower map from A_0 to A_0 is the identity, so the lower A_0 modules may be canceled to obtain

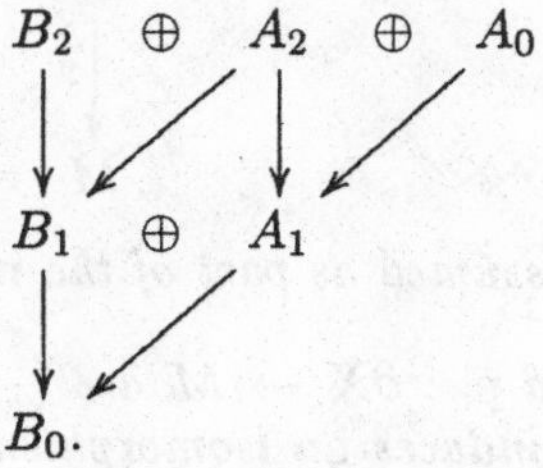

Repeat this process, and define $A'_{\#}$ so that $B_{\#} \oplus A'_{\#}$ is the resulting chain complex. □

Lemma 6.2. *Let $X \to M$ be a bounded CW complex which is 0-connected with bounded fundamental group π, so that the cellular chain complex over $\mathbb{Z}\pi$ is decomposed as (based) modules $B_{\#} \oplus A_{\#}$ for which the boundary map*

has the form

$$\begin{array}{ccc} B_\# & \oplus & A_\# \\ \downarrow & \swarrow & \downarrow \\ B_{\#-1} & \oplus & A_{\#-1}. \end{array}$$

If there is a bounded chain homotopy s with $(s|B_\#) = 0$, from the identity to a morphism which is 0 on $A_\#$ for $\# < k$, and if $A_\# = 0$ for $\# \leq 2$, then X may be changed by a bounded simple homotopy equivalence to X', so that the cellular chains have the form $B_\# \oplus A'_\#$ where $A'_\# = 0$ for $\# < k$ and $A'_\# = A_\#$ for $\# \geq k+2$.

Proof. Perform the same operations as above, but do them geometrically, using handle additions, rather than algebraically. □

In the above lemma, if $k > \dim(A_\# \oplus B_\#)$, the cellular chain complex becomes $B_\#$ in low dimensions together with modules in some pair of adjacent high dimensions with ∂ an isomorphism between them. The hypothesis that $A_\# = 0$ for $\# \leq 2$ is necessary to avoid π_1 problems.

7. The bounded π-π Theorem

As in [43, Chapter 9], the π-π Theorem is the key theorem in setting up a geometric version of bounded surgery theory.

Theorem 7.1. *(Bounded π-π Theorem) Let $(X^n, \partial X)$, $n \geq 6$, be a bounded Poincaré duality space over an allowable control space M. Consider a bounded surgery problem*

$$\begin{array}{ccc} (W, \partial W) & \xrightarrow{\phi} & (X^n, \partial X) \\ & & \downarrow{\scriptstyle p} \\ & & M \end{array}$$

with bundle information assumed as part of the notation.

If both $p : X \to M$ and $p| : \partial X \to M$ are (-1)-, and 0-connected and if the inclusion $\partial X \to X$ induces an isomorphism of bounded fundamental groups π, then we may do surgery to obtain a bounded normal bordism from $(W, \partial W) \to (X, \partial X)$ to $(W', \partial W') \to (X, \partial X)$, where the second map is a bounded simple homotopy equivalence of pairs.

Proof. We begin with the case $n = 2k$. By Theorem 5.3 we may do surgery below the middle dimension. We obtain a surgery problem $W' \xrightarrow{\phi'} X$ so that ϕ' is an inclusion which is the identity through dimension k.

This means that cancelling cells in $K_{\#}(W', \partial W')$ yields a complex which is 0 through dimension $k-1$. Abusing the notation, we will assume that the chain complex $K_{\#}(W', \partial W')$ is 0 for $\# \leq k-1$. The generators of $K_{k-1}(W)$ correspond to k-cells in $\partial X - W$. Cancelling these against the k-cells described in Remark 5.4(ii), and leaving out the primes for notational convenience, we have

$$\begin{aligned} K_{\#}(W, \partial W) &= 0 \qquad \# \leq k-1 \\ K_{\#}(W) &= 0 \qquad \# \leq k-1. \end{aligned}$$

Since

$$K^{n-\#}(W, \partial W) \simeq K_{\#}(W)$$

there is a bounded algebraic homotopy σ on $K^{\#}(W, \partial W)$ satisfying $\sigma\delta + \delta\sigma = 1$ for $\# \geq k+1$. Taking duals as in Definition 2.2, there is an algebraic homotopy s on $K_{\#}(W, \partial W)$ such that $s\partial + \partial s = 1$ for $\# \geq k+1$. Since $K_{\#} = K_{\#}(W, \partial W)$ is 0 in high dimensions, the "cell trading" procedure may be applied upside down, so that the $K_{\#}$ is changed to

$$0 \to K'_{k+2} \xrightarrow{\partial} K'_{k+1} \xrightarrow{\partial} K_k \to 0$$

together with a homotopy s so that $s\partial + \partial s = 1$ except at degree k. We leave out the primes for notational convenience. Corresponding to each generator of K_{k+2} (and at a point near where the generator sits in the control space) we introduce a pair of cancelling $(k-1)$- and k-handles and excise the interior of the $(k-1)$-handle from $(W, \partial W)$. The chain complex for this modified W is

$$\begin{array}{ccccccccc} 0 & \longrightarrow & K_{k+2} & \longrightarrow & K_{k+1} & \longrightarrow & K_k & \longrightarrow & 0 \\ & & & & & & \oplus & & \\ & & & & & & K_{k+1} & & \end{array}$$

All generators of $K_k \oplus K_{k+1}$ are represented by discs. We may represent any linear combination of these discs by an embedded disc, and these embedded discs may be assumed to be disjoint by the usual piping argument. See [43, p. 39]. This uses the surjective part of the π-π condition. We do surgery on the following elements: For each generator x of K_k, we do surgery on $(x - \partial sx, sx)$ and for each generator y of K_{k+2}, we do surgery on $(0, \partial y)$. This time, we can think of the process as introducing pairs of cancelling k- and $(k+1)$-handles, performing handle additions with the k-handles, and then excising the k-handles from $(W, \partial W)$. The resulting chain complex is:

$$\begin{array}{ccccccc}
0 \longrightarrow K_{k+2} & \xrightarrow{\partial} & K_{k+1} & \longrightarrow & K_k & & \\
& & \oplus & \nearrow_{1-\partial s} & \oplus & & \\
& & K_k & \xrightarrow{s} & K_{k+1} & \longrightarrow & 0 \\
& & \oplus & \nearrow_{\partial} & & & \\
& & K_{k+2} & & & &
\end{array}$$

which is easily seen to be contractible, the contraction being

$$\begin{array}{ccccccc}
0 \longleftarrow K_{k+2} & \xleftarrow{s} & K_{k+1} & \longleftarrow & K_k & & \\
& & \oplus & \swarrow_{1-\partial s} & \oplus & & \\
& & K_k & \xleftarrow{\partial} & K_{k+1} & \longleftarrow & 0 \\
& & \oplus & \swarrow_{s} & & & \\
& & K_{k+2} & & & &
\end{array}$$

Dualizing, we see that after surgery, $K_{\#}(W, \partial W)$ is boundedly chain contractible. Poincaré duality shows that $K_{\#}(W)$ is boundedly chain contractible. Together, these imply the bounded chain contractibility of $K_{\#}(\partial W)$. Using Theorem 2.15 now shows that $\partial X \to \partial W$ and $X \to W$ are bounded homotopy equivalences. This is where the hypothesis of allowability and the full π-π condition are used. An easy argument composing deformations in the mapping cylinder of $(W, \partial W) \to (X, \partial X)$ completes the proof that $(W, \partial W) \to (X, \partial X)$ is a controlled homotopy equivalence.

Having obtained a homotopy equivalence of pairs, we can vary by an h-cobordism of pairs to obtain a simple homotopy equivalence of pairs. The argument for this is easy, using only standard facts about torsion of h-cobordisms and the fact that if $\phi : (M^n, \partial M) \to (N^n, \partial N)$ is a homotopy equivalence of pairs, then $\tau(f|\partial M) = \tau(f) + (-1)^n \tau(f)^*$. This is a straightforward consequence of the *simplicity* of Poincaré duality at the chain level. This completes the even-dimensional case.

To obtain the π-π-Theorem in the odd-dimensional case we resort to a trick.

(i) Cross with S^1 to get back to an even dimension and do the simple surgery.

(ii) Go to the cyclic cover and use the simplicity of the above homotopy equivalence to split and obtain a homotopy equivalence of the ends. See Theorem 7.2 below.

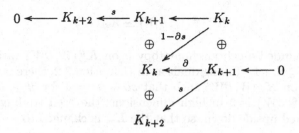

(iii) Vary by an h-cobordism of pairs to get a simple homotopy equivalence of pairs.

This completes the proof. □

In the above, we used the following splitting theorem from [9], which is essentially a bounded version of Quinn's End Theorem [29, 30]. In section 12 we give a proof based on the algebraic theory of surgery [36].

Theorem 7.2. *Let $(X^n, \partial X)$, $n \geq 6$, be a bounded Poincaré duality space over M. If both $p : X \to M$ and $p| : \partial X \to M$ are (-1), and 0-connected and if $\phi : (W^n, \partial W) \to (X^n, \partial X) \times \mathbb{R}$ is a bounded simple homotopy equivalence of pairs over $M \times \mathbb{R}$, where M is allowable, then $(W^n, \partial W) \cong (N^n, \partial N) \times \mathbb{R}$ and ϕ is boundedly homotopic to $\phi' \times id$, where $\phi' : (N^n, \partial N) \to (X^n, \partial X)$ is a bounded homotopy equivalence over M.*

8. Manifold 1-skeleton

Our construction of bounded surgery groups is modeled on Ch. IX in Wall [43]. An essential ingredient there is Wall's Lemma 2.8, which says that Poincaré duality spaces have manifold 1-skeleta.

In this and the following section we specialize to allowable metric spaces.

Proposition 8.1. *Suppose that M is an allowable metric space. Given a finitely presented group π and an integer $n \geq 4$, there exists a (-1)- and 0-connected n-dimensional manifold $W \to M$ with bounded fundamental group equal to π.*

Proof. Let N be a compact 4-manifold with fundamental group π, and let W be a regular neighborhood of a proper embedding of the 2-skeleton of M (or rather of a 0- and 1-connected PL-complex mapping to M see Definition 2.11) in $\mathbb{R}^5$. Now $\partial W \times N \to M$ satisfies the conditions except for dimension.

By general position once again, we may embed the 2-skeleton of $\partial W \times N$ properly in $\mathbb{R}^{n+1}$. The boundary of a regular neighborhood mapped to M now satisfies the conditions. □

Proposition 8.2. *A bounded Poincaré duality space $X \to M$ of dimension ≥ 5, with bounded fundamental group π has a manifold 1-skeleton, i.e., there exist a manifold with boundary $(W, \partial W) \to M$ and a bounded homotopy equivalence $X \sim W \cup_{\partial W} Y$ where Y is obtained from ∂W by attaching cells of dimension 2 and higher, and $(Y, \partial W)$ is a bounded Poincaré pair.*

Proof of proposition. Our proof, which is modeled on Wall's, consists of changing the CW structure on X to make it similar to the CW structure of the dual chains in high dimensions. We then exploit the fact that the boundary map from the $(n-1)$- to the n-cells in the dual complex has a very special form.

Let $A_\# = D^{n-\#}(X)$, a chain complex in $\mathcal{C}_M(\mathbb{Z}\pi)$. Denote $D_\#(X)$ by $B_\#$. Poincaré duality gives a homotopy equivalence

$$f : A_\# \to B_\#, \qquad g : B_\# \to A_\#.$$

We start by constructing an algebraic model for the new cell structure on X. The complex we construct will be equal to $B_\#$ for $\# = 0, 1, 2$ and $A_\#$ for $\# \geq 5$.

Consider the 2-skeleton of $B_\#$ and the mapping cylinder chain complex of $g|B_\#^{(2)} : B_\#^{(2)} \to A$.

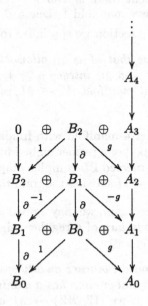

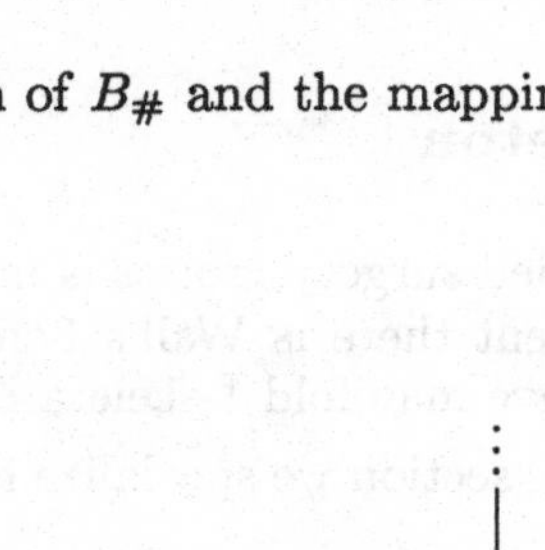

This is boundedly chain homotopy equivalent to $A_\#$. Since g is a homotopy equivalence, we can trade cells to get a new chain complex which looks like $B_\#$ through dimension 2. Calling the resulting chain complex $A_\#$ again, we begin to work geometrically. Introducing cancelling cells, we get a space bounded homotopy equivalent to X with chain complex changed as follows:

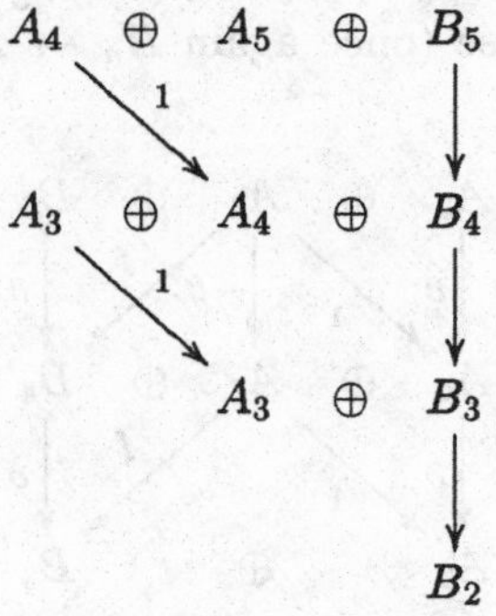

Adding A_n-cells to A_{n-1} via ∂ for $n \geq 4$ and A_n-cells to B_n via f for $n \geq 3$ results in the chain complex

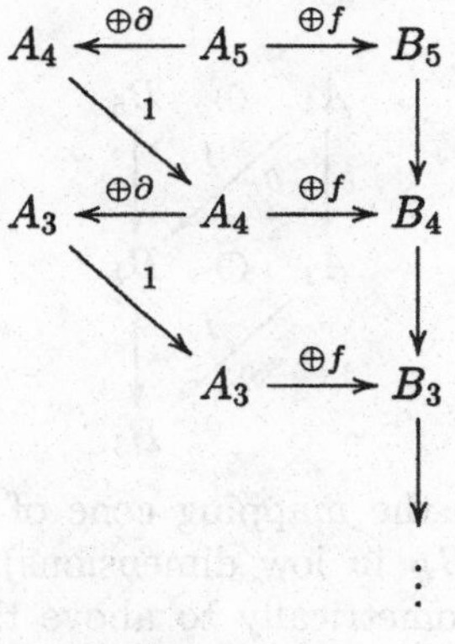

which gives the following boundary maps:

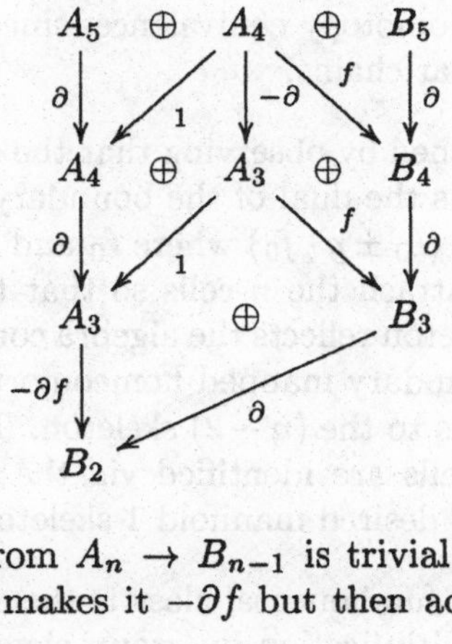

Note that the map from $A_n \to B_{n-1}$ is trivial for a nontrivial reason, adding A_n to B_n via f makes it $-\partial f$ but then adding A_n to A_{n-1} via ∂ makes it $-\partial f + f\partial = 0$.

Finally, we add B_n to A_n via $-g$. A short computation shows that this changes the boundary map $B_3 \to B_2$ to 0 (since $g : B_2 \to A_2$ is the identity) and it changes nothing else (once again $B_n \to A_{n-1}$ is 0 for a nontrivial reason). We now have

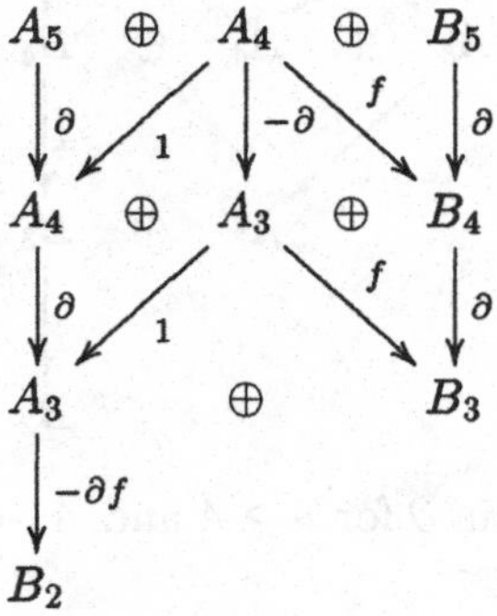

The chain complex

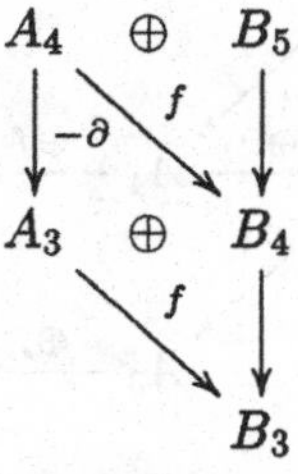

is contractible, since it is the mapping cone of a homotopy equivalence. (Remember that $A_\# = B_\#$ in low dimensions). We may now trade up this subchain complex geometrically to above the dimension of X. The n-skeleton of the resulting CW complex has cellular chains equal to $A_\#$ and the map to $D_\#(X)$ is the given map. It follows from Theorem 2.15 that this is a bounded homotopy equivalence, since it induces a homotopy equivalence on the cellular chains.

The proof is now finished by observing that the algebraic boundary map from n- to $(n-1)$-cells is the dual of the boundary map $D_1(X) \to D_0(X)$, and that each 1-cell hits $(\ell_0 \pm g \cdot f_0)$ where ℓ_0 and f_0 are 0-cells and $g \in \pi$. This means that if we attach the n-cells so that the map to the $(n-1)$-skeleton mod $(n-2)$-skeleton reflects the algebra completely, then the n-cells have patches on the boundary mapped homeomorphically to their images, and the complement goes to the $(n-2)$ skeleton. These patches are paired off 2 and 2, so the n-cells are identified via the $(n-1)$-cells to build a manifold W. This is the desired manifold 1-skeleton.

The point is that the fundamental class is the sum of all the n-handles suitably modified by multiplication by group elements. This is seen from the special nature of the boundary map. Using this, let W be the union of slightly shrunken n-cells with their identifications along $(n-1)$-cells. In the diagram

$$\begin{array}{ccccccccc} 0 & \longrightarrow & D_{\#}W & \hookrightarrow & D_{\#}X & \longrightarrow & D_{\#}(\overline{X-W}, \partial W) & \longrightarrow & 0 \\ & & & & \uparrow{\scriptstyle [X]\cap -} & & & & \\ 0 & \longrightarrow & D^{\#}(W, \partial W) & \hookrightarrow & D^{\#}(X) & \longrightarrow & D^{\#}(\overline{X-W}) & \longrightarrow & 0 \end{array}$$

$[X]\cap$ restricted to $D^{\#}(W,\partial W)$ is seen to be an isomorphism, and it follows from arguments similar to above (rolling up in the mapping cones) that $[X]\cap -$ induces a homotopy equivalence of $D_{\#}(\overline{X-W}) \leftarrow D^{\#}(\overline{X-W})$, so we have split off a Poincaré duality complex. The reader is referred to [42, Corollary 2.3.2] for further details. □

We finally need the following:

Lemma 8.3. *Let $X \xrightarrow{p} M$ be a bounded Poincaré duality complex, and let $W \xrightarrow{f} X$ be a degree 1 normal map of a manifold W to X. This means that $W \to X \to M$ is proper, there is a bundle ξ over X with a framing of $\tau_W \oplus f'\xi$, and that the fundamental class $[W] \in H_n^{\ell f}(W;\mathbb{Z})$ is sent to $[X] \in H_n^{\ell f}(X;\mathbb{Z})$. Then there is a normal bordism of $W \xrightarrow{f} X$ to a manifold (which we will again call W) so that f is a homeomorphism over the 1-skeleton of X found in Proposition* 8.2, *i.e., so that $f|f^{-1}$ (regular neighborhood of 1-skeleton) is a homeomorphism.*

Proof. We start with $F = f \circ \mathrm{proj} : W \times I \to X$ and make $F|W \times 1$ transverse to the barycenter of each n-cell of X. Since f is degree 1, the inverse image of this point counted with signs must be 1. On pairs of points of opposite sign we attach a 1-handle to $W \times 1$, and extend F over the resulting bordism, sending the core of the 1-handle to the point and the normal bundle of the core to normal bundle of the point. The restriction of F to the new boundary has 2 fewer double points. Continuing this process, $F|$(new boundary) becomes a homeomorphism over the 0-handles of our 1-skeleton.

Now consider the 1-handles. Assuming $F|$ transverse to the core of a 1-handle, the inverse image must be a union of finitely many S^1's and one interval. After a homotopy the interval may be assumed to map homeomorphically onto its image, and we only need to eliminate the S^1's. But each S^1 maps to the interior of an interval so this map is homotopy trivial. Attaching a 2-handle to each S^1 in the induced framing, extending the map to the core D^2 by the null homotopy, and extending to the normal direction by the framing removes S^1 from the inverse image of the 1-handle. Doing this to all of the 1-handles completes the process. □

9. The surgery groups

Our surgery groups are bordism groups patterned on Wall's Chapter 9. As usual, we restrict ourselves to allowable metric spaces. Following [43, p. 86], we define an "n-dimensional unrestricted object" to consist of:

(i) A bounded Poincaré pair, i.e. a pair (Y, X) with control map to M, such that each component has bounded fundamental group and is a bounded Poincaré duality complex in the sense of §3.
(ii) A proper map $\varphi : (W, \partial W) \to (Y, X)$ of pairs of degree 1, where W is a manifold and $\varphi| : \partial W \to X$ is a simple homotopy equivalence, simplicity being measured in $Wh_M(\pi_1(X))$.
(iii) A stable framing F of $\tau_W \oplus \varphi^*(\tau)$.
(iv) A map $\omega : Y \to K$, where K is a pointed CW complex which is fixed with a fixed pointed double cover $\widehat{K}$. It is required that the pullback of this double cover to Y be the orientation covering.

Bordism of these objects is defined similarly. See the reference above for details. We denote the bordism group of unrestricted objects by $L^1_{n,M}(K)$. Note that, as in Wall, (Y, X) is allowed to vary along with $(W, \partial W)$. There is a natural group structure on $L^1_{n,M}(K)$ with the empty set as the 0 element and the sum represented by disjoint union. As is usual in bordism theories, the groups are functorial in K, in the sense that a pointed map from K_1 to K_2 which is covered by a pointed map of double covers induces a group homomorphism. Note that Y may have infinitely many components, but that since $Y \to M$ is proper, there are only finitely many components locally.

Still following Wall, we define "restricted objects" by requiring that $X \to M$ be (-1)- and 0-connected and have bounded fundamental group and that the map $Y \to K$ induce a π_1-isomorphism. We define $L^2_{n,M}(K)$ to be the set of restricted bordism classes of restricted objects, i.e., we require objects as well as bordisms to be (-1)- and 0-connected and to have the same bounded fundamental group as K. See [43, pp. 86–88]. Note that $L^2_{n,M}(K)$ is only a set – we have no zero object and no sum, since the empty set is not allowed and since disjoint union destroys 0- connectedness.

The π-π Theorem shows, just as in the classical case, that we may do surgery with a fixed restricted target if and only if the invariant in $L^2_{n,M}(K)$ vanishes ([43, Theorem 9.3]).

Theorem 9.1. *Let $\phi : (W, \partial W) \to (Y, X)$ be a restricted surgery problem, i.e., a (-1)- and 0-connected surgery problem with bounded fundamental group π and reference map $Y \to K$ inducing an isomorphism of fundamental groups and pulling back the orientation double cover of Y. We assume $n = dim(X) \geq 5$. Then there is a normal cobordism rel ∂W of W to a bounded* *homotopy equivalence if and only if the equivalence class of ϕ in $L^2_{n,M}(K)$ vanishes.*

We also have the analogue of [43, Theorem 9.4].

Theorem 9.2. *The natural map*

$$L^2_{n,M}(K) \longrightarrow L^1_{n,M}(K)$$

is a bijection if $n \geq 5$ and K has a finite 2-skeleton.

Proof. By Proposition 8.1, there exists $W \to M$ so that the fundamental group is bounded $\pi_1 K$ and W is (-1)-connected. The existence of a map $W \to K$ inducing an isomorphism on π_1 is assured by the construction of W. The surgery problem $V \xrightarrow{f} X$ is equivalent to $V \amalg W \xrightarrow{f \amalg 1} X \amalg W$, since crossing with I may be considered a bordism (one is allowed to forget components that are homotopy equivalences).

Each component of X has a manifold 1-skeleton, and after a bordism of V we may assume f to be a homeomorphism of these 1-skeleta, by Lemma 8.3, so the stage is set for simultaneous surgery. Attach a 1-handle from every 0-handle of X to a 0-handle of W so that the image of this 1-handle in M is small.

Now mimic this construction in the domain by attaching a 1-handle from V to W. For each 1-handle of X we get a path from a point in W to a point in W through the one handles we attached above. Join up these points in W so that the loop created maps trivially to K. This is possible, since W has bounded π_1 isomorphic to $\pi_1 K$. Now attach 2-handles simultaneously in the domain and range to kill these loops. Since every 0- and 1-handle of X has been equated with some element of W, the result is a target which is (-1)-connected and which has bounded fundamental group equal to $\pi_1 K$ induced by the map to K. We have thus constructed a bordism from any object to a restricted object. Injectivity is proved using the same argument on the bordism. □

The group $L^1_{n,M}(K)$ and the set $L^2_{n,M}(K)$ can thus be given the common name $L_{n,M}(K)$, or better $L^s_{n,M}(K)$ because we require simple homotopy equivalence. Naturally, the definition may be varied by only requiring homotopy equivalence. These groups are denoted $L^h_{n,M}(K)$. We may, of course, also specify a $*$-invariant subgroup G of $Wh_M(\pi)$ defining $L^G_{n,M}(K)$. We shall sometimes suppress the upper index.

It is very important to notice that the functor $L_{n,M}(K)$, as a functor in M, does not send restricted objects to restricted objects, and that it is only for restricted objects that the invariant measures whether or not one may do surgery to obtain a bounded homotopy equivalence. We discuss this in the following:

Example 9.3. Consider the inclusion $L \subset L_+$, where L_+ is L union a disjoint basepoint. Assume that a restricted surgery problem has an obstruction in $L_{n,O(L)}(K)$ which vanishes in $L_{n,O(L_+)}(K)$. To understand this geometrically, we first have to replace the image in $L_{n,O(L_+)}(K)$ by a restricted object. The image is not (-1)-connected, but this may be corrected by doing simultaneous surgery – adding a tail to the surgery problem. This doesn't change the surgery problem away from a compact subset of the target. This means that if the surgery obstruction vanishes in the new problem, then it is possible to solve the original surgery problem "near infinity."

Similarly, if one considers the map from L_+ to L sending the extra point to some point of L, the induced map from $L_{n,O(L_+)}(K)$ to $L_{n,O(L)}(K)$ will hit an element which is not 0-connected, so some simultaneous surgery has to be done before the vanishing of this invariant implies that one can surger the source to a homotopy equivalence. Again, it is possible to do this via a small modification of the original problem, only changing the target along a ray out to infinity, rather than by using the more general construction of the proof of Theorem 9.2.

Notice that an analogous phenomenon occurs in classical compact surgery. A surgery problem comes equipped with a reference map, usually to $B\pi$ where π is the fundamental group of the target. Given a group homomorphism $\pi \to \rho$, the vanishing of the image $L^h_n(\mathbb{Z}\pi) \to L^h_n(\mathbb{Z}\rho)$ means that after simultaneous surgery on source and target to make the fundamental groups equal to ρ, one can do surgery to obtain a homotopy equivalence.

Theorem 9.4. *A map $K_1 \to K_2$ which induces an isomorphism of fundamental groups and which is covered by a map of based covering spaces induces an isomorphism $L_{n,M}(K_1)$ to $L_{n,M}(K_2)$ for $n \geq 5$.*

Proof. Given $K_1 \to K_2$ inducing an isomorphism on π_1, etc., we get a long exact sequence of bordism groups, as in any bordism theory. The relative groups are 0 by the π-π Theorem. □

As noted by Quinn [32], the proof of Theorem 9.5 of [43] can be used to prove the following:

Theorem 9.5. *Given a (-1)- and 0-connected manifold $V^{(n-1)}$ with bounded $\pi_1 = \pi$ and an element of $\alpha \in L_{n,M}(\pi)$, $n \geq 6$, then α may be represented with $V \times I$ as target.*

Proof. We may assume that α is realized by $\phi : (W, \partial W) \to (Y, X)$ where X and Y are (-1)- and 0-connected with bounded fundamental group π, and $\partial W \to X$ is a homotopy equivalence. We may glue ∂W to X by a mapping cylinder and assume that $\partial W \to Y$ is the identity. There is a bordism from $\partial W \to \partial W$ to $V \to V$ by equating 0- and 1-handles. Attaching this to ∂W, we may change the representative above to be of the form $\phi : (W, V) \to (X', V)$, where $\phi|V = id$. By the π-π Theorem, there is a bordism $(P, \partial P)$ *not* rel V from ϕ to $\phi' : (W', V) \to (X', V)$ which is a homotopy equivalence of pairs. $(P, \partial P)$ may be reinterpreted to be a bordism from ϕ to a surgery problem with $V \times I$ as target. □

Finally, we should mention that there is an important variation of the theory where we only ask for conditions to be satisfied near ∞. Thus, we only ask for bounded homotopy equivalence and Poincaré duality over $O(K) = \{t \cdot x | t \in [0, \infty), x \in K\}$ for t large. The groups $L^{s,\infty}_{n,O(K)}(\mathbb{Z}\pi)$ are defined as in the beginning of this section, but everything is only done near ∞. Spaces and maps only have to be defined near ∞, as well. It is shown in [9] that $\mathcal{C}_{O(K),\infty}(\mathcal{A}) \to \mathcal{C}_{O(K_+)}(\mathcal{A})$ is a homotopy equivalence, so Whitehead torsion makes sense in this framework. We obtain the following quite useful:

Theorem 9.6.

$$L^s_{n,O(K_+)}(\mathbb{Z}\pi) \to L^{s,\infty}_{n,O(K)}(\mathbb{Z}\pi)$$

is an isomorphism if $n \geq 5$.

Proof. The map is the forgetful map taking a problem over $O(K_+)$ to its germ near infinity over $O(K)$.

For simplicity, assume first that $n \gg dim(K)$. In this case we can form a manifold P with a map to $O(K_+)$ as follows. Embed K in S^{n-5} and let P' be the boundary of a regular neighborhood of K in S^{n-5}. Form P by gluing together the part of $O(P')$ outside the unit sphere and a copy of $(-\infty, 1] \times P'$. Map $[0,1] \times P'$ to the part of $O(K)$ inside the unit sphere and send $(-\infty, 0] \times P'$ to $O(+)$. Form a manifold V^{n-1} by taking the product of P with a closed manifold Q^4 which has fundamental group π. Note that it is reasonable to talk about "levels" in V, just as in $O(K_+)$.

To show that the forgetful map is onto, we first use a version of Theorem 9.5 near infinity to show that we can represent a given $\alpha \in L^{s,\infty}_{n,O(K)}(\mathbb{Z}\pi)$ by a map $\phi : (W, \partial W) \to (V \times I, \partial(V \times I))$ such that $\phi|\partial W$ is a simple homotopy equivalence near infinity. Using the simplicity, we can split ϕ over the part of V at level T for some large T, obtaining $\phi| : (W_0, \partial W_0) \to (V \times I, \partial(V \times I)) \cap (level\ T)$ with $\phi|\partial W_0$ a homotopy equivalence. Adding a copy of $(W_0, \partial W_0) \times (-\infty, T]$ to the part of W outside of level T and mapping to $V \times I$ in the obvious way produces an element of $L^s_{n,O(K_+)}(\mathbb{Z}\pi)$ whose image in $L^{s,\infty}_{n,O(K)}(\mathbb{Z}\pi)$ is α.

To show that the map is monic, let $\alpha \in L^s_{n,O(K_+)}(\mathbb{Z}\pi)$ be an element which becomes trivial in $L^{s,\infty}_{n,O(K)}(\mathbb{Z}\pi)$. Representing α by a map $\phi : (W, \partial W) \to (V \times I, \partial(V \times I))$ as above, ϕ is bordant rel ∂ to a map $\phi' : (W', \partial W') \to (V \times I, \partial(V \times I))$ which is a simple homotopy equivalence near infinity. Using the simplicity, we can split the homotopy equivalence over some level T. Expanding a collar around this level in the domain and sliding the remainder of the manifold down towards the cone point and out the "tail", $O(+)$, produces a bordism from ϕ' to a simple homotopy equivalence. Note that we actually *gain* control through this sliding process and that the bordism from ϕ to ϕ' need only be defined on a neighborhood of infinity for this process to succeed, since an easy transversality argument allows us to construct a bordism from $(W, \partial W)$ to a manifold which equals $(W', \partial W')$ near infinity in $O(K)$ and $(W, \partial W)$ over $O(+)$.

The general case, where we do not have $n \gg dim(K)$ is similar. The manifold target is less homogeneous, though, so we represent the problem with manifold target $(V \times I, \partial(V \times I))$, but here V has no good homogeneity properties, split at countably many levels going out the tail $O(+)$, and use the compact π-π-Theorem between the splittings to solve the surgery problem. Alternatively, we could use the periodicity result of §12 to deduce the low-dimensional case from the high-dimensional case treated above. □

Remark 9.7. Of course, we are not limited to defining absolute surgery groups. The same definition may be varied as in pp. 91–93 of [43] to define relative, or even n-ad, surgery groups.

10. Ranicki-Rothenberg sequences, and $L^{-\infty}$

In this section we study the properties of $L^s_{n,M}(\mathbb{Z}\pi)$ in the special case where $M = O(K)$.

Proposition 10.1. *Assume $n \geq 5$. There is a long exact Ranicki-Rothenberg sequence*

$$\to L^s_{n,M}(\mathbb{Z}\pi) \to L^h_{n,M}(\mathbb{Z}\pi) \to \widehat{H}^n(\mathbb{Z}_2 ; \widetilde{K}_1(\mathcal{C}_M(\mathbb{Z}))) \to$$

Proof. The proof is formal, the sequence is a bordism long exact sequence where the Tate cohomology groups are identified with the relative bordism groups of surgery problems with simple boundaries (see [26]). □

Note that $O(\Sigma K) = O(K) \times \mathbb{R}$ *as a metric space.* This leads to the following useful proposition.

Proposition 10.2. *Assume $n \geq 5$. Crossing with $\mathbb{R}$ produces an isomorphism*

$$L^h_{n,O(K)}(\mathbb{Z}\pi) \to L^s_{n+1,O(\Sigma K)}(\mathbb{Z}\pi).$$

Proof. First, consider the case where $K = \emptyset$. Then $O(K) = pt$ and $O(\Sigma K) = \mathbb{R}$. To see that the map is monic, let $\phi : (W, \partial W) \to (X, \partial X)$ be a compact surgery problem. Crossing with $\mathbb{R}$ gives $\phi \times id : (W, \partial W) \times \mathbb{R} \to (X, \partial X) \times \mathbb{R}$. If $\phi \times id$ represents 0 in $L^s_{n+1,R}(\mathbb{Z}\pi)$, then $\phi \times id$ is normally bordant rel ∂W to a simple homotopy equivalence $\phi' \times id : (W', \partial W') \times \mathbb{R} \to (X, \partial X) \times \mathbb{R}$. Since the homotopy equivalence is simple, W' has trivial end obstruction, so $W' = M' \times \mathbb{R}$ and by transversality we get W bordant rel ∂W to $M' \simeq X$, showing that ϕ represents 0 in $L^h_{n,pt}(\mathbb{Z}\pi)$.

To see that the map is an epimorphism, represent $\alpha \in L^s_{n+1,R}(\mathbb{Z}\pi)$ by a problem $\phi : (W, \partial W) \to ((V \times I) \times \mathbb{R}, \partial(V \times I) \times \mathbb{R})$ with V^{n-1} a closed manifold with fundamental group π. Since the homotopy equivalence on the boundary is simple, we can split the homotopy equivalence over $(V \times \partial I) \times \{T\}$ and continue the splitting over $(V \times I) \times \{T\}$ by transversality, getting a compact surgery problem $(M, \partial M) \to ((V \times I), \partial(V \times I))$. Letting a collar around M grow in both directions gives a bordism from the original problem to $(M, \partial M) \times \mathbb{R} \to ((V \times I), \partial(V \times I)) \times \mathbb{R}$.

In case $K \neq \emptyset$, the argument is similar. One does the same things boundedly over $O(K)$. □

Definition 10.3. By the groups $L^{2-k}_{n-k}(\mathbb{Z}\pi)$ we shall mean $L^s_{n-k}(\mathbb{Z}\pi)$ when $k = 0$, $L^h_{n-k}(\mathbb{Z}\pi)$ when $k = 1$, $L^p_{n-k}(\mathbb{Z}\pi)$ when $k = 2$, and the negative L-groups of [34, 35] for $k > 2$. It is well known that these groups are 4-periodic in the lower index.

Theorem 10.4. *When* $n \geq 5$

$$L^s_{n,R^k}(\mathbb{Z}\pi) \cong L^{2-k}_{n-k}(\mathbb{Z}\pi)$$
$$L^h_{n,R^{k-1}}(\mathbb{Z}\pi) \cong L^{2-k}_{n-k}(\mathbb{Z}\pi).$$

Proof. First, we consider the case $n \geq k+5$. The general case will follow from 4-periodicity which is proved algebraically in §12. We use induction on the Ranicki-Rothenberg exact sequence.

We have an algebraically defined inclusion $L^{-i}_n(\mathbb{Z}\pi) \subset L^s_{n+i+2}(\mathbb{Z}(\pi \times \mathbb{Z}^{i+2}))$. There is also a map $L^s_{n+i+2}(\mathbb{Z}(\pi \times \mathbb{Z}^{i+2})) \to L^s_{n+i+2,\mathbb{R}^{i+2}}(\mathbb{Z}\pi)$, which is defined geometrically by taking cyclic covers. Combining these maps, we get a map of exact sequences:

$$\begin{array}{ccccc}
L^{-i+1}_n(\mathbb{Z}\pi) & \longrightarrow & L^{-i}_n(\mathbb{Z}\pi) & \longrightarrow & \widehat{H}^n(\mathbb{Z}_2, K_{-i}(\mathbb{Z}\pi)) \\
\downarrow & & \downarrow & & \downarrow \\
L^s_{n+i+1,\mathbb{R}^{i+1}}(\mathbb{Z}\pi) & \longrightarrow & L^h_{n+i+1,\mathbb{R}^{i+1}}(\mathbb{Z}\pi) & \longrightarrow & \widehat{H}^n(\mathbb{Z}_2, K_{-i}(\mathbb{Z}\pi)).
\end{array}$$

Here, we are using that $K_1(\mathcal{C}_{\mathbb{R}^{i+1}}(\mathbb{Z}\pi)) \cong K_{-i}(\mathbb{Z}\pi)$ ([23]). Combining with the isomorphisms

$$L^h_{i,\mathbb{R}^n}(\mathbb{Z}\pi) \cong L^s_{i+1,\mathbb{R}^{n+1}}(\mathbb{Z}\pi)$$

this inductively proves that

$$L^{-i}_n(\mathbb{Z}\pi) \cong L^s_{n+i+2,\mathbb{R}^{i+2}}(\mathbb{Z}\pi).$$

□

Note that this proves that the L-groups are 4-periodic, at least when $K = S^i$, i.e. $O(K) = \mathbb{R}^{i+1}$ and $n \geq i+6$. We shall now investigate $L^s_{n,O(K)}(\mathbb{Z}\pi)$ as a functor of K, from the category of finite complexes and Lipschitz morphisms.

Theorem 10.5. *$L^s_{n,O(K)}(\mathbb{Z}\pi)$ is homotopy invariant, $n \geq 5$.*

Proof. We have to show that

$$L^s_{n,O(K)}(\mathbb{Z}\pi) \to L^s_{n,O(K\times I)}(\mathbb{Z}\pi)$$

is an isomorphism. By functoriality, it is a split monomorphism. To see that it is onto, we note that a homotopy can always be viewed as an unrestricted bordism. Thus a surgery problem parameterized by $O(K \times I)$ becomes bordant, and hence equivalent in the unrestricted bordism group of §9, to the induced problem parameterized by projecting to one end. □

Continuing to investigate $L^s_{n,O(K)}(\mathbb{Z}\pi)$ as a functor in K, we define

$$L^{-\infty}_{n,O(K)}(\mathbb{Z}\pi) = \lim_i L^s_{n+i,O(\Sigma^i K)}(\mathbb{Z}\pi)$$

where the maps are given by crossing with the reals.

Theorem 10.6. *$L^{-\infty}_{n,O(K)}(\mathbb{Z}\pi)$ is a reduced homology theory in the variable K.*

Remark 10.7. This is a geometric version of Theorem 3.4 of the thesis of Yamasaki [45]. There, $L^{-\infty}(\mathbb{Z}\pi)$ is defined abstractly as a spectrum.

To prove the theorem we need the following:

Lemma 10.8. *Let $C(L)$ be the cone on L. Then $L^s_{n,O(CL)}(\mathbb{Z}\pi) = 0$.*

Proof. First, note that there is an isometry $O(C(L)) \cong O(L) \times [0,\infty)$. Let $\alpha \in L^s_{n,O(CL)}(\mathbb{Z}\pi)$ be represented by a surgery problem:

$$\begin{array}{ccc} (W, \partial W)) & \longrightarrow & (X, \partial X) \\ & & \big\downarrow p \\ & & O(L) \times [0,\infty) \end{array}$$

This is the boundary of:

$$\begin{array}{ccc} (W,\partial W) & \longrightarrow & (X,\partial X) \\ & & \downarrow \\ & & O(L)\times[0,\infty) \end{array}$$

where $q(x,t) = p(x) + t$. This shows that $\alpha = 0$. $\square$

Proof of theorem. To show that the functor is half exact in the variable K, consider a cofibration

$$L \subset K \to K \cup CL.$$

Applying O we get:

$$O(L) \to O(K) \to O(K \cup C(L)) \cong O(K) \cup_{O(L)} O(L) \times [0,\infty).$$

The composite is the trivial map, since it factors through $O(CL)$, so consider a surgery problem:

$$\begin{array}{ccc} (W,\partial W) & \longrightarrow & (X,\partial X) \\ & & \downarrow \\ & & O(K) \end{array}$$

which goes to 0 in $L^{-\infty}_{n,O(K\cup C(L))}(\mathbb{Z}\pi)$. As usual, we may assume that X is a manifold and that ϕ is (-1)- and 0-connected. The vanishing of $[\phi]$ over $O(K \cup C(L)) \cong O(K) \cup_{O(L)} O(L) \times [0,\infty)$ means that after simultaneous surgery on the domain and range to obtain a (-1)- and 0-connected $\phi' : W' \to X'$, ϕ' is bordant to a bounded homotopy equivalence. Clearly, we can do the simultaneous surgery in such a way that the parts of W' and X' over $O(L) \times [1,\infty)$ are products.

If ϕ' is bordant to a bounded *simple* homotopy equivalence ϕ'', we can split the bordism over $O(L)\times\{T\}$ for some large T using controlled splitting over the boundary and at ϕ'' [9]. Projecting back to $O(K)$, the split bordism becomes an unrestricted bordism from the original problem to a bounded homotopy equivalence together with the inverse image of $O(L)\times\{T\}$, showing that the original problem was in the image of $L^{-\infty}_{n,O(L)}(\mathbb{Z}\pi)$.

If ϕ' is bordant to a bounded *non-simple* homotopy equivalence, we cross with $\mathbb{R}$ to kill the torsion and proceed as above. This torsion problem is the reason that we have to stabilize to obtain a homology theory. In fact, this argument shows that $L^{s}_{n,O(K)}(\mathbb{Z}\pi)$ is *not* half exact.

To finish the proof, note that we have already shown that there is an isomorphism $L^h_{n,O(K)}(\mathbb{Z}\pi) \cong L^s_{n+1,O(\Sigma K)}(\mathbb{Z}\pi)$. Since $L^{-\infty}_{n,O(S^k)}(\mathbb{Z}\pi)$ is naturally 4-periodic, it follows from half exactness that $L^{-\infty}_{n,O(-)}(\mathbb{Z}\pi)$ is a 4-periodic homology theory. □

Remark 10.9. We get a *periodic* homology theory and thus not a *connective* homology theory. A geometric interpretation of this is that a (-1)-connected $M^k \to \mathbb{R}^{m+k}$ behaves like a $-m$-dimensional manifold.

11. The surgery exact sequence

In this section we consider the following: Let M be an allowable metric space and $X \to M$ a 0- and -1-connected bounded Poincaré duality space with bounded fundamental group $= \pi$, and assume there is a given lift of

$$\begin{array}{ccc} & & B\,\mathrm{CAT} \\ & & \downarrow \\ X & \longrightarrow & BG \end{array}$$

where CAT = TOP, PL or O. We define the bounded structure set $\mathcal{S}_b(X)$ as usual in surgery theory: An element consists of a manifold W and a bounded homotopy equivalence

$$\begin{array}{ccc} W & \simeq & X \\ & & \downarrow \\ & & M \end{array}$$

two such being equivalent if there is a homeomorphism $h : W_1 \to W_2$ such that the diagram

$$\begin{array}{ccc} W_1 & & \\ \downarrow{\scriptstyle h} & \searrow & \\ & & X \\ & \nearrow & \\ W_2 & & \end{array}$$

is bounded homotopy commutative. As usual we get a surgery exact sequence.

Theorem 11.1. *For $n \geq 5$ there is an exact sequence of surgery*

$$\cdots \to \mathcal{S}_b(X \times I, \delta(X \times I)) \to [\Sigma X, F/\mathrm{CAT}] \to$$
$$\to L^s_{n+1,M}(\mathbb{Z}\pi) \to \mathcal{S}_b(X) \longrightarrow [X, F/\mathrm{CAT}] \to L^s_{n,M}(\mathbb{Z}\pi)$$

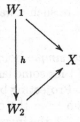

and relative versions hereof.

Proof. The proof is standard as in [40]. Given a lift of $X \to BF$ to $B\,\mathrm{CAT}$ there is a surgery problem obtained by transversality. If the obstruction to doing surgery vanishes, we obtain a bounded homotopy equivalence. Given two elements in the structure set, they determine two lifts to $B\,\mathrm{CAT}$. If the lifts are fibre homotopic we obtain a normal cobordism by transversality as in standard surgery theory. □

Given a finitely dominated Poincaré complex X, we obtain a surgery exact sequence for $X \times \mathbb{R}^k \longrightarrow \mathbb{R}^k$ as follows. Cross X with S^1 to obtain a finite Poincaré complex over $\mathbb{R}$, pass to the cyclic cover, and cross with $\mathbb{R}^{k-1}$ to obtain a bounded Poincaré model for $X \times \mathbb{R}^k \longrightarrow \mathbb{R}^k$. A radial homeomorphism $f : \mathbb{R}^k \longrightarrow \mathbb{R}^k$ which is Lipschitz (but whose inverse is not necessarily Lipschitz) induces a map of the surgery exact sequences, the point being that if a homotopy is bounded with respect to a map p to $\mathbb{R}^k$, it is certainly also bounded with respect to $f \cdot p$.

Theorem 11.2. *Reparameterization by a radial homeomorphism which is Lipschitz induces the identity on the surgery exact sequence.*

Proof. Since f is a radial homeomorphism, $\bar{x} + t \cdot f(\bar{x})$ is a homotopy of f through Lipschitz maps to the identity, and it follows easily that f induces the identity of L-groups and normal invariants. To see that f induces the identity on the structure set we need the result of Chapman [7] that a map from a manifold to $\mathbb{R}^k$ which has the bounded homotopy lifting property can be boundedly approximated by a map with the epsilon homotopy lifting property for all epsilon. If $W \longrightarrow X \times \mathbb{R}^k$ is a bounded equivalence, then the composition $W \longrightarrow \mathbb{R}^k$ is boundedly approximated by an approximate fibration. There is then an approximate fiber homotopy equivalence $W \longrightarrow X \times \mathbb{R}^k$ boundedly homotopic to the original map. This approximate fiber homotopy equivalence remains bounded under arbitrary radial reparameterization. □

12. The algebraic surgery theory

The algebraic surgery theory has been developed over a number of years by Ranicki see [34, 35] . The extension to additive categories with involution has also been developed by A. Ranicki (see [36]). This section depends strongly on [36]. Here are some of the basic definitions:

An involution on an additive category $\mathcal{A}$ is a contravariant functor

$$* : \mathcal{A} \to \mathcal{A};\ A \to A^*$$

$$(f : A \to B) \to (f^* : B^* \to A^*)$$

together with a natural equivalence

$$e : id_{\mathcal{A}} \longrightarrow **$$

such that the coherence condition

$$e(A^*) = (e(A)^{-1})^* : A^* \longrightarrow A^{***}$$

is satisfied.

Example 12.1. Let R be an associative ring with an antiinvolution, $R = \mathbb{Z}G$ with $\overline{\Sigma n_g g} = \Sigma w(g) n_g g^{-1}$, for example. Consider the category of finitely-generated projective R-modules. Then duality induces an involution on the category.

Of more interest to us is the following:

Example 12.2. Let $\mathcal{R}$ be the category of finitely generated free R-modules with involution as above. We get an induced involution on $\mathcal{C}_M(\mathcal{R})$ for M a metric space by the prescription $(A^*)_x = (A_x)^*$.

Ranicki has shown that his theory of algebraic surgery extends to additive categories with involution, so, in particular, he has defined $L_n^t(\mathcal{C}_M(\mathbb{Z}\pi))$, where the decoration t is h or corresponds to any involution-invariant subgroup of $Wh_M(\pi)$, as is usual in L-theory. In this setup L^p is the composite of L^h with idempotent completion of the additive category. To be able to treat the simple L-groups L^s corresponding to the 0-subgroup of $Wh_M(\pi)$ one needs a system of stable isomorphisms of the objects so that composites that are automorphisms have trivial torsion. This is obtained from an Eilenberg swindle on the *objects* in case M is unbounded (and as usual a specific choice of basis in case M is bounded). In this section we prove that these algebraically defined L-groups are the obstruction groups for bounded surgery problems in the case where there is no boundary or that there is a homotopy equivalence on the boundary. First recall from [36, p. 169] the basic definitions.

Let $\mathcal{A}$ be an additive category with involution. A sequence of objects and morphisms $0 \to A \xrightarrow{f} B \xrightarrow{g} C \to 0$ is split exact if g is split by a morphism h such that $(f, h) : A \oplus C \to B$ is an isomorphism. Let ϵ denote ± 1. An ϵ-quadratic form in $\mathcal{A}$ is an equivalence class of maps $\psi : A \to A^*$ two such being equivalent if they differ by a morphism of the form $\phi - \epsilon\phi^*$. It is nonsingular if $\psi + \epsilon\psi^*$ (which only depends on the equivalence class) is an isomorphism. A Lagrangian in a non-singular form (A, ψ) is a morphism $i : B \to A$ such that $\psi \cdot i = 0$ and $0 \to B \to A \to B^* \to 0$ is split exact. Ranicki then proves that a non-singular ϵ quadratic form is equivalent to the hyperbolic form $(B \oplus B^*, \{\begin{smallmatrix} 0 & 1 \\ 0 & 0 \end{smallmatrix}\})$ if and only if it admits a Lagrangian. The even L-groups are now defined as the Grothendieck construction on isomorphism classes of non-singular quadratic forms with $\epsilon = 1$ in dimensions $\equiv 0(4)$ and $\epsilon = -1$ when the dimension is $\equiv 2(4)$.

To define the odd L-groups one needs formations. A nonsingular ϵ-quadratic formation in $\mathcal{A}$, (A, ψ, F, G) is a non-singular ϵ-quadratic form (A, ψ) together with an ordered pair of Lagrangians F and G. (H_ϵ, P, P^*) is considered a trivial formation where H_ϵ is the hyperbolic form on $P \oplus P^*$. With the obvious notion of isomorphism Ranicki defines the odd L-groups to be the Grothendieck construction on isomorphism classes of formations modulo trivial formations and the relation $(A, \psi; F, G) + (A, \psi; G, H) = (A, \psi; F, H)$, with $\epsilon = 1$ in dimensions $\equiv 3(4)$ and $\epsilon = -1$ in dimensions $\equiv 1(4)$. Given this we now proceed along the lines of Wall's original method.

Theorem 12.3. *Consider a bounded surgery problem*

$$\begin{array}{ccc}(M^n, \partial M) & \longrightarrow & (X, \partial X) \\ & & \downarrow \\ & & Z\end{array}$$

where $\partial M \to \partial X$ is a bounded simple homotopy equivalence, X is 0 and -1-connected with bounded fundamental group π, and $n \geq 5$. Then one can do surgery rel boundary to produce a bounded simple homotopy equivalence if and only if an invariant in

$$L^s_n(\mathcal{C}_Z(\mathbb{Z}\pi))$$

vanishes. Moreover every element of $L^s_n(\mathcal{C}_Z(\mathbb{Z}\pi))$ is realized as the obstruction on a surgery problem with target $N \times I$ and homotopy equivalence on the boundary for an arbitrary $n-1$-dimensional manifold $N \to Z$ which is -1 and 0-connected with bounded fundamental group π.

Proof. First consider the even-dimensional case. We proceed as in §7 and obtain a highly connected surgery problem. We obtain a chain complex homotopy equivalent to $K_\#(M)$ which is concentrated in 3 dimensions:

$$0 \to K_{k+2} \to K_{k+1} \to K_k \to 0$$

and a contracting homotopy s (except in dimension k) which is obtained from Poincaré duality. Introducing cancelling $k+1$ and $k+2$ handles, we may change this to

$$\begin{array}{ccccccc} 0 & \longrightarrow & K_{k+2} & \longrightarrow & K_{k+1} & \longrightarrow & K_k \\ & & & & \oplus & & \oplus \\ & & & & K_{k+2} & \xrightarrow{1} & K_{k+2} \end{array}$$

after adding $k+1$-handles to K_{k+2} along s we may cancel the K_{k+2}-handles to shorten this chain complex to a 2-term chain complex which we write $0 \to K'_{k+1} \to K'_k \to 0$. Abusing notation we omit the primes. Notice that all generators of K_k are still represented by immersed spheres. For each generator of K_{k+1} we do a trivial surgery to get a chain complex

$$0 \longrightarrow K_{k+1} \longrightarrow \begin{array}{c} K_k \\ \oplus \\ K_{k+1} \\ \oplus \\ K_{k+1} \end{array}$$

We still have the contraction s in dimension $k+1$, so we may add handles along s and cancel K_{k+1} to obtain a chain-complex concentrated in one degree. Recall there is a similar need to do trivial surgeries in compact surgery theory because the homology modules are only stably free. Denote the remaining module by A. Poincare duality produces an isomorphism $\phi : A \to A^*$ which determines the intersections of *different generators* i. e. $\phi(e_i)(e_j)$ determines the intersections of e_i and e_j when e_i and e_j are different. Now total order the basis and define a map $\nu : A \to A^*$ so that $\nu(e_i)(e_j)$ is 0 when $i > j$ and the intersection counted with sign in $\mathbb{Z}\pi$ when $i \leq j$. By symmetrization $\nu + \epsilon\nu^* = \phi$, hence an isomorphism. This represents the surgery obstruction. If this obstruction is zero, [36, Proposition 2.6] shows us how to find a Lagrangian, and doing surgery on this Lagrangian will produce a homotopy equivalence. More specifically [36, Proposition 2.6] tells us that after stabilization with a hyperbolic form, we may find a Lagrangian. Using -1-connectedness we may do trivial surgeries at points chosen such that this hyperbolic form is added to A. Once we have a Lagrangian each basis element in the Lagrangian is a linear combination of generators in A, so we find representations by immersed spheres by tubing up the generators in A. This uses 0-connectedness. Using the assumption that we have bounded fundamental group π we may do the Whitney tricks to cancel double points so that the geometric intersections correspond to the algebraic intersections meaning that the generators of the Lagrangian are represented by framed, imbedded spheres. After surgery on these spheres an easy calculation shows that the new $K_{\#}(M)$ is contractible.

To see the obstruction is well defined it suffices to show that the obstruction is zero on a boundary, but doing surgery on the bounding manifold to make it highly connected will produce a Lagrangian as in classical surgery surgery theory. Plumbing shows that all algebraically defined surgery obstructions are realized by some surgery problem with boundary. This is done as follows: Let (A, ν) be an element of L_n, $n \geq 5$. Choose a -1, 0 connected $2k-1$-dimensional manifold $N \to Z$, with bounded fundamental group π, and trivially imbedded $k-1$-spheres corresponding to the generators of A.

Consider

$$(\cup S^{k-1}) \times I \subset N \times I$$

Piping against the boundary $N \times 1$ we change these imbeddings to immersions having selfintersection form defined by ν. Do surgery on the spheres imbedded in $M \times 1$, the non-singularity of ν implies that the trace of this surgery $W \to N \times I$ is a homotopy equivalence on the boundary and realizes the given surgery obstruction (A, ν)

In odd dimensions we can dodge the problem by crossing with the reals at both the manifold and parameter space level. Now we have a surgery problem parameterized by $O(\Sigma K)$. Since we now have an even-dimensional problem, we can translate to algebra as above, and use the algebraic fact that

$$L^s_{n+1}(\mathcal{C}_{O(\Sigma K)}(\mathbb{Z}\pi)) \cong L^h_n(\mathcal{C}_{O(K)}(\mathbb{Z}\pi))$$

[37] to finish off the proof by an application of Theorem 7.2. This gives the L^h result, but not the L^s result. An argument that solves the odd-dimensional case directly was shown to us by A. Ranicki. It goes as follows:

Doing surgery below the mid-dimension and furthermore proceeding as above we may obtain a length 2 chain complex

$$0 \to K_{k+1} \to K_k \to 0.$$

Now do surgeries on embedded $S^k \times D^{k+1}$'s in such a fashion that, denoting the trace of the surgery by W, the chain complexes $K_\#(W, M)$, $K_\#(W)$ and $K_\#(W, M')$ are homotopy equivalent to chain complexes which are zero except in dimension $k+1$. One way to do this could be to do surgeries to all the generators of K_k. Denote the resulting manifold by M'. The surgery obstruction is now defined to be the following formation

$$(K_{k+1}(W, M) \oplus K_{k+1}(W, M'), K_{k+1}(W, M), K_{k+1}(W))$$

where the first Lagrangian is the inclusion on the first factor, and the second Lagrangian is induced by the pair of inclusions. Poincaré duality shows that these are indeed Lagrangians. We need to see this is a well-defined element in the odd L-group. First, the choice of imbeddings of $S^k \times D^{k+1}$ may be changed by a regular homotopy, but that changes the formation by an isomorphism. Next we need to compare the effect of choosing a different set of spheres. Let W_1 and W_2 be two traces satisfying the conditions above. Let W_{12} denote the result of surgery by both sets of spheres. After attaching the first set of handles the second set is attached by homotopically trivial spheres so after a regular homotopy we have that W_{12} is W_1 with further trivial surgeries done. It is easy to see that trivial surgeries do not change the equivalence class of the formation, so W_1 and W_{12} define equivalent formations. Similarly W_2 and W_{12} define equivalent formations and we are done. We need to see this is a normal cobordism invariant, but given 2 highly connected normally cobordant surgery problems, we may do surgery on the normal cobordism and cancel handles as in the even dimensional case to obtain a normal cobordism which is just a trace of surgeries as described. This means we have a well defined element in the L-group. If $K_{\#}(M)$ is contractible we may choose to do no surgeries and thus get the 0-formation which does represent 0 in the L-group. Since the operations that are allowed on Lagrangians in the odd L-groups [36] can be mimicked geometrically, using the -1, 0-connectedness, and bounded fundamental group assumptions we see that surgery can be done if and only if the element is 0 in the L-group. Showing all algebraically defined elements are realized geometrically is done by plumbing: Given a non-singular formation we may think of it as $(H \oplus H^*, H, K)$, using the first Lagrangian to identify the form with a hyperbolic form. Start out with a -1, 0-connected $2k$-manifold $N \to Z$, with bounded fundamental group π and do trivial surgeries to a set of generators corresponding to generators of H. In the resulting manifold M' we have the kernel $K_k = H \oplus H^*$. Now do surgeries to spheres corresponding to generators of K to obtain a homotopy equivalence again. The union of the traces of these surgeries along M', $W \to N \times I$ will have surgery obstruction given by $(H \oplus H^*, H, K)$. □

Corollary 12.4. *The groups $L_{n,O(K)}(\mathbb{Z}\pi)$ are 4-periodic for $n \geq 5$. The isomorphism is given by multiplication by $\mathbb{CP}^2$.*

Proof. First, note that it suffices to prove periodicity in L^h, since the Ranicki-Rothenberg sequence and the 5 Lemma then give L^s periodicity. (Multiplication by $\mathbb{CP}^2$ is an isomorphism on Tate cohomology since $\mathbb{CP}^2$ has odd Euler characteristic.) The L^h groups are 4-periodic because the algebraically defined groups only depend on n mod 4. To see that the isomorphism is given by multiplication by $\mathbb{CP}^2$, one has to go through steps analogous to the compact proof [43, Theorem 9.9, p. 96]. □

Remark 12.5. As remarked above, it would be nicer to have a direct description of a map from the geometrically defined bordism groups to the algebraically defined bordism groups. Note, however, that the identification of $L^s_{n,O(S^i)}(\mathbb{Z}\pi)$ with $L^{1-i}_{n-i-1}(\mathbb{Z}\pi)$ is independent of the algebra of this section.

We now give a proof of theorem 7.2 based on the material in this section.

Proof of Theorem 7.2. Given the algebraic description of the surgery groups (in the case without boundary) we may establish the surgery exact sequences of the last section without reference to a [42, Chapter 9] type definition of the L-groups. First assume ∂X is empty. Consider the diagram of surgery exact sequences

$$\begin{array}{ccccccc} \cdots \longrightarrow & L^h_n(\mathcal{C}_M(\mathbb{Z}\pi)) & \longrightarrow & \mathcal{S}^h_b(X) & \longrightarrow & [X, F/\mathrm{TOP}] & \longrightarrow \cdots \\ & \downarrow & & \downarrow & & \downarrow & \\ \cdots \to & L^s_{n+1}(\mathcal{C}_{M\times\mathbb{R}}(\mathbb{Z}\pi)) & \to & \mathcal{S}^s_b(X\times\mathbb{R}) & \to & [X\times\mathbb{R}, F/\mathrm{TOP}] & \to \cdots \end{array}$$

where the vertical maps are induced by crossing with $\mathbb{R}$. On the normal invariant we clearly get an isomorphism, and it is proved in [37] that

$$L^h(\mathcal{C}_M(\mathbb{Z}\pi)) \to L^s(\mathcal{C}_{M\times\mathbb{R}}(\mathbb{Z}\pi))$$

is an isomorphism (see also [6]) hence an element in the simple structure set parameterized over $M \times \mathbb{R}$ is the product with $\mathbb{R}$ with an element in $\mathcal{S}^h_b(X)$. The reader should note that, as usual in surgery theory, the surgery exact sequence is *not* a sequence of abelian groups and homomorphisms. The L-groups act on the structure set and exactness at the structure set means that two elements having the same normal invariant differ by an action of the L-group. It is however easy to see that a version of the 5-lemma sufficient for our purposes is valid, so we do get a 1-1 correspondence of structure sets. To get the splitting, we finally need to refer to the bounded s-cobordism theorem 2.17.

The relative case is treated by first splitting the boundary then working relative to the boundary. □

13. The annulus theorem, CE approximation, and triangulation

In this section we show how bounded surgery theory can be applied to give direct proofs of Kirby's annulus theorem and Siebenmann's CE approximation Theorem. We also take a look at triangulation theory through the lens of bounded topology.

Theorem 13.1. (Kirby [16]) *If C^n, $n \geq 5$, is a bicollared ball in $\mathbb{R}^n$ containing a bicollared ball D^n in its interior, then $C - \mathring{D}$ is homeomorphic to $S^{n-1} \times [0,1]$.*

Proof. By the generalized Schönflies Theorem [5], [19], there is a homeomorphism $h : \mathbb{R}^n \to \mathbb{R}^n$ with $h(D^n) = B^n$, where B^n is the standard ball. Now $h : \mathbb{R}^n \to \mathbb{R}^n$ is certainly a controlled homotopy equivalence, so h defines an element of the bounded structure set

$$\mathcal{S}^{\mathrm{PL}}_b \begin{pmatrix} \mathbb{R}^n \\ \downarrow id \\ \mathbb{R}^n \end{pmatrix}.$$

The surgery exact sequence in this case is

$$\to L_{n,\mathbb{R}^{n+1}}(e) \to \mathcal{S}_b^{\mathrm{PL}}\begin{pmatrix}\mathbb{R}^n\\ \downarrow id\\ \mathbb{R}^n\end{pmatrix} \to [\mathbb{R}^n, F/\mathrm{PL}] \to L_{n,\mathbb{R}^n}(e)$$

where there are no decorations on the L-groups because π_1 is trivial and all homotopy equivalences are therefore simple. This uses Bass-Heller-Swan and [24]. By Theorem 10.4, $L_{n,\mathbb{R}^{n+1}}(e) = L_1(e)$, which is zero by Kervaire-Milnor [4, p. 49]. The space F/PL is connected, so $[\mathbb{R}^n, F/\mathrm{PL}]$ is trivial.

Thus, $\mathcal{S}_b^{\mathrm{PL}}\begin{pmatrix}\mathbb{R}^n\\ \downarrow id\\ \mathbb{R}^n\end{pmatrix}$ is trivial, which means that there is a PL homeomorphism $k : \mathbb{R}^n \to \mathbb{R}^n$ which is boundedly close to h. Let $\mathbb{R}^n$ be compactified to D^n by adding a sphere at infinity. We use the notation LB^n to denote the ball of radius L centered at the origin in $\mathbb{R}^n$. Since $k^{-1} \circ h : \mathbb{R}^n \to \mathbb{R}^n$ is a homeomorphism which extends to a homeomorphism $\overline{k^{-1} \circ h} : D^n \to D^n$, we see

(i) $D^n - L\mathring{B}^n$ is an annulus, so

$$\overline{k^{-1} \circ h}(D^n - L\mathring{B}^n) = D^n - (k^{-1} \circ h)(L\mathring{B}^n)$$

is an annulus for all L.

(ii) This implies that $(k^{-1} \circ h)(LB^n) - M\mathring{B}^n$ is an annulus for $L \gg M$, since the annulus $D^n - M\mathring{B}^n$ is $(k^{-1} \circ h)(LB^n) - M\mathring{B}^n$ plus the collar $D^n - (k^{-1} \circ h)(L\mathring{B}^n)$ and adding a collar on the boundary of a manifold leaves the homeomorphism type unchanged.

(iii) Applying k, we see that $h(LB^n) - k(M\mathring{B}^n)$ is an annulus for $L \gg M$.

(iv) Since k is PL, $k(M\mathring{B}^n) - \mathring{B}^n$ is an annulus for M large [38, p. 36] and the collaring trick shows that $h(LB^n) - \mathring{B}^n$ is an annulus for very large L.

(v) Applying the collaring trick yet again shows that $h(B^n) - \mathring{B}^n$ is an annulus.

□

Here is a geometric restatement of the surgery theory involved in this argument: Bundle theory over $\mathbb{R}^n$ is trivial, so a transversality argument shows that the bounded PL structure given by h is normally bordant to the identity. Repeated splitting shows that the obstruction to surgering this bordism to a bounded h-cobordism over $\mathbb{R}^n$ is the codimension n surgery obstruction over the transverse inverse images of points in $\mathbb{R}^n$. This uses the π-π theorem and periodicity, since we need to multiply the original problem by $\mathbb{CP}^2$ to keep the dimensions from dropping below 5. All of the surgery groups that we use here are rel boundary, so the problem of transferring between the geometry and algebra alluded to in the last section does not arise in this connection. The codimension n surgery obstruction for surgering the bordism is an odd-dimensional simply connected (ordinary) surgery obstruction and is therefore zero. We can therefore surger to a bounded PL h-cobordism, at which point we can apply the bounded h-cobordism theorem over $\mathbb{R}^n$ (see [25]) to produce the bounded PL approximation k.

Corollary 13.2. (of the proof) *Orientation-preserving homeomorphisms of* $\mathbb{R}^n$, $n \geq 5$ *are stable.*

Proof. We assume that the reader is familiar with [16]. The homeomorphism k is stable because it is orientation-preserving and PL, while the homeomorphism $k^{-1} \circ h$ is stable because it is bounded. Compositions of stable homeomorphisms are stable, so $h = k \circ (k^{-1} \circ h)$ is stable. □

Remark 13.3. This is the lone surgical ingredient in the proof of the Kirby-Siebenmann Product Structure Theorem, which says that M^n has a PL structure if and only if $M \times \mathbb{R}^k$ has a PL structure for some k. See [15, p. 33]. We could also prove the product structure theorem directly using Theorem 7.2. The existence of handlebody decompositions for high-dimensional TOP manifolds is a direct consequence. See [15, pp. 104 ff.]. It also follows immediately by a general bundle theory argument [21] that M^n has a PL structure if and only if the stable tangent bundle of M has a PL reduction. Thus, triangulation is a lifting problem and the triangulation problem is reduced to determining the structure of TOP / PL.

The same lemma gives a proof of Siebenmann's CE approximation theorem.

Theorem 13.4. *If* $n \geq 5$ *and* $f : M \to N$ *is a* CE *map, then* f *is a uniform limit of homeomorphisms.*

Proof. Let $U \subset N$ be the interior of a bicollared ball in N. Then $f : f^{-1}(U) \to U \cong \mathbb{R}^n$ is a bounded structure on $\mathbb{R}^n$. The manifold $f^{-1}(U)$ is contractible, so by the Product Structure Theorem, $f^{-1}(U)$ has a PL structure and the argument above shows that there is a PL homeomorphism $k : f^{-1}(U) \to U$ approximating f so closely that the map $\bar{f} : M \to N$ defined by

$$\bar{f}(x) = \begin{cases} f(x) & x \notin f^{-1}(U) \\ k(x) & x \in f^{-1}(U) \end{cases}$$

is continuous. Performing similar modifications over all of the sets U in an open cover of N gives a homeomorphism homotopic to f. If the open sets U are taken to be small, the homeomorphism approximates f. □

We can approach Kirby-Siebenmann's triangulation theory similarly. A topological homeomorphism $h : V^{\mathrm{PL}} \to D^k \times \mathbb{R}^m$, $m + k = n \geq 5$, which is a

PL homeomorphism over a neighborhood of the boundary gives an element of $\mathcal{S}_b^{\mathrm{PL}}\begin{pmatrix} D^k\times\mathbb{R}^m \\ \downarrow id \\ O(D^k\times S^{m-1})\end{pmatrix}$. The surgery exact sequence is

$$\ldots \to L_{n,O(D^k\times S^{m-1})}(e) \to \mathcal{S}_b^{\mathrm{PL}}\begin{pmatrix} D^k\times\mathbb{R}^m \\ \downarrow id \\ O(D^k\times S^{m-1})\end{pmatrix}$$
$$\to [D^k\times\mathbb{R}^m, \partial; F/\mathrm{PL}] \to L_{n,O(D^k\times S^{m-1})}(e).$$

By homotopy invariance, this is

$$\cdots \to L_{n,O(S^{m-1})}(e) \to \mathcal{S}_b^{\mathrm{PL}}\begin{pmatrix} D^k\times\mathbb{R}^m \\ \downarrow id \\ O(D^k\times S^{m-1})\end{pmatrix} \to \pi_k(F/\mathrm{PL}) \to L_{n,O(S^{m-1})}(e)$$

which is

$$\pi_{k+1}(F/\mathrm{PL}) \to L_{k+1}(e) \to \mathcal{S}_b^{\mathrm{PL}}\begin{pmatrix} D^k\times\mathbb{R}^m \\ \downarrow id \\ O(D^k\times S^{m-1})\end{pmatrix} \to \pi_k(F/\mathrm{PL}) \to L_k(e).$$

The usual plumbing argument shows that the maps $\pi_k(F/\mathrm{PL}) \to L_k(e)$ are isomorphisms for $k \neq 4$, in which case Rochlin's Theorem shows that the map is multiplication by 2. This shows that such structures are trivial for $k \neq 3$ and allows the straightening of all but 3-handles. The one nontrivial structure on $D^3 \times \mathbb{R}^m$ comes from a homotopy equivalence

$$f : V \to D^k \times \mathbb{R}^m$$

which is a PL homeomorphism near the boundary and which is bounded over $O(D^k \times S^{m-1})$. Add a boundary to V to form $\bar{V}$ and extend the map. This uses Quinn's end theorem or our bounded modification thereof and requires $m+k \geq 6$. By the Generalized Poincaré Conjecture, $\bar{V}$ is a disk. The limiting map is CE and we approximate by a homeomorphism. Coning produces a TOP homeomorphism h proper homotopic to the original f. Comparing the bounded and proper surgery exact sequences shows that the bounded structure given by h is equivalent to the original f, so composing h with an appropriate PL homeomorphism which is the identity on the boundary gives a TOP homeomorphism boundedly close to f. Thus, the nonstraightenable "bounded homotopy handle" comes from a TOP homeomorphism, $\pi_3(\mathrm{TOP}/\mathrm{PL}) \cong \mathbb{Z}/2\mathbb{Z}$, and the development of the theory proceeds as in [15].

14. Extending the algebra

In this section we extend the bounded algebraic theory in two directions. First, we consider the equivariant case, i.e., we extend the theory to allow non-bounded fundamental groups coming from group action. Second, we introduce germ methods, which allow us to disregard what happens in a bounded neighborhood of a subset of the metric space.

In the following, suppose that M is a metric space with a group G acting by quasi-isometries.

Definition 14.1. An object of $\mathcal{C}_{M,G}(R)$ is a left RG-module A together with a set map $f : A \to F(M)$, where $F(M)$ is the finite subsets of M such that

(i) f is G-equivariant.
(ii) $A_x = \{a \in A \mid f(a) \subseteq \{x\}\}$ is a finitely generated free sub R-module.
(iii) As an R-module $A = \oplus_{x \in M} A_x$.
(iv) $f(a + b) \subseteq f(a) \cup f(b)$.
(v) The set $\{x \in M \mid A_x \neq 0\}$ is locally finite.

A morphism $\varphi : A \to B$ is a morphism of RG-modules so that there is a $k = k(\varphi)$ so that $\varphi_n^m : A_m \to B_n$ is 0 for $d(m, n) > k$.

Remark 14.2. In case G is the trivial group, $\mathcal{C}_{M,e}(R)$ and $\mathcal{C}_M(R)$ are identified by sending an object A in $\mathcal{C}_M(R)$ to $\oplus_{x \in M} A_x$ together with the map $f : \oplus_{x \in M} A_x \to F(M)$ picking out non-zero coefficients. Similarly when the action of G on M is trivial, the categories $\mathcal{C}_{M,G}(R)$ and $\mathcal{C}_M(RG)$ may be identified.

Definition 14.3. If R is a ring with involution, the category $\mathcal{C}_{M,G}(R)$ has an involution given by $A^* = \mathrm{Hom}_R^{lf}(A, R)$, the set of locally finite R-homomorphisms. We define $f^* : A^* \to F(M)$ by $f^*(\phi) = \{x \mid \phi(A_x) \neq 0\}$, which is finite by assumption.

Given a metric space M with an action by G and an equivariant submetric space $N \subset M$, let us denote the k-neighborhood of N by N^k. We shall now develop germ methods "away from N".

Definition 14.4. The category $\mathcal{C}_{M,G}^{>N}(R)$ has the same objects as $\mathcal{C}_{M,G}(R)$, but morphisms $\varphi_1, \varphi_2 : A \to B$ are identified if there exists k such that $\varphi_{1y}^x = \varphi_{2y}^x$ for $x \notin N^k$.

Using the methods of [27] , [33], and [6] we get the following:

Theorem 14.5. *Let $M \cup N \times [0, \infty)$ have the metric included from $M \times [0, \infty)$. The forgetful map (functor!)*

$$\begin{array}{ccc} \mathcal{C}_{M \cup N \times [0,\infty),G}(R) & \longrightarrow & \mathcal{C}^{>N \times [0,\infty)}_{M \cup N \times [0,\infty),G}(R) \\ & & \| \\ & & \mathcal{C}^{>N}_{M,G}(R) \end{array}$$

induces isomorphisms on algebraic K-theory and (if R is a ring with involution) on algebraic L-theory.

Proof. Let $\mathcal{A}$ be the full subcategory of $\mathcal{U} = \mathcal{C}_{M \cup N \times [0,\infty), G}(R)$ with objects 0 except for a bounded neighborhood of $N \times [0, \infty)$. Then $\mathcal{U}$ is $\mathcal{A}$-filtered in the sense of Karoubi and the result follows from [27] since $\mathcal{A}$ has an obvious Eilenberg swindle making the K-theory trivial. □

Arguing as above with Karoubi filtrations we get the following from [27], see also [6]:

Theorem 14.6. *Assume that M is a metric space with a group G acting by quasi-isometries, and let N be an invariant subspace. Form $M \cup N \times [0, \infty)$ with metric induced from $M \times \mathbb{R}$ and the induced G-action. Then the sequence of categories (with morphisms restricted to isomorphisms)*

$$\mathcal{C}_{N,G}(R) \to \mathcal{C}_{M,G}(R) \to \mathcal{C}_{M \cup N \times [0,\infty),G}(R)$$

induces a fibration of K-theory spectra, and hence a long exact sequence in K-theory.

In the important special case where the metric space M is $O(K)$ for some finite complex K with a cellular action on K, the combination of these two theorems allows the computation (in the sense of providing exact sequences) of $K_*(\mathcal{C}_{O(K),G}(R))$. Computations are further facilitated by the fact [13] that these functors are Mackey functors in the variable G.

When R is a ring with involution $\mathcal{C}_{M,G}(R)$ is a category with involution, so following Ranicki the algebraic L-theory is defined. There are exact sequences similar to the above sequences for computing L-theory. See Remark 19.4, and [6].

15. Extending the geometry

In this section, the basic setup is going to be a group G acting on a metric space M by quasi-isometries and freely, cellularly, on a bounded CW-complex X such that the reference map $p : X \to M$ is equivariant. We call this a *free bounded $G - CW$ complex.* The cellular chains take values in the category $\mathcal{C}_{M,G}(\mathbb{Z})$ and will be denoted $D_{\#}(X)$. Thus, the basic point of view is equivariant instead of working with a fundamental group. We do however have to worry about interference from the fundamental group of X.

Let N be an equivariant subset of M. We shall use the following language:

Definition 15.1. Let $p : X \to M$ be a bounded G-CW complex. The term *away from N* means "when restricted to a subset of X whose complement under p maps to a bounded neighborhood of N." Similarly, *in a bounded neighborhood of N* means a subset of X mapping to a bounded neighborhood of N under p. Similarly,

(i) $p: X \to M$ is (-1)-*connected away from* N if there exists k so that for every point x in M except for a bounded neighborhood of N there exists $y \in X$ such that $d(x, p(y)) < k$.

(ii) The bounded CW complex (X, p) is (-1)-*connected away from* N if there are $k, l \in \mathbb{R}_+$ so that for each point $m \in M$ either there is a point $x \in X$ such that $d(p(x), m) < k$ or $d(m, N) < l$.

(iii) (X, p) is 0-*connected away from* N if for every $d > 0$ there exist k and l depending on d so that if $x, y \in X$ and $d(p(x), p(y)) \leq d$, then either x and y may be joined by a path in X whose image in M has diameter $< k(d)$ or $d(x, N) < l(d)$ or $d(y, N) < l(d)$. Notice that we have set up our definitions so that 0-connected does not imply (-1)-connected.

(iv) (X, p) is 1-*connected away from* N if for every $d > 0$, there exist $k = k(d)$ and $l = l(d)$ so that for every loop $\alpha : S^1 \to X$ with $d(\alpha(1), N) > l$ and $\mathrm{diam}(p \circ \alpha(S^1)) < d$, there is a map $\bar{\alpha} : D^2 \to X$ so that the diameter of $p \circ (D^2)$ is smaller than k. We also require $p : X \to M$ to be 0-connected away from N, but not (-1)-connected.

(v) (X, p) has *bounded fundamental group* π *away from* N if there exists a π-covering of X away from N which is 0- and 1-connected away from N. We do not require (X, p) to be (-1)-connected away from N.

Definition 15.2. A free bounded G-CW complex $X \to M$ is a *G-Poincaré duality complex away from* N if $X \to M$ is 0- and 1-connected away from N, and there is a class $[X] \in H^{lf}(X/G; C)$ so that a transfer of $[X]$ induces a bounded homotopy equivalence $[X] \cap - : D^{\#}(X) \to D_{\#}(X)$ as chain complexes in $\mathcal{C}^{>N}_{M,G}(\mathbb{Z})$.

Definition 15.3. A G-metric space M (i.e., a metric space M with a group G acting by quasi-isometries) is allowable if there exists a finite dimensional complex K with a free cellular G-action and a map $p : K \to M$ making K a free bounded (-1)-, 0- and 1- connected free bounded $G - CW$-complex.

With these definitions, the theory detailed in the preceding sections for the case of a boundedly constant fundamental group carries through, so we obtain the analogue of the main theorem of this paper, the surgery exact sequence.

Theorem 15.4. *Let $X \to M$ be a (-1), 0, and 1-connected n-dimensional G-Poincaré duality complex away from N. For $n \geq 5$, there is a surgery exact sequence*

$$\cdots \to \mathcal{S}_b((X \times I), \delta(X \times I), G)^{>N} \to [\Sigma X/G, F/\mathrm{CAT}]^{>N} \to$$
$$\to L^s_{n+1}(\mathcal{C}^{>N}_{M,G}(\mathbb{Z})) \to \mathcal{S}_b(X, G) \to [X, F/\mathrm{CAT}]^{>N} \to L^s_n(\mathcal{C}^{>N}_{M,G}(\mathbb{Z})).$$

Here simpleness is measured in $K_1(\mathcal{C}^{>N}_{M,G})/G$, $\mathcal{S}_b(X,G)^{>N}$ denotes bounded equivariant structures away from N, and $[X, F/\mathrm{CAT}]^{>N}$ denotes germs of homotopy classes of maps away from N.

Of course, we have L^h-groups as well as L^s-groups and these groups are connected via the usual Ranicki-Rothenberg exact sequence.

Proposition 15.5. *Assume $n \geq 5$. There is a long exact Ranicki-Rothenberg sequence*

$$\to L^s_n(\mathcal{C}^{>N}_{M,G}(\mathbb{Z})) \to L^h_n(\mathcal{C}^{>N}_{M,G}(\mathbb{Z})) \to \widehat{H}^n(\mathbb{Z}_2, \widetilde{K}_1(\mathcal{C}^{>N}_{M,G}(\mathbb{Z}))) \to .$$

16. Spectra and resolution of ANR homology manifolds

Following the tradition of Quinn, Ranicki, and Nicas, we spacify our bounded surgery groups, producing spectra such that the surgery groups are the homotopy groups of these spectra.

Theorem 16.1. *Let (M,G) be an allowable G-metric space. There is an infinite loop space $\mathbb{L}^s_{M,G}(\mathbb{Z})$ depending functorially on (M,G), such that*

$$\pi_i \mathbb{L}^s_{M,G}(\mathbb{Z}) = L^s_i(\mathcal{C}_{M,G}(\mathbb{Z})).$$

Proof. We construct a Δ-set whose n-simplices are n-ads of (M,G)-surgery problems. The realization is an infinite loop space, as in the classical case. See [28] and Nicas [22] for details. □

In the special case where G is the trivial group, i.e., the case of simply-connected bounded surgery, we can improve a bit on the situation, getting an analogue of the main theorem of Pedersen-Weibel [27].

Theorem 16.2. *The functor sending a finite complex K to $\mathbb{L}_{O(K)}(\mathbb{Z})$ sends cofibrations to fibrations.*

Proof. Let $L \hookrightarrow K \to K \cup CL$ be a cofibration. The composite $\mathbb{L}_{O(L)}(\mathbb{Z}) \to \mathbb{L}_{O(K)}(\mathbb{Z}) \to \mathbb{L}_{O(K\cup CL)}(\mathbb{Z})$ is the zero map, since it factors through $\mathbb{L}_{O(CL)}(\mathbb{Z})$, which is contractible. On homotopy groups we get an exact sequence (by Theorem 10.5) and thus the Five Lemma shows that $\mathbb{L}_{O(L)}(\mathbb{Z})$ is homotopy equivalent to the homotopy fibre of $\mathbb{L}_{O(K)}(\mathbb{Z}) \to \mathbb{L}_{O(K\cup CL)}(\mathbb{Z})$. □

Following [27], this identifies the homology theory. Denoting the four-periodic simply connected surgery spectrum by $\mathbb{L}$ we have:

Theorem 16.3.

$$L_n(\mathcal{C}_{O(K)}(\mathbb{Z})) \cong h_{n-1}(K, \mathbb{L})$$

where $h(-, \mathbb{L})$ denotes the reduced homology theory associated with the spectrum $\mathbb{L}$.

Proof. It is shown in [27], Theorem 3.1 that the spectrum for the homology theory $L_n(\mathcal{C}_{O(K)})$ is given by the spectrum whose n'th space is $\mathbb{L}_{\mathbb{R}^n}(\mathbb{Z})$, but that is exactly four-periodic simply connected L-theory. □

Let k denote the (unreduced) homology theory with coefficients in connective simply connected $\mathbb{L}$-theory, and h the (unreduced) homology theory with coefficients in 4-periodic simply connected $\mathbb{L}$-theory. As always, there is a natural transformation from h to k, sending the periodic spectrum to the connective version. We define a natural transformation α from k to h as follows:

Let K be a finite complex, W a regular neighborhood of K. Now $k_*(K) \cong k_*(W) \cong k^{|W|-*}(W/\partial W) \cong k^0(\Sigma^{*-|W|}W/\partial W) = [\Sigma^{*-|W|}W/\partial W, F/\mathrm{TOP}]$. Consider $W \times [0, \infty) \to O(K)$, a simply connected bounded Poincaré duality complex away from 0, and with boundary. The normal invariant of $W \times [0, \infty)$ away from 0 relative to the boundary is given by $[\Sigma^{*-|W|}W/\partial W, F/\mathrm{TOP}]$ and the surgery exact sequence maps from there to $L_{(*-|W|)+|W|}(\mathcal{C}_{O(K)}(\mathbb{Z})) = h_*(K)$.

Theorem 16.4. *The composite of natural transformations $k_* \xrightarrow{\alpha} h_* \to k_*$ is an isomorphism.*

Proof. It is enough to verify this for spheres. What we need to prove is that the bounded structure space $\mathcal{S}_b\begin{pmatrix} S^n \times [0,\infty) \\ \downarrow \\ O(S^n) \end{pmatrix}^{>0}$ is contractible. The classical structure space of a sphere is a point (by the high-dimensional Poincaré conjecture). Crossing with $\mathbb{R}^k$ into bounded $\mathbb{L}$-theory is an isomorphism both on normal invariants and L-groups, so $[\Sigma^i(S^n \times \mathbb{R}^k); F/\mathrm{TOP}] \cong L(\mathcal{C}_{\mathbb{R}^k}(\mathbb{Z}))$. Away from 0, the Poincaré complex $S^n \times \mathbb{R}^k \to \mathbb{R}^k$ is $S^n \times S^{k-1} \times [0, \infty) \to O(S^{k-1})$, but that, on the other hand, is the image of $S^n \times S^{k-1} \times [0, \infty) \to O(S^n \times S^{k-1})$ away from 0 induced by the projection $S^n \times S^{k-1} \to S^{k-1}$. By naturality, we obtain that α is an isomorphism on one of the summands when applied to $S^n \times S^{k-1}$, but then by naturality it must be an isomorphism on spheres. □

The point of the proof above is to relate the obvious isomorphism for the case $S^n \times \mathbb{R}^k \to \mathbb{R}^k$ to the definition of α. This theorem can also be proved using Chapman and Ferry's α-approximation theorem [8].

We get a new proof of Quinn's obstruction to resolution.

Theorem 16.5. *Let X be an* ANR *homology manifold. Then there is an integral obstruction to producing a resolution of X.*

Proof. First, assume that X admits a TOP reduction of its Spivak normal fibre space. Theorem 16.6 below shows that such a reduction always exists. Consider the bounded surgery exact sequence

$$\mathcal{S}_b\begin{pmatrix} X\times[0,\infty) \\ \downarrow \\ O(X) \end{pmatrix}^{>0} \to [X, F/\mathrm{TOP}] \to L_{|X|}(\mathcal{C}^{>0}_{O(X)}(\mathbb{Z})).$$

By the theorem above and Theorem 16.7 below,

$$L_{|X|}(\mathcal{C}^{>0}_{O(X)}(\mathbb{Z})) \cong [X, F/\mathrm{TOP}\times\mathbb{Z}],$$

and the map $[X, F/\mathrm{TOP}] \to [X, F/\mathrm{TOP}\times\mathbb{Z}]$ followed by the map to connective L-theory, i.e., to $[X, F/\mathrm{TOP}]$, is an isomorphism. Hence it is only the component in $F/\mathrm{TOP}\times\mathbb{Z}$ that X maps into which is the obstruction to the nonemptiness of the bounded structure set $\mathcal{S}_b\begin{pmatrix} X\times[0,\infty) \\ \downarrow \\ O(X) \end{pmatrix}^{>0}$. Assume that this integral obstruction vanishes. Choose an element in the bounded structure set

$$\begin{array}{ccc} \phi : W & \cong & X\times[0,\infty) \\ & & \downarrow \\ & & O(X). \end{array}$$

ϕ is a bounded homotopy equivalence away from $0 \in O(X)$. A neighborhood of infinity in W maps to X, and since $X\times[0,\infty) \to O(X)$ is the identity away from 0, and ϕ is a bounded homotopy equivalence, the end of W mapping to X is tame and simply connected, so we may add an end M to W and extend the map $W \to X$ to M. The map $M \to X$ is a resolution because it is an arbitrarily small homotopy equivalence. The theorem will now follow from: □

Theorem 16.6. *An* ANR *homology manifold X has a canonical* TOP *reduction.*

Preparing for the proof, first notice that by [1] $L_n(\mathcal{C}_{O(-)}(\mathbb{Z}))$ is a functor defined on compact subsets of S^N, N large, and all continuous maps, not only Lipschitz maps. We now have

Theorem 16.7. *$L_n(\mathcal{C}_{O(-)}(\mathbb{Z}))$ satisfies Milnor's wedge axiom.*

Proof. Consider $\bigvee X_\alpha \subset S^N$. We have

$$L_n(\mathcal{C}_{O(\bigvee X_\alpha)}(\mathbb{Z})) \cong L_n(\mathcal{C}^{>O(*)}_{O(\bigvee X_\alpha)}(\mathbb{Z})).$$

By suspension, we may assume n divisible by 4, so an element is given by a self-intersection form ν, which is bounded, so when we disregard a neighborhood of $O(*)$ we get a self-intersection form on each $L^{>O(*)}_{O(K_\alpha)}(\mathbb{Z})) \cong L_n(\mathcal{C}_{O(X_\alpha)}(\mathbb{Z}))$. To combine an element in $\Pi_\alpha L_n(\mathcal{C}^{>O(*)}_{O(X_\alpha)}(\mathbb{Z}))$ to get an element in $L_n(\mathcal{C}_{O(\bigvee X_\alpha)}(\mathbb{Z}))$, all we need to do is reparameterize radially so that all components have the same bound. $\square$

This means that the identification of bounded L-theory over open cones with homology theory extends beyond finite complexes as a Steenrod homology theory, and that homology with locally finite coefficients may be defined as reduced homology of the one-point compactification.

Proof of Theorem 16.6. Cover X by open sets U_α so that the Spivak normal fibration restricted to U_α is trivial. On U_α we obviously have a TOP reduction, giving rise to a surgery exact sequence as above, denoting the *topological* boundary of U_α by ∂U_α,

$$\mathcal{S}_b\begin{pmatrix} U_\alpha \times [0,\infty) \\ \downarrow \\ O(\bar{U}_\alpha) \end{pmatrix}^{>\partial U_\alpha} \to [U_\alpha, F/\mathrm{TOP}] \xrightarrow{\phi} L(\mathcal{C}^{>O(\partial \bar{U}_\alpha)}_{O(\bar{U}_\alpha)}(\mathbb{Z})).$$

By Poincaré duality, ϕ may be identified with

$$[U_\alpha, F/\mathrm{TOP}] \to [U_\alpha, F/\mathrm{TOP} \times \mathbb{Z}],$$

so by changing the lift of U_α we may ensure that the surgery obstruction is just an integer, and assuming this integer vanishes, we can produce a resolution over U_α as above. This surgery exact sequence is natural with respect to restriction to smaller open sets, so the lifts combine to give a lift over the whole of X. $\square$

Remark 16.8. Strictly speaking, the argument above is flawed in that arbitrary wedges of polyhedra cannot be embedded in a single finite-dimensional sphere. This can be cured by using the unit sphere in a Hilbert space or, better, by embedding X isometrically into the bounded functions from X to $\mathbb{R}$ and taking a cone there.

Next, we want to understand assembly from the point of view of bounded surgery. Given a boundedly simply-connected surgery problem away from 0 parametrized by $O(K)$, the induced map from K to a point gives a surgery problem parametrized by $[0,\infty)$ away from 0. We can turn this problem into a simply-connected surgery problem by doing simultaneous surgery on source and target, giving the usual functorial property with respect to K, but avoiding that, we obtain a simple surgery problem (simplicity measured in $Wh(\pi_1(K))$) together with a reference map to K, in other words, an element in $L^s(\mathcal{C}^{>0}_{[0,\infty)}(\mathbb{Z}\pi_1(K))) \cong L^s(\mathcal{C}_{\mathbb{R}}(\mathbb{Z}\pi_1(K))) \cong L^h(\mathbb{Z}\pi_1(K))$. We claim this forget control map is the assembly map.

Theorem 16.9. *Let M be a manifold. Then the forgetful map*

$$F : L(\mathcal{C}^{>0}_{O(M)}(\mathbb{Z})) \to L^h(\mathbb{Z}(\pi_1(M)))$$

is the assembly map.

Proof. Consider the following diagram

$$\begin{array}{ccccc} \cdots \longrightarrow & [\Sigma^i M, F/\mathrm{TOP}] & \xrightarrow{\alpha} & L_{m+i+1}(\mathcal{C}_{O(M)}(\mathbb{Z})) \\ & \| & & \downarrow F \\ \cdots \longrightarrow & [\Sigma^i M, F/\mathrm{TOP}] & \xrightarrow{A} & L^h_{m+i}(\mathbb{Z}(\pi_1(M))) \end{array}$$

where the lower row is the classical surgery exact sequence with the assembly map. We have just proved that α is an isomorphism for $i > 0$ and the inclusion of a direct summand for $i = 0$. For $i > 0$, this identifies the map with assembly. Since the algebraically defined groups are 4-periodic, this also identifies F with the higher assembly maps when $i = 0$. □

This gives a curious relation between the resolution problem and the Novikov Conjecture.

Theorem 16.10. *Let M be a closed $K(\pi, 1)$-manifold such that the assembly map is an integral monomorphism. Then an* ANR *homology manifold X homotopy equivalent to M admits a resolution.*

Proof. Consider the diagram

$$\begin{array}{ccccc} \mathcal{S}_b\begin{pmatrix} X\times[0,\infty) \\ \downarrow \\ O(X) \end{pmatrix}^{>0} & \longrightarrow & [X, F/\mathrm{TOP}] & \xrightarrow{\alpha} & L_{m+1}(\mathcal{C}^{>0}_{O(X)}(\mathbb{Z})) \\ \downarrow & & \| & & \downarrow F \\ \mathcal{S}(X) & \longrightarrow & [X, F/\mathrm{TOP}] & \xrightarrow{A} & L^h_m(\mathbb{Z}\pi_1(X)). \end{array}$$

Since $\mathcal{S}(X)$ is nonempty, there is a $\sigma \in [X, F/\mathrm{TOP}]$ so that $A(\sigma) = 0$. But then $\alpha(\sigma) = 0$ and $\mathcal{S}_b\begin{pmatrix} X\times[0,\infty) \\ \downarrow \\ O(X) \end{pmatrix}^{>0}$ is nonempty, showing that X admits a resolution. □

Bob Daverman has pointed out that there is an easy geometric proof that there is no nonresolvable ANR homology manifold X homotopy equivalent to the n-torus. The universal cover of such an ANR homology manifold could be compactified by adding a sphere at infinity. Adding an external collar would then give an ANR homology manifold with both manifold points and points with neighborhoods from X, showing that the resolution obstruction for X was trivial. Many of the classes of groups for which the map α is known to be 1-1 admit similarly nice compactifications. In fact,

Ferry and Weinberger have recently announced a proof of the Novikov conjecture for all Γ such that $K = K(\Gamma, 1)$ is a finite complex (*not* necessarily a manifold) and $\widetilde{K}$ admits a sufficiently nice compactification [11].

17. Geometric constructions

In this section we prepare for the applications in the next section by describing some geometric constructions. Consider a continuous proper map $X \to O(K)$.

Definition 17.1. The K-completion $\widehat{X}^K$, of X is defined as follows: As a set $\widehat{X}^K$ is the disjoint union of X and K. The open sets of $\widehat{X}^K$ have as a basis:

(i) all open sets of X
(ii) for every open set U of K, and every $k \in \mathbb{R}_+$, the set
$$\{p^{-1}\ (x,t) \in O(K) | x \in U, t > k\}\ \cup U.$$

This construction is sometimes called the tear drop construction. It is easy to see that $\widehat{X}^K$ is a compact metric space. This construction generalizes one-point compactification.

Theorem 17.2. *Let W_1 and W_2 be manifolds properly parameterized by $O(K)$, and assume that h is a bounded homotopy equivalence from W_1 to W_2. Then $\widehat{W}_1^K$ is a manifold if and only if $\widehat{W}_2^K$ is a manifold.*

Proof. Assume that $\widehat{W}_1^K$ is a manifold. It is easy to see that $\widehat{W}_2^K$ is an ANR homology manifold. Using the homotopy equivalence, the disjoint two-disc property is also carried over, so the result follows by the Manifold Recognition Theorem [31]. See [10] for a detailed proof of the disjoint two-disc property in the case where $K = S^1$. □

Remark 17.3. This result is very useful in proving the existence of group actions by varying the complement of the singular set. For studying group actions we also need the following:

Proposition 17.4. *Let G be a finite group with a stratum-preserving action on a finite complex K. Assume that W_1 and W_2 are manifolds parameterized by $O(K)$ on which G acts freely and compatibly with the action on K. Then W_1 is boundedly equivariantly homotopy equivalent to W_2 over $O(K)$ if and only if W_1/G is boundedly homotopy equivalent to W_2/G over $O(K/G)$.*

Proof. One way is trivial, so assume that $h : W_1/G \to W_2/G$ is a bounded homotopy equivalence. Certainly we get an equivariant homotopy equivalence $\tilde{h} : W_1 \to W_2$. The length of the path in the homotopy can be at most $|G|$ times the length of the path measured in $O(K/G)$, so we are done. □

18. More applications

We begin with an application to group actions.

Consider the standard $n+k$ sphere S^{n+k} with the standard $k-1$ subsphere $S^{k-1} \subset S^{n+k}$ so that $S^{n+k} = S^n * S^{k-1}$. Let G be a finite group and assume that G acts semifreely (topologically) on S^{n+k} fixing S^{k-1}. It was proved in [3] that G has to be a periodic group, since $(S^{n+k} - S^{k-1})/G$ is finitely dominated and $S^{n+k} - S^{k-1}$ has the homotopy type of a sphere. Hence, the homotopy type of $(S^{n+k} - S^{k-1})/G$ is given by G and a single k-invariant which is a unit in $\mathbb{Z}/|G|$. It was further proved that such actions exist if and only if a certain surgery problem has a solution, i.e., if and only if a certain Spivak normal bundle has a reduction and the resulting surgery problem can be solved.

This surgery program was completed in [12] and [20], and also the maps $L^h_n(\mathbb{Z}G) \to L^p_n(\mathbb{Z}G) \to L^{-1}_n(\mathbb{Z}G)$ were computed for the relevant groups, but the computation did not give the classification one usually obtains from surgery theory. It is the purpose of this section to show how ε-surgery can turn the computations of [12] and [20] into such a classification.

Let X be a Swan complex. The surgery exact sequence of §11 takes the following form:

$$\cdots \to L^s_{n+k+1,\mathbb{R}^k}(\mathbb{Z}G) \to \mathcal{S}_b\begin{pmatrix} X\times\mathbb{R}^k \\ \downarrow \\ \mathbb{R}^k \end{pmatrix} \to [X, F/\mathrm{TOP}] \to L^s_{n+k,\mathbb{R}^k}(\mathbb{Z}G).$$

We proved in §10 that $L^s_{n+k,\mathbb{R}^k}(\mathbb{Z}G) = L^{2-k}_n(\mathbb{R}G)$, so we get a surgery exact sequence:

$$\cdots \to L^{1-k}_{n+1}(\mathbb{Z}G) \to \mathcal{S}_b\begin{pmatrix} X\times\mathbb{R}^k \\ \downarrow \\ \mathbb{R}^k \end{pmatrix} \to [X, F/\mathrm{TOP}] \to L^{2-k}_n(\mathbb{Z}G) \to \cdots .$$

Comparing this with [12] and [20] we see that what is actually being computed is $\mathcal{S}_b\begin{pmatrix} X\times\mathbb{R}^k \\ \downarrow \\ \mathbb{R}^k \end{pmatrix}$.

In [3] a map is defined:

$$\mathcal{S}_b\begin{pmatrix} X\times\mathbb{R}^k \\ \downarrow \\ \mathbb{R}^k \end{pmatrix} \xrightarrow{h} \left\{ \begin{array}{c} \text{Conjugacy classes of semifree group ac-} \\ \text{tions of } G \text{ on } S^{n+k} \text{ fixing } S^{k+1} \text{ with } k\text{-} \\ \text{invariant of } S^{n+k} - S^{k-1}/G \text{ given by } X, \\ \text{so that } (S^{n+k} - S^{k-1})/G \simeq X. \end{array} \right\}$$

The construction in [3] starts with a compact manifold homotopy equivalent to $X \times T^n$, and then passes to the $\mathbb{Z}^n$ cover, but it is clear that the same construction gives a map as above. The right-hand side is a classification of semifree group actions, so we will be done once we prove:

Theorem 18.1. *h is an isomorphism.*

Proof. The map is well-defined because a bounded homeomorphism extends to a completion by the identity on the fixed sphere. To see that h is onto, consider a semifree action on S^{n+k} fixing S^{k-1} and let $W = S^{n-k} - S^{k-1}/G$. It is shown in [3] that W has a tame end at S^{k-1}. Killing the obstruction to completing the end by multiplying with a torus, we obtain that $W \times T^n \simeq W' \times \mathbb{R}^n$, where the radial directions in $\mathbb{R}^n$ point to the points of S^{n-1}. Going to the cyclic cover we have $W \simeq W \times \mathbb{R}^n \simeq \widetilde{W'} \times \mathbb{R}^n$ and $\widetilde{W'} \simeq X$ so we are done. That the map is monic follows from Theorem 11.2 that radial reparameterization induces the identity on the structure set. □

Remark 18.2. Note that this surgery theory is not restricted to the category of manifolds. All that is needed is that the objects be manifolds away from the singular set. It thus makes perfectly good sense to suspend group actions. Suspension is just crossing with the reals in the nonsingular part and suspending on the singular part, at least if one assumes nonempty singular sets. This means that questions such as the above may be treated in two stages:

(i) Suspend enough times that $K_{-i}(\mathbb{Z}[\pi]) = 0$, and apply $L^{-\infty}$.
(ii) Try to split off real factors to get back to the manifold situation.

As a second example, consider a closed PL manifold $M^n \subset S^{m-1} \subset \mathbb{R}^m$ which contains a simply-connected polyhedron Y. Let p:$M \to M/Y$ be the projection map. As in the proof of Theorem 9.6, form a two-ended manifold W which looks like $O(M)$ near $+\infty$ and like $M \times \mathbb{R}$ near $-\infty$ and parameterize W over $O(M_+/Y)$. The map $id : W \to W$ is a bounded structure on $W \to O(M_+/Y)$, so we have an exact surgery sequence:

$$\cdots \to L_{n+1,O(M_+/Y)}(\mathbb{Z}) \to \mathcal{S}_b\begin{pmatrix} W \\ \downarrow \\ O(M_+/Y)\end{pmatrix} \to [M, F/\mathrm{TOP}]$$
$$\to L_{n,O(M_+/Y)}(\mathbb{Z})$$

where there are no decorations on the L's because of the simple connectivity. In this case $L^s_{n,O(M_+/Y)}(\mathbb{Z}) = L^{-\infty}_{n,O(M_+/Y)}(\mathbb{Z})$, so the obstructions lie in $h_n(M/Y; F/\mathrm{TOP})$. This is unreduced homology.

An element of $\mathcal{S}_b\begin{pmatrix} W \\ \downarrow \\ O(M_+/Y)\end{pmatrix}$ is an equivalence class of bounded homotopy equivalences $\phi : W' \to W$. Splitting such a ϕ over $M \times \{T\}$ for some large T produces a homotopy equivalence $\phi| : M' \to M$ which is arbitrarily small over M/Y. By the thin h-cobordism Theorem, this splitting is well-defined up to small homeomorphism over M/Y.

If $N \supset Y$ is a regular neighborhood of Y in M, the main theorem of [8] shows that $\phi|$ is close to a homeomorphism over $M - \mathrm{int}(N)$. Thus, M' is the union of a copy of $M - \text{int}(N)$ and a copy of $(\phi|)^{-1}(N) = N'$. Since $M - Y \cong M - N \cong M' - N'$, we see that M' is a compactification of $M - Y$ by a polyhedron homotopy equivalent to Y. If Y has codimension-three or greater in M, then a polyhedron Y' homotopy equivalent to Y and having the same dimension as Y embeds in N' and $M' - N' \cong M' - Y'$.

There is, of course, a related existence question. If M is an open manifold and we wish to compactify M by adding a complex K at ∞, we can proceed by constructing a nonmanifold "Poincaré completion" and then try to solve the resulting bounded surgery problem over the open cone on the one-point compactification of M. Note that the use of bounded surgery here is the reverse of the group actions application above. There, we started with a manifold and a control map and used our theory to vary the complement. Here, we control over the complement and allow the theory to construct the manifold completion. One interesting aspect of this theory is that, except for predicting the dimension of Y', it works well for Y of any codimension.

Another way of exploiting the same control map $M \to M/Y$ is to start with a homotopy equivalence $\phi : N \to M$ and try to solve the resulting controlled surgery problem over $O(M_+/Y)$, as above. As before, we encounter obstructions lying in $h_n(M/Y; F/\text{TOP})$. If we succeed in solving this surgery problem, we obtain a bordism from N to a manifold N' which is controlled homotopy equivalent to M over M/Y. As above, such a manifold splits into a copy of $M - Y$ and a polyhedron homotopy equivalent to Y. The bordism comes equipped with a degree one normal map to $M \times I$, so there is a further ordinary surgery obstruction to surgering the bordism to an s-cobordism from N' to N. Note that this is a nonsimply connected surgery obstruction, since M is not required to be simply connected. In the case $Y = pt$, the resulting exact sequence is the ordinary surgery exact sequence. In the general case, we have obtained a 2-stage obstruction to splitting N into a manifold homeomorphic to $M - Y$ and a complex homotopy equivalent to Y.

As a final example, consider a manifold M homotopy equivalent to the total space of a bundle (or quasifibration or approximate fibration) of manifolds:

$$\begin{array}{ccc} & & F \\ & & \downarrow \\ M & \simeq & E \\ & & \downarrow \\ & & B \end{array}$$

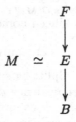

We may ask whether M can be turned into a bundle of some sort over B. Assuming that the bundle splits at fundamental group level , we obtain a surgery problem

$$\begin{array}{ccc} M \times \mathbb{R} & \longrightarrow & E \times \mathbb{R} \\ & & \downarrow \\ & & O(B_+) \end{array}$$

with fundamental group $\pi = \pi_1(F)$. The obstruction lies in

$$L^s_{n,O(B_+)}(\mathbb{Z}[\pi_1 F]).$$

Assuming that the obstruction vanishes, we obtain a manifold N normally cobordant to M and a homotopy equivalence of N to E which is arbitrarily small when measured in B. But this means that $N \to B$ is an approximate fibration, so we have obtained a normal cobordism from M to a manifold which approximate fibers over B. As before, we now have an ordinary surgery obstruction to turning this cobordism into an s-cobordism. The result is an obstruction theory for homotoping a map to an approximate fibration. Note that the fact that $E \to B$ was a bundle was barely used. If $E \to B$ is any map from a manifold to a polyhedron which is a "trivial bundle on π_1," and $M \to E$ is a homotopy equivalence, then solving the same sequence of problems would produce a map $M \to B$ with the same "shape fiber structure" as $E \to B$.

19. A variant L-theory

It is sometimes a problem that the L-theory described in §9 is not a homology theory as a functor of the control space. This is unlike K-theory [27]. Inspired by discussions with Quinn, we give a a variant definition of L-theory which is (at least) a half exact functor in the control space. This, on the other hand, means that it cannot degenerate to usual L-theory when the control space is a point. The idea is to mix the torsion requirements. As in §9, our definition is modeled on [43, Ch. 9].

Given a space K, an object is a surgery problem

$$\begin{array}{ccc} (M, \partial M) & \longrightarrow & (X, \partial X) \\ & & \downarrow \\ & & O(K) \end{array}$$

where $(X, \partial X)$ is a bounded Poincaré pair with bounded fundamental group π and a specific CW structure. Thus, we have a specific simple type of $(X, \partial X)$ parameterized by $O(K)$, but we only require $(X, \partial X)$ to be a Poincaré pair. We do not require Poincaré torsion to vanish in $Wh_{O(K)}(\mathbb{Z}\pi)$. We assume $\partial M \to \partial X$ to be a simple homotopy equivalence.

Associated to such an object we have a Poincaré torsion $\tau(X)$. We use sign conventions for Poincaré torsion as in [23].

The usual equivalence of bordism is to say $(M_1, \partial M_1) \to (X_1, \partial X_1)$ is bordant to $(M_2, \partial M_2) \to (X_2, \partial X_2)$ if there is a triad surgery problem

$$(W, M_1, M_2) \to (Y, X_1, X_2).$$

We refine this relation by requiring that $\tau(Y, X_1) = 0$.

We claim that this refined type of bordism is an equivalence relation on the set of surgery problems. The condition $\tau(Y, X_1) = 0$ is equivalent to the condition $\tau(Y) = \tau(X_1) = \tau(X_2)$ (see e.g. [23]). We have $\tau(X \times I, X \times 0) = 0$ showing that an object is equivalent to itself. Symmetry follows from $\tau(Y, X_1) = \pm\tau(Y, X_2)$. Finally, if Y is a bordism from X_1 to X_2 and Z is a bordism from X_2 to X_3, then $\tau(Y \cup Z) = \tau(Y) + \tau(Z) - \tau(X_2)$ showing that $\tau(Y \cup Z) = \tau(X_1) = \tau(X_3)$.

All constructions involving simultaneous surgery as in §9 are allowed, since manifolds have trivial Poincaré torsion, and these are manifold constructions.

The Grothendieck construction on the set of surgery problems with fundamental group π, and only requiring homotopy equivalence of the Poincaré duality map, not simple homotopy equivalence, parameterized by $O(K)$, modulo the above equivalence relation, we shall denote by

$$L^{h,s}_{n,O(K)}(\mathbb{Z}\pi).$$

The basic idea is to require relations to be simpler than generators.

Theorem 19.1. *The functor $L^{h,s}_{n,O(K)}(\mathbb{Z}\pi)$, $n \geq 5$, is half exact in the variable K.*

Proof. In §9 we studied the functor $L^h_{O(K),n}(\mathbb{Z}\pi)$ as a functor in K. In trying to prove half exactness, there was a splitting obstruction, but this splitting obstruction must vanish because of the assumption of simpler relations. □

Let

$$\begin{aligned} H^h_n(\mathbb{Z}_2; K_1) &= \{\sigma \in K_1 \,|\, \sigma^* = (-1)^n \sigma\}, \\ H^s_n(\mathbb{Z}_2; K_1) &= \{\sigma \in K_1 \,|\, \sigma = \tau + (-1)^n \tau^*\} \end{aligned}$$

where $K_1 = K_1(\mathcal{C}_{O(K)}(\mathbb{Z}\pi))$.

Theorem 19.2. *For $n \geq 5$, there are exact sequences*

$$H^s_n(\mathbb{Z}_2; K_1) \to L^{h,s}_{n,O(K)}(\mathbb{Z}\pi) \to L^h_{n,O(K)}(\mathbb{Z}\pi) \to 0$$
$$0 \to L^s_{n,O(K)}(\mathbb{Z}\pi) \to L^{h,s}_{n,O(K)}(\mathbb{Z}\pi) \to H^h_n(\mathbb{Z}_2; K_1)$$

which together with the usual Ranicki-Rothenberg exact sequence fit into a commutative braid

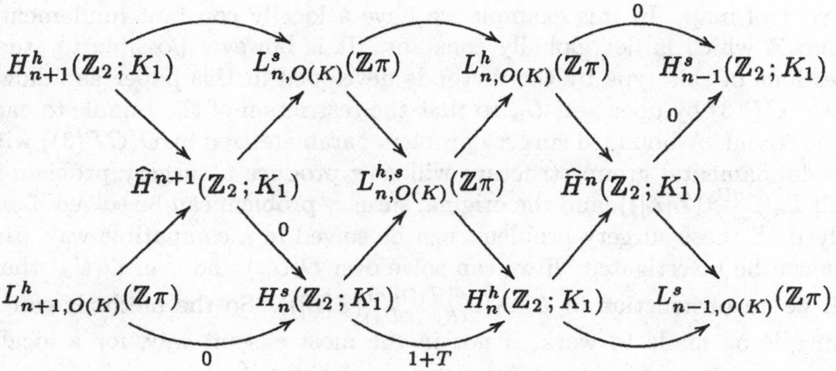

Proof. In L^h we allow more relations than in $L^{h,s}$, so clearly there is an epimorphism. Similarly, in L^s we allow fewer generators than in $L^{h,s}$, so there is a monomorphism. The proof is now completed by a slight modification of the main argument in [26], realization of h-cobordisms, and the π-π theorem. □

Remark 19.3. The authors believe that the $L^{h,s}$ groups coincide with the diagonal L-groups as proposed by Quinn in various lectures. The notation is chosen to indicate that one may always define L-groups with two upper decorations instead of only one, corresponding to a $*$-invariant subgroup of the Whitehead group containing another $*$-invariant subgroup.

Remark 19.4. Ranicki has recently proved the existence of a useful exact sequence [37]. Given a cofibration $A \to X \to X \cup_A CA$ there is a long exact sequence:

$$\ldots \to L^h_n(\mathcal{C}_{O(A)}(R)) \to L^h_n(\mathcal{C}_{O(X)}(R)) \to L^K_n(\mathcal{C}_{O(X\cup_A CA)}(R)) \to \\ \to L^h_{n-1}(\mathcal{C}_{O(A)}(R)) \to \ldots$$

where $K = \mathrm{Im}(K_1(\mathcal{C}_{O(X)}(R)) \to K_1(\mathcal{C}_{O(X\cup_A CA)}(R)))$. This seems to be an adequate substitute for being a homology theory. See also the extensions of Ranicki's results given in [6, Section 4] which give a general result of the above mentioned type in the language of Karoubi–filtered categories.

20. Final Comments

Throughout this paper we have been working under the assumption of a constant fundamental group or a group action. This does exclude some examples one might want to study, for example

$$(S^7 \to CP(3)) \rightsquigarrow \{S^7 \times \mathbb{R} \to O(CP(3)_+))\}$$

as control map. In this example we have a locally constant fundamental group $\mathbb{Z}$ which is not globally constant. It is however possible to study questions of this type by the methods developed in this paper as follows: Cover $CP(3)$ by open sets U_α so that the restriction of the bundle to each U_α is trivial. A bounded surgery problem parameterized by $O(CP(3))$ with this fundamental group structure will now produce a surgery problem in each $L_n(\mathcal{C}_{\bar{U}_\alpha}^{\partial U_\alpha}(\mathbb{Z}[\mathbb{Z}]))$, and the original surgery problem can be solved if and only if all these surgery problems can be solved in a compatible way. But this can be investigated: If we can solve over $O(U_\alpha)$ and over $O(U_\beta)$ there will be an obstruction in $L_{n+1}(\mathcal{C}_{O(\bar{U}_\alpha \cap \bar{U}_\beta)}^{\partial(U_\alpha \cap U_\beta)}(\mathbb{Z}[\mathbb{Z}]))$. So the methods can in principle be made to work, if not in the most elegant way, for a locally constant system of fundamental groups in the stratified sense.

References

1. D. R. Anderson, F. Connolly, S. C. Ferry, and E. K. Pedersen, *Algebraic K-theory with continuous control at infinity*, J. Pure Appl. Algebra **94** (1994), 25–47.
2. D. R. Anderson and H. J. Munkholm, *Foundations of Boundedly Controlled Algebraic and Geometric Topology*, Lecture Notes in Mathematics, vol. 1323, Springer-Verlag, Berlin-New York, 1988.
3. D. R. Anderson and E. K. Pedersen, *Semifree topological actions of finite groups on spheres*, Math. Ann. **265** (1983), 23–44.
4. W. Browder, *Surgery on simply connected Manifolds*, Springer-Verlag, Berlin-New York, 1972.
5. M. Brown, *A proof of the generalized Schoenflies theorem*, Bull. Amer. Math. Soc. **66** (1960), 74 - 76.
6. G. Carlsson and E. K. Pedersen, *Controlled algebra and the Novikov conjectures for K- and L-theory*, Topology (1995), (To appear).
7. T. A. Chapman, *Approximation results in Topological Manifolds*, Mem. Amer. Math. Soc., no. 251 (1981).
8. T. A. Chapman and S. C. Ferry, *Approximating homotopy equivalences by homeomorphisms*, Amer. J. Math. **101** (1979), 583–607.
9. S. C. Ferry and E. K. Pedersen, *Controlled algebraic K-theory*, In preparation.
10. ———, *Some mildly wild circles in S^n arising from algebraic K-theory*, *K*-theory **4** (1991), 479–499.
11. S. C. Ferry and S. Weinberger, *A coarse approach to the Novikov conjecture*, Proceedings of the 1993 Oberwolfach Conference on the Novikov Conjecture (S. Ferry, A. Ranicki, and J. Rosenberg, eds.), London Mathematical Society Lecture Notes, 1994.
12. I. Hambleton and I. Madsen, *Actions of finite groups on $\mathbb{R}^{n+k}$ with fixed set R^k*, Canad. J. Math. **38** (1986), 781–860.
13. I. Hambleton and E. K. Pedersen, *Bounded surgery and dihedral group actions on spheres*, J. Amer. Math. Soc. **4** (1991), 105–126.
14. C. B. Hughes, L. R. Taylor, and B. Williams, *Controlled surgery over manifolds*, preprint.

15. R. Kirby and L. Siebenmann, *Foundational Essays on Topological Manifolds, Smoothings and Triangulations*, Princeton University Press, 1977.
16. R. C. Kirby, *Stable homeomorphisms and the annulus conjecture*, Ann. of Math. (1969), 575–582.
17. I. Madsen and M. Rothenberg, *On the homotopy theory of equivariant automorphism groups*, Invent. Math. **94** (1988), 623–638.
18. S. Maumary, *Proper surgery groups and Wall-Novikov groups*, Proc. 1972 Battelle Seattle Algebraic K-theory Conference (Berlin-New York) (H. Bass, ed.), Springer-Verlag, 1973, Lecture Notes in Mathematics 343, pp. 526–539.
19. B. Mazur, *On embeddings of spheres*, Bul. Amer. Math. Soc. **65** (1959), 59–65.
20. J. Milgram, *Patching techniques in surgery and the solution of the compact space form problem*, Stanford mimeo (1981).
21. J. Milnor, *Microbundles and smoothing*, Princeton University Notes.
22. A. Nicas, *Induction theorems for groups of manifold homotopy structure sets*, Mem. Amer. Math. Soc., no. 267 (1982).
23. E. K. Pedersen, *Geometrically defined transfers, comparisons*, Math. Z. **180** (1982), 535–544.
24. ———, *On the K_{-i} functors*, J. Algebra **90** (1984), 461–475.
25. ———, *On the bounded and thin h-cobordism theorem parameterized by $\mathbb{R}^n$*, Transformation Groups, Poznań 1985, Proceedings (Berlin-New York) (S. Jackowski and K. Pawałowski, eds.), Springer-Verlag, 1986, Lecture Notes in Mathematics 1217, pp. 306–320.
26. E. K. Pedersen and A. A. Ranicki, *Projective surgery theory*, Topology **19** (1980), 239–254.
27. E. K. Pedersen and C. Weibel, *K-theory homology of spaces*, Algebraic Topology, Proceedings Arcata 1986 (Berlin-New York) (G. Carlsson, R. L. Cohen, H. R. Miller, and D. C. Ravenel, eds.), Springer-Verlag, 1989, Lecture Notes in Mathematics 1370, pp. 346–361.
28. F. Quinn, *A geometric formulation of surgery*, Ph.D. thesis, Princeton University, 1969.
29. ———, *Ends of maps, I*, Ann. of Math. **110** (1979), 275–331.
30. ———, *Ends of maps, II*, Invent. Math. **68** (1982), 353–424.
31. ———, *An obstruction to the resolution of homology manifolds*, Michigan Math. J. **301** (1987), 285–292.
32. ———, *Resolutions of homology manifolds, and the topological characteriztation of manifolds*, Invent. Math. **72** (1987), 267–284.
33. A. A. Ranicki, *Algebraic L-theory II, Laurent extensions*, Proc. Lond. Math. Soc. (3) **27** (1973), 126–158.
34. ———, *The algebraic theory of surgery, I. Foundations*, Proc. Lond. Math. Soc. (3) **40** (1980), 87–192.
35. ———, *The algebraic theory of surgery II. Applications to topology*, Proc. Lond. Math. Soc. (3) **40** (1980), 163–287.
36. ———, *Additive L-theory*, K-theory **3** (1989), 163–195.
37. ———, *Lower K- and L-theory*, Lond. Math. Soc. Lect. Notes, vol. 178, Cambridge Univ. Press, Cambridge, 1992.
38. C. Rourke and B. Sanderson, *Introduction to PL topology*, Springer-Verlag, Berlin-New York, 1972.
39. L. C. Siebenmann, *Infinite simple homotopy types*, Indag. Math. **32** (1970), 479–495.
40. D. Sullivan, *Triangulating homotopy equivalences*, Ph.D. thesis, Princeton University, 1965.
41. L. Taylor, *Surgery on paracompact manifolds*, Ph.D. thesis, UC at Berkeley, 1972.
42. C. T. C. Wall, *Poincaré complexes*, Ann. of Math. **86** (1970), 213–245.
43. ———, *Surgery on Compact Manifolds*, Academic Press, 1971.

44. S. Weinberger, *The Topological Classification of Stratified Spaces*, University of Chicago Press, 1994.
45. M. Yamasaki, *L-groups of crystallographic groups*, Invent. Math. **88** (1987), 571–602.

DEPARTMENT OF MATHEMATICAL SCIENCES, SUNY AT BINGHAMTON, BINGHAMTON, NEW YORK 13901, USA

email: steve@math.binghamton.edu

email: erik@math.binghamton.edu

On the coarse Baum-Connes conjecture

Nigel Higson and John Roe

1. Introduction

The *Baum-Connes conjecture* [2, 3] concerns the K-theory of the reduced group C^*-algebra $C_r^*(G)$ for a locally compact group G. One can define a map from the equivariant K-homology of the universal proper G-space $\underline{E}G$ to $K_*(C_r^*(G))$: each K-homology class defines an index problem, and the map associates to each such problem its analytic index. The conjecture is that this map is an isomorphism. The injectivity of the map has geometric and topological consequences, implying the Novikov conjecture for example; the surjectivity has consequences for C^*-algebra theory and is related to problems in harmonic analysis.

In geometric topology it has proved to be very useful to move from studying classical surgery problems on a compact manifold M to studying bounded surgery problems over its universal cover (see [8, 24] for example). In terms of L-theory, one replaces the classical L-theory of $\mathbf{Z}\pi$ by the L-theory, bounded over $|\pi|$, of $\mathbf{Z}$ (here $|\pi|$ denotes π considered as a metric space, with a word length metric). Now the authors, motivated by considerations of index theory on open manifolds, have studied a C^*-algebra $C^*(X)$ associated to any proper metric space X, and it has recently become quite clear that the passage from $C_r^*(\pi)$ to $C^*(|\pi|)$ is an analytic version of the same geometric idea. Moreover, various descent arguments have been given [4, 9, 5, 17, 27], both in the topological and analytic contexts, which show that a 'sufficiently canonical' proof of an analogue of the Baum-Connes conjecture in the bounded category will imply the classical version of the Novikov conjecture.

The purpose of this paper is to give a precise formulation of the Baum-Connes conjecture for the C^*-algebras $C^*(X)$ (filling in the details of the hints in the last section of [26]) and to prove the conjecture for spaces which are non-positively curved in some sense, including affine buildings and hyperbolic metric spaces in the sense of Gromov. Notice that while the classical Novikov conjecture has been established for the analogous class of groups, the Baum-Connes conjecture has not. Unfortunately there does not seem to be any descent principle for the surjectivity side of the Baum-Connes conjecture as there is for the injectivity side.

The main tool that we will use in this paper is the invariance of $K_*(C^*(X))$ under *coarse homotopy*, established by the authors in [15]. Coarse homotopy is a rather weak equivalence relation on metric spaces, weak enough that (for example) many spaces are coarse homotopy equivalent to open cones $\mathcal{O}Y$ on compact spaces Y.

The idea of 'reduction to a cone on an ideal boundary' is also used in some topological approaches to the Novikov conjecture, but the notion of coarse homotopy appears at present[1] to be peculiar to our analytic set-up. Certainly the proofs in [15] make heavy use of C^*-algebraic machinery.

A different and very interesting approach to the coarse Baum-Connes conjecture has been proposed by G. Yu [31].

2. Coarse homology theories

The *coarse category* UBB was defined in [26] to be the category whose objects are proper metric spaces (that is, metric spaces in which closed bounded sets are compact) and whose maps are proper Borel maps f satisfying the growth condition

$$\forall R > 0 \quad \exists S > 0 \quad \text{such that} \quad d(x_0, x_1) < R \quad \Rightarrow \quad d(f(x_0), f(x_1)) < S.$$

Two morphisms f and g are called *bornotopic* if there is a constant C such that $d(f(x), g(x)) < C$ for all x.

A substantial part of the paper [26] was devoted to the explicit construction of a bornotopy-invariant cohomology theory on the coarse category. Now let M_* be any generalized homology theory on the category of locally compact spaces and proper maps. We will show how it is possible to 'coarsen' M to a 'coarse homology theory' MX_*, functorial on the coarse category and invariant under bornotopy.

To do this recall from [26], Definition 3.13, that an *anti-Čech system* for a proper metric space X is a sequence $\mathcal{U}_1, \mathcal{U}_2$ of successively coarser open covers of X, with the property that the diameter of each set in $\mathcal{U}_n$ is bounded by a constant R_n which is less than the Lebesgue number of $\mathcal{U}_{n+1}$, and the constants R_n tend to infinity. It follows that each member of the cover $\mathcal{U}_n$ is contained within a member of $\mathcal{U}_{n+1}$, and from now on we shall include a choice of such as part of the structure of an anti-Čech system. Passing to the nerves of the covers $\mathcal{U}_n$, this extra data determines 'coarsening' maps

$$|\mathcal{U}_1| \to |\mathcal{U}_2| \to |\mathcal{U}_3| \to \dots$$

by associating to each member of the cover $\mathcal{U}_n$ the member of the cover $\mathcal{U}_{n+1}$ chosen to contain it.[2]

---

[1] **Added in proof:** Since writing this paper we have learned of unpublished calculations of Ferry and Pedersen which make use of similar ideas.

[2] Recall that the nerve of a cover $\mathcal{U} = \{U_\alpha\}$ is the simplicial complex with vertices $[U_\alpha]$ labelled by members of $\mathcal{U}$, and a p-simplex $[U_{\alpha_0} \dots U_{\alpha_p}]$ for each $(p+1)$-tuple in $\mathcal{U}$ with $U_{\alpha_0} \cap \dots \cap U_{\alpha_p} \neq \emptyset$.

(2.1) DEFINITION: *We define the coarse M-homology of X to be the direct limit*

$$MX_*(X) = \lim_{\rightarrow} M_*(|\mathcal{U}_n|).$$

If $f\colon X \to Y$ is a coarse map, and if $\mathcal{U}_n$ and $\mathcal{V}_m$ are anti-Čech systems for X and Y respectively, then for each n there is an m_n such that if $U \in \mathcal{U}_n$ then $f[U]$ is contained in a member of $\mathcal{U}_{m_n}$. After selecting one such member of $\mathcal{U}_{m_n}$ for each member of $\mathcal{U}_n$ we get a proper map

$$f_n\colon |\mathcal{U}_n| \to |\mathcal{V}_{m_n}|.$$

Defining the maps f_n inductively, we can arrange that the diagrams

$$\begin{array}{ccc} |\mathcal{U}_n| & \to & |\mathcal{U}_{n+1}| \\ f_n \downarrow & & \downarrow f_{n+1} \\ |\mathcal{V}_{m_n}| & \to & |\mathcal{U}_{m_{n+1}}| \end{array}$$

commute. Passing to the direct limit we obtain an induced map

$$f_*\colon MX_*(X) \to MX_*(Y).$$

It is independent of the choices involved in the definition of the maps f_n.

Applying this observation to the identity map on X we note that two different choices of anti-Čech system on X will give rise to canonically isomorphic direct limits $MX_*(X)$, which justifies our omission of the anti-Čech system $\mathcal{U}$ in the notation $MX_*(X)$.

We have made MX_* into a functor on the category UBB.

(2.2) PROPOSITION: *Two bornotopic maps $f, g\colon X \to Y$ induce the same transformation on MX_*.*

PROOF: For each n one can choose an m_n such that the maps f_n and g_n in the above construction both map to the same $|\mathcal{V}_{m_n}|$ and are linearly homotopic. The induced maps on homology are therefore the same. □

We shall occasionally find it useful to confine our attention to coarse maps which are continuous. Imposing this requirement on morphisms we obtain the *continuous coarse category* UBC, a subcategory of UBB. The following definition is best viewed within this context.

(2.3) DEFINITION: *Let X and Y be proper metric spaces. A* coarse homotopy *from X to Y is a continuous and proper map*

$$h\colon X \times [0,1] \to Y$$

such that for every $R > 0$ there exists $S > 0$ with

$$d(x,x') \le R \quad \Rightarrow \quad d(h(x,t),h(x',t)) \le S, \quad \textit{for all } t \in [0,1].$$

Two coarse maps $f_0, f_1 : X \to Y$ are coarsely homotopic if there is a coarse homotopy $h : X \times [0,1] \to Y$ such that

$$f_0(x) = h(x,0) \quad \text{and} \quad f_1(x) = h(x,1), \quad \text{for all } x \in X.$$

A coarse map $f : X \to Y$ is a coarse homotopy equivalence if there is a coarse map $g : Y \to X$ such that $f \circ g$ and $g \circ f$ are coarsely homotopic to the identity maps on Y and X, respectively.

REMARK: It is possible to relax the continuity requirement in this definition to *pseudocontinuity*, that is continuity 'on a certain scale'; see the remarks to Section 1 in [15]. This point will be of some importance when we discuss hyperbolic spaces later in this paper.

(2.4) THEOREM: *Let M be a generalized homology theory. Then coarse M-homology, MX_*, is also a coarse homotopy invariant functor on UBC.*

PROOF: Let $h : X \times [0,1] \to Y$ be a coarse homotopy, and let $\mathcal{U}_n$ and $\mathcal{V}_m$ be anti-Čech systems for X and Y respectively. From the definition of coarse homotopy it follows that for each n there is an m_n such that h induces a proper homotopy $h_n : |\mathcal{U}_n| \to |\mathcal{V}_{m_n}|$. The result follows. □

3. Comparison of homology and coarse homology

Let X be a proper metric space, let $\mathcal{U} = \{U_\alpha\}$ be a locally finite open cover and let $\{\varphi_\alpha\}$ be a partition of unity subordinate to this cover. Define a map $\kappa : X \to |\mathcal{U}|$ by

$$\kappa(x) = \sum_\alpha \varphi_\alpha(x)[U_\alpha].$$

To explain this formula, note that for each x we have $\sum \varphi_\alpha(x) = 1$, and those finitely many vertices $[U_\alpha]$ for which $\varphi(x) \neq 0$ span a simplex in $|\mathcal{U}|$. So $\kappa(x)$ describes, in barycentric coordinates, a point of $|\mathcal{U}|$.

Suppose that we apply this construction to the first cover $\mathcal{U}_1$ in an anti-Čech system. Two different choices of partition of unity will give rise to maps which are properly homotopic, and so induce the same map on homology. Passing to the direct limit we obtain a canonical *coarsening map*

$$c : M_*(X) \to MX_*(X).$$

(This is dual to the map c considered in [26] in a cohomology context.) We wish to inquire when c is an isomorphism.

Since the passage from X to $|\mathcal{U}_1|$, and then $|\mathcal{U}_2|$, and so on, obliterates the 'small scale' topology of X, it is natural to confine our attention to *uniformly contractible* spaces. These are defined by the requirement that for each $R > 0$ there be some $S > R$ such that $B(x; R)$ is contractible within $B(x; S)$, for every $x \in X$. One might conjecture that if X is uniformly contractible then the coarsening map c is an isomorphism. However, this is not so: Dranishnikov, Ferry and Weinberger [6] have constructed an example of a uniformly contractible space X for which the coarsening map in K-homology is not an isomorphism. But we shall show that if X is a *bounded geometry complex* (defined below) then uniform contractibility does imply that the coarsening map is an isomorphism.

Recall that a *path metric* on a space X is a metric such that the distance between any two points of X is the infimum of the lengths of the continuous paths connecting them. A *path metric space* is a metric space whose metric has this property.

(3.1) DEFINITION: *A path metric space X is called a* metric simplicial complex *if it is a simplicial complex and its metric coincides on each simplex with the usual spherical metric.*

The spherical metric on the standard n-simplex Δ^n is obtained by regarding it as the set of points of $S^n \subseteq \mathbf{R}^{n+1}$ with nonnegative coordinates. Any locally finite simplicial complex can be given a complete metric that makes it into a metric simplicial complex.

The following result is proved in [25, Section 3].

(3.2) PROPOSITION: *Let X be a complete path metric space, $\mathcal{U}$ an open cover of X that has positive Lebesgue number and such that the sets of $\mathcal{U}$ have bounded diameter. Then the map $\kappa\colon X \to |\mathcal{U}|$, defined above, is a bornotopy-equivalence.*

(3.3) LEMMA: *let $f\colon X \to Y$ be a coarse map. Suppose that X is a finite-dimensional metric simplicial complex and that Y is uniformly contractible. Then there exists a continuous map $g\colon X \to Y$ that is bornotopic to f. Moreover, if f is already continuous on a subcomplex X', then we may take $g = f$ on X'.*

PROOF: We construct g by induction over the skeleta X_k of (X, X'). The base step is provided by setting $g = f$ on $X' \cup X_0$. Assume then that g has been defined on $X' \cup X_k$. Then g is defined on the boundary of each $k+1$-simplex Δ of (X, X'), and as Y is uniformly contractible, $g|\partial\Delta$ can be extended to a map $\Delta \to Y$ whose image lies within a bounded distance of the image of the vertex set of Δ. Proceeding thus inductively, after finitely many stages we obtain a continuous map $g\colon X \to Y$ which coincides with f on $X' \cup X_0$, and which has the property that there is a constant $C > 0$ such that $d(g(x), g(x')) < C$ whenever $x \in X_0$ is a vertex of a simplex containing x'. Since X_0 is coarsely dense, g is bornotopic to f. $\square$

(3.4) LEMMA: *Let X be a finite-dimensional metric simplicial complex and*

Y a uniformly contractible space; then any two bornotopic continuous coarse maps from X to Y are properly homotopic.

PROOF: Let $h\colon X \times [0,1] \to Y$ be a bornotopy, and apply Lemma 3.3 to h which is continuous on the subcomplex $X \times \{0,1\}$ of $X \times [0,1]$. □

(3.5) COROLLARY: *If two uniformly contractible, finite dimensional metric simplicial complexes are bornotopy equivalent then they are proper homotopy equivalent.*

The following notion is due to Fan [7].

(3.6) DEFINITION: *A proper metric space X has* bounded coarse geometry *if there is some $\varepsilon > 0$ such that for each $R > 0$ there is $C > 0$ such that the ε-capacity of any ball of radius R is at most C.*

Recall [21] that the ε-capacity of a set Y is the maximum number of elements in an ε-separated subset of Y.

One can show that bounded coarse geometry implies that for all sufficiently large ρ there is a universal bound on the ρ-entropy and the ρ-capacity of any subset of X in terms of its diameter.

Bounded coarse geometry has the following consequence which will be important for us.

(3.7) LEMMA: *If X is a space of bounded coarse geometry, then for any $R > 0$ there is $S > 0$ such that X has an open cover $\mathcal{U}$ with:*

- *The Lebesgue number of $\mathcal{U}$ is at least R;*
- *The cover $\mathcal{U}$ is of finite order (that is, its nerve is finite-dimensional);*
- *The sets of $\mathcal{U}$ have diameter less than S.*

The proof is straightforward.

For brevity, we will abbreviate the phrase 'metric simplicial complex with bounded coarse geometry' to 'bounded geometry complex.' It is easy to check that *every bounded geometry complex is finite dimensional.*

(3.8) PROPOSITION: *Let X be a uniformly contractible, bounded geometry complex, and let MX_* be the coarse homology theory associated to a generalized homology theory M_* as above. Then the natural map $c\colon M_*(X) \to MX_*(X)$ is an isomorphism.*

PROOF: We will construct an anti-Čech system by induction as follows. Let $\mathcal{U}_1$ be any cover of X of the kind described in Lemma 3.7, and let $f_1\colon X \to |\mathcal{U}_1|$ be the map $\kappa_{\mathcal{U}_1}$. By 3.2, f_1 is a bornotopy equivalence; so it admits a bornotopy inverse $g_1\colon |\mathcal{U}_1| \to X$. Since X is flabby and $|\mathcal{U}_1|$ is finite-dimensional, Lemmas 3.3 and 3.4 show that g_1 may be assumed to be continuous and to be a left proper homotopy inverse of f_1.

The map $f_1 \circ g_1$ is bornotopic to the identity map on $|\mathcal{U}_1|$. It is therefore possible to find a second cover $\mathcal{U}_2$ of the kind described in Lemma 3.7, which coarsens $\mathcal{U}_1$ with coarsening map $f_2 \colon |\mathcal{U}_1| \to |\mathcal{U}_2|$ and which has $f_2 \circ f_1 \circ g_1$ properly homotopic to f_2 by a linear homotopy. Proceeding inductively we may obtain an anti-Čech system

$$X \xrightarrow{f_1} |\mathcal{U}_1| \xrightarrow{f_2} |\mathcal{U}_2| \xrightarrow{f_3} \cdots$$

which has the following properties:

- The maps f_i are continuous;
- The maps $h_i = f_i \circ \cdots \circ f_1$ admit left proper homotopy inverses g_i;
- The maps f_{i+1} and $h_{i+1} \circ g_i$ are properly homotopic.

It follows (using the proper homotopy invariance of M-homology) that $(h_i)_* : M_*(X) \to M_*(|\mathcal{U}_i|)$ is an isomorphism onto the image of $(f_i)_*$. Thus the induced map to the direct limit is an isomorphism. □

REMARK: Let X be a proper metric space. We define a *coarsening* of X to be a uniformly contractible, bounded geometry complex Y equipped with a bornotopy-equivalence $X \to Y$. A space X that has a coarsening might be called *coarsenable*. It follows from Corollary 3.5 that a coarsening of X, if it exists, is unique up to a proper homotopy equivalence (which is at the same time a bornotopy equivalence).

Moreover, by Lemmas 3.3 and 3.4, coarsening is functorial: a coarse map between two spaces induces a unique proper homotopy class of continuous coarse maps between their coarsenings. Thus one may define the coarse M-homology of a space X simply to be the ordinary M-homology of a coarsening of X, and indeed one may make the analogous definition for cohomology also. This definition has the disadvantage of applying only to the category of coarsenable spaces, which seems to be rather hard to characterize by an internal description.

One situation in which the notion of coarsening can be made concrete, however, is that in which X is (the underlying metric space of) a finitely generated discrete group Γ. The usual Baum-Connes conjecture for Γ relates to the equivariant K-homology of a certain space $\underline{E}\Gamma$, the universal space for proper Γ-actions. A model for $\underline{E}\Gamma$ as a Γ-finite simplicial complex, if one exists, will automatically be a bounded geometry complex in our sense, and will in fact be a coarsening of Γ. Thus the coarse K-homology of $|\Gamma|$ is a 'nonequivariant' version of the left-hand side of the ordinary Baum-Connes conjecture for Γ.

4. Cones

In this section we shall analyze the coarse homology of an *open cone*. Recall that if Y is a compact subset of the unit sphere in a normed space then the open cone on Y, denoted $\mathcal{O}Y$, is the set of all non-negative multiples of points in Y.

(4.1) DEFINITION: *Let $r\colon \mathbf{R}^+ \to \mathbf{R}^+$ be a contractive map*[3] *such that $r(0) = 0$ and $r(\infty) = \infty$, and let $\mathcal{O}Y$ be an open cone. The* radial contraction *associated to r is the map $\rho\colon \mathcal{O}Y \to \mathcal{O}Y$ defined by*

$$\rho(ty) = r(t)y, \quad y \in Y.$$

Any radial contraction is coarsely homotopic to the identity map, and therefore induces the identity on coarse homology. On the other hand, radial contractions can be used to force more or less arbitrary maps from open cones to obey a growth condition; this is the content of the next lemma. These two properties taken together make radial contractions extremely useful in computations involving coarse homology.

(4.2) LEMMA: *Let $\mathcal{O}Y$ be an open cone, as above, and let Z be any metric space. Let $f\colon \mathcal{O}Y \to Z$ be a continuous (or pseudocontinuous) and proper map. Then there exists a radial contraction $\rho\colon \mathcal{O}Y \to \mathcal{O}Y$ such that $f \circ \rho$ is a coarse map.*

PROOF: Fix $\varepsilon > 0$ such that, for any $x \in \mathcal{O}Y$, the inverse image $f^{-1}(B(f(x);\varepsilon))$ is a neighbourhood of x. (Any ε will do if f is continuous; pseudocontinuity means, by definition, that there exists an ε with this property.) For each $s \geq 0$ let $K_s = \{x \in \mathcal{O}Y : \|x\| \leq s\}$ and let $\delta(s)$ be a Lebesgue number for the covering

$$\{f^{-1}(B(f(x);\varepsilon)) : x \in K_s\}$$

of K_s. Then $\delta(s)$ is a monotone decreasing function of s. Define a function $r(t)$ by the equation

$$t = \int_0^{r(t)} \frac{(s+1)}{\min\{\delta(s), 1\}}\, ds.$$

By elementary calculus, r is a contractive map. Consider the radial contraction ρ defined by r. By construction, for any $R > 0$ there exists $r_0 > 0$ such that if $x, x' \in \mathcal{O}Y$ with $d(x,x') < R$ and $\|x\| > r_0$, then $d(\rho(x), \rho(x')) < 1/\delta(\max\{\|x\|, \|x'\|\})$, and hence $d(f \circ \rho(x), f \circ \rho(x')) < \varepsilon$. Thus, whenever $d(x,x') < R$, one has $d(f \circ \rho(x), f \circ \rho(x')) < S$, where

$$S = \varepsilon + \operatorname{diam} f(K_{r_0}).$$

[3]That is, a Lipschitz map with Lipschitz constant less than or equal to 1.

Hence $f \circ \rho$ is a coarse map. □

For example, suppose that $h: Y \to Y'$ is a homeomorphism. The induced map $\mathcal{O}h: \mathcal{O}Y \to \mathcal{O}Y'$ need not be a coarse map; in fact, it will be so if and only if the original h is Lipschitz. Nevertheless, by the above result there exists a radial contraction ρ such that both $\rho \circ h$ and $h^{-1} \circ \rho$ are coarse maps. Since ρ is coarsely homotopic to the identity, these maps are coarse homotopy equivalences[4] between $\mathcal{O}Y$ and $\mathcal{O}Y'$. This allows one to extend the scope of the definition of 'open cone'. Let Y be any finite-dimensional compact metrizable space; then it is a classical theorem that Y is homeomorphic to a subset of a sphere in a Euclidean space (see Theorem V.2 in [18]). The open cone $\mathcal{O}Y$ can be defined as the open cone on any such homeomorphic image, and will be well defined up to coarse homotopy equivalence.

We should now like to calculate the coarse M-homology of such an open cone. If Y is a finite complex, then $\mathcal{O}Y$ can be triangulated as a uniformly contractible bounded geometry complex, and so the calculation will follow from 3.8. However, we will need to consider open cones on more unpleasant metrizable spaces also; our proof of the Baum-Connes conjecture for a hyperbolic space will proceed *via* the open cone on its Gromov boundary.

It will be necessary to assume that the homology theory M satisfies the *strong excision axiom* [29, 22, 19]. We recall that this axiom states that for any pair (X, A) of compact metric spaces, the natural map $M_*(X, A) \to M_*(X \setminus A)$ is an isomorphism. For example, Steenrod homology satisfies strong excision, but singular homology does not. Of more direct relevance to this paper is the fact that Kasparov's K-homology [20] satisfies the strong excision axiom.

(4.3) PROPOSITION: *If M_* is a generalized homology theory satisfying the strong excision axiom, then the coarsening map*

$$c: M_*(\mathcal{O}Y) \to MX_*(\mathcal{O}Y)$$

is an isomorphism for any finite-dimensional compact metric space Y.

PROOF: We consider $\mathcal{O}Y$ to be embedded in $\mathbf{R}^n$. Form an anti-Čech system as follows: the cover $\mathcal{U}_i$ is made up of all the nonempty intersections $B \cap \mathcal{O}Y$, where B runs over the set of open balls in $\mathbf{R}^n$ with centres at the integer lattice points and radius 3^i. Let X_i be the geometric realization of the nerve of $\mathcal{U}_i$. We also let $X_0 = \mathcal{O}Y$ itself. There are obvious coarsening maps

$$X_0 \to X_1 \to X_2 \to X_3 \to \ldots$$

---

[4]In one or two places in the literature one can find statements which might mislead the unwary into believing that $\mathcal{O}Y$ and $\mathcal{O}Y'$ are *bornotopy* equivalent. This is not, in general, the case.

and $MX_*(\mathcal{O}Y)$ is by definition the direct limit of the M-homology of this sequence. We make the following claims about this construction.

- *Claim 1:* Each X_i can be compactified to a space $\overline{X_i}$ obtained by adding a copy of Y as a set of points at infinity.[5]
- *Claim 2:* The coarsening maps can be extended by the identity on Y to give continuous maps $\overline{X_i} \to \overline{X_{i+1}}$.
- *Claim 3:* The extended map $\overline{X_i} \to \overline{X_{i+1}}$ is nullhomotopic.

Granted these three claims, the result follows. For consider the commutative diagram

$$\begin{array}{ccccccc}
\tilde{M}_*(Y) & \longrightarrow & \tilde{M}_*(\overline{X_i}) & \longrightarrow & M_*(X_i) & \xrightarrow{\partial} & \tilde{M}_{*-1}(Y) \\
\downarrow{=} & & \downarrow & & \downarrow & & \downarrow{=} \\
\tilde{M}_*(Y) & \longrightarrow & \tilde{M}_*(\overline{X_{i+1}}) & \longrightarrow & M_*(X_{i+1}) & \xrightarrow{\partial} & \tilde{M}_{*-1}(Y)
\end{array}$$

in which the rows are the long exact sequences in reduced M-homology arising from the pairs $(\overline{X_i}, Y)$ and the columns are the coarsening maps. Because we are using reduced homology, Claim 3 implies that the second vertical map is in fact zero. A diagram chase then shows that ∂ gives an isomorphism between $\mathrm{Im}(M_*(X_i) \to M_*(X_{i+1}))$ and $\tilde{M}_{*-1}(Y)$. Passing to the direct limit we obtain another commutative diagram

$$\begin{array}{ccc}
M_*(\mathcal{O}Y) & \xrightarrow{\partial} & \tilde{M}_{*-1}(Y) \\
c\downarrow & & =\downarrow \\
MX_*(\mathcal{O}Y) & \xrightarrow{\partial} & \tilde{M}_{*-1}(Y)
\end{array}$$

where both ∂'s are isomorphisms, and therefore c is an isomorphism too.

It remains to prove the claims. Claims 1 and 2 are straightforward; to check Claim 3, let us notice that there are continuous maps $\alpha_i \colon X_i \to \mathrm{Pen}(\mathcal{O}Y; 2\cdot 3^i)$ and $\beta_i \colon \mathrm{Pen}(\mathcal{O}Y; 2\cdot 3^i) \to X_{i+1}$ defined as follows: α_i sends each vertex of X_i to the centre of the corresponding ball in $\mathbf{R}^n$ and extends by linearity, and $\beta_i(x)$ is defined to be

$$\frac{\sum_p \varphi(|x-p|)\cdot p}{\sum_p \varphi(|x-p|)}$$

where the sum ranges over all integer lattice points p and $\varphi(r)$ is a positive continuous function equal to 1 for r small and equal to 0 for $r > 3^i$. (Notice that, because $x \in \mathrm{Pen}(\mathcal{O}Y; 2\cdot 3^i)$, the points p for which $\varphi(x-p) > 0$ do indeed define the vertices of a simplex in X_{i+1}.) Clearly $\mathrm{Pen}(\mathcal{O}Y; 2\cdot 3^i)$ can

---

[5] In the case of X_0, the compactification is simply the usual closed cone on Y, and is therefore contractible. The point of the construction is to obtain a 'pro'-version of this fact for the sequence of coarsenings X_i.

be compactified by adding Y at infinity, and both α_i and β_i extend continuously by the identity. Moreover, $\beta_i\alpha_i$ is homotopic to the coarsening map (by a linear homotopy). It is therefore enough to prove that the continuous extension of α_i is nullhomotopic, and this is so because $\overline{\text{Pen}(\mathcal{O}Y; 2\cdot 3^i)}$ is star-shaped about 0. □

5. *K*-Homology and Paschke duality

In this section we will recall the basic definitions of Kasparov's K-homology theory, together with the duality theory of Paschke that relates K-homology and K-theory.

Let X be a locally compact metrizable space. By an *X-module*, we will mean a separable Hilbert space equipped with a representation of the C^*-algebra $C_0(X)$ of continuous functions (tending to zero at infinity) on X. We shall say that an X-module is *non-degenerate* if the representation of $C_0(X)$ is non-degenerate, and as in [16] we shall say that an X-module is *standard* if it is non-degenerate and no non-zero function in $C_0(X)$ acts as a compact operator.

Let T be a bounded operator on an X-module H_X. By definition, T is *locally compact* if the operators $T\varphi$ and φT are compact for every $\varphi \in C_0(X)$. We say that T is T *pseudolocal* if $\varphi T\psi$ is a compact operator for all pairs of continuous functions on X with compact and disjoint supports. As Kasparov remarks (see Proposition 3.4 in [20]), T is pseudolocal if and only if the commutator $\varphi T - T\varphi$ is compact, for every $\varphi \in C_0(X)$. This makes it clear that the set of all pseudolocal operators on H_X is a C^*-algebra containing the set of all locally compact operators as a closed two-sided ideal. We shall use the notation

$$\Psi_0(X, H_X) = \text{pseudolocal operators on } H_X$$

and

$$\Psi_{-1}(X, H_X) = \text{locally compact operators on } H_X,$$

which is meant to suggest that pseudolocal operators should be thought of as abstract pseudodifferential operators of order ≤ 0, and locally compact operators as abstract pseudodifferential operators of order ≤ -1. Compare [1, 20].

Kasparov's K-homology groups $K_i(X)$ are generated by certain cycles modulo a certain equivalence relation. A cycle for $K_0(X)$ is given by a pair (H_X, F), comprised of an X-module H_X and a pseudolocal operator F on H_X such that $FF^* - I$ and $F^*F - I$ are locally compact.[6] A cycle for $K_1(X)$

---

[6] **Kasparov's theory allows for more a complicated sort of cycle, comprised of a pseudolocal operator $F\colon H_X \to H'_X$ between two different Hilbert spaces, but there is a simple trick to convert such a cycle to one of the simpler ones we are considering. See**

is given by a similar pair (H_X, F), but with the additional requirement that F be self-adjoint. In both cases the equivalence relation on cycles is given by homotopy of the operator F, unitary equivalence, and direct sum of 'degenerate' cycles, these being cycles for which $F\varphi - \varphi F$, $\varphi(F^*F - 1)$, and so on, are not merely compact but actually zero. See [20] for further details.

The K-homology of X may be related to the K-theory of the algebras of abstract pseudodifferential operators on X. Let H_X be an X-module. Then for $i = 0, 1$ there are maps

$$K_i(\Psi_0(X, H_X)/\Psi_{-1}(X, H_X)) \to K_{1-i}(X) \tag{5.1}$$

defined as follows. In the odd case ($i = 1$), we map a unitary U in $\Psi_0(X)/\Psi_{-1}(X)$, representing an odd K-theory class, to the even K-homology cycle (H_X, F), where F is any lifting of U to $\Psi_0(X)$. Similarly in the even case ($i = 0$), we map a projection P, representing an even K-theory class, to the odd K-homology cycle $(H_X, 2Q - 1)$, where Q is any self-adjoint lifting of P. It is readily checked that these operations respect the various equivalence relations and give a well-defined homomorphism of abelian groups. Now Paschke [23] (see also [14]) proved the following result:

(5.2) PROPOSITION: *If H_X is a standard X-module, then the map* 5.1 *is an isomorphism.*

Suppose now that a standard X-module H_X has been fixed. One can in fact identify not only the K-theory of the quotient $\Psi_0(X)/\Psi_{-1}(X)$ but also the K-theory of the algebras $\Psi_0(X)$ and $\Psi_{-1}(X)$ individually. It turns out that $K_i(\Psi_{-1}(X))$ is always zero, except when $i = 0$ and X is compact in which case it is $\mathbf{Z}$, and from this and Paschke's duality theorem it follows that $K_i(\Psi_0(X))$ is isomorphic to $\tilde{K}_{1-i}(X)$, the *reduced* K-homology of X.

It is also possible to interpret the boundary map in K-homology in terms of Paschke duality. We will summarize the results of [14] in a special case which suffices for our purposes. Suppose that X is locally compact but not compact, and that $\overline{X} = X \cup Y$ is a compactification of X, that is a compact space containing X as a dense open subspace. Then H_X can be thought of as an $\overline{X}$-module, because any representation of $C_0(X)$ on a Hilbert space has a unique extension to a representation of $C(\overline{X})$. It therefore makes sense to consider the algebra $\Psi_0(\overline{X}) \cap \Psi_{-1}(X)$, and one easily sees that this is an ideal in $\Psi_0(\overline{X})$. The relative form of Paschke's duality is then the following

(5.3) PROPOSITION: *The K-theory groups of the algebras $\Psi_0(\overline{X}) \cap \Psi_{-1}(X)$, $\Psi_0(\overline{X})$, and $\Psi_0(\overline{X})/(\Psi_0(\overline{X}) \cap \Psi_{-1}(X))$ are isomorphic to the reduced K-homology groups of Y, $\overline{X}$, and X, with a dimension shift in each case.*

Section 1 of [14].

Moreover, the isomorphisms transform the six-term exact sequence in K-theory arising from the C^*-algebra extension

$$0 \to \Psi_0(\overline{X}) \cap \Psi_{-1}(X) \to \Psi_0(\overline{X}) \to \Psi_0(\overline{X})/(\Psi_0(\overline{X}) \cap \Psi_{-1}(X)) \to 0$$

into the six-term exact sequence in reduced K-homology arising from the pair $(\overline{X}, Y)$.

6. Formulation of the Baum-Connes conjecture

In previous papers we have associated to each proper metric space X a C^*-algebra $C^*(X)$ [16, 15] whose K-theory is functorial for maps in UBB and is a bornotopy invariant. Our objective in this section is to construct an 'analytic index' map

$$\mu: K_*(X) \to K_*(C^*(X)). \tag{6.1}$$

This map is analogous to the assembly map in bounded L-theory.

We begin by briefly recalling some relevant definitions. Let X and Y be proper metric spaces. We refer the reader to Section 4 of [16] for the definition of the *support* of a bounded linear operator from an X-module to a Y-module. It is a closed subset of $X \times Y$, and generalizes the support of the distributional kernel in the C^∞ context. We recall that a bounded linear operator T on an X-module has *finite propagation* if there exists some $R > 0$ such that

$$(x, x') \in \operatorname{Supp}(T) \quad \Rightarrow \quad d(x, x') \leq R.$$

The least such R is called the *propagation* of T. The set of all locally compact, finite propagation operators on a non-degenerate X-module H_X is a $*$-algebra, and we denote by $C^*(X, H_X)$ the C^*-algebra obtained by closing it in the operator norm.

It is shown in Section 4 of [16] that if H_X is any non-degenerate X-module and H'_X is a standard X-module then there is a canonical homomorphism of K-theory groups

$$K_*(C^*(X, H_X)) \to K_*(C^*(X, H'_X)).$$

Moreover, if H_X is also standard, then this map is an isomorphism. So, at the level of K-theory at least, the C^*-algebra $C^*(X, H_X)$ does not depend on the choice of standard X-module H_X. For this reason we shall often suppose that that a particular standard X-module has been chosen, and write $C^*(X)$ in place of $C^*(X, H_X)$.

To define our index map (6.1) it will be convenient to introduce one more C^*-algebra. The set of all pseudolocal, finite propagation operators on a non-degenerate X-module H_X is a $*$-algebra, and we define

$D^*(X, H_X) =$ norm closure of the pseudolocal, finite propagation operators.

As is the case for $C^*(X, H_X)$, the K-theory of $D^*(X, H_X)$ depends only slightly on the choice of H_X. We shall discuss this point in the next section. Here we make the following important observation.

(6.2) LEMMA: *Let X be any proper metric space. The inclusion of $D^*(X, H_X)$ into the C^*-algebra of pseudolocal operators $\Psi_0(X, H_X)$ induces an isomorphism of quotient C^*-algebras*

$$D^*(X, H_X)/C^*(X, H_X) \cong \Psi_0(X, H_X)/\Psi_{-1}(X, H_X).$$

This should be compared with the simple fact that every properly supported pseudodifferential operator can be perturbed by a properly supported smoothing operator so as to have support confined to a strip near the diagonal in $X \times X$.

PROOF: It suffices to show that every pseudolocal operator T can be written as a sum of a finite propagation operator and a locally compact operator.

Choose a partition of unity ψ_j^2 subordinate to a locally finite open cover of X by sets of uniformly bounded diameter. The series

$$T' = \sum_j \psi_j T \psi_j$$

converges in the strong topology. Indeed the partial sums are uniformly bounded in norm and $\sum_j \psi_j T \psi_j v$ converges (in fact it is a finite sum) for any vector v in the dense subset $C_c(X)H \subset H$. Clearly T' is an operator of finite propagation. On the other hand, if φ is a function of compact support, then

$$(T' - T)\varphi = \sum_j [\psi_j, T]\psi_j \varphi$$

is a finite sum of compact operators, hence is compact. Similarly, $\varphi(T - T')$ is compact, and thus $T - T'$ is locally compact. □

Now fix a standard X-module H_X, and consider the long exact K-theory sequence associated to the extension

$$0 \to C^*(X, H_X) \to D^*(X, H_X) \to D^*(X, H_X)/C^*(X, H_X) \to 0. \quad (6.3)$$

The boundary map in this sequence is a map

$$K_{i-1}(D^*(X, H_X)/C^*(X, H_X)) \to K_i(C^*(X, H_X)).$$

But by the lemma above together with Paschke duality (5.2), the first group that appears here is simply $K_i(X)$. Thus we have obtained a homomorphism

$$\mu\colon K_i(X) \to K_i(C^*(X))$$

which is our analytic index map.

If X is compact, then $C^*(X)$ is just the algebra of compact operators and so $K_0(C^*(X)) \cong \mathbf{Z}$. In this case the map μ associates to each operator in $K_0(X)$ its usual Fredholm index in $K_0(C^*(X)) \cong \mathbf{Z}$. On the other hand, if X is a complete Riemannian manifold, D is a Dirac operator on a Clifford bundle S over X, and $[D]$ denotes its K-homology class, then $\mu[D]$ is the index of D in $K_*(C^*(X))$ as defined in [26]. To prove this one needs to verify that if Ψ is a 'chopping function' (as defined in [26]), then $\Psi(D)$ belongs to the algebra $D^*(X, L^2(S))$. This can be accomplished by a straightforward finite propagation speed argument.

We can now formulate a first version of the coarse Baum-Connes conjecture.

(6.4) CONJECTURE: *If X is a uniformly contractible bounded geometry complex then the map $\mu\colon K_*(X) \to K_*(C^*(X))$ is an isomorphism.*

In [26] this was conjectured for all uniformly contractible spaces, but the example of Dranishnikov, Ferry and Weinberger cited earlier shows that this more wide-ranging conjecture is false.

A more general version of the conjecture removes the hypothesis of uniform contractibility. To formulate it, recall that in the paper [16], the group $K^*(C^*(X))$ is made into a bornotopy invariant functor on UBB. Now let X be a complete path metric space. By 3.2, X admits an anti-Čech system $\mathcal{U}_i$ consisting of covers whose geometric realizations $|\mathcal{U}_i|$ are all metric simplicial complexes bornotopy-equivalent to X. The maps

$$\mu_i\colon K_*(|\mathcal{U}_i|) \to K_*(C^*(|\mathcal{U}_i|)) \cong K_*(C^*(X))$$

therefore give in the direct limit a map

$$\mu_\infty\colon KX_*(X) \to K_*(C^*(X)). \tag{6.5}$$

(6.6) CONJECTURE: *For any complete path metric space of bounded coarse geometry, the map* (6.5) *is an isomorphism.*

This is essentially Conjecture 6.30 of [26]. Using Proposition 3.8, it implies Conjecture 6.4.

It is clear from the definitions that

$$\mu_\infty \circ c = \mu$$

where c is the coarsening map. In the Dranishnikov-Ferry-Weinberger example alluded to above, the map c fails to be injective, and therefore the map μ cannot be injective either. This example therefore leaves intact the possibility that Conjecture 6.6 is true for all spaces whether or not they are of bounded geometry, and indeed we will prove the conjecture for certain spaces of non-bounded geometry in this paper. Nevertheless, it seems safer to restrict the general statement to spaces of bounded geometry for the time being.

REMARK: Note that conjecture 6.6 is coarse homotopy invariant, in the following sense. It is easy to see that the maps μ and μ_∞ are natural under coarse maps of X. Now the functors $X \mapsto KX_*(X)$ and $X \mapsto K_*(C^*(X))$ are both coarse homotopy invariant. It follows (by considering the obvious commutative diagram) that if X is coarse homotopy equivalent to X', and the conjecture holds for X, then it holds for X' also. We will make use of a version of this principle in our discussion of hyperbolic metric spaces.

REMARK: Suppose that Γ is a finitely generated discrete group; how is the coarse Baum-Connes conjecture for the underlying metric space $|\Gamma|$ of Γ related to the usual Baum-Connes conjecture for Γ? One answer is as follows: we have seen that $KX_*(|\Gamma|) = K_*(\underline{E}\Gamma)$, the K-homology of the universal space for proper actions of Γ. On the analytic side there is a natural action of Γ on $C^*(|\Gamma|)$, and it is not hard to show that the fixed subalgebra $C^*(|\Gamma|)^\Gamma$ is Morita equivalent to the reduced group C^*-algebra $C^*_r(\Gamma)$. In fact there is a commutative diagram

$$\begin{array}{ccc} K^\Gamma_*(\underline{E}\Gamma) & \longrightarrow & K_*(C^*_r(\Gamma)) \\ \downarrow & & \downarrow \\ K_*(\underline{E}\Gamma) & \longrightarrow & K_*(C^*(|\Gamma|)) \end{array}$$

where the top line is the conjectured isomorphism of the usual Baum-Connes conjecture, the bottom line is the conjectured isomorphism of our coarse version, and the vertical arrows represent a process of forgetting the Γ-equivariance of the situation.

REMARK: We have identified the boundary map in the K-theory exact sequence corresponding to the extension 6.3 with an assembly map. This suggests very strongly that the whole of this K-theory exact sequence should be thought of as an analytic analogue of the (simple, bounded) surgery exact sequence. In particular, $K_*(D^*(X))$ should be thought of as an analytic analogue of the simple structure set for X bounded over itself, and this can be made precise by relating it to the boundedly controlled model for the structure set discussed in [4]. It is an intriguing problem to determine what kind of 'structures' this 'structure set' is classifying.

7. Proof of the conjecture for open cones, nonpositively curved manifolds, and affine buildings

In this section we shall work in the category UBC.

The following is a slight weakening of the notion of 'rescaleable space' in [26].

(7.1) DEFINITION: *A proper metric space X is* scaleable[7] *if there is a continuous and proper map $f: X \to X$, coarsely homotopic to the identity map, such that*

$$d(f(x), f(x')) \leq \frac{1}{2} d(x, x')$$

for all $x, x' \in X$.

Every cone is scaleable, as is every complete, simply connected, non-positively curved Riemannian manifold, every tree and affine building (so long as the trees and buildings are locally finite).

Here is the main result of this section.

(7.2) THEOREM: *If X is a scaleable space then the the index map*

$$\mu: K_*(X) \to K_*(C^*(X))$$

is an isomorphism.

This gives an immediate proof of the following cases of Conjecture 6.6.

(7.3) COROLLARY: *If X is an open cone on a finite-dimensional compact metric space Y then the coarse Baum-Connes conjecture* 6.6 *holds for X.*

PROOF: This follows immediately from the fact that μ is an isomorphism together with the result that coarsening gives an isomorphism on the K-homology of a cone (4.3). □

(7.4) COROLLARY: *If X is a complete, simply connected, non-positively curved Riemannian manifold then the coarse Baum-Connes conjecture (*6.6*) holds for X.*

PROOF: This argument was already given in [15] and [16]; we use the fact that the exponential map is a coarse homotopy equivalence to reduce to the case of Euclidean space, which (for example) may be thought of as a cone on a sphere. □

---

[7] This definition is closely related to 'Lipschitz contractibility' in the sense of Gromov [12, page 25].

(7.5) COROLLARY: *If X is a bounded geometry complex which is either a tree or an affine Bruhat-Tits building then the coarse Baum-Connes conjecture holds for X.*

PROOF: Every tree or affine building is uniformly contractible. So the result follows from 3.8. □

Affine buildings and trees need not be of bounded geometry. However it is not difficult to extend the previous corollary to all buildings and trees by using non-positive curvature to broaden Proposition 3.8 to these cases. Alternatively, one can adapt the argument of the next section so as to apply to buildings. But we shall not pursue this matter here.

We begin by looking the dependence of $K_*(D^*(X, H_X))$ on H_X, and its functoriality in UBC. We shall use the following well-known result of Voiculescu [30].

(7.6) THEOREM: *Let A be a separable C^*-algebra and let H and H' be two separable Hilbert spaces equipped with non-degenerate representations of A. If the representation of A on H' has the property that no non-zero element of A acts as a compact operator on H then there is an isometry $V : H \to H'$ such that $Va - aV$ is a compact operator for every $a \in A$.* □

In the concrete cases of interest to us it is usually possible to construct such isometries V explicitly, but Voiculescu's theorem provides a convenient general framework.

(7.7) LEMMA: *Let X and Y be proper metric spaces, let H_X be a non-degenerate X-module and let and H_Y be a standard Y-module. Let*

$$H_Y^\infty = H_Y \oplus H_Y \oplus H_Y \oplus \dots .$$

If $f : X \to Y$ is a continuous coarse map then there is an isometry

$$W : H_X \to H_Y^\infty$$

such that

$$\mathrm{Supp}(W) \subseteq \{(x, y) : d(f(x), y) < R\},$$

for some $R > 0$, and $\varphi W - W\varphi \circ f$ is compact, for every $\varphi \in C_0(Y)$.

PROOF: Let $\mathcal{U} = \{U_j\}$ and $\mathcal{V} = \{V_k\}$ be locally finite open covers of X and Y by sets of uniformly bounded diameter with the property that f maps each U_j into some V_{k_j}. Let

$$H_X^j = \overline{C_0(U_j)H_X} \quad \text{and} \quad H_Y^k = \overline{C_0(V_k)H_Y}.$$

By Voiculescu's theorem (applied to $A = C(\overline{V_{k_j}})$), for each j there is an isometry $W_j : H_X^j \to H_Y^{k_j}$ such that $\varphi W_j - W_j\varphi \circ f$ is compact, for every $\varphi \in C_0(Y)$.

If $\psi_1^2, \psi_2^2, \ldots$ is any partition of unity subordinate to $\mathcal{U}$. then we can define W by

$$Wf = W_1\psi_1 f \oplus W_2\psi_2 f \oplus \ldots \quad ,$$

which has all the required properties. □

Conjugation with W gives a map

$$\mathrm{Ad}(W)\colon D^*(X, H_X) \to D^*(Y, H_Y^\infty).$$

If W and W' are two isometries satisfying the conclusion of the lemma then $W'W^* \in D^*(Y, H_Y^\infty)$. It follows that the induced maps $\mathrm{Ad}(W)_*$ and $\mathrm{Ad}(W')_*$ on K-theory are equal (see Lemma 3.2 of [16], for example).

Let us apply these remarks to the identity map $1\colon X \to X$. If H and H' are two standard X-modules then we obtain canonical maps

$$K_*(D^*(X, H^\infty)) \leftrightarrow K_*(D^*(X, H'^\infty))$$

which are inverse to one another. In view of this, when convenient we shall drop the term H_X^∞ from our notation, writing $D^*(X)$ in place of $D^*(X, H_X^\infty)$. Selecting one standard module for each X we obtain a functor on UBC, and the same construction makes $K_*(D^*(X)/C^*(X))$ a functor on UBC. By Paschke duality (5.2), this latter functor is simply Kasparov's K-homology, $K_{*-1}(X)$.

(7.8) LEMMA: *The functor $K_*(D^*(X))$ is coarse homotopy invariant.*

PROOF: Let $h\colon X \times [0,1] \to Y$ be a coarse homotopy, and denote by $Z = X \times_h [0,1]$ the product space $X \times [0,1]$ endowed with the warped product metric introduced in [15]. As noted in that paper, to prove that h_0 and h_1 induce the same map on K-theory it suffices to show that the projection map $Z \to X$ induces an isomorphism on K-theory. But consider the diagram

$$\begin{array}{ccccccccc} 0 & \to & C^*(Z) & \to & D^*(Z) & \to & D^*(Z)/C^*(Z) & \to & 0 \\ & & \downarrow & & \downarrow & & \downarrow & & \\ 0 & \to & C^*(X) & \to & D^*(X) & \to & D^*(X)/C^*(X) & \to & 0 \end{array}$$

where the vertical maps are given by conjugation with an isometry chosen as in Lemma 7.7. By the main theorem in [15], the first vertical map induces an isomorphism on K-theory, and by the identifications

$$K_*(D^*(Z)/C^*(Z)) \cong K_{*-1}(Z) \quad \text{and} \quad K_*(D^*(X)/C^*(X)) \cong K_{*-1}(X),$$

the third vertical map also induces an isomorphism on K-theory. So by the Five Lemma the middle vertical map also induces an isomorphism on K-theory. □

(7.9) LEMMA: *The map $\mu\colon K_*(X) \to K_*(C^*(X))$ is an isomorphism if and only if $K_*(D^*(X)) = 0$.*

PROOF: The map μ, as defined in the previous section, is precisely the identification $K_{*-1}(X) \cong K_*(D^*(X)/C^*(X))$ followed by the boundary map in K-theory for the short exact sequence

$$0 \to C^*(X) \to D^*(X) \to D^*(X)/C^*(X) \to 0.$$

So the lemma follows from the long exact sequence in K-theory. □

PROOF OF THEOREM 7.2: By the previous lemma it suffices to show that if X is a scaleable space then $K_*(D^*(X)) = 0$. Let $f\colon X \to X$ be a contraction of X, as in Definition 7.1. For simplicity, let us suppose that f is one-to-one (as it is for cones, if we make the obvious choice for f). We shall deal with the slight extra complications of the general case at the end of this proof. Let S be a countable dense subset of X which is invariant under f. Form the standard X-module

$$H = \ell^2 S \oplus \ell^2 S \oplus \ldots,$$

and let

$$H' = H \oplus H \oplus \ldots.$$

There is an obvious inclusion of H as the first summand in H', and a corresponding inclusion $I\colon D^*(X,H) \to D^*(X,H')$. It suffices to show that this inclusion induces the zero map on K-theory, since by the remarks following Lemma 7.7 it also induces an identification $K_*(D^*(X,H)) \cong K_*(D^*(X,H'))$. Define an isometry on $\ell^2 S$ by mapping the basis vector corresponding to $s \in S$ to the basis vector corresponding to $f(s) \in S$ (it is here that we are using our assumption that f be one-to-one). Taking the direct sum of this isometry with itself infinitely many times we obtain an isometry V of H, and an isometry V' of $\mathcal{H}'$. Conjugation with V' induces the map f_* on $K_*(D^*(X,H'))$, and thanks to coarse homotopy invariance this is the identity map.

Consider now the following map from operators on H to operators on H':

$$I \oplus \mathrm{Ad}(V) \oplus \mathrm{Ad}(V^2) \oplus \cdots : T \mapsto T \oplus \mathrm{Ad}(V)(T) \oplus \mathrm{Ad}(V^2)(T) \oplus \cdots.$$

We observe that this maps $D^*(X,H)$ into $D^*(X,H')$. The reason is that if T has propagation R then $\mathrm{Ad}(V)(T)$ has propagation at most $R/2$, thanks to the fact that f contracts distances by at least $1/2$. So if φ and ψ are functions on X with disjoint compact supports then $\varphi\,\mathrm{Ad}(V^n)(T)\psi = 0$ for large enough n. Having made this observation, the rest of the argument is routine. At the level of K-theory we have

$$\begin{aligned}(I \oplus \mathrm{Ad}(V) \oplus \mathrm{Ad}(V^2) \oplus \cdots)_* \\ &= I_* + (0 \oplus \mathrm{Ad}(V) \oplus \mathrm{Ad}(V^2) \oplus \cdots)_* \\ &= I_* + (\mathrm{Ad}(V) \oplus \mathrm{Ad}(V^2) \oplus \cdots)_* \\ &= I_* + \mathrm{Ad}(V')_*(I \oplus \mathrm{Ad}(V) \oplus \mathrm{Ad}(V^2) \oplus \cdots)_* \\ &= I_* + (I \oplus \mathrm{Ad}(V) \oplus \mathrm{Ad}(V^2) \oplus \cdots)_*.\end{aligned}$$

We subtract $(I \oplus \mathrm{Ad}(V) \oplus \mathrm{Ad}(V^2) \oplus \cdots)_*$ from everything to get $I_* = 0$.

To complete the proof of the theorem we consider the case where f is not necessarily one-to-one. In this case we replace $\ell^2 S$ with $\ell^2 S \times A$, where A is any countably infinite set, and define an isometry on $\ell^2 S \times A$ by mapping the basis vector labelled by (a, s) to a basis vector corresponding to some $(a', f(s))$, where a' is chosen so that this map on basis vectors is one-to-one. Of course $\ell^2 S \times A$ is an X-module in the obvious way, and we proceed as above. □

REMARK: One can give a somewhat different proof of the conjecture for open cones, more in the spirit of certain topological approaches to the Novikov conjecture, as follows. One proves first that the functor $Y \mapsto K_*(C^*(\mathcal{O}Y))$ defines a generalized homology theory on the category of compact metrizable spaces, and that there is a natural transformation from this functor to the functor $Y \mapsto K_{*-1}(Y)$ which is an isomorphism on spheres. One shows further that this functor has the continuity properties summarized in the Steenrod axioms [22, 19], and one then appeals to a uniqueness theorem for Steenrod homology theories to complete the proof.

8. Proof of the conjecture for hyperbolic metric spaces

Gromov [11] has introduced a notion of ***hyperbolicity*** for general metric spaces. A metric space is hyperbolic if it is 'negatively curved on the large scale'. In this section we will show that the coarse Baum-Connes conjecture is true for any (geodesic and locally compact) hyperbolic metric space X.

Our proof will proceed by way of the *Gromov boundary* $Y = \partial_g X$ of the hyperbolic space X. This finite-dimensional compact metrizable space is the boundary of a 'radial compactification' of X, and there are natural 'exponential' and 'logarithm' maps between X and the open cone $\mathcal{O}Y$ on its Gromov boundary. We intend to show that these maps define coarse homotopy equivalences in a certain sense. Since the coarse Baum-Connes conjecture is coarse homotopy invariant, this will prove the conjecture for X by reducing it to the conjecture for $\mathcal{O}Y$, which was proved in the preceding section.

We recall a definition of the Gromov boundary. Fix a basepoint $*$ in X and consider the set of geodesic rays in X originating from $*$. Two such rays are said to be *equivalent* if they lie within a finite distance of one another; it can be shown that there is an absolute constant 8δ such that if γ_1 and γ_2 are equivalent, then $d(\gamma_1(t), \gamma_2(t)) < 8\delta$ for all t. The Gromov product of two such rays may be defined by

$$(\gamma_1|\gamma_2) = \lim_{s,t\to\infty} (\gamma_1(s)|\gamma_2(t))$$

where we note that the expression $(\gamma_1(s)|\gamma_2(t))$ is an increasing function of s and t, and the metric on Y has the property that there are constants $A > 0$ and $\varepsilon > 0$ for which

$$A^{-1}d([\gamma_1],[\gamma_2]) < e^{-\varepsilon(\gamma_1|\gamma_2)} < Ad([\gamma_1],[\gamma_2]).$$

Details of these constructions may be found in [10, 11].

Let $\mathcal{O}Y$ denote the open cone on the Gromov boundary Y. We define the *exponential map* $\exp\colon \mathcal{O}Y \to X$ in the natural way: to a pair $([\gamma], t)$ assign the point $\gamma(t) \in X$. (In order to have a well-defined map we choose one representative from each equivalence class; the choice of representative cannot affect the exponential map by more than 8δ.) The exponential map is highly expansive, but by 4.2 it is always possible to choose a radial shrinking ρ of $\mathcal{O}Y$ so that $\exp\circ\rho$ is a coarse map. The main result of this section is then

(8.1) PROPOSITION: *The map*

$$\exp\circ\rho\colon \mathcal{O}Y \to X$$

induces an isomorphism on coarse K-homology and on the K-theory of the corresponding C^*-algebras.

To avoid wearisome repetition, let us simply say that a coarse map is a *weak equivalence*[8] if it has the property described in proposition 8.1, namely that it induces an isomorphism on coarse K-homology and on the K-theory of the corresponding C^*-algebras.

(8.2) COROLLARY: *The coarse Baum-Connes conjecture 6.6 is true for any hyperbolic metric space X.*

PROOF: Consider the diagram

$$\begin{array}{ccc} KX_*(\mathcal{O}Y) & \longrightarrow & K_*(C^*(\mathcal{O}Y)) \\ \downarrow & & \downarrow \\ KX_*(X) & \longrightarrow & K_*(C^*(X)) \end{array}$$

[8] This is really the analogue of the notion of *homology* equivalence in ordinary topology, rather than of weak *homotopy* equivalence. But since fundamental group problems are not really relevant here, we ask the reader to permit us this convenient abuse of language.

in which the vertical maps are induced by $\exp \circ \rho$, the horizontal maps are the assembly maps μ_∞, and the first horizontal map is an isomorphism by the results of the previous section. □

We will prove proposition 8.1 in two stages: first we will prove that the exponential map is a weak equivalence onto its range, and secondly we will prove that the range of the exponential map is weakly equivalent to the whole space X.. (Notice that the exponential map need not in general be surjective; consider the example of a tree formed from the real line by attaching a branch of length $|n|$ to each integer point n.)

Let $R \subseteq X$ denote the range of the exponential map. A map $\log\colon R \to \mathcal{O}Y$ can be defined by sending a point $x \in R$ to the pair $([\gamma], |x|)$, where γ is some choice of geodesic ray that passes within 8δ of x. Let us check that the choices involved here do not affect the map by more than a bounded amount. If γ_1 and γ_2 are two rays both of which pass within 8δ of x, then by definition, $(\gamma_1|\gamma_2) > |x| - 8\delta$ and so the distance in the open cone between $([\gamma_1], x)$ and $([\gamma_2], x)$ is bounded above by

$$A|x|e^{-\varepsilon(|x|-8\delta)}$$

which is bounded by some constant independent of $|x|$.

(8.3) LEMMA: *The map* $\log\colon R \to \mathcal{O}Y$ *defined above is a coarse map.*

PROOF: It is sufficient to prove that if x_1 and x_2 belong to R and have $|x_1| = |x_2| = r$ say, then there is an estimate of the form

$$d(\log x_1, \log x_2) \le \Phi(d(x_1, x_2))$$

for some universal function Φ. Let γ_1 and γ_2 be rays passing within 8δ of x_1 and x_2 respectively; then

$$(\gamma_1|\gamma_2) \ge (x_1|x_2) - 16\delta \ge r - 16\delta - d(x_1, x_2)/2.$$

Therefore

$$d(\log x_1, \log x_2) \le Are^{-\varepsilon r - 16\delta} e^{\varepsilon d(x_1,x_2)/2}$$

and this is bounded by $Ae^{\varepsilon(16\delta + d(x_1,x_2)/2)}$. □

(8.4) LEMMA: *The composite map* $\log \circ \exp \circ \rho\colon \mathcal{O}Y \to \mathcal{O}Y$ *is weakly equivalent to the identity.*

PROOF: The map $\log \circ \exp$ is bornotopy equivalent to the identity, and the map ρ is coarsely homotopy equivalent to the identity. □

Now recall from [15] the notion of a *generalized coarse homotopy*, this being a map which satisfies all the requirements of definition 2.3 except that it need only be pseudocontinuous rather than continuous. If the space Y is a path metric space, then any generalized coarse homotopy from X to Y can be factored into a coarse homotopy followed by a bornotopy-equivalence.

Thus a generalized coarse homotopy whose range is a path metric space will be a weak equivalence.

(8.5) LEMMA: *The composite map* $\exp\circ\rho\circ\log\colon R \to X$ *is weakly equivalent to the inclusion* $R \hookrightarrow X$.

PROOF: Let ρ_t, with $\rho_0 = \rho$ and $\rho_1 =$ identity, be a linear coarse homotopy of ρ to the identity map. It is not hard to check that $\exp\circ\rho_t\circ\log$ is then a generalized coarse homotopy of $\exp\circ\rho\circ\log$ to $\exp\circ\log$, which in turn is bornotopic to the inclusion map. □

We must now prove that the range R of the exponential map is weakly equivalent to the whole space X. Let us begin by noticing that R is *coarsely convex*[9], that is, there is a constant 16δ such that given any two points x_0 and x_1 in R, any geodesic segment $[x_0, x_1]$ lies within 16δ of R. Now we have

(8.6) LEMMA: *For any coarsely convex subset W of a hyperbolic metric space X, there is a map $\pi\colon X \to W$ such that $d(x,\pi(x)) \leq d(x,W) + \delta$; moreover, this map is unique up to bornotopy.*

Note that such a map π is 'coarsely contractive', that is, $d(\pi(x_0),\pi(x_1)) < d(x_0,x_1) + 2\delta$.

PROOF: It suffices to prove that there is a constant $a > 0$ such that if w_1 and w_2 are points of W within distance $d(x,W)+\delta$ of x, then $d(w_1,w_2) < a$. Let p be the point on the geodesic segment $[w_1,w_2]$ that is closest to x. Then, for $i = 1,2$,

$$d(x,w_i) < d(x,W) + \delta < d(x,p) + \delta + c$$

where c is the constant implied in the statement that W is coarsely convex. However, by Lemma 17 and Proposition 21 of Chapter 2 of [10],

$$d(x,p) < (w_1|w_2)_x + 4\delta < d(x,p) + 5\delta + c - d(w_1,w_2)/2.$$

Therefore

$$d(w_1,w_2) < 10\delta + 2c$$

as required. □

Now take $W = R$ in the above lemma. Define a family of maps $\pi_t\colon X \to X$ by interpolating 'linearly' between π_0, the identity map, and $\pi_1 = \pi$; in other words, $\pi_t(x)$ is the point at distance $td(x,\pi(x))$ along a geodesic segment from x to $\pi(x)$. By the approximate convexity of hyperbolic spaces (Proposition 25 in Chapter 2 of [10]), the maps π_t are uniformly coarsely contractive. In order to show that the π_t give a generalized coarse homotopy

[9] The importance of this point was indicated by M. Gromov in a helpful conversation with one of the authors, for which we are grateful.

between π and the identity map (and thereby to show that the inclusion of R into X is a weak equivalence) it will suffice to prove

(8.7) LEMMA: *The map* $\Pi: X \times [0,1] \to X$ *representing the family* π_t *defined above is proper.*

PROOF: Suppose not. Then there is a sequence of points (x_j, t_j) in $X \times [0,1]$ such that the points (x_j, t_j) are all at mutual distance at least 2 whereas the $\pi_{t_j}(x_j)$ remain in a compact set. Extracting a subsequence and using the contractiveness of the π's, we can suppose that there is a sequence x_j which tends to a point $x_\infty \in \partial X$ while $\pi(x_j)$ remains in a compact set. This, however, is impossible. For the point x_∞ is represented by a geodesic ray γ which belongs to R. Let $y_j = \gamma((x_j|\gamma)) \in R$. By the construction of y_j and the thinness of geodesic triangles, there is a constant 16δ such that $d(x_j, y_j) < |x_j| - |y_j| + 16\delta$. Since $\pi(x_j)$ remains in a compact set, there is a constant C such that $d(x_j, \pi(x_j)) > |x_j| - C$. Since $d(x_j, \pi(x_j))$ is (up to an error of 2δ) the shortest distance from x_j to any point of R, $d(x_j, \pi(x_j)) < d(x_j, y_j) + 2\delta$, and it follows that

$$|y_j| < 18\delta + C.$$

But $|y_j| = (x_j|\gamma) \to \infty$ as $j \to \infty$, because x_j tends to the point at infinity represented by the geodesic ray γ, and this contradiction completes the proof. □

Combining Lemmas 8.4, 8.5, and 8.7, we complete the proof of Proposition 8.1, and therefore of the coarse Baum-Connes conjecture for hyperbolic metric spaces.

9. Appendix: Remarks on ideal boundaries

In this section we want to show how the machinery developed in this paper may be applied to clarify the relationship between the analytic index map μ and the K-homology theory of suitable 'coronas' of a space X. This idea was of some importance in the paper [26], and we would like to discuss it here both because it is rather simple from the present point of view and because the article [13] by one of us to which reference was made at the relevant point of [26] is no longer available in the form in which it was cited.

The situation of interest is the following. X is a proper metric space as usual, and $\overline{X} = X \cup Y$ is a *corona compactification* of X, which means that the continuous functions on $\overline{X}$ restrict to functions on X whose gradient, measured relative to the metric on X, tends to zero at infinity. In this situation proposition 5.18 of [26] (compare also section **3** of [28]) states in our notation that $C^*(X) \subseteq \Psi_0(\overline{X}) \cap \Psi_{-1}(X)$, and thus by relative Paschke duality (5.3) we get a map

$$b : K_i(C^*(X)) \to \tilde{K}_{i-1}(Y).$$

The result that we would like to prove states that $b \circ \mu$ is just the boundary map $\partial\colon K_i(X) \to \tilde{K}_{i-1}(Y)$ in reduced K-homology; this result was used in the proof of Proposition 5.29 in [26]. To prove it, we need to note first that $D^*(X) \subseteq \Psi_0(\overline{X})$; this can be proved by exactly the same methods as those used for $C^*(X)$. Now consider the diagram of six-term exact sequences in K-theory associated to the diagram of extensions

$$\begin{array}{ccccc} C^*(X) & \longrightarrow & D^*(X) & \longrightarrow & D^*(X)/C^*(X) \\ \downarrow & & \downarrow & & \downarrow \\ \Psi_0(\overline{X}) \cap \Psi_{-1}(X) & \longrightarrow & \Psi_0(\overline{X}) & \longrightarrow & \Psi_0(\overline{X})/(\Psi_0(\overline{X}) \cap \Psi_{-1}(X)) \end{array}$$

The connecting map associated to the lower sequence is ∂ by the relative Paschke duality theorem, and that associated to the upper sequence is μ by definition. The two quotient algebras both have K-theory isomorphic to the K-homology of X, and the right-hand vertical map induces the identity[10] on $K_*(X)$. The desired result follows immediately.

References

1. M.F. Atiyah. Global theory of elliptic operators. In *Proceedings of the International Symposium on Functional Analysis*, pages 21–30. University of Tokyo Press, 1969.
2. P. Baum and A. Connes. K-theory for discrete groups. In D. Evans and M. Takesaki, editors, *Operator Algebras and Applications*, pages 1–20. Cambridge University Press, 1989.
3. P. Baum, A. Connes, and N. Higson. Classifying space for proper G-actions and K-theory of group C^*-algebras. In R. S. Doran, editor, *C^*-Algebras: 1943–1993, A Fifty-Year Celebration*, Contemporary Mathematics, vol. 167, pages 241–291. Amer. Math. Soc., Providence, RI, 1994.
4. G. Carlsson and E. K. Pedersen. Controlled algebra and the Novikov conjectures for K and L theory. *Topology*, 1994. To appear.
5. A. Connes, M. Gromov, and H. Moscovici. Group cohomology with Lipschitz control and higher signatures. *Geometric and Functional Analysis*, 3:1–78, 1993.
6. A.N. Dranishnikov, S.C. Ferry, and S. Weinberger. Large Riemannian manifolds which are flexible. Preprint, 1993.
7. P. Fan. *Coarse ℓ_p Geometric Invariants.* PhD thesis, University of Chicago, 1993.
8. S.C. Ferry and E.K. Pedersen. Epsilon surgery theory. In S. Ferry, A. Ranicki, and J. Rosenberg, editors, *Proceedings of the 1993 Oberwolfach Conference on the Novikov Conjecture.* To appear.
9. S.C. Ferry and S.Weinberger. A coarse approach to the Novikov conjecture. In S. Ferry, A. Ranicki, and J. Rosenberg, editors, *Proceedings of the 1993 Oberwolfach Conference on the Novikov Conjecture.* To appear.

[10] One can show that the map is actually an isomorphism, but we do not need this fact.

10. E. Ghys and P. de la Harpe. *Sur les Groupes Hyperboliques d'après Mikhael Gromov*, volume 83 of *Progress in Mathematics*. Birkhäuser, Basel, 1990.
11. M. Gromov. Hyperbolic groups. In S.M. Gersten, editor, *Essays in Group Theory*, pages 75–263. Springer, 1987. Mathematical Sciences Research Institute Publications 8.
12. M. Gromov. Asymptotic invariants for infinite groups. In G.A. Niblo and M.A. Roller, editors, *Geometric Group Theory*, volume 182 of *LMS Lecture Notes*, pages 1–295. Cambridge University Press, 1993.
13. N. Higson. On the relative K-homology theory of Baum and Douglas. Unpublished preprint, 1990.
14. N. Higson. C^*-algebra extension theory and duality. *Journal of Functional Analysis*, 129:349–363, 1995.
15. N. Higson and J. Roe. A homotopy invariance theorem in coarse geometry and topology. *Transactions of the American Mathematical Society*, 345:347–365, 1994.
16. N. Higson, J. Roe, and G. Yu. A coarse Mayer-Vietoris principle. *Mathematical Proceedings of the Cambridge Philosophical Society*, 114:85–97, 1993.
17. S. Hurder. Exotic index theory and the Novikov conjecture. In S. Ferry, A. Ranicki, and J. Rosenberg, editors, *Proceedings of the 1993 Oberwolfach Conference on the Novikov Conjecture*. To appear.
18. W. Hurewicz and H. Wallman. *Dimension Theory*. Princeton, 1941.
19. J. Kaminker and C. Schochet. K-theory and Steenrod homology: applications to the Brown-Douglas-Fillmore theory of operator algebras. *Transactions of the American Mathematical Society*, 227:63–107, 1977.
20. G.G. Kasparov. Topological invariants of elliptic operators I:K-homology. *Mathematics of the USSR — Izvestija*, 9:751–792, 1975.
21. A.N. Kolmogorov and V. M. Tihonirov. ε-entropy and ε-capacity of sets in functional space. *AMS Translations*, 17:277–364, 1961.
22. J.W. Milnor. On the Steenrod homology theory. In S. Ferry, A. Ranicki, and J. Rosenberg, editors, *Proceedings of the 1993 Oberwolfach Conference on the Novikov Conjecture*. Berkeley notes from 1961, reprinted.
23. W. Paschke. K-theory for commutants in the Calkin algebra. *Pacific Journal of Mathematics*, 95:427–437, 1981.
24. A. Ranicki. *Algebraic L-Theory and Topological Manifolds*. Cambridge, 1992.
25. J. Roe. Hyperbolic metric spaces and the exotic cohomology Novikov conjecture. *K-Theory*, 4:501–512, 1991. See also the erratum published in the next volume.
26. J. Roe. Coarse cohomology and index theory on complete Riemannian manifolds. *Memoirs of the American Mathematical Society*, 497, 1993.
27. J. Rosenberg. Analytic Novikov for topologists. In S. Ferry, A. Ranicki, and J. Rosenberg, editors, *Proceedings of the 1993 Oberwolfach Conference on the Novikov Conjecture*. To appear.
28. J. Rosenberg. K and KK: Topology and operator algebras. In W.B. Arveson and R.G. Douglas, editors, *Operator Theory/Operator Algebras and Applications*, volume 51 of *Proceedings of Symposia in Pure Mathematics*, pages 445–480. American Mathematical Society, 1990.
29. E. Spanier. *Algebraic Topology*. McGraw-Hill, 1966.
30. D. Voiculescu. A non-commutative Weyl-von Neumann theorem. *Revue Roumaine des Mathématiques Pures et Appliqués*, 21:97–113, 1976.
31. G. Yu. Baum-Connes conjecture and coarse geometry. Preprint, 1993.

Department of Mathematics, Penn State University, University Park, PA 16802, USA

email: higson@math.psu.edu

Jesus College, Oxford OX1 3DW, England

email: jroe@spinoza.jesus.ox.ac.uk

Exotic index theory and the Novikov conjecture

Steven Hurder[1]

1 Introduction

The *Novikov conjecture*, that the higher signatures of a compact oriented manifold are homotopy invariant, is well-known both for its depth and for the breadth of the research it has spawned. There are two approaches to proving the conjecture for various classes of groups — one via geometric surgery techniques [8, 16, 9, 17], and the other via index theory methods. The index method appeared first in the work of Lusztig [33], which was greatly extended by Miščenko [35, 36] and Kasparov [28, 29, 30, 32]. The two methods are directly compared in Rosenberg's article [42]. Connes introduced an intermediate approach which combines index methods with the more geometric techniques of cyclic cohomology [4, 10, 11, 12, 13, 15].

The index theory approach for a compact oriented manifold with fundamental group Γ traditionally has two steps (cf. the survey by Kasparov [31]). The first is the Kasparov-Miščenko construction of the "parametrized signature class" $\sigma(M) \in K_0(C_r^*(\Gamma))$, and the key point is that this class is a homotopy invariant [25, 27, 34, 35]. Secondly, one shows that operator assembly map $\beta: K_*(B\Gamma) \to K_*(C_r^*\Gamma)$ from the K-homology of $B\Gamma$ to the K-theory of the reduced group C^*-algebra is rationally injective and the image of the K-homology class determined by the signature operator coincides with $\sigma(M)$. Hence, the signature K-homology class is also homotopy invariant and its pairing with group cohomology classes will be invariant — which is the conclusion of the Novikov conjecture.

The Γ-invariance of the lifted operator $\tilde{\mathcal{D}}$ on the universal cover $\widetilde{M}$ of $M \cong B\Gamma$ is used to define a class $\beta[\mathcal{D}] \in K_*(C_r^*\Gamma)$, basically given by the differences of the homotopy classes of the representation of Γ on the kernel and cokernel of $\tilde{\mathcal{D}}$. The "index data" of an operator is localized in its spectrum around 0, while the $C_r^*(\Gamma)$-index contains information related to all the unitary representations of Γ weakly contained in the regular representation, so *a priori* has in it more structure than is necessary for the study of the Novikov conjecture. The idea of index theory in coarse geometry (cf. Roe [40]) is to localize the C^*-index in a neighborhood of infinity spacially,

[1]Supported in part by NSF grant DMS 91-03297.

which by a non-commutative uncertainty principle corresponds to localizing the index at 0 spectrally. The applications of this fundamental idea are still being developed.

The purpose of this note is to prove the injectivity of the map β for a large class of groups using the methods of exotic index theory; that is, index theory for families of coarse metric spaces. The exotic index approach has the advantage that "coarsening the index class" results in technical simplifications which allow the Miščenko-Kasparov method to be applied more broadly. As an example, exotic index methods yield new cases of the Foliation Novikov Conjecture [26, 19].

The topological context for the "index data localized at infinity" of a Γ-invariant elliptic operator is the geometric structure given by a family of parametrized metric spaces with base a model for $B\Gamma$, and fibers the coarse metric type of Γ. The basic idea is to consider such a family as a generalized vector bundle and introduce its *coarse Bott class* which is the analogue in coarse theory of the Thom class (cf. [1]). The data given by a coarse Bott class corresponds to the "dual Dirac" KK-class of Miščenko-Kasparov, but formulated within the coarse geometry category. Our most general result using this technique is that the operator assembly map is injective on K-theory for a group Γ whose associated field of metric spaces admits a coarse Bott class. This note contains a complete proof of this assertion for such Γ when $B\Gamma$ represented by a complete Riemannian manifold.

A Riemannian n-manifold $\widetilde{M}$ is *ultraspherical* (Definition 6.1, [40]) if it admits a proper map $f: \widetilde{M} \to \mathbf{R}^n$ where f has non-zero degree and the gradient ∇f has uniformly bounded pointwise norm on $\widetilde{M}$. Given a complete Riemannian manifold M with fundamental group Γ and universal covering $\widetilde{M}$, we say that M is Γ-*ultraspherical* if the balanced product $\widetilde{M\Gamma} = (\widetilde{M} \times \widetilde{M})/\Gamma$ admits a map $F: \widetilde{M\Gamma} \to TM$ which preserves fibers, and the restriction of F to each fiber $F_x: \{x\} \times \widetilde{M} \to T_xM$ is ultraspherical. The coarse index approach yields the following:

THEOREM 1.1 *Let Γ be a group so that the classifying space $B\Gamma$ is represented by a spin manifold M which admits a complete Riemannian metric so that M is Γ-ultraspherical. Then the operator assembly map $\beta: K_*(B\Gamma) \to K_*(C_r^*\Gamma)$ is injective.*

The proof of Theorem 1.1 is especially transparent when M is compact without boundary. For this reason, we give the proof in the closed case first, in sections 2 through 6. Section 7 discusses the non-compact case.

Note that the hypotheses of Theorem 1.1 do not require that M be of the homotopy type of a CW-complex of finite type. Also note, if $B\Gamma$ is represented by a finite CW complex, then it is also represented by an open complete spin manifold M: embed $B\Gamma$ in $\mathbf{R}^\ell$ for appropriate $\ell \gg 0$ and take M to be a regular neighborhood. Modify the restriction of the Euclidean metric to M so that it rapidly vanishes near the boundary of the closure of M, and one obtains a complete Riemannian metric on TM as well. The key hypothesis which must be verified, is that the universal covering $\widetilde{M}$ of M is ultraspherical for the particular choice of complete metric on TM.

There is a version of the theorem for non-spin manifolds, where the signature operator is used in place of the Dirac operator in the application of Poincaré duality.

The coarse index approach can also be used to prove:

THEOREM 1.2 *Let Γ be a finitely presented group whose classifying space $B\Gamma$ is homotopy equivalent to a finite CW complex. Suppose the universal covering $E\Gamma$ of $B\Gamma$ admits a metrizable, contractible compactification $X\Gamma$ such that*

- *The deck translation action of Γ on $E\Gamma$ extends continuously to $X\Gamma$.*
- *For any compact subset $K \subset E\Gamma$ and sequence of group elements $\{\gamma_n\} \subset \Gamma$ which tend to infinity in the word norm, the translates $K \cdot \gamma_n$ have diameter tending to zero.*

Then the operator assembly map $\beta\colon K_*(B\Gamma) \to K_*(C^*_r\Gamma)$ is injective.

Theorem 1.2 follows by proving that for Γ as in the theorem, there is a complete open manifold V and a free action of Γ on V so that $(V \times V)/\Gamma \to V/\Gamma \cong B\Gamma$ admits a coarse Bott class. This follows via a geometric method similar to that used in [12] to establish the theorem for word hyperbolic groups, which will be discussed in detail in a later paper. Example 8.3 gives the proof of Theorem 1.2 for the very particular case where $\Gamma \cong \pi_1(M)$ and M is a complete open manifold with restrictions on its geometry. The most general case will be treated in a subsequent paper.

The hypotheses of Theorem 1.2 correspond with those assumed in Carlsson-Pedersen [9] and Ferry-Weinberger [17] to prove injectivity of the integral assembly map via controlled surgery theory. The results of Bestvina-Mess [6] show that the hypotheses of Theorem 1.2 are satisfied for groups which are word hyperbolic.

Theorems 1.1, 1.2, 6.2 and 7.6 have similar conclusions to all approaches to the Novikov conjecture using the KK-approach. Kasparov and Skandalis [32] construct a Dirac and dual Dirac class for C^*-algebras associated to Bruhat-Tits buildings, and use this to prove the conjecture for many classes of arithmetic groups. A. Connes, M. Gromov, and H. Moscovici [12] developed the theory of proper Lipschitz cohomology (a work that directly inspired this note) and used it to give the first proof of the Novikov Conjecture for word hyperbolic groups. Our notion of a "coarse Bott class" for a group is clearly related to their

Lipschitz-Poincaré dual of a group (cf. Epilogue to [12].) The non-finite type case of Theorem 1.1 yields a new class of examples where the conjecture has been verified, for the case where $B\Gamma$ is represented by a complete manifold not of finite type.

Here is an outline of this paper. The coarse index class is introduced in section 2. The parametrized corona $\partial_\pi\Gamma$ of the fundamental group is introduced in section 3. In section 4 we develop a pairing between the K-theory $K^1(\partial_\pi\Gamma)$ and the exotic indices. Section 5 considers the exotic index for operators on compact manifolds. The coarse Bott class hypotheses is introduced in section 6, where we prove that the μ-map is injective for a group with this condition. It is immediate that a group satisfying the hypotheses of Theorem 1.1 admits a coarse Bott class, so its proof is completed. Section 7 details the modifications needed in the open case. Finally, in section 8 we show that the groups satisfying the hypotheses of Theorem 1.2 satisfy the hypotheses of Theorem 1.1.

The results of this paper were presented to the meeting on *Novikov Conjectures, Index Theorems and Rigidity* at Oberwolfach, September 1993. There are close parallels between the operator methods of this paper and those used in the most recent controlled K-theory approaches to the Novikov Conjecture [9, 17]. This manuscript is a substantial revision of an earlier preprint (dated April 16, 1993) with these parallels accentuated.

This author is indebted to Jonathan Block, Alain Connes, Steve Ferry, James Heitsch, Nigel Higson, Jonathan Rosenberg, Mel Rothenberg and Shmuel Weinberger for discussions on the Novikov conjecture and related topics, and especially to John Roe for sharing his insights on exotic index theory and its application to the Novikov conjecture.

2 The coarsening map

We begin by recalling some basic ideas of coarse geometry. A *pseudometric* on a set X is a non-negative, symmetric pairing $d(\cdot,\cdot) : X \times X \to [0,\infty)$ satisfying the triangle inequality

$$d(x,z) \leq d(x,y) + d(y,z) \qquad \text{for all } x,y,z \in X.$$

A map $f : X_1 \to X_2$ is said to be *coarsely quasi-isometric* with respect to pseudometrics $d_i(\cdot,\cdot)$ on the X_i if there exists constants $A, B > 0$ so that for all $y, y' \in X_1$

$$A^{-1} \cdot (d_1(y, y') - B) \leq d_2(f(y), f(y')) \leq A \cdot (d_1(y, y') + B) . \tag{2.1}$$

This notion is stricter than Roe's notion of *uniformly bornologous* (cf. Definition 2.1, [40]), where the estimator $A = S(R)$ need not be a linear function of $R = d(y, y')$.

A subset $\mathcal{Z} \subset X$ is called *C-dense* if for each $x \in X$, there exists $n(x) \in \mathcal{Z}$ so that $d(x, n(x)) \leq C$.

A map $f : X_1 \to X_2$ is said to be a *coarse isometry* with respect to pseudometrics $d_i(\cdot, \cdot)$ if f is a coarse quasi-isometry and the image $f(X_1)$ is C-dense in X_2 for some C.

A *net*, or *quasi-lattice*, is a collection of points $\mathcal{N} = \{x_\alpha \mid \alpha \in \mathcal{A}\} \subset X$ so that there are C, $D > 0$ with $\mathcal{N}$ a C-dense set, and distinct points of $\mathcal{N}$ are at least distance D apart. The inclusion of a net $\mathcal{N} \subset X$ is a coarse isometry for the restricted metric on $\mathcal{N}$.

For sections 2 through 6, we will assume that M is a compact spin manifold without boundary, and we fix a Riemannian metric on M. Let $\widetilde{M} \to M$ denote the universal covering of M, with the fundamental group $\Gamma = \pi_1(M)$ acting via translations $T\colon \widetilde{M} \times \Gamma \to \widetilde{M}$ on the right. The Riemannian metric on TM lifts to a Γ-invariant complete Riemannian metric on $T\widetilde{M}$, and $\widetilde{M}$ is endowed with the Γ-invariant path length metric. Introduce the *balanced product* $\widetilde{M}\Gamma = (\widetilde{M} \times \widetilde{M})/\Gamma$, with projections $\pi_i\colon \widetilde{M}\Gamma \to M$ for $i = 1, 2$ onto the first and second factors, respectively. The Γ-invariant metric on $\widetilde{M}$ induces a Γ-invariant metric on the second factor of $\widetilde{M} \times \widetilde{M}$, which descends to a fiberwise metric on $\widetilde{M}\Gamma$: for $x \in M$ and $y, y' \in \widetilde{M}_x$ then $d_x(y, y')$ denotes the "fiberwise" distance from y to y'. Note that that $\widetilde{M}_x = \pi_1^{-1}(x)$ is naturally isometric to $\widetilde{M}$.

For $x \in \widetilde{M}$, let $\Delta(x) \in \widetilde{M}\Gamma$ denote the equivalence class of $x \times x$, with $\pi_1(\Delta(x)) = \pi_2(\Delta(x)) = x$. This factors through a map, also denoted by $\Delta\colon M \to \widetilde{M}\Gamma$, which includes M along the diagonal.

Let $\mathcal{H}(\widetilde{M}, N)$ denote the Hilbert space of L^2-sections of the trivial bundle $\widetilde{M} \times \mathbf{C}^N$, where the inner product is determined by the Riemannian volume form on $\widetilde{M}$. We will often abuse notation and write $\mathcal{H}(\widetilde{M}) = \mathcal{H}(\widetilde{M}, N)$ where $N \gg 0$ is a sufficiently large integer determined as needed. Introduce the continuous field of Hilbert spaces over M,

$$\mathcal{H}_M = \{\mathcal{H}_x = \mathcal{H}(\widetilde{M}_x) \mid x \in M\}$$

where each fiber $\mathcal{H}_x$ is isomorphic to $\mathcal{H}(\widetilde{M})$. Note the inner product of two sections of $\mathcal{H}_M$ is a continuous function on M by definition.

The *Roe algebra* $C^*(X)$ of a complete metric space X equipped with a Radon measure is the norm closure of the $*$-algebra $\mathcal{B}^*_\rho(X)$ of bounded propagation, locally compact, bounded operators on $\mathcal{H}(X)$ (cf. section 4.1 [40]). The compact operators on $\mathcal{H}(X)$ satisfy $\mathcal{K}(\mathcal{H}(X)) \subset C^*(X)$. The inclusion is strict if and only if X is non-compact.

We next extend the construction of the Roe algebra from the open manifold $\widetilde{M}$ to the fibers of the map $\pi: \widetilde{M\Gamma} \to M$. This technical step is one of the main ideas of the coarse approach to the Novikov Conjecture. We begin with a non-equivariant version of the extension. Let $\mathcal{B}^*_\rho(\widetilde{M} \times \widetilde{M}, \pi)$ be the $*$-algebra generated by the continuous families of operators $A = \{A_x \in \mathcal{B}^*_\rho(\widetilde{M}_x) \mid x \in \widetilde{M}\}$ with uniformly bounded propagation. That is, we require that there exists a constant C so that for each $x \in \widetilde{M}$, the operator A_x can be represented by a kernel on the fiber $\widetilde{M}_x$ supported in a C-neighborhood of the diagonal in $\widetilde{M}_x \times \widetilde{M}_x$, and the function $x \mapsto A_x$ is continuous in x for the operator norm on $\mathcal{B}^*_\rho(\widetilde{M}_x) \cong \mathcal{B}^*_\rho(\widetilde{M}) \subset \mathcal{B}(\mathcal{H}(\widetilde{M}))$. Using this last identification we have a natural isomorphism

$$\mathcal{B}^*_\rho(\widetilde{M} \times \widetilde{M}, \pi) \cong C_{up}(\widetilde{M}, \mathcal{B}_\rho(\widetilde{M})) \tag{2.2}$$

where the subscript "*up*" indicates the uniform control on propagation.

The right translation action of Γ induces a right action on operators $\mathcal{B}_\rho(\widetilde{M})$. Given $A = \{A_x\}$ and $\gamma \in \Gamma$ we obtain an action

$$\gamma^* A = (T\gamma)^* A_{T\gamma^{-1}(x)}.$$

Let $\mathcal{B}^*_\rho(\widetilde{M} \times \widetilde{M}, \pi)^\Gamma$ denote the $*$-subalgebra of Γ-invariant elements of $C_{up}(\widetilde{M}, \mathcal{B}_\rho(\widetilde{M}))$, and $C^*(\widetilde{M\Gamma}, \pi) \subset \mathcal{B}(\mathcal{H}_M)$ its C^*-closure for the $*$-representation on $\mathcal{H}(\widetilde{M})$. An element of $C^*(\widetilde{M\Gamma}, \pi)$ is a continuous field of operators $\{A_x \mid x \in M\}$ where for each $x \in M$, $A_x \in C^*(\widetilde{M}_x)$. The C^*-completion commutes with the Γ-action, hence:

PROPOSITION 2.1 *There is a natural isomorphism*

$$C^*(\widetilde{M\Gamma}, \pi) \cong C(\widetilde{M}, C^*(\widetilde{M}))^\Gamma.$$

Let $C^*_r(\Gamma)$ denote the reduced C^*-algebra of Γ, and $\mathcal{B}(\mathcal{H}(\widetilde{M}))^\Gamma \subset \mathcal{B}(\mathcal{H}(\widetilde{M}))$ the C^*-subalgebra of Γ-invariant bounded operators.

PROPOSITION 2.2 *There is an injective map of C^*-algebras, $\mathcal{C}: C^*_r(\Gamma) \to C^*(\widetilde{M\Gamma}, \pi)$. The induced map on K-theory, called the* coarsening map *for Γ,*

$$\chi: K_*(C^*_r(\Gamma)) \to K_*(C^*(\widetilde{M\Gamma}, \pi)), \tag{2.3}$$

is independent of the choices made in defining $\mathcal{C}$.

Proof: Let $\mathcal{K} \subset \mathcal{B}(\mathcal{H}(\widetilde{M}))$ denote the C^*-subalgebra of compact operators. Let $P_\xi \in \mathcal{K}$ denote the projection onto a unit vector $\xi \in \mathcal{H}(\widetilde{M})$. Choose ξ with compact support, so that $P_\xi \in \mathcal{B}_\rho(\mathcal{H}(\widetilde{M}))$. Define the inclusion of C^*-algebras, $P_\xi^* : C_r^*(\Gamma) \to C_r^*(\Gamma)\hat{\otimes}\mathcal{K}$ mapping the unit element $\mathbf{1}_e \mapsto \mathbf{1}_e \otimes P_\xi$.

PROPOSITION 2.3 [13] *There is an isomorphism of C^*-algebras*

$$C_r^*(\Gamma)\hat{\otimes}\mathcal{K} \cong C^*(\widetilde{M})^\Gamma \subset \mathcal{B}(\mathcal{H}(\widetilde{M}))^\Gamma$$

where the image of the projection $\mathbf{1}_e \otimes P_\xi$ is contained in $\mathcal{B}_\rho^*(\widetilde{M}\Gamma, \pi)^\Gamma$.

The natural inclusion of $C^*(\widetilde{M})^\Gamma$ into $C^*(\widetilde{M}\Gamma, \pi)$ is induced from the inclusion of the constant functions in terms of the isomorphism (2.2). The composition

$$\mathcal{C}: C_r^*(\Gamma) \to C_r^*(\Gamma)\hat{\otimes}\mathcal{K} \cong C^*(\widetilde{M})^\Gamma \subset C^*(\widetilde{M}\Gamma, \pi)$$

is the desired map. Note that the choice of ξ affects $\mathcal{C}$ but not χ. □

Here is a simple but typical example of the above construction, for the group $\Gamma = \mathbf{Z}^n$ with model $M = \mathbf{T}^n$ the n-torus, and $\widetilde{M} = \mathbf{R}^n$. Fourier transform gives an isomorphism $C_r^*(\mathbf{Z}^n) \cong C_0(\mathbf{T}^n)$, so that the K-theory $K_*(C_r^*(\mathbf{Z}^n)) \cong K^*(\mathbf{T}^n)$. Higson and Roe [24, 23] have calculated the K-theory $K_*(C^*(X))$ for a wide range of complete metric spaces X, which in the case of $\mathbf{R}^n$ yields $K_*(C^*(\mathbf{R}^n)) \cong K^*(\mathbf{R}^n)$. There is a natural *unparametrized* coarsening map $K_*(C_r^*(\mathbf{Z}^n)) \to K_*(C^*(\mathbf{R}^n))$, which assigns to a $\mathbf{Z}^n$-invariant elliptic operator on $\mathbf{R}^n$ its index in the coarse group $K_*(C_\rho^*(\mathbf{Z}^n)) \cong K_*(C^*(\mathbf{R}^n)) \cong K^*(\mathbf{R}^n)$. This coarsening process retains only the information contained in the top degree in K-homology.

The Higson-Roe results and standard algebraic topology yield

$$K_*(C^*(\mathbf{R}^n\mathbf{Z}^n, \pi)) \cong K^*(\mathbf{T}^n \times \mathbf{R}^n) \cong K^{*+n}(\mathbf{T}^n)$$

where the last isomorphism is cup product with the Thom class for the bundle $\widetilde{M}\Gamma = \mathbf{R}^n\mathbf{Z}^n = (\mathbf{R}^n \times \mathbf{R}^n)/\mathbf{Z}^n \to \mathbf{T}^n$. In this case the parametrized coarsening map

$$\chi: K_*(C(\mathbf{T}^n)) \cong K_*(C_r^*(\mathbf{Z}^n)) \to K_*(C^*(\mathbf{R}^n\mathbf{Z}^n, \pi)) \cong K^*(\mathbf{T}^n \times \mathbf{R}^n)$$

is an isomorphism.

3 Parametrized coronas

In this section, the parametrized (Higson) corona $\partial_\pi \widetilde{M}\Gamma$ is defined, which encapsulates the topological data relevant to the coarse indices of Γ-invariant operators on $\widetilde{M}$. Begin by introducing subalgebras of the bounded continuous functions on $\widetilde{M}\Gamma$:

• $C^b(\widetilde{M}\Gamma)$ denotes the algebra of bounded continuous functions on $\widetilde{M}\Gamma$, with the usual *sup*-norm

$$|h| = \sup_{y \in \widetilde{M}\Gamma} |h(y)|.$$

• $C^b_u(\widetilde{M}\Gamma)$ is the unital algebra of uniformly continuous bounded functions on $\widetilde{M}\Gamma$.

• $C_0(\widetilde{M}\Gamma)$ is the uniform closure of the compactly supported functions $C_c(\widetilde{M}\Gamma)$.

For $x \in M$ and $r > 0$, introduce the *fiberwise variation function*

$$\begin{aligned} \mathbf{Var}_{(x,r)} : C_u(\widetilde{M}_x) &\rightarrow C(\widetilde{M}_x, [0,\infty)) \\ \mathbf{Var}_{(x,r)}(h)(y) &= \sup\{|h(y') - h(y)| \text{ such that } d_x(y,y') \leq r\} \end{aligned}$$

and also set

$$\begin{aligned} \mathbf{Var}_r : C_u(\widetilde{M}\Gamma) &\rightarrow C(\widetilde{M}\Gamma, [0,\infty)) \\ \mathbf{Var}_r(h)(y) &= \mathbf{Var}_{(\pi(y),r)}(h)(y). \end{aligned}$$

We say $f \in C^b_u(\widetilde{M}\Gamma)$ has *uniformly vanishing variation at infinity* if for each $r > 0$ there exists a function $D(f,r) : [0,\infty) \rightarrow [0,\infty)$ so that $d_x(y, \Delta(x)) \geq D(f,r)(\epsilon) \Longrightarrow \mathbf{Var}(x,r)(f)(y) \leq \epsilon$. When M is compact, this condition is equivalent to saying that $\mathbf{Var}_r(f) \in C_0(\widetilde{M}\Gamma)$ for all $r > 0$. Let $C_h(\widetilde{M}\Gamma, \pi) \subset C^b_u(\widetilde{M}\Gamma)$ denote the subspace of functions which have uniformly vanishing variation at infinity.

LEMMA 3.1 (cf. 5.3, [40]) *$C_h(\widetilde{M}\Gamma, \pi)$ is a commutative C^*-algebra.* □

The spectrum of the C^*-algebra $C_h(\widetilde{M}\Gamma, \pi)$, denoted by $\overline{\widetilde{M}\Gamma}$, is a compactification of $\widetilde{M}\Gamma$, where the inclusion of the ideal $C_0(\widetilde{M}\Gamma) \hookrightarrow C_h(\widetilde{M}\Gamma, \pi)$ induces a topological inclusion $\widetilde{M}\Gamma \subset \overline{\widetilde{M}\Gamma}$ as an open dense subset.

Define the *parametrized* (Higson) *corona* of $\widetilde{M}\Gamma$ as the boundary

$$\partial_\pi \widetilde{M}\Gamma = \overline{\widetilde{M}\Gamma} - \widetilde{M}\Gamma,$$

which is homeomorphic to the spectrum of the quotient C^*-algebra

$$C_h(\widetilde{M}\Gamma, \pi)/C_0(\widetilde{M}\Gamma).$$

The functions in $C(M)$ act as multipliers on $C_h(\widetilde{M}\Gamma, \pi)/C_0(\widetilde{M}\Gamma)$, hence the projection $\pi = \pi_1$ extends to continuous maps $\widehat{\pi}\colon \overline{\widetilde{M}\Gamma} \to M$ and $\partial\pi\colon \partial_\pi \widetilde{M}\Gamma \to M$. One can show, using that $C_h(\widetilde{M}\Gamma, \pi)$ is a subalgebra of the uniformly continuous functions on $\widetilde{M}\Gamma$, that both maps $\widehat{\pi}$ and $\partial\pi$ are fibrations. The typical fiber of $\partial\pi\colon \partial_\pi\widetilde{M}\Gamma \to M$ is not metrizable, even in the simplest case where $M = \mathbf{S}^1$ (cf. page 504, [38]).

4 The boundary pairing

The K-theory of the parametrized corona $\partial_\pi\widetilde{M}\Gamma$ naturally pairs with the coarse indices of Γ-invariant operators on $\widetilde{M}$. This pairing is a parametrized extension of N. Higson's observation [20, 21] that the vanishing condition on gradients for C^1-functions on a open complete Riemannian manifold X is the exact analytic condition required to form a pairing between the K-theory of its corona $\partial_h X$ and the K-homology of X. Roe extended this idea to complete metric spaces [40], and a parametric form of this construction was introduced in [26]. We require the following result:

THEOREM 4.1 *There is a natural pairing, for* $p, q = 0, 1$,

$$B_e\colon K_q(C_r^*(\Gamma)) \otimes K^p(\partial_\pi\widetilde{M}\Gamma) \longrightarrow K^{p+q+1}(M). \tag{4.1}$$

For each $[u] \in K^p(\partial_\pi\widetilde{M}\Gamma)$ *evaluation of the pairing on* $[u]$ *yields the* exotic index map

$$B_e[u]\colon K_q(C_r^*(\Gamma)) \longrightarrow K^{q+p+1}(M). \tag{4.2}$$

Proof: The idea of the proof is to construct a natural map

$$\partial_e\colon K^p(\partial_\pi\widetilde{M}\Gamma) \to KK_{p+1}(C^*(\widetilde{M}\Gamma, \pi), C(M)). \tag{4.3}$$

The map (4.1) is obtained as the Kasparov product between the images of the maps (2.3) and (4.3):

$$\begin{aligned}
\chi \otimes \partial_e \quad &: \quad K_q(C_r^*(\Gamma)) \otimes K^p(\partial_\pi\widetilde{M}\Gamma) \\
&\longrightarrow \quad KK_q(\mathbf{C}, C^*(\widetilde{M}\Gamma, \pi)) \otimes KK_{p+1}(C^*(\widetilde{M}\Gamma, \pi), C(M)) \\
&\longrightarrow \quad KK_{p+q+1}(\mathbf{C}, C(M)) \cong K^{p+q+1}(M).
\end{aligned}$$

The map (4.3) is exactly the coarse analogue of the dual Dirac construction. We give the essential points of the construction below — further details can be found in section 6, [26]. For example, one must use care because both the full Roe algebra $C^*(\widetilde{M}\Gamma, \pi)$ and the Higson corona $\partial_\pi\widetilde{M}\Gamma$ are not separable, so it is actually necessary to work with separable subalgebras and direct limits to define the pairing (4.3). This poses no problem, as every class in $K^p(\partial_\pi \widetilde{M}\Gamma)$ factors through a separable quotient of the corona $\partial_\pi \widetilde{M}\Gamma$, and the image of a projection in $C_r^*(\Gamma)$ lies in a separable subalgebra of $C^*(\widetilde{M}\Gamma, \pi)$.

Suppose $p = 1$. Let $[u] \in K^1(\partial_\pi \widetilde{M}\Gamma)$; then we will define $\partial_e[u]$. Represent $[u]$ by a continuous map $u : \partial_\pi \widetilde{M}\Gamma \to U(N)$ for some $N > 0$. Let $j : U(N) \subset GL(N, \mathbf{C}) \subset \mathbf{C}^{N^2}$ be the embedding obtained by the standard coordinates on matrices. Apply the Tietze Extension Theorem to the map from the boundary of $\widetilde{M}\Gamma$ into a regular neighborhood retract of $U(N)$, $u: \partial_\pi \widetilde{M}\Gamma \to U(N) \subset N(U(N)) \subset \mathbf{C}^{N^2}$, to obtain a continuous extension $\widetilde{M}\Gamma \to \mathrm{End}(\mathbf{C}^{N^2})$, and then use the retraction from $N(U(N))$ back to $U(N)$ to obtain a map $\hat{u}: \overline{\widetilde{M}\Gamma} \to \mathrm{End}(\mathbf{C}^{N^2})$ such that $\hat{u}(x)$ is a unitary matrix for x in an open neighborhood of $\partial_\pi \widetilde{M}\Gamma$ in $\overline{\widetilde{M}\Gamma}$.

The Kasparov $(C^*(\widetilde{M}\Gamma, \pi), C(M))$-bimodule that represents $\partial_e[u]$ is constructed from $\hat{u}$ following the method of G. Yu [44], as adapted to the bivariant context. (Yu's method implements a K-theory duality in the index theory of coarse spaces, in the sense of Higson's original paper introducing these ideas [20].) Introduce the KK-cycle:

$$(\mathcal{E}_0 \oplus \mathcal{E}_1, \Phi = \begin{bmatrix} 0 & F^* \\ F & 0 \end{bmatrix}, \phi_0 \oplus \phi_1, \psi) \tag{4.4}$$

whose components are defined as follows:

• Let the integer N be determined by the representative $u : \partial_\pi \widetilde{M}\Gamma \to U(N)$. Recall that $\mathcal{H}_\Gamma$ consists of fiberwise sections of the Hermitian vector bundle $\mathbf{E} = \overline{\widetilde{M}\Gamma} \times \mathbf{C}^N$, so there is a natural module action ϕ_i of $C(M)$ on $\mathcal{H}_\Gamma$ via the induced map $\pi_1^*: C(M) \to C(\widetilde{M}\Gamma)$. Set $\mathcal{E}_i = \mathcal{H}_\Gamma$ for $i = 0, 1$.

• The matrix-valued function $\hat{u}$ induces a map of bundles $F: \mathcal{E}_0 \to \mathcal{E}_1$ which is an Hermitian isomorphism outside of a compact set in $\widetilde{M}\Gamma$.

• Let ψ be the diagonal representation of $C^*(\widetilde{M}\Gamma, \pi)$ on $\mathcal{E}_0 \oplus \mathcal{E}_1$. Note that ψ is a $C(M)$-representation, as the operators in $C^*(\widetilde{M}\Gamma, \pi)$ are fiberwise, and the module action of $C(M)$ is via fiberwise constant multipliers.

It is now routine to check

PROPOSITION 4.2 (cf. Lemma 3 [44]) $(\mathcal{E}_0 \oplus \mathcal{E}_1, \Phi, \phi_0 \oplus \phi_1, \psi)$ *defines a Kasparov* $(C^*(\widetilde{M}\Gamma, \pi), C(M))$*-bimodule. Its* KK*-class*

$$\partial_e[u] \in KK(C^*(\widetilde{M}\Gamma, \pi), C(M))$$

depends only on the class $[u] \in K^1(\partial_\pi \widetilde{M}\Gamma)$.

Proof: To establish that this yields a KK-bimodule, it only remains to check that for all $a \in C^*(\widetilde{M}\Gamma, \pi)$, the graded commutator $[\Phi, \psi(a)]$ is a uniformly fiberwise compact operator. This follows from the compact support of Φ and the bounded propagation property of a; the calculation follows exactly as that for a single complete open manifold by Higson and Roe (cf. Proposition 5.18, [40]). The homotopy invariance of the KK-class follows by standard methods (cf. Chapter VIII, [7]).

A similar construction applies in the even case, where we represent a K-theory class $[\mathbf{E}] \in K^0(\partial_\pi \widetilde{M}\Gamma)$ by a map to a Grassmannian embedded in Euclidean space, then follow the above outline to obtain

$$\partial_e[\mathbf{E}] \in KK_1(C^*(\widetilde{M}\Gamma, \pi), C(M)).$$

5 Coarse geometry and the exotic index

The exotic indices for an elliptic differential operator $\mathcal{D}$ on M are obtained from its Γ-index class by the pairing (4.1) with boundary classes.

Let $\mathcal{D}$ be a first order differential operator defined on the smooth sections $C^\infty(M, \mathbf{E}_0)$ of an Hermitian bundle $\mathbf{E}_0 \to M$, determining a K-homology class $[\mathcal{D}] \in K_*(M)$. The cap product $\cap: K_*(M) \otimes K^*(M) \to K_*(M)$ is realized by an operator pairing: given $[\mathbf{E}] \in K(M)$ the choice of an Hermitian metric on a representative $\mathbf{E} = \mathbf{E}_+ - \mathbf{C}^\ell$ determines an extension of $\mathcal{D}$ to an elliptic first order operator $\mathcal{D} \otimes \mathbf{E}$ acting on $C_c^\infty(M, \mathbf{E}_0 \otimes \mathbf{E})$. The K-homology class $[\mathcal{D} \otimes \mathbf{E}] \in K_*(M)$ represents $[\mathcal{D}] \cap \mathbf{E}$. We recall a fundamental result:

THEOREM 5.1 ([5]; Corollary 4.11 [30]) *Let M be a closed spin-manifold and $\mathcal{D} = \partial$ the Dirac operator on spinors. Then $[\mathcal{D}] \cap : K^*(M) \to K_*(M)$ is the Poincaré duality isomorphism on K-theory.*

We define the Baum-Connes map [3] $\mu[\mathcal{D}]: K^*(M) \to K_*(C_r^*(\Gamma))$ in terms of the lifts of operators to coverings: For $[\mathbf{E}] \in K^*(M)$, the lift of $\mathcal{D} \otimes \mathbf{E}$ to a Γ-invariant differential operator $\widetilde{\mathcal{D} \otimes \mathbf{E}}$ on the compactly supported sections $C_c^\infty(\widetilde{M}, \widetilde{\mathbf{E}_0 \otimes \mathbf{E}})$ has a Γ-index

$$\mu[\mathcal{D}]\,[\mathbf{E}] = [\widetilde{\mathcal{D} \otimes \mathbf{E}}] \in K_*(C_r^*(\Gamma)).$$

When $M = B\Gamma$ the Baum-Connes map can be written simply as $\mu[\mathcal{D}]([\mathbf{E}]) = \beta\,([\mathcal{D}] \cap [\mathbf{E}])$.

DEFINITION 5.2 *For each class $[u] \in K^p(\partial_\pi \widetilde{M}\Gamma)$ and K-homology class $[\mathcal{D}] \in K_q(M)$, define the exotic index map*

$$\mathrm{Ind}_e([u],[\mathcal{D}]) = B_e[u] \circ \mu[\mathcal{D}] : K^*(M) \longrightarrow K^{*+p+q+1}(M). \tag{5.1}$$

The exotic index map (5.1) can be evaluated in terms of ordinary indices of a family [14]. Let $\tilde{\mathcal{D}}_\pi$ denote the differential operator along the fibers of $\pi\colon \widetilde{M}\Gamma \to M$, obtained from the suspension of the Γ-invariant operator $\tilde{\mathcal{D}}$. It determines a Connes-Skandalis index class $\mathrm{Ind}_\pi(\tilde{\mathcal{D}}_\pi) \in KK_*(C_0(\widetilde{M}\Gamma), C(M))$, where in the case of graded operators we have suppressed notation of the grading. Use the boundary map δ in K-theory for $(\widetilde{M}\Gamma, \partial_\pi \widetilde{M}\Gamma)$ and the KK-external product

$$KK(\mathbf{C}, C_0(\widetilde{M}\Gamma)) \otimes KK_*(C_0(\widetilde{M}\Gamma), C(M)) \longrightarrow KK(\mathbf{C}, C(M)) \cong K^*(M)$$

to pair $\delta[u] \in KK(\mathbf{C}, C_0(\widetilde{M}\Gamma))$ with $\mathrm{Ind}_\pi(\tilde{\mathcal{D}}_\pi)$ to obtain a map

$$\mathrm{Ind}_\pi\left(\delta[u] \otimes \tilde{\mathcal{D}}_\pi\right) \in K^*(M).$$

The Kasparov pairing $\delta[u] \otimes (\tilde{\mathcal{D}}_\pi)$ is operator homotopic to a family of fiberwise operators over M which are invertible off a compact set — exactly the class of operators considered by Gromov and Lawson [18, 39]. An elegant homotopy argument of G. Yu for the indices of special vector bundles on an open complete manifold (Theorem 2, [44]) adapts to the parametrized case to relate these two indices in $K^*(M)$:

PROPOSITION 5.3 (Exotic families index theorem)

$$\mathrm{Ind}_e([u],[\mathcal{D}])[\mathbf{E}] = \mathrm{Ind}_\pi\left(\delta[u] \otimes (\widetilde{\mathcal{D} \otimes \mathbf{E}})_\pi\right) \in K^*(M). \tag{5.2}$$

REMARK 5.4 Formula (5.2) gives an expression for the exotic indices as the indices of a family of operators — a decidedly non-exotic index. The notation that $\mathrm{Ind}_e([u],[\mathcal{D}])[\mathbf{E}]$ is an "exotic" index is retained because it results from pairing with coefficients $\delta[u]$ that are the transgression of a K-theory class "at infinity". The description of the index invariants as "exotic" thus parallels exactly the usage in characteristic class theory, where classes arising from boundary constructions are usually called exotic or secondary. It should be noted that almost all other authors now refer to such constructions as *coarse* invariants, due to the role of coarse geometry in their construction.

6 Coarse Bott classes

The notion of a *coarse Bott class* in K-theory for a family of pseudo-metric spaces is introduced, and we prove the map $\mu: K_*(B\Gamma) \to K_*(C_r^*\Gamma)$ is injective for a group Γ with a coarse Bott class coming from the corona. There is a close analogy between the coarse techniques of this section and the constructions of "proper Lipschitz cocycles" introduced by Connes-Gromov-Moscovici [12].

The definition of coarse Bott classes is motivated by the Bott class for vector bundles:

DEFINITION 6.1 *Let $M = \widetilde{M}/\Gamma$ be a connected, compact spin-manifold, and ∂ the Dirac operator on spinors for M. We say that $\Theta \in K^*(\widetilde{M}\Gamma)$ is a* coarse Bott class *if there exists $[u_\Theta] \in K^*(\partial_\pi \widetilde{M}\Gamma)$ so that $\Theta = \delta[u_\Theta]$, and for some $x \in M$, hence for all x, the index of the operator $\partial_x \otimes (\Theta|\widetilde{M}_x)$ on the fiber $\widetilde{M}_x$ satisfies* $\mathrm{Ind}(\partial_x \otimes (\Theta|\widetilde{M}_x)) = \pm 1$.

Here is the main result of this note for the case where $B\Gamma \cong M$ is compact:

THEOREM 6.2 *Let Γ be a group so that the classifying space $B\Gamma$ is represented by a complete, orientable Riemannian spin manifold $M = \widetilde{M}/\Gamma$ such that there is a coarse Bott class $\Theta \in K^*(\widetilde{M}\Gamma)$. Then $\beta: K_*(B\Gamma) \to K_*(C_r^*\Gamma)$ is injective.*

Proof: Let $u_\Theta \in K^*(\partial_\pi \widetilde{M}\Gamma)$ with $\Theta = \delta u_\Theta$ be a coarse Bott class, and let ∂ be the Dirac operator on M. By Theorem 5.1 it will suffice to show that the exotic index map

$$\mathrm{Ind}_e([\mathcal{D}], [u_\Theta]) = B_e[u_\Theta] \circ \mu[\mathcal{D}] : K^*(M) \longrightarrow K^*(M)$$

is injective. This in turn follows from a simple topological observation and a calculation. The observation is contained in the following lemma, which is a simple reformulation of Kasparov's "homotopy lemma" (cf. the proof of **Theorem**, section 6.5, page 193 [30]).

LEMMA 6.3 *For each* $[\mathrm{E}] \in K^*(M)$, $\Theta \otimes \pi_1^*[\mathrm{E}] = \Theta \otimes \pi_2^*[\mathrm{E}] \in K^*(\widetilde{M}\Gamma)$.

Proof: The projections $\pi_i: \widetilde{M}\Gamma \to M$ for $i = 1, 2$ are homotopic, so the induced module actions π_1^* and π_2^* of $K^*(M)$ on $K(\widetilde{M}\Gamma)$ coincide. □

Theorem 6.2 now follows from a calculation using Proposition 5.3 (the exotic families index theorem) applied to the fibration $\pi = \pi_1: \widetilde{M}\Gamma \to M$.

LEMMA 6.4 *For each non-zero* $[\mathbf{E}] \in K^*(M)$*, the exotic index* $\mathrm{Ind}_e([u_\Theta], [\partial])[\mathbf{E}] \in K^*(M)$ *is non-zero.*

Proof:

$$\begin{aligned}
\mathrm{Ind}_e([u_\Theta],[\partial])[\mathbf{E}] &= \mathrm{Ind}_\pi\Big(\Theta \otimes (\widetilde{\partial \otimes \mathbf{E}})_\pi\Big) \\
&= \mathrm{Ind}_\pi\Big(\Theta \otimes \pi_2^*\mathbf{E} \otimes \widetilde{\partial}_\pi\Big) \\
&= \mathrm{Ind}_\pi\Big(\Theta \otimes \pi_1^*\mathbf{E} \otimes \widetilde{\partial}_\pi\Big) \\
&= \mathrm{Ind}_\pi\Big(\Theta \otimes \widetilde{\partial}_\pi\Big) \otimes [\mathbf{E}] \\
&\neq 0.
\end{aligned}$$

The last conclusion follows as $\mathrm{Ind}(\partial \otimes (\Theta|\widetilde{M}_x)) = \pm 1$ for all $x \in M$ implies $\mathrm{Ind}_\pi\Big(\Theta \otimes \widetilde{\partial}_\pi\Big) \in K^*(M)$ is invertible in $K^*(M)$.

Note that one can weakened the notion of a coarse Bott class, requiring only that the indices of the operators $\partial_x \otimes (\Theta|\widetilde{M}_x)$ on the fibers $\widetilde{M}_x$ satisfy $\mathrm{Ind}(\partial_x \otimes (\Theta|\widetilde{M}_x)) \neq 0$. Then in the above proof, $\mathrm{Ind}_\pi\Big(\Theta \otimes \widetilde{\partial}_\pi\Big) \in K^*(M)$ will be invertible in $K^*(M) \otimes \mathbf{Q}$, which implies the rational injectivity of the operator assembly map β in this case. We expect this more general case will prove useful for the study of further classes of groups.

7 The relative case

We extend Theorem 6.2 to the case where $B\Gamma \cong M$ is a complete open manifold. We do not assume that M has finite type, so the main result of this section, Theorem 7.6, implies the Novikov Conjecture for certain classes groups $\Gamma = \pi_1(M)$ which need not be of finite type.

Fix a complete Riemannian metric on TM. As in the compact case, the metric lifts to a Γ-invariant Riemannian metric on $T\widetilde{M}$, and the induced path-length metric on $\widetilde{M}$ is Γ-invariant and complete. Endow the fibers of $\widetilde{M\Gamma} \to M$ with the quotient metric obtained from that on $\widetilde{M}$ — for each $x \in M$ the fiber $\widetilde{M}_x = \pi_1^{-1}(x)$ is isometric to $\widetilde{M}$. Recall that $\Delta\colon M \to \widetilde{M\Gamma}$ is the quotient of the diagonal mapping $\Delta\colon \widetilde{M} \to \widetilde{M} \times \widetilde{M}$.

$\mathcal{B}^*_\rho(\widetilde{M\Gamma}, \pi)$ is the C^*-algebra with typical element a family of operators $A = \{A_x \in \mathcal{B}^*_\rho(\widetilde{M}_x) \mid x \in M\}$ with uniformly bounded propagation, exactly as in § 2 for the compact case, with a natural isomorphism

$$\mathcal{B}^*_\rho(\widetilde{M\Gamma}, \pi) \cong C_{up}(\widetilde{M}, \mathcal{B}_\rho(\widetilde{M}))^\Gamma. \tag{7.1}$$

A uniformly continuous function $\psi: \widetilde{M\Gamma} \to \mathbf{C}$ has Δ-*compact support* if there exists a constant $R_\psi > 0$ so that the support of ψ is contained in the fiberwise R_ψ-tube $N(\Delta, R_\psi)$ around the diagonal $\Delta(M) \subset \widetilde{M\Gamma}$:

$$N(\Delta, R_\psi) = \{y \in \widetilde{M\Gamma} \mid d_x(y, \Delta(\pi_1(y))) \leq R_\psi\}.$$

The construction of the parametrized corona for $\widetilde{M\Gamma}$ remains unchanged in the case where M is non-compact: Γ acts naturally on the Higson corona $\partial_h \widetilde{M}$, and we set

$$\partial_\pi \widetilde{M\Gamma} \equiv (\widetilde{M} \times \partial_h \widetilde{M})/\Gamma.$$

Let $\pi: \partial_\pi \widetilde{M\Gamma} \to M$ be induced by projection onto the first factor.

A key difference in the non-compact case is that we must introduce the "locally finite" K-theory $K^*_{lf}(\partial_\pi \widetilde{M\Gamma})$ of $\partial_\pi \widetilde{M\Gamma}$. An *exhaustion for M* is a nested increasing sequence $\{M_n\}$ of compact submanifolds (with boundary) $M_0 \subset M_1 \subset \cdots M$ whose union is all of M. For each $n \geq 1$ define $\widetilde{M_n\Gamma} = \pi^{-1}(M_n)$ and define its corona to be $\partial_\pi \widetilde{M_n\Gamma} = \partial\pi^{-1}(M_n)$.

We will take for an operating definition

$$\begin{aligned} K^*_{lf}(\partial_\pi \widetilde{M\Gamma}) = \{u \quad &= \quad \lim_{\leftarrow}[u_n] \in \lim_{\leftarrow} K^*(\partial_\pi \widetilde{M_n\Gamma}) \mid \exists R > 0 \;\; \forall n > 0 \\ & \qquad \delta[u_n] \text{ has support in } N(\Delta, n, R)\}, \end{aligned}$$

where $N(\Delta, n, R) = \{y \in \widetilde{M_n\Gamma} \mid d_{\pi(y)}(y, \Delta\pi(y)) \leq R\}$.

The support condition on an element $u = \lim_{\leftarrow}[u_n]$ has a subtle aspect that we comment on. The sequence of representatives $\{[u_n]\}$ are supported in a uniform tube $N(\Delta, R) \subset \widetilde{M\Gamma}$ around $\Delta(M)$. If we suppose that M admits a compactification to a manifold with boundary, $\overline{M}$, then $\widetilde{M\Gamma}$ can be trivialized on a collar neighborhood of $\overline{M} - M$ hence $\widetilde{M\Gamma} \to M$ extends to a fibration over $\overline{M}$, with a section

$$\tau: \overline{M} \to (\widetilde{\overline{M} \times \widetilde{M}})/\Gamma.$$

With respect to the "intuitively natural" section τ, the tube $N(\Delta, R)$ intersects the fiber $\widetilde{M}_x$ in a section which becomes infinitely far from $\tau(x)$ as x tends to the boundary $\overline{M} - M$.

In particular, given any compact set $K \subset \widetilde{M\Gamma}$, the distance between $\{y \in \widetilde{M}_x | d_x(y, \Delta(x)) \leq R\}$ and K tends to infinity as x tends to the boundary $\overline{M} - M$. This latter condition makes sense whether M admits a manifold compactification or not, and suggests that the proper way to consider the fiberwise supports of the sections $\{[u_n]\}$ is that they lie in supports tending to infinity with respect to a geometric "cross-section".

Another unique aspect of the open manifold case, is the need to introduce the algebra of fiberwise operators $A = \{A_x \mid x \in M\}$ on $\widetilde{M}\Gamma$ whose support in a fiber $\widetilde{M}_x$ tends to infinity with respect to the basepoint $\Delta(x)$ as x tends to infinity in M. To be precise, introduce the subalgebra $\mathcal{B}^*_{c\rho}(\widetilde{M}\Gamma, \pi) \subset \mathcal{B}^*_\rho(\widetilde{M}\Gamma, \pi)$ generated by the continuous fields of operators $A = \{A_x \mid x \in M\}$ such that for each Δ-compactly supported function $\psi: \widetilde{M}\Gamma \to \mathbf{C}$, the field $\psi \cdot A = \{\psi \cdot A_x \mid x \in M\}$ has compact support in M. Let $C^*_{c\rho}(\widetilde{M}\Gamma, \pi) \subset C^*_\rho(\widetilde{M}\Gamma, \pi)$ denote the C^*-closure of $\mathcal{B}^*_{c\rho}(\widetilde{M}\Gamma, \pi)$.

PROPOSITION 7.1 *There is an injective map of C^*-algebras, $\mathcal{C}: C^*_r(\Gamma) \to C^*_{c\rho}(\widetilde{M}\Gamma, \pi)$, and the coarsening map for Γ given by the induced map on K-theory,*

$$\chi: K_*(C^*_r(\Gamma)) \to K_*(C^*_{c\rho}(\widetilde{M}\Gamma, \pi)),$$

is natural.

Proof: Choose a compactly supported function $\xi \in C_0(M) \subset L^2(M)$ with L^2-norm one, and compact support $K_\xi \subset M$. The function ξ defines a projection $P_\xi \in \mathcal{K} \cong \mathcal{K}(M)$ which lifts to a Γ-invariant operator $\widetilde{P_\xi}$ on $\widetilde{M}$. Let P_ξ also denote the fiberwise operator in $\mathcal{B}^*_\rho(\widetilde{M}\Gamma, \pi)$. The product $\psi \cdot P_\xi$ then has compact support in the R-tube over the support K_ξ of ξ.

As in the proof of Proposition 2.2, $\widetilde{P_\xi}$ determines a map

$$P^*_\xi: C^*_r(\Gamma) \to C^*_r(\Gamma)\hat{\otimes}\mathcal{K} \subset C^*(\widetilde{M}\Gamma, \pi)$$

whose image similarly consists of classes represented by operators with supports contained in the fibers over the support of ξ, hence lie in $C^*_{c\rho}(\widetilde{M}\Gamma, \pi)$. □

The construction of the boundary pairing B_e proceeds exactly as before, where we note the important detail that by Proposition 7.1 its range is the K-theory with compact supports of M:

THEOREM 7.2 *There is a natural pairing, for $p, q = 0, 1$,*

$$B_e: K_q(C^*_r(\Gamma)) \otimes K^p_{lf}(\partial_\pi \widetilde{M}\Gamma) \longrightarrow K^{p+q+1}(M). \tag{7.2}$$

The definition of coarse Bott classes is modified as follows:

DEFINITION 7.3 *Let $M = \widetilde{M}/\Gamma$ be a connected, complete spin-manifold, and ∂ the Dirac operator on spinors for M. We say that $\Theta \in \lim\limits_{\leftarrow} K^*(\widetilde{M}_n\Gamma)$ is a* coarse Bott class *if there exists $[u_\Theta] \in K^*_{lf}(\partial_\pi \widetilde{M}\Gamma)$ so that $\Theta = \delta[u_\Theta]$, and for some $x \in M$ hence for all x, the fiber index* $\mathrm{Ind}(\partial \otimes (\Theta | V_x)) = \pm 1$.

With the previous set-up, we now have the following extension of Lemma 6.3 within the framework of the fiberwise foliation index theorem applied to $\pi: \widetilde{M}\Gamma \to M$:

LEMMA 7.4 *Let $\Theta \in \varprojlim K^*(\widetilde{M}_n\Gamma)$ be a coarse Bott class. For each $[\mathbf{E}] \in K^*(M)$,*

$$\mathrm{Ind}_\pi\left(\Theta \otimes \widetilde{\partial}_\pi \otimes \pi_2^*\mathbf{E}\right) = \mathrm{Ind}_\pi\left(\Theta \otimes \widetilde{\partial}_\pi \otimes \pi_1^*\mathbf{E}\right). \tag{7.3}$$

Proof: By Theorem 7.2 the index class of the operator $\Theta \otimes \widetilde{\partial}_\pi$ is supported in a compact subset of $\widetilde{M}\Gamma$, so we can apply the homotopy argument from the proof of Lemma 6.3. □

Poincaré duality for a manifold with boundary extends to exhaustion sequences:

THEOREM 7.5 (Corollary 4.11 [30]) *Let M be a complete spin-manifold and ∂ the Dirac operator on spinors. Then $[\partial] \cap : K^*(M) \to K_*(M)$ is the Poincaré duality isomorphism on K-theory.*

Finally, we have established the preliminaries needed to prove the main result of this note, the version of Theorem 6.2 applicable for M an open complete manifold:

THEOREM 7.6 *Let Γ be a group so that the classifying space $B\Gamma$ is represented by a complete, orientable Riemannian spin manifold $M = \widetilde{M}/\Gamma$ such that there is a coarse Bott class $\Theta \in K^*(\widetilde{M}\Gamma)$. Then $\beta: K_*(B\Gamma) \to K_*(C_r^*\Gamma)$ is injective.*

Proof: This follows exactly the outline of the proof of Theorem 6.2, where we use the natural pairing

$$\varprojlim K^*(M_n) \otimes K^*(M) \longrightarrow K^*(M),$$

Poincaré duality as in Theorem 7.5, and the fact that the class

$$\mathrm{Ind}_\pi\left(\Theta \otimes \widetilde{\partial}_\pi\right) \in \varprojlim K^*(M_n)$$

is invertible. □

8 Applications

In this final section, we give three illustrations of the use of Theorem 6.2.

EXAMPLE 8.1 (cocompact lattices) Let $\Gamma \subset G$ be a torsion-free uniform lattice in a connected semi-simple Lie group G with finite center. Choose a maximal compact subgroup $K \subset G$; then the double quotient $M = K\backslash G/\Gamma$ is a compact manifold, as Γ is torsion-free and discrete. The inverse of the geodesic exponential map is a degree-one proper Lipschitz map $\log: V \to \mathbf{R}^n$. The geodesic ray compactification of V, corresponding to the spherical compactification of $\mathbf{R}^n$ via the map log, is a Γ-equivariant quotient of the corona $\partial_h V$ (this is the usual Miščenko calculation), so there is a fiberwise degree-one map $\widetilde{M\Gamma} \to TM$. Thus the usual Bott class in $TM \to M$ pulls back to a Bott class for $\widetilde{M\Gamma}$.

For M a spin manifold, we have satisfied all of the hypotheses of Theorem 6.2 so the operator assembly map $\beta: K_*(B\Gamma) \to K_*(C^*_r\Gamma)$ is injective.

The case where G is simply connected nilpotent Lie group follows similarly, except that the construction of the Γ-equivariant spherical quotient of the corona $\partial_h V$ uses the ***special vector field*** technique of Rees [37].

This example reproduces the Miščenko method [35, 36] in the "coarse language"; the cases discussed below are seen to be just successive embellishments of it.

EXAMPLE 8.2 (non-uniform lattices) Consider a discrete, non-uniform
torsion-free subgroup $\Gamma \subset G$ of a connected semi-simple Lie group G with finite center. Choose a maximal compact subgroup $K \subset G$; then the quotient $\widetilde{M} = K\backslash G$ is ultra-spherical for a K-G bi-invariant Riemannian metric on TG. The double quotient $M = K\backslash G/\Gamma$ is a complete open manifold, to which we apply the methods of section 7.

This calculation reproduces the Kasparov's method [28, 29, 30] for handling non-uniform lattices. Kasparov's construction of the "realizable K-functor for C^*-algebras" has been replaced with the usual inverse limit construction on the K-theory of the corona.

EXAMPLE 8.3 (word hyperbolic groups) We give a sketch of the proof of Theorem 1.2 for the case where Γ satisfies the following hypotheses: there is an open manifold M representing $B\Gamma$ such that $\widetilde{M\Gamma}$ has a coarse Bott class. Assume

- the classifying space $B\Gamma$ has finite type.

• The universal covering $\widetilde{M}$ of M admits a metrizable, contractible compactification $\overline{\widetilde{M}}$.

• The right deck translation action of Γ on V extends continuously to $\overline{\widetilde{M}}$.

• For any compact subset $K \subset E\Gamma$ and sequence of group elements $\{\gamma_n\} \subset \Gamma$ which tend to infinity in the word norm, the translates $K \cdot \gamma_n$ have diameter tending to zero.

These assumptions imply that $\overline{\widetilde{M}}$ is an equivariant quotient of the Higson compactification of $\widetilde{M}$, so it will suffice to construct a class $\Theta \in K^*(\widetilde{M}\Gamma)$ which is the boundary of a class $\theta \in K^*((\widetilde{M} \times \overline{\widetilde{M}})/\Gamma)$.

Use the exponential map to define a local diffeomorphism from an open δ-neighborhood $N(TM, \delta) \subset TM$ of the zero section in TM to an open neighborhood of $*M \subset \widetilde{M}\Gamma$. Choose a Bott class in $H^*(TM)$ supported near the diagonal. It will pull back to a compactly supported class around the diagonal of $\widetilde{M}\Gamma$. This is induced from a relative class in $H^*((\widetilde{M} \times \overline{\widetilde{M}})/\Gamma, (\widetilde{M} \times \delta\overline{\widetilde{M}})/\Gamma)$. Replacing M with $M \times S^1$, we can assume that the restriction of the Bott class to the diagonal is trivial in cohomology, hence the image of the pull-back to $H^*((\widetilde{M} \times \overline{\widetilde{M}})/\Gamma, (\widetilde{M} \times \delta\overline{\widetilde{M}})/\Gamma)$ maps to the trivial class in $H^*((\widetilde{M} \times \overline{\widetilde{M}}))$. (This uses that $(\widetilde{M} \times \overline{\widetilde{M}})/\Gamma$ retracts to the diagonal, or that the added boundary to $\widetilde{M}\Gamma$ is a Z-set.) Choose $\omega \in H^{*-1}((\widetilde{M} \times \delta\overline{\widetilde{M}})/\Gamma)$ which maps to the pull-back class. Now use the isomorphism between rational K-theory and rational cohomology to pull these cohomology classes on the pair $((\widetilde{M} \times \overline{\widetilde{M}})/\Gamma, (\widetilde{M} \times \delta\overline{\widetilde{M}})/\Gamma)$ back to K-theory classes. This yields the θ and Θ required to have a coarse Bott class.

The more general case, where $B\Gamma$ is simply a finite CW complex, requires embedding $B\Gamma$ into Euclidean space, taking a regular neighborhood M of $B\Gamma$, then repeating this argument for the open manifold M. However, one must show that $\widetilde{M}$ admits a Z-set compactification, given that $\widetilde{M}\Gamma$ admits one. The proof of this uses an engulfing technique similar to that in [12].

References

[1] M. F. Atiyah. Bott periodicity and the index of elliptic operators. *Quart. J. Math. Oxford*, 19:113–140, 1968.

[2] M. F. Atiyah. Elliptic operators, discrete groups and von Neumann algebras. pages 43–72, 1976. Astérisque 32-33.

[3] P. Baum and A. Connes. Geometric K-theory for Lie groups and foliations. 1982. IHES preprint.

[4] P. Baum, A. Connes, and N. Higson. Classifying space for proper actions and K-theory of group C^*-algebras. In R. S. Doran, editor, *C^*-algebras: a 50-year Celebration*, pages 241–291, Providence, 1994. Amer. Math. Soc. Contemp. Math., vol. 167.

[5] P. Baum and R. G. Douglas. Toeplitz operators and Poincaré duality. In *Proc. Toeplitz Memorial Conf., Tel Aviv 1981*, pages 137–166, Basel, 1982. Birkhäuser.

[6] M. Bestvina and G. Mess. The boundary of negatively curved groups. *Journal Amer. Math. Soc.*, 4:469–481, 1991.

[7] B. Blackadar. *K-Theory for Operator Algebras.* Springer-Verlag, New York & Berlin, 1986.

[8] S. Cappell. On homotopy invariance of higher signatures. *Invent. Math.*, 33:171–179, 1976.

[9] G. Carlsson and E. Pedersen. Controlled algebra and the Novikov conjectures for K- and L-theory. *Topology.* to appear.

[10] A. Connes. Cyclic cohomology and the transverse fundamental class of a foliation. In H. Araki and E. G. Effros, editors, *Geometric methods in Operator Algebras*, pages 52–144. Pitman, 1986. Research Notes in Math. Series 123.

[11] A. Connes, M. Gromov, and H. Moscovici. Conjecture de Novikov et fibrés presque plats. *C. R. Acad. Sci., Paris*, 310:273–277, 1990.

[12] A. Connes, M. Gromov, and H. Moscovici. Group cohomology with Lipschitz control and higher signatures. *Geometric and Functional Analysis*, 3:1–78, 1993.

[13] A. Connes and H. Moscovici. Cyclic cohomology, the Novikov conjecture and hyperbolic groups. *Topology*, 29:345, 1990.

[14] A. Connes and G. Skandalis. The longitudinal index theorem for foliations. *Publ. Res. Inst. Math. Sci. Kyoto Univ.*, 20:1139–1183, 1984.

[15] T. Fack. Sur la conjecture de Novikov. In *Index Theory of Elliptic Operators, Foliations, and Operator Algebras*, pages 43–102, Providence, 1988. Amer. Math. Soc. Contemp. Math. vol. 70.

[16] F. T. Farrell and W.-C. Hsiang. On Novikov's Conjecture for non-positively curved manifolds, I. *Annals of Math.*, 113:199–209, 1981.

[17] S. Ferry and S. Weinberger. A coarse approach to the Novikov conjecture. *these proceedings*, 1995.

[18] M. Gromov and H. B. Lawson, Jr. Positive scalar curvature and the Dirac operator on complete Riemannian manifolds. *Publ. Math. Inst. Hautes Etudes Sci.*, 58:83–196, 1983.

[19] G. Hector and S. Hurder. The Foliation Novikov Conjecture for totally proper foliations. *in preparation*, 1995.

[20] N. Higson. On the relative K-homology theory of Baum and Douglas. *preprint*, 1989.

[21] N. Higson. C^*-algebra extension theory and duality. *Jour. Func. Anal.*, 129:349–363, 1995.

[22] N. Higson and J. Roe. A homotopy invariance theorem in coarse cohomology and K-theory. *Trans. Amer. Math. Soc.*, 345:347–365, 1994.

[23] N. Higson and J. Roe. On the coarse Baum-Connes conjecture. *these proceedings*, 1995.

[24] N. Higson, J. Roe, and G. Yu. A coarse Mayer-Vietoris principle. *Math. Proc. Camb. Phil. Soc.*, 114:85–97, 1993.

[25] M. Hilsum and G. Skandalis. Invariance par homotopie de la signature à coefficients dans un fibré presque plat. *Jour. Reine Angew. Math.*, 423:73–99, 1992.

[26] S. Hurder. Exotic index theory for foliations. *preprint*, 1993.

[27] J. Kaminker and J. Miller. Homotopy invariance of the analytic signature operator over C^*-algebras. *Jour. Operator Theory*, 14:113–127, 1985.

[28] G. G. Kasparov. K-theory, group C^*-algebras and higher signatures. I, II. 1981. Conspectus, Chernogolovka. Published in these proceedings.

[29] G. G. Kasparov. The operator K-functor and extensions of C^*-algebras. *Math. U.S.S.R.-Izv.*, 16:513–572, 1981.

[30] G. G. Kasparov. Equivariant KK-theory and the Novikov conjecture. *Invent. Math.*, 91:147–201, 1988.

[31] G. G. Kasparov. Novikov's conjecture on higher signatures: the operator K-Theory approach. In *Contemp. Math.*, volume 145, pages 79–99. Amer. Math. Soc., 1993.

[32] G.G. Kasparov and G. Skandalis. Groups acting on buildings, operator K-theory, and Novikov's conjecture. *K-Theory*, 4:303–337, 1991.

[33] G. Lusztig. Novikov's higher signatures and families of elliptic operators. *Jour. Differential Geom.*, 7:229–256, 1972.

[34] A. S. Miščenko. Homotopy invariants of non-simply connected manifold. I: rational invariants. *Math. U.S.S.R.-Izv.*, 4:506–519, 1970.

[35] A. S. Miščenko. Infinite-dimensional representations of discrete groups and higher signature. *Math. U.S.S.R.-Izv.*, 8:85–111, 1974.

[36] A. S. Miščenko. Controlled Fredholm representations. *these proceedings*, 1995.

[37] H. Rees. Special manifolds and Novikov's conjecture. *Topology*, 22:365–378, 1983.

[38] J. Roe. Hyperbolic metric spaces and the exotic cohomology Novikov conjecture. *K-Theory*, 4:501–512, 1991.

[39] J. Roe. A note on the relative index theorem. *Quart. J. Math. Oxford*, 42:365–373, 1991.

[40] J. Roe. Coarse cohomology and index theory on complete Riemannian manifolds. *Memoirs Amer. Math. Soc.*, 104(497), July 1993.

[41] J. Rosenberg. C^*-algebras, positive scalar curvature and the Novikov conjecture. *Publ. Math. Inst. Hautes Etudes Sci.*, 58:197–212, 1983.

[42] J. Rosenberg. Analytic Novikov for topologists. *these proceedings*, 1995.

[43] S. Weinberger. Aspects of the Novikov conjecture. In J. Kaminker, editor, *Geometric and Topological Invariants of Elliptic Operators*, pages 281–297, Providence, 1990. Amer. Math. Soc. Contemp. Math., vol. 105.

[44] G. Yu. K-theoretic indices of Dirac type operators on complete manifolds and the Roe algebra. 1991. Math. Sci. Res. Inst., Berkeley, preprint.

DEPARTMENT OF MATHEMATICS (M/C 249), UNIVERSITY OF ILLINOIS AT CHICAGO, 851 S. MORGAN ST., CHICAGO IL 60607-7045, USA

email: **hurder@boss.math.uic.edu**

Bounded and continuous control

Erik K. Pedersen

The purpose of this note is to clarify the relation between bounded and continuous control as it occurs in various proofs of the Novikov conjecture [2, 3]. In [2] the bounded categories $\mathcal{C}$ of [8] are used to prove the algebraic K-theory Novikov conjecture for certain groups. In [3] the continuously controlled categories $\mathcal{B}$ of [1] are used to prove the algebraic K- and L-theory Novikov conjectures for a larger class of groups.

Since continuous control is finer than bounded control, there should be a forgetful functor from continuously controlled algebra to bounded algebra. This is only true, however, after delooping the categories. A functor from bounded algebra to continuously controlled algebra does not always exist, but if the metric space can be given an ideal boundary satisfying conditions of the type that "bounded is small at infinity", one gets a functor from bounded control to continuous control at infinity. To make this precise let us recall some definitions. We shall use the language of algebraic K- or L-theory, but it is relatively straightforward to modify these considerations to topological K-theory or A-theory, using [5, 4].

Let $\mathcal{A}$ be a small additive category, M a proper metric space.

Definition 1. The *bounded category* $\mathcal{C}(M;\mathcal{A})$ has objects $A = \{A_x\}_{x\in M}$, a collection of objects from $\mathcal{A}$ indexed by points of M, satisfying $\{x|A_x \neq 0\}$ is locally finite in M. A morphism $\phi : A \to B$ is a collection of morphisms $\phi^x_y : A_x \to B_y$ so that there exists $k = k(\phi)$ so $\phi^x_y = 0$ if $d_M(x,y) > k$.

Composition is defined as matrix multiplication. Given a subspace $N \subset M$, we denote the full subcategory with objects A so that $\{x|A_x \neq 0\}$ is contained in a bounded neighborhood of N by $\mathcal{C}(M;\mathcal{A})_N$. It is easy to see that $\mathcal{C}(M;\mathcal{A})$ is $\mathcal{C}(M,\mathcal{A})_N$-filtered in the sense of Karoubi [6]. We denote the quotient category by $\mathcal{C}(M,\mathcal{A})^{>N}$, the category of germs away from N. $\mathcal{C}(M;\mathcal{A})^{>N}$ thus has the same objects as $\mathcal{C}(M,\mathcal{A})$, but morphisms are identified if they agree except in a bounded neighborhood of N.

Next we recall the definition of continuous control. Let (X,Y) be a pair of Hausdorff spaces, Y closed in X, and contained in $\overline{X-Y}$, $\mathcal{A}$ a small additive category.

The author was partially supported by NSF grant DMS 9104026

Definition 2. The *continuously controlled category* $\mathcal{B}(X,Y;\mathcal{A})$ has objects $A=\{A_x\}_{x\in X-Y}$ satisfying $\{x|A_x\neq 0\}$ is locally finite in $X-Y$. A morphism $\phi: A\to B$ is a collection of morphisms $\phi^x_y: A_x\to B_y$ satisfying that for every $z\in Y$ it is continuously controlled i. e. for every neighborhood U of z in X, there is a neighborhood V of z in X so that $x\in V$, $y\notin U$ implies $\phi^x_y=0$ and $\phi^y_x=0$.

Again composition is defined by matrix multiplication. If Z is a closed subset of Y, we denote the full subcategory of $\mathcal{B}(X,Y;\mathcal{A})$ with objects A so that $\partial(\{x|A_x\neq 0\}\subseteq Z$, by $\mathcal{B}(X,Y;\mathcal{A})_Z$. Here ∂ denotes the *topological* boundary. It is easy to see that $\mathcal{B}(X,Y;\mathcal{A})$ is $\mathcal{B}(X,Y\mathcal{A})_Z$-filtered in the sense of Karoubi. We denote the quotient category by $\mathcal{B}(X,Y;\mathcal{A})^{Y-Z}$ or $\mathcal{B}(X,Y;\mathcal{A})^{>Z}$. The objects are the same as in $\mathcal{B}(X,Y;\mathcal{A})$ but morphisms are identified if they agree in a neighborhood of $Y-Z$.

Recall the following from [1] and [2, 3]. Given a small additive category $\mathcal{A}$, we associate a spectrum $K^{-\infty}(\mathcal{A})$ as follows: restricting the morphism to isomorphisms we get a symmetric monoidal category to which there is a functorially associated connective spectrum $K(\mathcal{A})$. The inclusion

$$K(\mathcal{C}(\mathbb{R}^i;\mathcal{A}))\to K(\mathcal{C}(\mathbb{R}^{i+1};\mathcal{A}))$$

is canonically null homotopic in two ways by Eilenberg swindle towards $+\infty$ and $-\infty$ respectively. Hence we get a map

$$\Sigma K(\mathcal{C}(\mathbb{R}^i;\mathcal{A}))\to K(\mathcal{C}(\mathbb{R}^{i+1};\mathcal{A}))$$

or adjointly

$$K(\mathcal{C}(\mathbb{R}^i;\mathcal{A}))\to \Omega K(\mathcal{C}(\mathbb{R}^{i+1};\mathcal{A}))$$

and we define

$$K^{-\infty}(\mathcal{A})=\operatorname{hocolim}(\Omega^i K(\mathcal{C}(\mathbb{R}^i;\mathcal{A})))\ .$$

Similarly if $\mathcal{A}$ is a small additive category with involution, we have the functor $\mathbb{L}^{-\infty}$ to spectra [10]. In this note we shall use spt to denote $K^{-\infty}$ or $\mathbb{L}^{-\infty}$. Using [5] the considerations in here generalize to topological K-theory, replacing additive categories by controlled C^*-algebras, and using [4] they generalize to A-theory, replacing additive categories by appropriate categories with cofibrations and weak equivalences.

Definition 3. Let spt be any one of $K^{-\infty}$, $\mathbb{L}^{-\infty}$, K^{top} or $A^{-\infty}$.

We shall use the language appropriate for $K^{-\infty}$ throughout.

Theorem 4. ([3, 2, 12, 11, 10]) *For any proper metric space M and any subspace $N\subseteq M$ there is a fibration of spectra*

$$\operatorname{spt}(\mathcal{C}(M;\mathcal{A})_N)\to \operatorname{spt}(\mathcal{C}(M;\mathcal{A}))\to \operatorname{spt}(\mathcal{C}(M,\mathcal{A})^{>N})$$

Theorem 5. ([1, 3]) *For any compact Hausdorff pair (X,Y) such that $Y \subset \overline{X-Y}$ and a closed subset $Z \subseteq Y$ there is a fibration of spectra*

$$\mathrm{spt}(\mathcal{B}(X,Y;\mathcal{A})_Z) \to \mathrm{spt}(\mathcal{B}(X,Y;\mathcal{A})) \to \mathrm{spt}(\mathcal{B}(X,Y;\mathcal{A})^{Y-Z}) .$$

Definition 6. A *Steenrod functor* is a homotopy invariant functor

$$\mathrm{St} : \{\text{compact metrizable spaces}\} \to \{\text{spectra}\} ; \; Y \to \mathrm{St}(Y)$$

with the following two properties :

(i) St is strongly excisive, i.e. if $Z \subseteq Y$ there is a fibration

$$\mathrm{St}(Z) \to \mathrm{St}(Y) \to \mathrm{St}(Y/Z) ,$$

(ii) St satisfies the wedge axiom, i.e. for any countable collection $\{X_i\}$ of compact metrizable spaces with wedge $\bigvee\limits_i X_i \subset \prod_i X_i$ the natural map is a homotopy equivalence

$$\mathrm{St}(\bigvee_i X_i) \simeq \prod_i \mathrm{St}(X_i) .$$

Milnor [7] showed that a natural transformation $\mathrm{St}_1 \to \mathrm{St}_2$ of Steenrod functors which induces a homotopy equivalence on S^0 induces a homotopy equivalence $\mathrm{St}_1(Y) \to \mathrm{St}_2(Y)$ for every compact metric space Y.

Theorem 5 is one of the key ingredients in proving:

Theorem 7. ([3]) *For any compact metrizable pair (X,Y) there is a homotopy equivalence*

$$\mathrm{spt}(\mathcal{B}(X,Y;\mathcal{A})) \simeq \mathrm{spt}(\mathcal{B}(CY,Y;\mathcal{A}))$$

where CY is the cone on Y. Moreover,

$$\begin{aligned} \mathrm{st} &: \{\textit{compact metrizable spaces}\} \to \{\textit{spectra}\} ; \\ Y &\to \mathrm{st}(Y) = \mathrm{spt}(\mathcal{B}(CY,Y;\mathcal{A})) \end{aligned}$$

is a Steenrod functor.

Definition 8. The functor

$$M \to h^{l.f.}(M; \Sigma\,\mathrm{spt}(\mathcal{A})) = \mathrm{st}(M_+)$$

is the *homology with locally finite coefficients in the spectrum* $\Sigma\,\mathrm{spt}(\mathcal{A})$ on the category of spaces M with metrizable one-point compactification M_+ and proper maps.

Recall that a metric space M is proper if closed balls are compact. The one-point compactification M_+ of a locally compact space M is metrizable if and only if M is second countable.

Definition 9. Let M be a proper metric space with a compactification $\overline{M}$, such that M is dense in $\overline{M}$. Thus $\partial M = \overline{M} - M$ is the topological boundary of M in $\overline{M}$. The compactification is *small at infinity* if for every $z \in \partial M$, $k \in \mathbb{N}$, and neighborhood U of z in $\overline{M}$, there is a neighborhood V of z in $\overline{M}$ so that $B(x,k) \subset U$ for all $x \in V \cap M$.

The one-point compactification M_+ of a proper metric space M is an obvious example of such a compactification. A more interesting class of examples is provided by radial compactifications of nonpositively curved manifolds.

Proposition 10. *If M is a proper metric space and $\overline{M}$ is a compactification which is small at infinity there is a "forget some control" map*

$$\mathcal{C}(M;\mathcal{A}) \to \mathcal{B}(\overline{M}, \partial M; \mathcal{A})$$

which is the identity on objects.

Proof. One simply observes that bounded morphisms are automatically continuously controlled at points of ∂M. □

We thus get an induced map of spectra

$$\operatorname{spt}(\mathcal{C}(M;\mathcal{A})) \to \operatorname{spt}(\mathcal{B}(\overline{M}, \partial M); \mathcal{A}) \simeq \operatorname{st}(\partial M)$$

Definition 11. The *continuously controlled assembly map* is the connecting map in the Steenrod theory

$$\partial : \operatorname{st}(M_+) = \operatorname{st}(\overline{M}/\partial M) \to \Sigma \operatorname{st}(\partial M)$$

The relation between bounded and continuous control is derived from the following :

Theorem 12. *If M is a proper metric space with a metrizable one-point compactification there is a* bounded assembly map

$$h^{l.f.}(M; \Sigma \operatorname{spt}(\mathcal{A})) \to \Sigma \operatorname{spt}(\mathcal{C}(M;\mathcal{A}))\ .$$

If $\overline{M}$ is a compactification of M which is small at infinity the diagram

$$\begin{array}{ccc} h^{l.f.}(M; \Sigma \operatorname{spt}(\mathcal{A})) & \longrightarrow & \Sigma \operatorname{spt}(\mathcal{C}(M;\mathcal{A})) \\ \| & & \downarrow \\ \operatorname{st}(\overline{M}/\partial M) & \xrightarrow{\ \partial\ } & \Sigma \operatorname{st}(\partial M) \end{array}$$

is commutative.

Proof. We need an appropriate categorical model for $\Sigma\,\mathrm{spt}(\mathcal{C}(M;\mathcal{A}))$ to be able to relate bounded control to continuous control, since continuous control cannot be discussed without the extra parameter in the cone. We shall use ad hoc notation. The following is a diagram of categories as above. The objects are parameterized by $M \times (0,1)$ in all cases, but the control conditions vary. As part of the proof we construct a map from $\mathrm{spt}(\mathcal{B}(C(M_+), M_+; \mathcal{A})) \to \mathrm{spt}(\mathcal{C}(M, \mathcal{A}))$ for any metric space M justifying the claim that there is always a forget control map from continuous to bounded control. Here is a diagram which we proceed to explain

$$\begin{array}{ccccc} \mathcal{A}_1 & \longrightarrow & \mathcal{A}_2 & \longrightarrow & \mathcal{A}_3 \\ \downarrow & & & & \uparrow \\ \mathcal{A}_4 & & \longrightarrow & & \mathcal{A}_5 \end{array}$$

Here

$$\mathcal{A}_4 = \mathcal{B}(C\overline{M}, C\partial M \cup \overline{M}; \mathcal{A})$$
$$\mathcal{A}_5 = \mathcal{B}(\Sigma\overline{M}, \Sigma\partial M; \mathcal{A})$$

and the map $\mathcal{A}_4 \to \mathcal{A}_5$ is induced by collapsing $\overline{M}$. This means that

$$\mathrm{spt}(\mathcal{A}_4) \to \mathrm{spt}(\mathcal{A}_5)$$

is the boundary map

$$\partial : \mathrm{st}(\overline{M} \cup \partial M) \to \mathrm{st}(\Sigma\partial M) \simeq \Sigma\,\mathrm{st}(\partial M)$$

The category

$$\mathcal{A}_1 = \mathcal{B}^b(C\overline{M}, C\partial M \cup \overline{M}; \mathcal{A})$$

is the subcategory of $\mathcal{A}_4$, where the morphisms are required also to have bounded control. The map $\mathcal{A}_1 \to \mathcal{A}_4$ forgets this extra control requirement. It induces a homotopy equivalence because in both cases the subcategories with support at $C(\partial M)$ allow Eilenberg swindles and the germ categories at M are isomorphic. The spectra $\mathrm{spt}(\mathcal{A}_1)$ and $\mathrm{spt}(\mathcal{A}_4)$ are thus models for $h^{l.f.}(M; \Sigma\,\mathrm{spt}(\mathcal{A}))$.

$\mathcal{A}_2$ is the categorical model for $\Sigma\,\mathrm{spt}(\mathcal{C}(M;\mathcal{A}))$. We think of $M \times (0,1)$ as compactified by $\Sigma\partial M$, and require continuous control at the suspension points and bounded control everywhere. The map $\mathcal{A}_1 \to \mathcal{A}_2$ is induced by collapsing $\overline{M}$ and forgetting the continuous control along $\partial M \times (0,1)$. The subcategories of $\mathcal{A}_2$ with support at 0 and 1 respectively, admit obvious Eilenberg swindles and intersect in a category isomorphic to $\mathcal{C}(M;\mathcal{A})$. Hence $\mathrm{spt}(\mathcal{A}_2)$ is a deloop of $\mathrm{spt}(\mathcal{C}(M;\mathcal{A}))$. Finally

$$\mathcal{A}_3 = \mathcal{B}(\Sigma\overline{M}, \Sigma\partial M, p_{\overline{M}}; \mathcal{A})$$

is the category with continuous control at the suspension points, but along $\partial M \times (0,1)$ we only require control in the ∂M-direction. Arguing as for

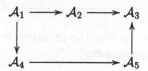

$\mathcal{A}_2$ we find spt($\mathcal{A}_3$) is a deloop of spt($\mathcal{B}(\overline{M}, \partial M; \mathcal{A})$). The map $\mathcal{A}_2 \to \mathcal{A}_3$ is a forget control map sending bounded control to control in the ∂M-direction. The map $\mathcal{A}_2 \to \mathcal{A}_3$ is precisely the deloop of the map $\mathcal{C}(M; \mathcal{A}) \to \mathcal{B}(\overline{M}, \partial M; \mathcal{A})$ mentioned earlier, the map forgetting bounded control, but remembering the control at infinity. To see this consider the diagram

$$\begin{array}{ccc} \mathcal{B}(I, S^0; \mathcal{C}(M; R)) & \longrightarrow & \mathcal{B}(I, S^0; \mathcal{B}(\overline{M}, \partial M; R)) \\ \downarrow & & \downarrow \\ \mathcal{A}_4 & \longrightarrow & \mathcal{A}_5 \end{array}$$

where all the maps are induced by the identity on objects, and the vertical maps induce homotopy equivalences. □

Looping the map

$$h^{l.f.}(M; \Sigma\,\mathrm{spt}(\mathcal{A})) = \mathrm{st}(\overline{M}/\partial M) \to \Sigma\,\mathrm{spt}(\mathcal{C}(M; \mathcal{A}))$$

gives a map from $h^{l.f.}(M; \mathrm{spt}(\mathcal{A}))$ to the bounded K-theory of M. This is the map used in [2], where $h^{l.f.}(M; \mathrm{spt}(\mathcal{A}))$ was given a more homotopy theoretic description.

Remark 13. The bounded L-theory assembly map of [9, p. 327]

$$A^{l.f.} : H^{l.f.}_*(X; \mathbb{L}.(R)) \to L_*(\mathcal{C}(X; R))$$

is the special case of the bounded assembly map of Theorem 12 when spt is the quadratic L-spectrum $\mathbb{L}.(R)$ of a ring with involution R.

In the applications to the Novikov conjecture in [3, 2] $M = E\Gamma$ is a contractible space with a free action of a discrete group Γ.

Definition 14. An *equivariant split injection* of spectra with Γ-actions $S \to T$ is an equivariant map of spectra for which there exists a spectrum R with Γ-action and an equivariant map of spectra $T \to R$ such that the composite $S \to T \to R$ is a homotopy equivalence (non-equivariantly).

Theorem 15. *Let Γ be a discrete group such that $B\Gamma$ is finite and the free Γ-action on $M = E\Gamma$ has a metrizable Γ-equivariant compactification $\overline{M}$ which is small at infinity.*

(i) *The continuously controlled assembly map factors through the bounded assembly map*

$$h^{l.f.}(E\Gamma; \Sigma\,\mathrm{spt}(\mathcal{A})) \to \Sigma\,\mathrm{spt}(\mathcal{C}(E\Gamma; \mathcal{A})) \to \mathrm{st}(\Sigma\partial M)\ .$$

(ii) *If the bounded assembly map*

$$\Sigma\,\mathrm{spt}(\mathcal{C}(E\Gamma;\mathcal{A})) \to \mathrm{st}(\Sigma\partial M)$$

is an equivariant split injection then the Novikov conjecture holds for Γ, *meaning that the assembly map*

$$h(B\Gamma;\mathrm{spt}(R)) \to \mathrm{spt}(R[\Gamma])$$

is a split injection of spectra for any commutative ring R *with* $K_{-i}(R) = 0$ *for sufficiently large* i.

(iii) *If the continuously controlled assembly map*

$$h^{l.f.}(E\Gamma;\Sigma\,\mathrm{spt}(\mathcal{A})) \to \mathrm{st}(\Sigma\partial M)$$

is an equivariant split injection the Novikov conjecture holds for Γ.

(iv) *If* $\overline{M}$ *is contractible then the Novikov conjecture holds for* Γ.

Proof. (i) Immediate from Theorem 12.
(ii) Immediate from [2].
(iii) Immediate from [3]. Alternatively, combine (i) and (ii).
(iv) Proved in [3]. Alternatively, note that if $\overline{M}$ is contractible then the continuously controlled assembly map is an equivariant map which is a homotopy equivalence. □

Conjecture 16. *If* M *is a proper metric space with a contractible compactification* $\overline{M}$ *which is small at infinity then the map* $\mathcal{C}(M;R) \to \mathcal{B}(\overline{M},\partial M;R)$ *induces a homotopy equivalence of spectra* $\mathrm{spt}(\mathcal{C}(M;R)) \to \mathrm{spt}(\mathcal{B}(\overline{M},\partial M;R))$ *for any ring* R *with* $K_{-i}(R) = 0$ *for sufficiently large* i.

Example 17. Conjecture 16 is true if $M = O(K)$ is the open cone on a finite CW complex K, since in this case $\mathrm{spt}(\mathcal{C}(M;R)) \to \mathrm{spt}(\mathcal{B}(\overline{M},\partial M;R))$ is a natural transformation of generalized homology theories which is a homotopy equivalence on the coefficients ([8, 1, 11, 4, 12]).

Example 18. Conjecture 16 is true if M is a complete nonpositively curved manifold ([5]).

Acknowledgment The author would like to thank A. Ranicki for his encouragement to write this paper, and for useful suggestions.

References

1. D. R. Anderson, F. Connolly, S. C. Ferry, and E. K. Pedersen, *Algebraic K-theory with continuous control at infinity.*, J. Pure Appl. Algebra **94** (1994), 25–47.
2. G. Carlsson, *Bounded K-theory and the assembly map in algebraic K-theory*, in these proceedings.
3. G. Carlsson and E. K. Pedersen, *Controlled algebra and the Novikov conjectures for K- and L-theory*, Topology (1993), (To appear).

4. G. Carlsson, E. K. Pedersen, and W. Vogell, *Continuously controlled algebraic K-theory of spaces and the Novikov conjecture*, Report No. 17 1993/1994, Institut Mittag-Leffler (1993), To appear.
5. N. Higson, E. K. Pedersen, and J. Roe, *C^*-algebras and controlled Topology*, preprint.
6. M. Karoubi, *Foncteurs derivees et K-theorie*, Lecture Notes in Mathematics, vol. 136, Springer-Verlag, Berlin-New York, 1970.
7. J. Milnor, *On the Steenrod homology theory*, in these proceedings.
8. E. K. Pedersen and C. Weibel, *A nonconnective delooping of algebraic K-theory*, Algebraic and Geometric Topology, Proceedings Rutgers 1983 (Berlin-New York) (A. Ranicki, N. Levitt, and F. Quinn, eds.), Springer-Verlag, 1985, Lecture Notes in Mathematics 1126, pp. 166–181.
9. A. A. Ranicki, *Algebraic L-theory and Topological Manifolds*, Cambridge Tracts in Mathematics, vol. 102, Cambridge University Press, Cambridge, UK, 1992.
10. ———, *Lower K- and L-theory*, Lond. Math. Soc. Lect. Notes, vol. 178, Cambridge Univ. Press, Cambridge, 1992.
11. J. Roe, *Coarse cohomology and index theory on complete Riemannian manifolds*, Memoirs, vol. 497, Amer. Math. Soc., Providence, R.I., 1993.
12. W. Vogell, *Algebraic K-theory of spaces with bounded control*, Acta Math. **165** (1990), 161–187.

Department of Mathematical Sciences, SUNY at Binghamton, Binghamton, New York 13901, USA

email: erik@math.binghamton.edu

On the homotopy invariance of the boundedly controlled analytic signature of a manifold over an open cone

Erik K. Pedersen, John Roe, and Shmuel Weinberger

1. Introduction

The theorem of Novikov [21], that the rational Pontrjagin classes of a smooth manifold are invariant under homeomorphisms, was a landmark in the development of the topology of manifolds. The geometric techniques introduced by Novikov were built upon by Kirby and Siebenmann [19] in their study of topological manifolds. At the same time the problem was posed by Singer [30] of developing an analytical proof of Novikov's original theorem.

The first such analytic proof was given by Sullivan and Teleman [33, 32, 34], building on deep geometric results of Sullivan [31] which showed the existence and uniqueness of Lipschitz structures on high-dimensional manifolds. (It is now known that this result is false in dimension 4 — see [10].) However, the geometric techniques needed to prove Sullivan's theorem are at least as powerful as those in Novikov's original proof[1]. For this reason, the Sullivan-Teleman argument (and the variants of it that have recently appeared) do not achieve the objective of *replacing* the geometry in Novikov's proof by analysis.

In an unpublished but widely circulated preprint [35], one of us (S.W.) suggested that this objective might be achieved by the employment of techniques from *coarse geometry*. A key part in the proposed proof is played by a certain homotopy invariance property of the 'coarse analytic signature' of a complete Riemannian manifold. We will explain in section 2 below what the coarse analytic signature is, in what sense it is conjectured to be homotopy invariant, and how Novikov's theorem should follow from the conjectured homotopy invariance. In section 3 we will prove the homotopy invariance modulo 2-torsion in the case that the control space is a cone on a finite polyhedron. This suffices for the proof of the Novikov theorem. In section 4 we will show how the methods of this paper can be improved to obtain homotopy invariance 'on the nose'.

[1] See the discussion on page 666 of [8].

Although the coarse signature is an index in a C^*-algebra, our proof is not a direct generalization of the standard proof of the homotopy invariance of signatures over C^*-algebras, as presented for example in [17]. (The assertion to the contrary in [35] is, unfortunately, not correct as it stands.) The problem is this: in the absence of any underlying uniformity such as might be provided by a group action, it becomes impossible to prove that the homotopies connecting two different signatures are represented by *bounded* operators on some Hilbert space. We circumvent this problem by comparing two theories, a 'bounded operator' theory and an 'unbounded' theory, by means of a Mayer-Vietoris argument. Homotopy invariance can be proved in the 'unbounded' theory, but since the two theories are isomorphic, it must hold in the 'bounded' theory as well. A somewhat similar argument was used by the first author in a different context [23].

Our 'unbounded' theory is just boundedly[2] controlled L-theory as defined in [25, 26], and to keep this paper to a reasonable length we will freely appeal to the results of this theory. We do not claim, therefore, that this paper gives a 'purely analytic' proof of Novikov's theorem; indeed, if one is prepared, as we are, to appeal to the homological properties of controlled L-theory, then one can prove Novikov's theorem quite directly and independently of any analysis (see [26], for example). Our point is rather the following. Conjecture 2.2 is a natural analogue of theorems about the homotopy invariance of appropriate kinds of symmetric signatures in other contexts. But those theorems have simple general proofs, whereas in our case the proof is indirect and depends strongly on the hypothesis that the control space possesses appropriate geometric properties, of the kind which can also be used to show the injectivity of the assembly map (compare [5]). Moreover, although 2.2 is a conjecture about C^*-algebras, it appears to be necessary to leave the world of C^*-algebras in order to prove it. It may be that conjecture 2.2 is in fact *false* for more general control spaces X, and, if this were so, then it would suggest the existence of some new kind of obstruction to making geometrically bounded problems analytically bounded also.

It is possible that the special case of conjecture 2.2 that is proved in this paper might be approachable by other, more direct, analytic methods, such as a modification of the almost flat bundle theory of [7, 16]; but it seems that similar questions about gaining appropriate analytic control would have to be addressed.

This paper provides a partial answer to a problem that was raised, in one form or another, by several of the participants at the Oberwolfach conference; see (in the problem session) Ferry-Weinberger, problem 1, Rosenberg, problem 2, and Roe, problems 1 and 2.

[2] Notice that there are two senses in which the word 'bounded' is used in this paper; we may distinguish them as *geometrically bounded* and *analytically bounded*.

We are grateful to Nigel Higson and Jonathan Rosenberg for helpful discussions and comments.

2. The coarse signature

Let X be a proper metric space. We refer to [28, 15, 14] for the construction of the C^*-algebra $C^*(X)$ of locally compact finite propagation operators and of the assembly map $\mu\colon K_*(X) \to K_*(C^*(X))$. We recall that the groups $K_*(C^*(X))$ are functorial under *coarse maps*, that is, proper maps f such that the distance between $f(x)$ and $f(x')$ is bounded by a function of the distance between x and x'. Such maps need not be continuous; but on the subcategory of continuous coarse maps the groups $K_*(X)$ are functorial also, and assembly becomes a natural transformation.

If X is a proper metric space, a *(smooth) manifold over* X is simply a manifold[3] M equipped with a control map $c\colon M \to X$; c must be proper but it need not be continuous. It is elementary that any such manifold M can be equipped with a complete Riemannian metric such that the control map c becomes a coarse map, and that any two such Riemannian metrics can be connected by a path of such metrics.

(2.1) DEFINITION: *Let (M, c) be a manifold over X. The* coarse analytic signature *of M over X is defined as follows: equip M with a Riemannian metric such that c becomes a coarse map, let D_M denote the signature operator on M. According to* [28] *this operator has a 'coarse index'* $\mathrm{Ind}(D_M) \in K_*(C^*(M))$, *which is in fact the image of the K-homology*[4] *class of D under the assembly map μ. We define*

$$\mathrm{Sign}_X(M) = c_*(\mathrm{Ind}\, D_M) \in K_*(C^*(X)).$$

REMARK: For clarity we should make explicit what is meant by the 'signature operator', especially on an odd-dimensional manifold. We use the language of Dirac operators on Clifford bundles (see [27], for example). Let C be the bundle of Clifford algebras associated to the tangent bundle TM, and let $\omega \in C$ be the volume form. Then $\omega^2 = \pm 1$, the sign depending on the dimension of M modulo 4, and so there is a decomposition of C into two eigenspaces $C^+ \oplus C^-$ of ω. If the dimension of M is even, ω anticommutes with the Clifford action of TM on C, and so C becomes a graded Clifford bundle, and we define the *signature operator* to be the associated graded Dirac operator. If the dimension of M is odd, then ω commutes with the action of TM, and so C^+ and C^- individually are Clifford bundles; we define

[3] All manifolds will be assumed to be oriented.

[4] In this paper, we define the *K-homology* of a locally compact metrizable space M to be the Kasparov group [18] $KK(C_0(M), \mathbb{C})$; this is the same as the *locally finite Steenrod K-homology* of X as defined in algebraic topology.

the *signature operator* in this case to be the (ungraded) Dirac operator of C^+.

It is implicit in the definition of the coarse analytic signature that $c_*(\text{Ind } D_M)$ is independent of the choice of Riemannian metric on M. This may be proved by the following development of the theory of [14]. Recall that in that paper the assembly map μ was defined to be the connecting map in the six-term K-theory exact sequence arising from an extension of C^*-algebras

$$0 \to C^*(X) \to D^*(X) \to D^*(X)/C^*(X) \to 0,$$

where $D^*(X)$ is the C^*-algebra of *pseudolocal* finite propagation operators. Using Pashcke's duality theory [22], it was shown that the K-theory of the quotient algebra $D^*(X)/C^*(X)$ was isomorphic to the (locally finite) K-homology of X. Now let us generalize the whole set-up to the case of a manifold M over X, which we write $\begin{pmatrix} M \\ \downarrow \\ X \end{pmatrix}$. We define algebras $C^*\begin{pmatrix} M \\ \downarrow \\ X \end{pmatrix}$ and $D^*\begin{pmatrix} M \\ \downarrow \\ X \end{pmatrix}$ to be the (completions of) the algebras of locally compact and pseudolocal operators, respectively, on M, that have finite propagation when measured in X. Then it is not hard to see on the one hand that the K-theory of $C^*\begin{pmatrix} M \\ \downarrow \\ X \end{pmatrix}$ maps canonically to the K-theory of $C^*(X)$ (in fact it is equal to it if the range of the control map is coarsely dense), and on the other hand that the K-theory of $D^*\begin{pmatrix} M \\ \downarrow \\ X \end{pmatrix} \Big/ C^*\begin{pmatrix} M \\ \downarrow \\ X \end{pmatrix}$ is canonically isomorphic to the (locally finite) K-homology of M. Thus we obtain an assembly map $\mu\colon K_*(M) \to K_*(C^*(X))$ which is independent of any choice of Riemannian metric on M; and naturality of the construction shows that $\mu(D_M)$ coincides with the coarse signature as defined above for any choice of metric.

The usual notions of algebraic topology may be formulated in the category of manifolds over X. In particular we have the concepts of ***boundedly controlled map***, ***boundedly controlled homotopy***, and ***boundedly controlled homotopy equivalence***. A map

$$\begin{array}{ccc} M_1 & \xrightarrow{\quad\varphi\quad} & M_2 \\ & {\scriptstyle c_1}\searrow \qquad \swarrow{\scriptstyle c_2} & \\ & X & \end{array}$$

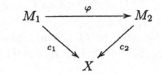

is thus boundedly controlled if φ is continuous, and c_1 is at most a uniformly bounded distance from $c_2 \cdot \varphi$. Similarly a boundedly controlled homotopy is a boundedly controlled map

$$\begin{array}{ccc} M_1 \times I & \xrightarrow{H} & M_2 \\ & {}_{c_1 \cdot p}\searrow \quad \swarrow_{c_2} & \\ & X & \end{array}$$

where p is projection on the first factor. Notice this means that $c_2(H(m \times I))$ has uniformly bounded diameter. The notion of a boundedly controlled homotopy equivalence now follows in an obvious manner.

The following is the homotopy invariance property that we wish to use:

(2.2) CONJECTURE: *If two smooth manifolds M and M' over X are homotopy equivalent by a boundedly controlled orientation-preserving homotopy equivalence, then their coarse analytic signatures agree:*

$$\mathrm{Sign}_X(M) = \mathrm{Sign}_X(M') \in K_*(C^*(X)).$$

We make a few comments on the difficulty in proving this along the lines of the analytic proof in [17]. One wants to construct chain homotopies which intertwine the L^2-de Rham complexes of M and M' (or some simplicial L^2-complexes constructed from an approximation procedure). Because we are working in the world of C^*-algebras, everything has to be a bounded operator on appropriate L^2-spaces. This means that one needs suitable estimates on the derivatives of the maps and homotopies involved, and such estimates do not seem automatically to be available unless one works in a 'bounded geometry' context. This would be appropriate for a proof of the *bi-Lipschitz* homeomorphism invariance of the Pontrjagin classes, but not, it seems, of the topological invariance.

We will now show that conjecture 2.2 implies Novikov's theorem. In fact, we will show a little more, namely that the conjecture implies that the K-homology class of the signature operator of a smooth manifold is invariant under homeomorphism. This is also the conclusion of Sullivan and Teleman's proof [32, 33] which uses Lipschitz approximation. To simplify the later proofs a little, we work away from the prime 2.

(2.3) PROPOSITION: *Suppose that Conjecture 2.2 is true modulo 2-torsion for control spaces X which are open cones on finite polyhedra. Then, if N and N' are homeomorphic compact smooth manifolds, the K-homology signatures of N and N' are equal in $K_*(N) \otimes \mathbb{Z}[\frac{1}{2}]$.*

(2.4) COROLLARY: *In the situation above, the rational Pontrjagin classes of M and M' agree.*

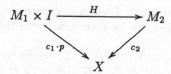

PROOF: By the Atiyah-Singer index theorem [2], the homology Chern character of the signature class is the Poincaré dual of the $\mathcal{L}$-class (which is the same as the Hirzebruch L-class except for some powers of 2); the rational Pontrjagin classes can be recovered from this. □

PROOF: (OF THE PROPOSITION): We begin by considering manifolds M and M' which are smoothly $N \times \mathbb{R}$ and $N' \times \mathbb{R}$ respectively. Let g_{ij} be a Riemannian metric on N. We equip M with a warped product metric of the form

$$dt^2 + \varphi(t)^2 g_{ij} dx^i dx^j,$$

where $\varphi(t)$ is a smooth function with $\varphi(t) = 1$ for $t < -1$ and $\varphi(t) = t$ for $t > 1$. The exact form of the metric is not especially important, provided that it has one cylinder-like and one cone-like end, so that N^+ (that is, N with a disjoint point added) is a natural Higson corona of M. Let $X = M$ considered as a metric space. Then M is obviously boundedly controlled over X (via the identity map!). We use the homeomorphism between N and N' to regard M' as boundedly controlled over X as well. A simple smoothing argument shows that M and M' are boundedly controlledly (smoothly) homotopy equivalent over X; thus by the conjecture their coarse analytic signatures agree.

We now identify the coarse analytic signature of M with the ordinary K-homology signature of N. To do this recall from [28, 14] that there is a natural map defined by Paschke duality

$$b\colon K_*(C^*(M)) \to \widetilde{K}_{*-1}(N^+) = K_{*-1}(N),$$

with the property that $b(\operatorname{Ind} D)$, for any Dirac-type operator D on M, is equal to $\partial[D]$, where $\partial\colon K_*(M) \to K_{*-1}(N)$ is the boundary map in K-homology. (In fact, b is an isomorphism for cone-like spaces such as M, but this fact will not be needed here.) On the other hand, it is a standard result in K-homology that 'the boundary of Dirac is Dirac' [13, 36]. It is not true that 'the boundary of signature is signature', but this *is* true up to powers of 2. In fact $\partial[D_M] = k[D_N]$, where $[D_N]$ is the class of the signature operator of N and k is 2 if M is even-dimensional, 1 if M is odd-dimensional. We will discuss the factor k in section 4.

By a similar argument we may identify the coarse analytic signature of M' with k times the ordinary K-homology signature of N', pulled back to $K_*(N)$ via the homeomorphism $N' \to N$. The desired result therefore follows from the equality of these two signatures. □

REMARK: With a little more effort, this argument might be made to work with the hypothesis that N and N' are ε-controlled homotopy equivalent for all ε, rather than homeomorphic. Of course one knows from the α-approximation theorem [6] that N and N' are in fact homeomorphic under this hypothesis, but the point is that one can avoid appealing to this geometric result.

In the next section we will need to know that the coarse analytic signature is bordism invariant. In other words, we will require

(2.5) PROPOSITION: *Suppose that N is an X-bounded manifold which is the boundary of an X-bounded manifold-with-boundary M. Then* $\mathrm{Sign}_X(N)$ *is a 2-torsion element in $K_*(C^*(X))$.*

PROOF: Let M° denote the interior of M. A portion of the exact sequence of K-homology is

$$K_{*+1}(M^\circ) \to K_*(N) \to K_*(M).$$

As remarked above, there is an integer k equal to 1 or 2 such that $\partial[D_{M^\circ}] = k[D_N]$, and it follows from exactness that the image of $[D_N]$ in $K_*(M)$ is a k-torsion element. But from the naturality of the assembly map there is a commutative diagram

$$\begin{array}{ccc} K_*(N) & \xrightarrow{\qquad\qquad} & K_*(M) \\ & \searrow \quad \swarrow & \\ & K_*(C^*(X)) & \end{array}$$

in which both vertical arrows are assembly maps. Therefore the image of $[D_N]$ under the assembly map, that is the coarse analytic signature, is k-torsion. □

3. Proof of homotopy invariance modulo 2-torsion

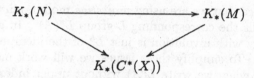

We begin by recalling the definition of the L^h-groups of an additive category with involution[25]. Given an additive category $\mathfrak{U}$ an involution on $\mathfrak{U}$ is a contravariant functor $*\colon \mathfrak{U} \to \mathfrak{U}$, sending U to U^*, and a natural equivalence $** \cong 1$. One of the defining properties of an additive category is that the Hom-sets are abelian groups, that is $\mathbb{Z}$-modules. All the categories that we will consider will have the property that the Hom-sets are in fact modules over the ring $\mathbb{Z}[i, \frac{1}{2}]$, and we will make this assumption from now on. This yields two simplifications in L-theory: the existence of $i = \sqrt{-1}$ makes L-theory 2-periodic, since dimensions n and $n+2$ get identified through scaling by i, and the existence of $\frac{1}{2}$ removes the difference between quadratic and symmetric L-theory. We therefore get the following description of L-theory. In degree 0 an element is given as an isomorphism $\varphi\colon A \to A^*$ satisfying $\varphi = \varphi^*$. Elements of the form $B \oplus B^* \cong B^* \oplus B$, with the obvious isomorphism, are considered trivial, and L_0^h is the Grothendieck construction determining whether a selfadjoint isomorphism is stably conjugate to a trivial isomorphism. In the definition of L^h_2 the condition $\varphi = \varphi^*$ is replaced by $\varphi = -\varphi^*$ but in the presence of i these groups become scale equivalent. In odd degrees the groups are given as automorphisms of trivial forms.

REMARK: Suppose the additive category $\mathfrak{U}$ is the category of finitely generated projective modules over a C^*-algebra A and the involution is given by the identity on objects and the $*$-operation on morphisms. One defines the projective L-groups $L^p_*(A)$ to equal $L^h_*(\mathfrak{U})$ for this category $\mathfrak{U}$. In this situation, the availability of the Spectral Theorem for C^*-algebras allows one to separate out the positive and negative eigenspaces of a nondegenerate quadratic form and thus to assign a signature in $K_*(A)$ (a formal difference of projections) to any element of $L^p_*(A)$. This construction goes back to Gelfand and Mischenko [11], and a very careful account may be found in Miller [20]; the exposition in Rosenberg [29] is couched in language similar to ours, and also includes a proof that one obtains in this way a homomorphism $L^p_*(A) \to K_*(A)$, which becomes an isomorphism after inverting 2.

REMARK: Notice that we are using *projective* modules in the above statement, so one calls the corresponding L-group $L^p(A)$. In general L^p of an additive category with involution is just L^h of the idempotent completion of the category. To simplify these issues we will work modulo 2-torsion, so from now on when we write $L(A)$ without upper index we shall mean $L^h(A)\otimes\mathbb{Z}[\frac{1}{2}]$, noting that by the Ranicki-Rothenberg exact sequences tensoring with $\mathbb{Z}[\frac{1}{2}]$ removes the dependency on the upper decoration. To retain the above mentioned isomorphism we obviously have to tensor K-theory with $\mathbb{Z}[\frac{1}{2}]$ as well.

We now recall the (geometrically) bounded additive categories defined in [24]. Let X be a metric space, and R a ring with anti-involution. This turns the category of left R-modules into an additive category with involution, since the usual dual of a left R-module is a right R-module, but by means of the anti-involution this may be turned into a left R-module.

The reader should keep in mind the model case in which X is the infinite open cone $\mathcal{O}(K)$ on a complex $K \subseteq S^n \subset \mathbb{R}^{n+1}$ and $R = \mathbb{C}$. The category $\mathfrak{C}_X(R)$ is defined as follows:

(3.1) DEFINITION: *An* object A *of* $\mathfrak{C}_X(R)$ *is a collection of finitely generated based free right R-modules* A_x*, one for each* $x \in X$*, such that for each ball* $C \subset X$ *of finite radius, only finitely many* A_x*,* $x \in C$*, are nonzero. A* morphism $\varphi\colon A \to B$ *is a collection of morphisms* $\varphi^x_y\colon A_x \to B_y$ *such that there exists* $k = k(\varphi)$ *such that* $\varphi^x_y = 0$ *for* $d(x,y) > k$.

The composition of $\varphi\colon A \to B$ *and* $\psi\colon B \to C$ *is given by* $(\psi\circ\varphi)^x_y = \sum_{z\in X}\psi^z_y\varphi^x_z$*. Note that* $(\psi\circ\varphi)$ *satisfies the local finiteness and boundedness* *conditions whenever ψ and φ do.*

(3.2) DEFINITION: *The* dual *of an object A of $\mathfrak{C}_X(R)$ is the object A^* with $(A^*)_x = A^*_x = \mathrm{Hom}_R(A_x, R)$ for each $x \in X$. A^*_x is naturally a left R-module, which we convert to a right R-module by means of the anti-involution. If $\varphi\colon A \to B$ is a morphism, then $\varphi^*\colon B^* \to A^*$ and $(\varphi^*)^x_y = h \circ \varphi^y_x$, where $h\colon B_x \to R$ and $\varphi^y_x\colon A_y \to B_x$. φ^* is bounded whenever φ is. Again, φ^* is naturally a left module homomorphism which induces a homomorphism of right modules $B^* \to A^*$ via the anti-involution.*

If we choose a countable set $E \subset X$ such that for some k the union of k-balls centered at points of E covers X, then it is easy to see that the categories $\mathfrak{C}_E(R)$ and $\mathfrak{C}_X(R)$ are equivalent.

It is convenient to assume that such a choice has been made once and for all. Then we may think of the objects of $\mathfrak{C}_X(\mathbb{C})$ as based complex vector spaces with basis a subset of $E \times \mathbb{N}$ satisfying certain finiteness conditions. Any based complex vector space has a natural inner product, and therefore a norm, and we define a morphism in $\mathfrak{C}_X(\mathbb{C})$ to be *analytically bounded* if it becomes a bounded operator when its domain and range are equipped with these natural ℓ^2 norms.

(3.3) DEFINITION: *The category $\mathfrak{C}^{b.o.}_X(\mathbb{C})$ has the same objects as $\mathfrak{C}_X(\mathbb{C})$, but the morphisms have to satisfy the further restriction that they define analytically bounded operators on $\ell^2(E \times \mathbb{N})$*

It is apparent that there is a close connection between the category $\mathfrak{C}^{b.o.}_X(\mathbb{C})$ and the C^*-algebra $C^*(X)$. In fact, the way we have arranged things any object A in the category $\mathfrak{C}^{b.o.}_X(\mathbb{C})$ can be thought of as a projection in $C^*(X)$ defined by the generating set for A and hence as a projective $C^*(X)$-module, and an endomorphism of A respects the $C^*(X)$-module structure. Since the involution on $\mathfrak{C}^{b.o.}_X(\mathbb{C})$ is given by duality, it corresponds to the $*$-operation on $C^*(X)$. Hence we get a map

$$L_*(\mathfrak{C}^{b.o}_X(\mathbb{C})) \to L_*(C^*(X)) = K_*(C^*(X)).$$

Similarly the forgetful functor $\mathfrak{C}^{b.o.}_X(\mathbb{C}) \to \mathfrak{C}_X(\mathbb{C})$ induces a map

$$L_*(\mathfrak{C}^{b.o}_X(\mathbb{C})) \to L_*(\mathfrak{C}_X(\mathbb{C})).$$

Notice that whenever we have a manifold $\begin{pmatrix} M \\ \downarrow \\ X \end{pmatrix}$ bounded over a metric space X, we may triangulate M in a bounded fashion so the cellular chain complex of M can be thought of as a chain complex in $\mathfrak{C}_X(\mathbb{Z})$ and, more relevantly, the chain complex with complex coefficients can be thought of as a chain complex in $\mathfrak{C}_X(\mathbb{C})$. Poincaré duality thus gives rise to a self-dual map and hence an element $\sigma_X[M] \in L_0(\mathfrak{C}_X(\mathbb{C}))$, the bounded symmetric signature of the manifold. The bounded symmetric signature is an invariant under bounded homotopy equivalence, since a bounded homotopy equivalence gives rise to a chain homotopy equivalence in the category $\mathfrak{C}_X(\mathbb{C})$ and the L-groups by their definition are chain homotopy invariant [25].

As mentioned above we get maps

$$K_*C^*(X) \xleftarrow{\alpha} L_*(\mathfrak{C}_X^{b.o.}(\mathbb{C})) \xrightarrow{\beta} L_*(\mathfrak{C}_X(\mathbb{C}))$$

(3.4) THEOREM: *In case $X = \mathcal{O}(K)$, the open cone on a finite complex, the map β is an isomorphism. Moreover,* $\operatorname{Sign}_{\mathcal{O}(K)} M = \alpha\beta^{-1}\sigma_{\mathcal{O}(K)}[M]$

PROOF: Let $\mathcal{F}$ be any of the functors

$$K \mapsto L_*(\mathfrak{C}^{b.o.}_{\mathcal{O}(K)}(\mathbb{C})), \quad K \mapsto L_*(\mathfrak{C}_{\mathcal{O}(K)}(\mathbb{C})).$$

Then $\mathcal{F}$ is a reduced generalized homology theory on the category of finite complexes. In case $\mathcal{F}(K) = L_*(\mathfrak{C}_{\mathcal{O}(K)}(\mathbb{C}))$ this is proved by Ranicki [25]. In the case $\mathcal{F}(K) = L_*(\mathfrak{C}^{b.o.}_{\mathcal{O}(K)}(\mathbb{C}))$ the proof needs the extensions to Ranicki's results provided in [5] but goes along exactly the same lines, noting that restricting the morphisms to the ones defining analytically bounded operators does not prevent Eilenberg swindles[5], and thus the basic Karoubi filtration technique goes through. Moreover, β is plainly a natural transformation of homology theories[6], and it is an isomorphism for $K = \emptyset$, so it is an isomorphism for all finite complexes. This proves the first statement.

To prove the second statement note that if M has a bounded triangulation of bounded geometry (meaning that for each $r > 0$ there is a number N_r such that the number of simplices meeting $c^{-1}(B(x;r))$, for any $x \in X$, is at most N_r), then the natural representative of $\sigma_{\mathcal{O}(K)}[M]$ is in fact an analytically bounded operator (since Poincaré duality is given by sending a cell to its dual cell combined with appropriate subdivision maps). Moreover, by following the line of proof given by Kaminker and Miller [17, 20], and using the de Rham theorem in the bounded geometry category [9], one may show that this bounded operator passes under α to the class of the signature operator in $K_*(C^*(\mathcal{O}(K))$ (compare [17], theorem 5.1). In case M is not of bounded geometry we need to notice that both $\sigma_{\mathcal{O}(K)}[M]$ and $\operatorname{Sign}_{\mathcal{O}(K)}[M]$ are $\mathcal{O}(K)$-bordism invariants modulo 2-torsion, the latter by proposition 2.5, and that any manifold $\begin{pmatrix} M \\ \downarrow \\ \mathcal{O}(K) \end{pmatrix}$ is $\mathcal{O}(K)$-bordant to a bounded geometry manifold. To see this latter statement make $M \to \mathcal{O}(K)$ transverse to a level $t \cdot K \subset \mathcal{O}(K)$, and let V be the inverse image of $(\geq t)\cdot K$, W the inverse image of $(\leq t) \cdot K$. We then get a bordism from M to $W \cup_{\partial W} \partial W \times [0,\infty)$ by $M \times I \cup_V V \times [0,\infty)$ and the map p extends to a proper map from the

[5]The key point is that the operator norm of an orthogonal direct sum is the *supremum* of the operator norms of its constituents. See [15] for the details of an Eilenberg swindle in the analytic situation.

[6]We do *not* assert that α is a natural transformation of homology theories.

bordism to $\mathcal{O}(K)$ by sending $(m,t) \in M \times I$ to $p(m)$ and $(v,s) \in V \times [0,\infty)$ to $(s+u)\cdot k$ where $u \cdot k = p(v)$. This is easily seen to be a proper map, and we do get a bordism over $\mathcal{O}(K)$ to a manifold of bounded geometry. □

REMARK: In the above argument we needed to reduce the manifold M to bounded geometry, and to do this we used the fact that it is always possible to split M over an open cone. If one could similarly reduce a homotopy equivalence bounded over an open cone to a bounded geometry homotopy equivalence, the proof of our theorem would be considerably simplified. However, it appears that the proof of such a result would require a lengthy excursion into bounded geometry surgery [3].

(3.5) COROLLARY: *In the situation above* $\operatorname{Sign}_{\mathcal{O}(K)}(M)$ *is an invariant modulo 2-torsion under boundedly controlled homotopy equivalence.*

As has already been explained, this suffices for a proof of Novikov's theorem.

4. Integral homotopy invariance

In the previous section we worked modulo 2-torsion, for simplicity. We will now justify the title of this paper by showing that it is not in fact necessary to invert 2 in corollary 3.5. In this section we will therefore, of course, suspend the convention made previously that all L and K groups are implicitly tensored with $\mathbb{Z}[\frac{1}{2}]$.

There are two points at which 2-torsion issues were neglected: the proof of the bordism invariance of the signature in section 2, and the identification of the various decorations on L-theory in section 3. We will address these in turn.

Bordism invariance. We begin by discussing in somewhat greater detail the reason for the appearance of the factor k in the formula for the boundary of the signature operator. The informal statement that 'the K-homology boundary of Dirac is Dirac' can be expressed more precisely as follows:

(4.1) PROPOSITION: *Let M be a manifold with boundary N, C a bundle of Clifford modules on M, D_C the corresponding Dirac operator. If M is even-dimensional, we assume that C is graded by a grading operator ε. Let n be a unit normal vector field to N. Define a bundle of Clifford modules ∂C on N as follows:*

(1) *If M is odd-dimensional, $\partial C = C$, graded by Clifford multiplication by in;*
(2) *If M is even-dimensional, ∂C is the $+1$ eigenspace of the involution $in\varepsilon$ on C (which commutes with the Clifford action of TN).*

Then the boundary, in K-homology, of the class of the Dirac operator $[D_C]$ *is the class of the Dirac operator of* ∂C.

While this particular statement does not appear to be in the literature, related results are proved in [12, 13, 36]. Now let us take C to be the bundle of Clifford modules that defines the signature operator on M (see the remark after definition 2.1). Then a simple calculation shows that if M is odd-dimensional, ∂C defines the signature operator on N, but that if M is even-dimensional, ∂C defines the direct sum of two copies of the signature operator on N. Hence the factor k in 2.3 and 2.5.

To get rid of the factor 2 in the even-dimensional case we employ an idea of Atiyah [1]. Suppose that the normal vector field n extends to a unit vector field (also called n) on M, and define an operator R_n on C by *right* Clifford multiplication by in. Then R_n is an involution and its ± 1 eigenspaces are bundles of (left) Clifford modules. Let D_{C^+} be the Dirac operator associated to the $+1$ eigenbundle of R_n. Using the result above, we find that $\partial[D_{C^+}]$ is exactly the signature operator of N. Thus, the same method of proof as that of 2.5 gives us

(4.2) PROPOSITION: *Suppose that N is an X-bounded manifold which is the boundary of an X-bounded manifold-with-boundary M. In addition, if M is even-dimensional, suppose that the unit normal vector field to N extends to a unit vector field on M. Then* $\mathrm{Sign}_X(N) = 0$ *in* $K_*(C^*(X))$.

Now we remark that there is no obstruction to extending the field n over any non-compact connected component of M. Moreover, provided that the control space X is non-compact and coarsely geodesic, there is no loss of generality in assuming that every connected component of M is non-compact; for, in any compact component, we may punch out a disc, replace it with an infinite cylinder, and control this cylinder over a ray in X. Thus we conclude that over any such space X, and in particular over an open cone on a finite polyhedron, the coarse signature is bordism invariant on the nose.

Decorations. With the issue of bordism invariance settled, the integral boundedly controlled homotopy invariance of the coarse analytic signature will follow (as in section 3) from:

(4.3) THEOREM: *The functors* $K \mapsto L^h_*(\mathfrak{C}^{b.o.}_{\mathcal{O}(K)}(\mathbb{C}))$ *and* $K \mapsto L^h_*(\mathfrak{C}_{\mathcal{O}(K)}(\mathbb{C}))$ *are isomorphic homology theories (under the forgetful map).*

PROOF: All we need to do is to prove that both functors are homology theories, since they agree on $\emptyset$. Since $\mathbb{C}$ is a field we have $K_{-i}(\mathbb{C}) = 0$ for $i > 0$ [4, Chap. XII]. Hence

$$L^h_*(\mathfrak{C}_{\mathcal{O}(K)}(\mathbb{C})) = L^{-\infty}_*(\mathfrak{C}_{\mathcal{O}(K)}(\mathbb{C}))$$

is a homology theory. To prove $L^h_*(\mathfrak{C}^{b.o.}_{\mathcal{O}(K)}(\mathbb{C}))$ is also a homology theory we use the excision result [5, Theorem 4.1]. Combining this with [5, Lemma 4.17] we only need to see that idempotent completing any of the categories $\mathfrak{C}^{b.o.}_{\mathcal{O}(K)}(\mathbb{C})$ does not change the value of L^h i.e. that K_0 of the idempotent completed categories is trivial. This is the object of the next proposition. □

(4.4) PROPOSITION: *With terminology as above we have*

$$K_0(\mathfrak{C}^{b.o.}_{\mathcal{O}(K)}(\mathbb{C})^\wedge) = 0$$

for K a non-empty finite complex.

PROOF: The proof follows the methods in [23] and [24] quite closely, and the reader is supposed to be familiar with these papers. Let L be a finite complex, $K = L \cup_\alpha D^n$. Consider the category $\mathfrak{U} = \mathfrak{C}^{b.o.}_{\mathcal{O}(K)}(\mathbb{C})$ and the full subcategory $\mathfrak{A} = \mathfrak{C}^{b.o.}_{\mathcal{O}(K)}(\mathbb{C})_{\mathcal{O}(L)}$ with objects having support in a bounded neighborhood of $\mathcal{O}(L)$. $\mathfrak{A}$ is isomorphic to $\mathfrak{C}^{b.o.}_{\mathcal{O}(L)}(\mathbb{C})$, and $\mathfrak{U}$ is $\mathfrak{A}$-filtered in the sense of Karoubi, so following [24], we get an exact sequence

$$\ldots \to K_1(\mathfrak{U}) \to K_1(\mathfrak{U}/\mathfrak{A}) \to K_0(\mathfrak{A}^\wedge) \to K_0(\mathfrak{U}^\wedge) \to \ldots$$

but $\mathfrak{U}/\mathfrak{A}$ is isomorphic to

$$\mathfrak{C}^{b.o.}_{\mathcal{O}(D^n)}(\mathbb{C})/\mathfrak{C}^{b.o.}_{\mathcal{O}(S^{n-1})}(\mathbb{C})$$

which has the same K-theory as $\mathfrak{C}^{b.o.}_{\mathbb{R}^n}(\mathbb{C})$. So by induction over the cells in K, it suffices to prove that

$$K_1(\mathfrak{C}^{b.o.}_{\mathbb{R}^n}(\mathbb{C})) = 0 \qquad n > 1$$

and

$$K_0(\mathfrak{C}^{b.o.}_{\mathbb{R}^{n-1}}(\mathbb{C})^\wedge) = 0 \qquad n > 1$$

but following the arguments in [23] it is easy to see these groups are equal. Now consider the ring $\mathbb{C}[t_1, t_1^{-1}, \ldots, t_k, t_k^{-1}]$. The category

$$\mathfrak{C}^{b.o.}_{\mathbb{R}^n}(\mathbb{C}[t_1, t_1^{-1}, \ldots, t_k, t_k^{-1}])$$

with geometrically bounded morphisms, inducing analytically bounded operators on the Hilbert space where the t_i powers are also used as basis has a subcategory

$$\mathfrak{C}^{b.o.,t_1,\ldots,t_k}_{\mathbb{R}^n}(\mathbb{C}[t_1, t_1^{-1}, \ldots, t_k, t_k^{-1}])$$

where the morphisms are required to use uniformly bounded powers of the t_i's. Turning t_i-powers into a grading produces a functor

$$\mathfrak{C}^{b.o.,t_1,\ldots,t_k}_{\mathbb{R}^n}(\mathbb{C}[t_1, t_1^{-1}, \ldots, t_k, t_k^{-1}]) \to \mathfrak{C}^{b.o.,t_1,\ldots,t_{i-1},t_{i+1},\ldots,t_k}_{\mathbb{R}^{n+1}}(\mathbb{C}[t_1, t_1^{-1}, \ldots, t_{i-1}, t_{i+1}, \ldots, t_k, t_k^{-1}]).$$

We claim this is a split epimorphism on K_1. Consider the automorphism β_{t_i} which is multiplication by t_i on the upper half of $\mathbb{R}^{n+1}$ and the identity on the lower half. Here upper and lower refers to the coordinate introduced when the t_i-powers were turned into a grading. The splitting is given by sending an automorphism α to the commutator $[\alpha, \beta_{t_i}]$ and restricting to a band. Since both the bounded operator and the bounded t-power conditions are responsive to the Eilenberg swindle arguments used in [23] the argument carries over to this present situation. From this it follows there is a monomorphism

$$K_1(\mathfrak{C}^{b.o.}_{\mathbb{R}^n}(\mathbb{C})) \to K_1(\mathfrak{C}^{b.o.,t_1,\dots,t_n}_{*}(\mathbb{C}[t_1, t_1^{-1}, \dots, t_n, t_n^{-1}])).$$

But the bounded t-power condition is vacuous, when the metric space is a point, and the uniformity given by the $\mathbb{Z}^n$-action renders the bounded operator condition vacuous too. Since the inclusion maps given by the commutator with β_{t_i} commute up to sign we find that the image of $K_1(\mathfrak{C}^{b.o.}_{\mathbb{R}^n}(\mathbb{C}))$ is contained in

$$K_{-i}(\mathbb{C}) \subset K_1(\mathbb{C}[t_1, t_1^{-1}, \dots, t_n, t_n^{-1}])$$

which is 0 since $\mathbb{C}$ is a field and we are done. □

References

1. M.F. Atiyah. Vector fields on manifolds. *Arbeitsgemeinschaft für Forschung des Landes Nordrhein Westfalen*, 200:7–24, 1970.
2. M.F. Atiyah and I.M. Singer. The index of elliptic operators I. *Annals of Mathematics*, 87:484–530, 1968.
3. O. Attie. *Quasi-Isometry Classification of Some Manifolds of Bounded Geometry.* PhD thesis, New York University, 1992.
4. H. Bass. *Algebraic K-Theory.* Benjamin, 1968.
5. G. Carlsson and E. K. Pedersen. Controlled algebra and the Novikov conjectures for K and L theory. *Topology*, 1995. To appear.
6. T.A. Chapman and S. Ferry. Approximating homotopy equivalences by homeomorphisms. *American Journal of Mathematics*, 101:583–607, 1979.
7. A. Connes, M. Gromov, and H. Moscovici. Conjecture de Novikov et fibrés presque plats. *Comptes Rendus de l'Académie des Sciences de Paris*, 310:273–277, 1990.
8. A. Connes, D. Sullivan, and N. Teleman. Quasiconformal mappings, operators on Hilbert space, and local formulae for characteristic classes. *Topology*, 33:663–681, 1994.
9. J. Dodziuk. L^2 harmonic forms on complete manifolds. In S.T. Yau, editor, *Seminar on Differential Geometry*, Annals of Mathematics Studies, no. 102, pages 291–302. Princeton, 1982.
10. S.K. Donaldson and D. Sullivan. Quasiconformal 4-manifolds. *Acta Mathematica*, 163:181–252, 1989.
11. I.M. Gelfand and A.S. Mischenko. Quadratic forms over commutative group rings and K-theory. *Functional Analysis and its Applications*, 3:277–281, 1969.
12. N. Higson. K-homology and operators on non-compact manifolds. Unpublished preprint, 1988.
13. N. Higson. C^*-algebra extension theory and duality. *Journal of Functional Analysis*, 129:349–363, 1995.

14. N. Higson and J. Roe. On the coarse Baum-Connes conjecture. these Proceedings.
15. N. Higson, J. Roe, and G. Yu. A coarse Mayer-Vietoris principle. *Mathematical Proceedings of the Cambridge Philosophical Society*, 114:85–97, 1993.
16. M. Hilsum and G. Skandalis. Invariance par homotopie de la signature à coefficients dans un fibré presque plat. *Journal für die reine und angewandte Mathematik*, 423:73–99, 1992.
17. J. Kaminker and J.G. Miller. Homotopy invariance of the index of signature operators over C^*-algebras. *Journal of Operator Theory*, 14:113–127, 1985.
18. G.G. Kasparov. The operator K-functor and extensions of C^*-algebras. *Mathematics of the USSR — Izvestija*, 16:513–572, 1981.
19. R.C. Kirby and L.C. Siebenmann. *Foundational Essays on Topological Manifolds, Smoothings, and Triangulations*, Annals of Mathematics Studies, no. 88. Princeton, 1976.
20. J. Miller. Signature operators and surgery groups over C^*-algebras. Preprint, 1992.
21. S.P. Novikov. The topological invariance of the rational Pontrjagin classes. *Mathematics of the USSR — Doklady*, 6:921–923, 1965.
22. W. Paschke. K-theory for commutants in the Calkin algebra. *Pacific Journal of Mathematics*, 95:427–437, 1981.
23. E.K. Pedersen. On the K_{-i}-functors. *Journal of Algebra*, 90:461–475, 1984.
24. E.K. Pedersen and C.A. Weibel. K-theory homology of spaces. In G. Carlsson et al, editor, *Algebraic Topology, Arcata 1986*, volume 1370 of *Springer Lecture Notes in Mathematics*, pages 346–361, 1989.
25. A. Ranicki. *Lower K- and L-Theory*, volume 178 of *London Mathematical Society Lecture Notes*. Cambridge, 1992.
26. A. Ranicki. *Algebraic L-Theory and Topological Manifolds*. Cambridge, 1993.
27. J. Roe. *Elliptic Operators, Topology, and Asymptotic Methods*, volume 179 of *Research Notes in Mathematics*. Longman, London, 1988.
28. J. Roe. Coarse cohomology and index theory on complete Riemannian manifolds. *Memoirs of the American Mathematical Society*, 497, 1993.
29. J. Rosenberg. Analytic Novikov for topologists. these Proceedings.
30. I.M. Singer. Future extensions of index theory and elliptic operators. In *Prospects in Mathematics*, Annals of Mathematics Studies, no. 70, pages 171–185. Princeton, 1971.
31. D. Sullivan. Hyperbolic geometry and homeomorphisms. In J.C. Cantrell, editor, *Geometric Topology*, pages 543–555. Academic Press, 1979.
32. D. Sullivan and N. Teleman. An analytic proof of Novikov's theorem on the rational Pontrjagin classes. *Publications Mathématiques de l'Institut des Hautes Études Scientifiques*, 58:39–78, 1983.
33. N. Teleman. The index of signature operators on Lipschitz manifolds. *Publications Mathématiques de l'Institut des Hautes Études Scientifiques*, 58:39–78, 1983.
34. N. Teleman. The index theorem for topological manifolds. *Acta Mathematica*, 153:117–152, 1984.
35. S. Weinberger. An analytic proof of the topological invariance of the rational Pontrjagin classes. Preprint, 1990.
36. F. Wu. The boundary map in KK-theory for manifolds with boundary. Preprint, 1990.

Department of Mathematics, State University of New York, Binghamton, Binghamton, NY 13901, USA

email: erik@math.binghamton.edu

Jesus College, Oxford OX1 3DW, England

email: jroe@spinoza.jesus.ox.ac.uk

Department of Mathematics, University of Pennsylvania, Philadelphia, PA 19104, USA

email: shmuel@archimedes.math.upenn.edu

3-Manifolds and PD(3)-Groups

C. B. Thomas

This article is a survey of what one knows about $PD(3)$-groups and related topics. Our aim is to show first how such groups are related to 3-manifold groups, and second to suggest why these groups should be of interest to mathematicians working on the Novikov conjecture. Recall that the discrete group Γ is said to be of type (FP) if there is a finite resolution of the trivial module $\mathbb{Z}$ by finitely generated projective module, and of type (FF) if each of these projective modules may be taken to be free. The group Γ is said to be an oriented PD^n-group if

(i) Γ is of type (FP) and

(ii) $H^i(\Gamma, \mathbb{Z}\Gamma) = \begin{cases} 0 & \text{if } i \neq n \\ \mathbb{Z} & \text{if } i = n. \end{cases}$

In the relative case let Δ be a subgroup of Γ, and give $\mathbb{Z}\Gamma/\Delta$ the natural structure of left Γ-module with the cosets $x\Delta$ as generators. There is a left module homomorphism $\epsilon : \mathbb{Z}\Gamma/\Delta \longrightarrow \mathbb{Z}$ which is such that $\epsilon(x\Delta) = 1$ for all x, and whose kernel B is a left Γ-module. Then (Γ, Δ) is a PD^n-pair if

(i) Γ is of type (FP) and

(ii) $H^i(\Gamma, \mathbb{Z}\Gamma) = \begin{cases} 0 & \text{if } i \neq n-1 \\ B & \text{if } i = n-1. \end{cases}$

Note that Δ is a PD^{n-1}-group, and that the definition can be generalised to allow Δ to have several components. Thus if $\Delta = \bigcup_i \Delta_i$ $\mathbb{Z}\Gamma/\Delta$ is to be interpreted as the free $\mathbb{Z}$-module on the cosets $\Delta_i g$. These conditions are equivalent to the more familiar ones in terms of homology/cohomology pairings, see [3], and by twisting the coefficients one can generalise the definitions to cover the non-oriented case. However in this paper we shall assume that all groups and manifolds are oriented. From the geometric point of view, if Γ satisfies the stronger conditions (FF), then Γ can be realized as the fundamental group of an n-dimensional Poincaré complex $K(\Gamma, 1)$, and in dimensions ≥ 5, subject perhaps to further restrictions on Γ one can hope to show that $K(\Gamma, 1)$ is homotopy equivalent to a topological manifold, see [15]. However, what is the situation in dimensions 2 and 3?

In [6] and [5] it is shown that PD^2-groups are surface groups, and building on this result it is clear there is a strong relation between PD^3-groups and the fundamental groups of irreducible 3-manifolds. (A connected, closed, compact oriented 3-manifold M^3 is said to be irreducible if every embedded 2-sphere bounds a 3-ball. Such manifolds either have finite fundamental group or $M^3 = K(\Gamma, 1)$, as above.) The classification of PD^3-groups now proceeds in parallel with Thurston's geometrisation programme for 3-manifolds. Indeed, one can mimic Thurston's "flow chart" [19, p.211] in the algebraic setting.

Notation: The discrete group Γ is said to be almost P for some property P if Γ contains a subgroup of finite index Γ_1 such that Γ_1 is P.

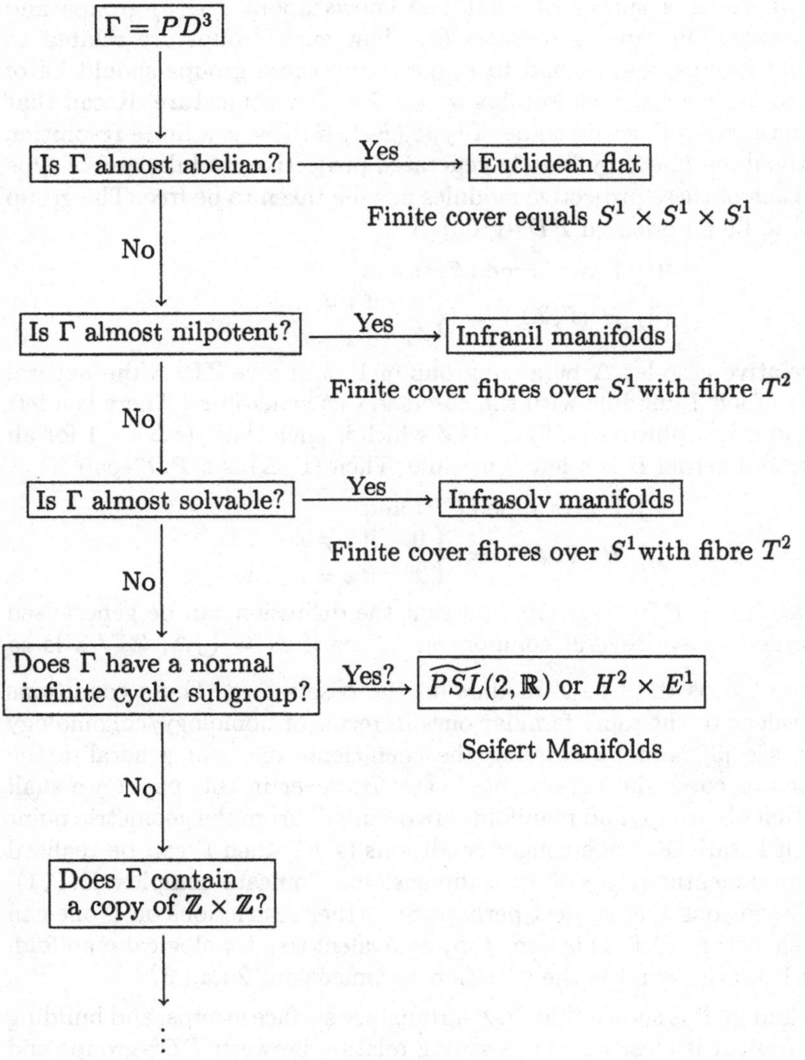

Remarks: The geometries S^3 and $S^2 \times E^1$ do not occur, since the corresponding complexes are not acyclic. But see Section 2 below. (Yes?) above refers to the question hanging over PD^3 (as opposed to manifold) groups with finite abelianisation.

We are left with the (large) residual class of groups containing those of hyperbolic type. As explained in the next section, assuming that Γ satisfies the maximal condition on centralizers *Max-c*, see [12], Γ can be split along embedded rank 2 abelian subgroups. This leaves a collection of PD^3-pairs $(\Gamma_i, \Delta_i = \bigcup_j \Delta_{ij})$, which are either covered by the relative version of some previous step or are *algebraically atoroidal* in the sense that any free abelian subgroup of rank 2 in Γ is conjugate to a subgroup of some "bounding" subgroup Δ_{ij}. At present this is as far as one can go, although the partial solution of the Novikov Conjecture suggests how to proceed.

The paper is organised as follows. Section 1 contains a commentary on the flow chart above, Section 2 summarises what is known about 3-complexes with finite fundamental group, and Section 3 is devoted to a previously unpublished argument for non-finitely generated solvable groups. This answers a question which was originally posed to me by David Epstein. However in a sense the entire paper shows the influence which he had on me as a graduate student.

1. Explanation of the flow chart

(a) If Γ is a solvable PD^3-group, then by [2, 2.3] Γ is a torsion-free polycyclic group, which contains a subgroup Γ_1 of finite index with a presentation of the form

$$1 \longrightarrow \mathbb{Z}^B \oplus \mathbb{Z}^C \longrightarrow \Gamma_1 \longrightarrow \mathbb{Z}^A \longrightarrow 1 .$$

The generator A acts on the subgroup $\langle B, C\rangle$ via some (2×2) unipotent integral matrix. It follows that Γ_1 is isomorphic to the fundamental group of a torus bundle over the circle S^1. Either by explicit enumeration of the possibilities or by an appeal to [1] we see that Γ is also a 3-manifold group. The abelian and nilpotent cases are included in this argument.

(b) Next suppose that Γ is given as an extension of the form

$$1 \longrightarrow \mathbb{Z}^A \longrightarrow \Gamma \longrightarrow Q \longrightarrow 1 .$$

The result which we would like to apply states that if $\mathrm{cd}(Q) < \infty$, then up to finite index (compare Γ_1 in (a)) Q is a PD^2-group. Unfortunately in general it is far from obvious when Q has finite cohomological dimension, and more roundabout methods must be used. At the time of writing the complete answer is not known, although it has recently been settled in the affirmative for 3-manifold groups by A.Casson and D.Jungreis, using earlier work by G.Mess, see [4] and also [9]. (I am grateful to C.T.C.Wall for directing my attention to the second reference.) Geometrically we distinguish between Haken and non-Haken manifolds; a sufficient condition for M^3 to be Haken is that $H_1(M;\mathbb{Z})$ to be infinite. That the Seifert fibre space conjecture holds for Haken manifolds has been known for a long time - see the references in [4]. The true interest of the Casson-Jungreis construction is that it does not start from the existence of a "good" embedded surface. Algebraically the analogue of this is infinite abelianisation, and subject to this restriction J. Hillman has proved the conjecture, [11]. The idea behind his argument is as follows: eliminating a number of easy cases we consider the following situation:

$$\begin{array}{ccccccccc} & & K' = \text{free} & & & & & & \\ & & \downarrow & & & & & & \\ 1 & \longrightarrow & K & \longrightarrow & \Gamma & \longrightarrow & \mathbb{Z} & \longrightarrow & 1 \\ & & \uparrow & & & & & & \\ & & \mathbb{Z}^A = \text{central} & & & & & & \end{array}$$

with $K' \cap \mathbb{Z}^A = \{1\}$. Using the structure of K/K' as a finitely generated $\mathbb{Z}[G/K]$-module we can find a subgroup H of K such that $H \equiv H/\mathbb{Z}^A \times \mathbb{Z}^A$. From this it follows that $\mathrm{cd}(H/\mathbb{Z}^A)$ is finite, and passing to Γ, that there is a subgroup Γ_1 of finite index defined as an extension

$$1 \longrightarrow \mathbb{Z}^A \longrightarrow \Gamma_1 \longrightarrow Q \longrightarrow 1\,,$$

with Q an HNN-extension of $H/\mathbb{Z}^A$, still having $\mathrm{cd}(Q) < \infty$. See also [18] for further results of this kind.

(c) Splitting a group Γ of residual type. The strongest result in this direction is:

Theorem 1. *If the PD^3-group Γ (i) has no infinite cyclic normal subgroup and (ii) does contain some rank 2 free abelian subgroup, then either (a) Γ is a non-trivial free product with amalgamation, $\Gamma = K *_H \Lambda$, or (b) Γ is an HNN-extension $\Gamma = B_{*H,t}$, where $\mathrm{cd}(H) = 2$ and H has an infinite cyclic subgroup which is commensurable with all of its conjugates.*

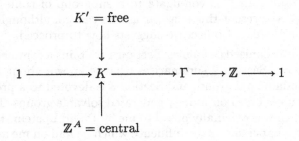

For a proof and discussion of the relation with the geometric Torus Theorem, see [13]. The latter is proved using transversality and 3-dimensional surgery, neither of which has an immediate analogue in the category of Poincaré complexes. This explains why, without some restriction, one cannot take the splitting subgroup H to be again abelian of rank 2. We have already mentioned *Max-c* (implying that every increasing chain of centralizers in Γ terminates) as one such, which also has the advantage of being satisfied by 3-manifold groups. Hence subject to *Max-c* the splitting theorem reduces the classification of PD^3-groups to those Γ which are neither of Seifert type, nor splittable. This brings us to the Novikov Conjecture: is such a Γ a group to which the known results, for example the final theorem in [15] apply? In particular does Γ satisfy the "no flat criterion" in [10, page 175 et seq.], which would suffice to make Γ word hyperbolic. Given that there is no embedding of $\mathbb{Z} \times \mathbb{Z}$ in Γ this amounts to finding a decomposition of $K(\Gamma, 1)$ by means of hyperbolic polyhedra, for example copies of the dodecahedron with suitable identifications. Together with the Casson-Jungreis theorem this possibility is plausible enough for one to formulate the

Conjecture. *If Γ is a PD^3-group satisfying Max-c, then either Γ is a 3-manifold group or Γ contains an algebraically atoroidal subgroup Γ_1 which is geometrically realisable by a manifold $X^{n+3} \simeq K(\Gamma_1, 1) \times S^n$, $n \geq 2$.*

Note that even if this conjecture is true it leaves open the possibility that there may exist groups Γ which are not manifold realisable in dimension 3, even though all algebraic obstructions to their existence vanish, This would be analogous to a phenomenon for finite groups discussed in the next section.

The essentially algebraic discussion above clarifies some problems with the geometrisation programme for 3-manifolds. Starting from the unique prime decomposition of M^3 one leaves on one side copies of $S^2 \times S^1$ and manifolds with finite fundamental group. The remaining prime manifolds are classified by our flow chart up to the final class of hyperbolic type. (Given [4] there is no need to make any Haken assumption until this point.) Any prime manifold which splits according to the Torus Theorem is amenable to geometrisation, so, apart from the unresolved elliptic question, it remains to clarify the status of closed atoroidal manifolds with finite first homology group.

2. Poincaré complexes with finite fundamental group

The 3-dimensional spherical space form problem concerns the classification of manifolds covered by S^3. The geometrisation conjecture says that such a manifold admits a metric of constant positive curvature, and hence is classified by representation theory, see [17] for what is known. In the category of Poincaré complexes one starts from the result of R. Swan which states that a finite group π with cohomological period 4 is geometrically realised as the fundamental group of a complex $Y^{4k-1} = \Sigma^{4k-1}/\pi$. Here $\Sigma^{4k-1} \simeq S^{4k-1}$ and k divides the order of π, see [16]. If we do not require Y to be finite, then there is no problem in taking $k = 1$, otherwise it is a question of determining the order of the so-called finiteness obstruction in the reduced projective class group $\widetilde{K}_0(\mathbb{Z}\pi)$. Although there is no systematic account of everything that is known in the literature techniques are available (a) to classify groups of period 4, (b) to determine the best value of k and (c) prove

Theorem 2. *Let π be a finite group of cohomological period 4 such that the optimal value k_0 of k above equals 1. Then the oriented homotopy classes of finite complexes Y^3 realising π are in 1-1 correspondence with a subset of the orbits of the induced* $\mathrm{Aut}(\pi)$*-action on* $H^4(\pi;\mathbb{Z})$.

A familiar example of this result is that the lens spaces $L(p,q)$ and $L(p,q')$ are *oriented* homotopy equivalent if and only if there exists some x with $q' \cong x^2 q (\mathrm{mod}\, p)$. Its interest from our point of view is that the class of Poincaré complexes Y^3 is certainly larger than that of manifolds S^3/π. Thus Swan gives an example in the Appendix to [16] with dihedral fundamental group D_6 which has since been generalise to the group D_{2n} for all values of n. Since D_{2n} contains non-central elements of order 2, there is a surgery obstruction to replacing Y^3 by a manifold.

More interesting are the groups $Q(8r,s) \cong \mathbb{Z}/rs \rtimes Q(8)$, where r and s are distinct odd primes, and $Q(8)$ is the quaternion group of order 8. The semidirect product is determined by the structural map $\phi : Q(8) \longrightarrow \mathrm{Aut}(\mathbb{Z}/rs)$ which has kernel of order 2. $Q(8r,s)$ has period 4 and acts freely and linearly on S^7. Furthermore there exist pairs (r_0,s_0) for which the finiteness obstruction multiple $k_0 = 1$. More interestingly one can specialise still further, and determine pairs (r_1,s_1) such that the surgery obstruction to replacing Y^3 by a manifold vanishes. Of course the dimension is so low that this tells us nothing about the actual existence of such a manifold, which would indeed be a counterexample to the full geometrisation conjecture. But it follows that if this is true, then there exist Poincaré complexes which are not homotopy equivalent to manifolds for specifically low-dimensional reasons. Put another way, by using a join-like construction on S^7 and the universal cover $\widetilde{Y}$, and allowing $Q(8r_1,s_1)$ to act on the resulting 11-dimensional complex, we obtain a quotient space which is homotopy equivalent to a manifold. This is analogous to the case of the residual groups of hyperbolic type in Section 1 - some of which may only be "stably" realisable as manifold groups. The reader is referred to [14] for more information on this question, and in particular for the arithmetic conditions which determine the pairs (r_0,s_0) and (r_1,s_1).

3. Non-finitely generated solvable groups

In [7] D. B. A. Epstein constructed an example of a non-compact 3-manifold with fundamental group isomorphic to $\mathbb{Q}^+$, the additive group of rationals. He went on to show that subgroups of $\mathbb{Q}^+$ were the only non-finitely generated abelian groups to arise as fundamental group in 3-manifold theory. In this section we prove algebraically that replacing the property "abelian" by "solvable" introduces no new groups. The result itself is not new, see for example Theorem 3.1 of [8]. Thus

Theorem 3. *If M^3 (oriented but possibly non-compact) has an infinite solvable fundamental group Γ, then either Γ is Thurston geometric or Γ is a subgroup of the additive group $\mathbb{Q}^+$.*

Proof. Let the solvable group Γ have the composition series

$$\Gamma = \Gamma_0 \triangleright \Gamma_1 \triangleright \Gamma_2 \triangleright \ldots \triangleright \Gamma_n = \{1\}$$

where Γ_{k-1}/Γ_k is abelian, and define

$$\mathrm{h}(\Gamma) = \sum_{k=1}^{n} \dim_{\mathbb{Q}}(\Gamma_{k-1}/\Gamma_k \otimes \mathbb{Q}) .$$

Then $\mathrm{h}(\Gamma) = \mathrm{hd}(\Gamma)$ and $\mathrm{cd}(\Gamma) \leq \mathrm{hd}(\Gamma) + 1$. (Indeed it can be shown that $\mathrm{cd}(\Gamma) = \mathrm{hd}(\Gamma)$ if and only if Γ is of type (FP).) Now let the non-compact manifold M^3 have fundamental group Γ; if Γ is finitely generated (hence as a 3-manifold group is also finitely presented) Γ has already been listed. Since Γ does not split as a free product, a non-compact variant of the Sphere Theorem shows that there exists an aspherical manifold with the same fundamental group as M, and non-compactness implies that $\mathrm{cd}(\Gamma) \leq 2$. Hence our problem reduces to classifying non-finitely generated groups of cohomological dimension 2. The restrictions show that if Γ is non-abelian, Γ must be a (split) extension of the form

$$1 \longrightarrow N \longrightarrow \Gamma \longrightarrow \mathbb{Z} \longrightarrow 1$$

where N is torsion free abelian of rank 1. If x denotes a generator of the quotient, then the finiteness of the cohomological dimension implies that N is of type $[p_1^\infty p_2^\infty \ldots p_s^\infty]$ and that x acts on N by multiplication by $\pm p_1^{\alpha_1} p_2^{\alpha_2} \ldots p_s^{\alpha_s}$, $0 \neq \alpha_i \in \mathbb{Z}$, $i = 1, 2, \ldots, s$, see [3, Section 7.4].
(Recall that the terminology means that $p_i^{-k} \in \mathbb{N}$ for all values of k, $i = 1, 2, \ldots, s$. For example the localised integers $\mathbb{Z}_{(q)} = [p^\infty : p \neq q]$ and $\mathbb{Z}[\frac{1}{p}] = [p^\infty]$.)
If all the exponents have the same sign then Γ is actually a finitely generated 1-relator group with $\mathrm{cd}(\Gamma) \leq 2$. This contradicts the assumption of non-finite generation. Otherwise $\mathrm{cd}(\Gamma) = 3$; the simplest example occurs when $N = \mathbb{Z}[\frac{1}{6}]$ and x acts by multiplication by $\frac{2}{3}$. In either case the form of the group extension shows that Γ has a non-compact 4-dimensional geometric realisation. Hence in our case Γ must be abelian.

References.

[1] Auslander, L., Johnson, F.E.A., *On a conjecture of C.T.C. Wall*, J. London Math. Soc. (2) 14, 331–332 (1976)

[2] Bieri, R., *Gruppen mit Poincaré Dualität*, Comm. Math. Helv. 47, 373–396 (1972)
[3] Bieri, R., *Homological dimension of discrete groups*, Queen Mary College Math. Notes, London, 1976
[4] Casson, A., Jungreis, D., *Convergence groups and Seifert fibred 3-manifolds* (preprint 1994)
[5] Eckmann, B., Linnell, P., *Poincaré duality groups of dimension 2, II*, Comm. Math. Helv. 58, 111–114 (1983)
[6] Eckmann, B., Müller, H., *Poincaré duality groups of dimension 2*, Comm. Math. Helv. 55, 510–520 (1980)
[7] Epstein, D.B.A., *Finite presentations of groups and 3-manifolds*, Quart. J. Math. (2) 12, 205–212 (1961)
[8] Evans, B. and Jaco, W., *Varieties of groups and three-manifolds*, Topology 12, 83–97 (1973)
[9] Gabai, D., *Convergence groups are Fuchsian groups*, Ann. of Math (2) 136, 447–510 (1992)
[10] Gromov, M., *Asymptotic invariants of infinite groups*, Geometric group theory II, LMS Lecture Notes 182, Cambridge University Press (1993)
[11] Hillman, J., *Seifert fibre spaces and Poincaré duality groups*, Math. Z. 190, 365–369 (1985)
[12] Kropholler, P., *An analogue of the torus decomposition theorem for certain Poincaré duality groups*, Proc. London Math. Soc. (3) 60, 503–529 (1990)
[13] Kropholler, P., *A group theoretic proof of the torus theorem*, Geometric group theory I, LMS Lecture Notes 181, Cambridge University Press, 138–158 (1993)
[14] Madsen, I., *Reidemeister torsion, surgery invariants and spherical space forms*, Proc. London Math. Soc. (3) 46, 193–240 (1983)
[15] Ranicki, A., *On the Novikov conjecture*, (these proceedings)
[16] Swan, R.G., *Periodic resolutions for finite groups*, Ann. of Maths. (2) 72, 267–291 (1960)
[17] Thomas, C.B., *Elliptic structures on 3-manifolds*, LMS Lecture Notes 104, Cambridge University Press (1986)
[18] Thomas, C.B., *Splitting theorems for certain PD^3-groups*, Math. Z. 186, 201–209 (1984)
[19] Thurston, W., *Geometry of 3-manifolds*, (mimeographed notes, Minnesota) (1993)

DEPT. OF PURE MATHEMATICS AND MATHEMATICAL STATISTICS, CAMBRIDGE UNIVERSITY, CAMBRIDGE CB2 1SB, ENGLAND, UK

email: C.B.Thomas@pmms.cam.ac.uk

Orthogonal complements and endomorphisms of Hilbert modules and C^*-elliptic complexes

Evgenii V. Troitsky

1. Introduction

In the present paper we discuss some properties of endomorphisms of C^*-Hilbert modules and C^*-elliptic complexes. The main results of this paper can be considered as an attempt to answer the question: what kinds of good properties can one expect for an operator on a Hilbert module, which represents an element of a compact group? These results are new, but we have to recall some first steps made by us before to make the present paper self-contained.

In §2 we define the Lefschetz numbers "of the first type" of C^*-elliptic complexes, taking values in $K_0(A) \otimes \mathbb{C}$, A being a complex C^*-algebra with unity, and prove some properties of them.

The averaging theorem 3.2 was discussed in brief in [15] and was used there for constructing an index theory for C^*-elliptic operators. In this theorem we do not restrict the operators to admit a conjugate, but after averaging they even become unitary. This raises the following question: is the condition on an operator on a Hilbert module to represent an element of a compact group so strong that it automatically has to admit a conjugate?

The example in section 4 gives a negative answer to this question. Also we get an example of closed submodule in Hilbert module which has a complement but has no orthogonal complement.

In §5 we define the Lefschetz numbers of the second type with values in $HC_0(A)$. We prove that these numbers are connected via the Chern character in algebraic K-theory. These results were discussed in [18] and we only recall them.

In §6 we get similar results for $HC_{2l}(A)$. We have to use in a crucial way the properties of representations.

Acknowledgment. I am indebted to A. A. Irmatov, V. M. Manuilov, A. S. Mishchenko and Yu. P. Solovyov for helpful discussions. This work was partially supported by the Russian Foundation for Fundamental Research Grant No. 94-01-00108-a and the International Science Foundation. I am grateful to the organizers of the Meeting "Novikov conjectures, index theorems and rigidity" at Oberwolfach for an extremely nice and helpful week in September 1993, especially to J. Rosenberg for very useful help in the preparation of the final form of the present paper.

2. Preliminaries

We consider the Hilbert C^*-module $l_2(P)$, where P is a projective module over a C^*-algebra A with unity (see [10, 4, 13, 15]).

2.1. Lemma. *Let $P^+(A)$ be the positive cone of the C^*-algebra A. For every bounded A-homomorphism $F : l_2(P) \to l_2(P)$ and every $u \in l_2(P)$ we have*

$$\langle Fu, Fu\rangle \leqslant \|F\|^2 \langle u, u\rangle$$

in $P^+(A)$.

Proof. For $c \in P^+(A)$ we have $c \leqslant \|c\| 1_A$. So if $\langle u, u\rangle = 1_A$, then

$$\langle Fu, Fu\rangle \leqslant \|Fu\|^2 1_A \leqslant \|F\|^2 \langle u, u\rangle .$$

Let now $\langle u, u\rangle$ be equal to $\alpha \in P^+(A)$, where α is an invertible element of A. We put $v = (\sqrt{\alpha})^{-1} u$. Then $u = \sqrt{\alpha}\, v$ and $\langle v, v\rangle = 1_A$. So

$$\langle Fv, Fv\rangle \leqslant \|F\|^2 \langle v, v\rangle ,$$

$$\langle Fu, Fu\rangle = \sqrt{\alpha}\, \langle Fv, Fv\rangle \left(\sqrt{\alpha}\right)^* \leqslant \sqrt{\alpha}\, \|F\|^2 \langle v, v\rangle \left(\sqrt{\alpha}\right)^* = \|F\|^2 \langle u, u\rangle .$$

Elements u with invertible $\langle u, u\rangle$ are dense in $l_2(A)$ (this is a consequence of Lemma 2 of [4]), so the continuity of the A-product gives the statement for $l_2(A)$. For $l_2(P)$ we have to use the stabilization theorem [10]. ∎

Let us recall the basic ideas of [16, 17].

2.2. Definition. Let $p : F \to X$ be a G-$\mathbb{C}$-bundle over a locally compact Hausdorff G-space X. Let $\Lambda(p^*F, s_F)$ be the well known complex of G-$\mathbb{C}$-bundles (see [5]) with, in general, non-compact support. Let a complex (E, α) represent an element $a \in K_G(X; A)$ (see [15, sect. 1.3]); then $(p^*E, p^*\alpha) \otimes \Lambda(p^*F, s_F)$ has compact support and defines an element of $K_G(F; A)$. We get the *Thom isomorphism* of $R(G)$-modules

$$\varphi = \varphi_A^F : K_G(X; A) \to K_G(F; A).$$

If we pass to K_G^1 by the Bott periodicity [15, 1.2.4], we can define

$$\varphi : K_G^*(X; A) \to K_G^*(F; A).$$

2.3. Theorem. *If X is separable and metrizable, then φ is an isomorphism.*

With the help of this theorem we can define the *Gysin homomorphism* $i_! : K_G(TX; A) \to K_G(TY; A)$ and the *topological index*

$$\text{t-ind}_G^X = \text{t-ind}_{G,A}^X : K_G(TX; A) \to K^G(A)$$

in a way similar to the case $A = \mathbb{C}$ [5]. Here $i : X \to Y$ is a G-inclusion of smooth manifolds and TX, TY are (co)tangent bundles.

We need the following property of the Gysin homomorphism.

2.4. Lemma. *Let $i : Z \to X$ be a G-inclusion of smooth manifolds, N its normal bundle. Then the homomorphism*

$$(di)^* i_! : K_G(TZ; A) \to K_G(TZ; A)$$

is the multiplication by

$$[\lambda_{-1}(N \otimes_{\mathbb{R}} \mathbb{C})] = \sum (-1)^i [\Lambda^i (N \otimes_{\mathbb{R}} \mathbb{C})] \in K_G(Z),$$

where Λ^i are the exterior powers, and we consider $K_G(TZ; A)$ as a $K_G(Z)$-module in the usual way.

2.5. Theorem. *Let $\text{a-ind}\, D \in K^G(A)$ be the analytic index of a pseudo differential equivariant C^*-elliptic operator [15] , $\sigma(D) \in K_G(TX; A)$ its symbol's class. Then*

$$\text{t-ind}_{G,A}^X \sigma(D) = \text{a-ind}\, D.$$

Now for the completeness of this text we recall a generalization of the result of [1]. Let, as above, G be a compact Lie group, X a G-space, X^g the set of fixed points of $g : X \to X$, $i : X^g \to X$ the inclusion.

2.6. Definition. Let E be a G-invariant A-complex on X, $\sigma(E)$ its sequence of symbols (see [15]) , $u = [\sigma(E)] \in K_G(TX; A)$, $\text{ind}_{G,A}^X(u) \in K_0(A) \otimes R(G)$. The *Lefschetz number of the first type* is

$$L_1(g, E) = \text{ind}_{G,A}^X(u)(g) \in K_0(A) \otimes \mathbb{C}.$$

2.7. Theorem. *Using the notation as above we have*

$$L_1(g, E) = (\text{ind}_{1,A}^{X^g} \otimes 1) \left(\frac{i_* u(g)}{\lambda_{-1}(N^g \otimes_{\mathbb{R}} \mathbb{C})(g)} \right).$$

Also we need the following theorem from [12].

2.8. Theorem. *Let M be a countably generated Hilbert A-module. Then we have a G-A-isomorphism*

$$M \cong \oplus_\pi \mathrm{Hom}_G(V_\pi, M) \otimes_{\mathbf{C}} V_\pi,$$

where $\{V_\pi\}$ is a complete family of irreducible unitary complex finite dimensional representations of G, non-isomorphic to each other. In

$$\mathrm{Hom}_G(V_\pi, M) \otimes_{\mathbf{C}} V_\pi,$$

the algebra A acts on the first factor and G on the second.

3. An averaging theorem

Let us recall some facts about the integration of operator-valued functions (see [9, §3]). Let X be a compact space, A be a C^*-algebra, $\varphi : C(X) \to A$ be an involutive homomorphism of algebras with unity, and $F : X \to A$ be a continuous map, such that for every $x \in X$ the element $F(x)$ commutes with the image of φ. In this case the integral

$$\int_X F(x)\, d\varphi \quad \in \quad A$$

can be defined in the following way. Let $X = \cup_{i=1}^n U_i$ be an open covering and

$$\sum_{i=1}^n \alpha_i(x) = 1$$

be a corresponding partition of unity. Let us choose the points $\xi_i \in U_i$ and compose the integral sum

$$\sum(F, \{U_i\}, \{\alpha_i\}, \{\xi_i\}) = \sum_{i=1}^n F(\xi_i)\varphi(\alpha_i).$$

If there is a limit of such integral sums then it is called the corresponding integral.

If $X = G$ then it is natural to take φ equal to the Haar measure

$$\varphi : C(X) \to \mathbf{C}, \quad \varphi(\alpha) = \int_G \alpha(g)\, dg$$

(though this is only a positive linear map, not a $*$-homomorphism) and to define for a norm-continuous $Q : G \to L(H)$

$$\int_G Q(g)\,dg = \lim \sum_i Q(\xi_i) \int_G \alpha_i(g)\,dg.$$

If we have $Q : G \to P^+(A) \subset L(H)$, then, since

$$\int_G \alpha_i(g)\,dg \geqslant 0,$$

we get

$$\sum_i Q(\xi_i) \cdot \int_G \alpha_i(g)\,dg \quad \in \quad P^+(A)$$

and

$$\int_G Q(g)\,dg \quad \in \quad P^+(A)$$

(the cone $P^+(A)$ is convex and closed). So we have proved the following lemma.

3.1. Lemma. *Let $Q : G \to P^+(A)$ be a continuous function. Then for the integral in the sense of* [9] *we have*

$$\int_G Q(g)\,dg \geqslant 0. \qquad \blacksquare$$

3.2. Theorem. *Let* GL *be the group of all bounded A-linear automorphisms of $l_2(A)$ (see* [14]*). Let $g \mapsto T_g \quad (g \in G, T_g \in \mathrm{GL})$ be a representation of G such that the map*

$$G \times l_2(A) \to l_2(A), \quad (g, u) \mapsto T_g u$$

is continuous. Then on $l_2(A)$ there is an A-product equivalent to the original one and such that $g \mapsto T_g$ is unitary with respect to it.

Proof. Let $\langle\ ,\ \rangle'$ be the original product. We have a continuous map

$$G \to A, \quad x \mapsto \langle T_x u, T_x v\rangle'$$

for every u and v from $l_2(A)$. We define the new product by

$$\langle u, v\rangle = \int_G \langle T_x u, T_x v\rangle'\,dx,$$

where the integral can be defined in the sense of either of the two definitions from [9, p. 810] because the map is continuous with the respect to the norm of the C^*-algebra. It is easy to see that this new product is a A-sesquilinear map $l_2(A) \times l_2(A) \to A$. Lemma 3.1 shows that $\langle u, u \rangle \geqslant 0$. Let us show that this map is continuous. Let us fix $u \in l_2(A)$. Then

$$x \mapsto T_x(u), \qquad G \to l_2(A)$$

is a continuous map defined on a compact space and so the set $\{T_x(u) | x \in G\}$ is bounded. Hence by the principle of uniform boundedness [2, v. 2, p. 309]

$$\lim_{v \to 0} T_x(v) = 0 \tag{1}$$

uniformly with respect to $x \in G$. If u is fixed then

$$\|T_x(u)\| \leq M_u = const$$

and by (1)

$$\| \langle u, v \rangle \| = \| \int_G \langle T_x(u), T_x(v) \rangle' \, dx \| \leq M_u \cdot vol\, G \cdot \sup_{x \in G} \|T_x(v)\| \to 0 \quad (v \to 0).$$

This gives the continuity at 0 and hence everywhere. For

$$T_x u = (a_1(x), a_2(x), \dots) \in l_2(A),$$

the equation $\langle u, u \rangle = 0$ takes the form

$$\int_G \sum_{i=1}^{\infty} a_i(x) a_i^*(x) \, dx = 0.$$

Let A be realized as a subalgebra of the algebra of all bounded operators in the Hilbert space L with inner product $(\ ,\ )_L$. For every $p \in L$ we have

$$\begin{aligned} 0 &= \left(\left(\int_G \sum_{i=1}^{\infty} a_i(x) a_i^*(x) \, dx \right) p, p \right)_L \\ &= \int_G \left(\sum_{i=1}^{\infty} a_i(x) a_i^*(x) p, \, p \right)_L \, dx = \int_G \left(\sum_{i=1}^{\infty} (a_i(x) p, \, a_i^*(x) p)_L \right) dx \end{aligned}$$

(cf. [9]). Hence $a_i(x) = 0$ almost everywhere, and thus $a_i(x) = 0$ for every x because of the continuity, and $T_x u = 0$. In particular, $u = 0$.

Since every T_y is an automorphism, we have (cf. [9])

$$\langle T_y u, T_y v\rangle = \int_G \langle T_{xy}u, T_{xy}v\rangle' \, dx = \int_G \langle T_z u, T_z v\rangle' \, dz = \langle u, v\rangle .$$

Now we will show the equivalence of the two norms and, in particular, the continuity of the representation. There is a number $N > 0$ such that $\|T_x\|' \leq N$ for every $x \in G$. So by [9]

$$\|u\|^2 = \| \langle u, u\rangle \|_A = \| \int_G \langle T_x u, T_x u\rangle' \, dx\|_A \leq \left(\sup_{x\in G} \|T_x u\|'\right)^2 \leq N^2(\|u\|')^2.$$

On the other hand, applying 2.1 and 3.1 we have

$$\begin{aligned}\langle u, u\rangle' &= \int_G \langle T_{g^{-1}}T_g u, T_{g^{-1}}T_g u\rangle' \, dg \leq \int_G \|T_{g^{-1}}\|^2 \langle T_g u, T_g u\rangle' \, dg \\ &\leq \int_G N^2 \langle T_g u, T_g u\rangle' \, dg = N^2 \int_G \langle T_g u, T_g u\rangle' \, dg = N^2 \langle u, u\rangle .\end{aligned}$$

Then

$$(\|u\|')^2 = \| \langle u, u\rangle' \|_A \leq N^2 \| \langle u, u\rangle \|_A = N^2 \|u\|^2. \quad \blacksquare$$

3.3. Remark. $l_2(P)$ is a direct summand in $l_2(A)$, so 3.2 holds for $l_2(P)$.

4. Complements and orthogonal complements

Let us recall some preliminary statements.

4.1. Lemma. *1. An A-linear operator $F : M \to H_A$ always admits a conjugate if $M \in \mathcal{P}(A)$ — the category of finitely generated projective modules.*

2. Let $0_A \leq \alpha < 1_A$. Then $\|\alpha\| < 1$.

3. Let $\alpha \geq 0$, $\alpha = \beta\beta^*$, $1 - \alpha > 0$. Then $1 - \beta$ is an isomorphism.

Here the strong inequality means that the spectrum of the operator is bounded away from zero.

4.2. Example. Let $A = C[0, 1]$, $\{e_i\}$ be the standard basis of H_A. Let

$$\varphi_i(x) = \begin{cases} 0 & \text{on } [0, \frac{1}{i}] \text{ and } [\frac{1}{i-1}, 1], \\ 1 & \text{at } x_i = \frac{1}{2}\left(\frac{1}{i} + \frac{1}{i-1}\right), \\ \text{linear on } [\frac{1}{i}, x_i] \text{ and } [x_i, \frac{1}{i-1}], \end{cases}$$

$i = 2, 3, \ldots$. Let

$$h_i = \frac{e_i + \varphi_i e_1}{(1 + \varphi_i{}^2)^{1/2}} \qquad (i = 2, 3, \ldots)$$

be an orthonormal system of vectors which generates $H_1 \subset H_A$, $H_1 \cong H_A$. Then $H_1 \oplus \operatorname{span}_A(e_1) = H_A$. Indeed, all $e_i \in H_1 + \operatorname{span}_A(e_1)$, and if

$$x = (\alpha_1, \alpha_2, \ldots) \in H_1 \cap \operatorname{span}_A(e_1),$$

$$x = (\alpha_1, 0, \ldots) = \sum_{i=2}^{\infty} \beta_i h_i,$$

then all $\beta_i = 0$, $x = 0$. However the module H_1 does not have an orthogonal complement. More precisely we have the following situation. Let $y = \sum_{j=1}^{\infty} \alpha_j e_j$ be in $H_1^{\perp}$. Then $\left\langle \sum_{j=1}^{\infty} \alpha_j e_j, h_i \right\rangle = 0$ for $i = 2, 3, \ldots$, so $\alpha_i + \alpha_1 \varphi_i = 0 \quad (i = 2, 3, \ldots)$, and $\alpha_i = -\alpha_1 \varphi_i$, hence

$$y = (\alpha_1, -\alpha_1 \varphi_2, -\alpha_1 \varphi_3, \ldots).$$

This is possible if and only if the function α_1 vanishes at 0: $\alpha_1(0) = 0$. If $H_1 \oplus H_1^{\perp} = H_A$, then for some α_1 we have $e_1 = y + \sum_{i=2}^{\infty} \beta_i h_i$. In particular the series $\sum_{i=2}^{\infty} \beta_i \bar{\beta}_i$ converges and

$$1 = \alpha_1 + \sum_{i=2}^{\infty} \frac{\beta_i \varphi_i}{(1 + \varphi_i{}^2)^{1/2}}.$$

But $\|\beta_i\|_A \to 0$, so for

$$\gamma = \sum_{i=2}^{\infty} \frac{\beta_i \varphi_i}{(1 + \varphi_i{}^2)^{1/2}}$$

we get $\gamma(0) = 0$, as well as for α_1. We come to a contradiction.

Let us investigate the involution J which determines a representation of $\mathbb{Z}_2$:

$$J(x) = \begin{cases} x & \text{if } x \in H_1, \\ -x & \text{if } x \in \operatorname{span}_A e_1, \end{cases}$$

This operator does not admit a conjugate. Indeed, let J^* exist. Then $(J^*)^2 = J^{2*} = \mathrm{Id}$, so J^* is also an involution.

$$\begin{aligned} J^* x = x \quad &\Leftrightarrow \quad (J^* x, y) = (x, y) \quad \forall y \quad \Leftrightarrow \quad (x, Jy) = (x, y) \quad \forall y \quad \Leftrightarrow \\ &\Leftrightarrow \quad (x, (J-1)y) = 0 \quad \forall y \quad \Leftrightarrow x \perp \operatorname{Im}(J-1) \quad \Leftrightarrow \\ &\Leftrightarrow \quad x \perp \operatorname{span}_A(e_1), \end{aligned}$$

and $J^*x = -x \quad \Leftrightarrow \quad x \perp H_1$. But H_1 has no orthogonal complement and so the involution J^* can not be defined. Nevertheless for the A-product averaged by the action of $\mathbb{Z}_2$

$$\langle x, y\rangle_2 = 1/2(\langle x, y\rangle + \langle Jx, Jy\rangle)$$

we get if $x \in H_1$, $y \in \mathrm{span}_A(e_1)$: $\quad \langle x, y\rangle_2 = 1/2(\langle x, y\rangle + \langle x, -y\rangle) = 0$, so the $+$ and $-$ subspaces of the involution are orthogonal to each other, and $J^*_{(2)} = J$.

Let us recall the definition of A-Fredholm operator [11, 13]. The theorem which will be proved is the crucial one for the possibility of construction of Sobolev chains in the C^*-case.

4.3. Definition. A bounded A-operator $F : H_A \to H_A$ admitting a conjugate is called Fredholm, if there exist decompositions of the domain of definition $H_A = M_1 \oplus N_1$ and the range $H_A = M_2 \oplus N_2$ where M_1, M_2, N_1, N_2 are closed A-modules, N_1, N_2 have a finite number of generators, and such that the operator F has in these decompositions the following form

$$F = \begin{pmatrix} F_1 & 0 \\ 0 & F_2 \end{pmatrix},$$

where $F_1 : M_1 \to M_2$ is an isomorphism.

4.4. Lemma. *Let $J : H_A \to H_A$ be a self adjoint injection. Then J is an isomorphism. Here injection means an injective A-homomorphism with closed range.*

Proof. Let us consider $J_1 = J : H_A \to J(H_A)$. It is an isomorphism of Hilbert modules admitting a conjugate $J_1^* = J^*|_{J(H_A)} = J|_{J(H_A)}$. Let $J_2 = J(J_1^*J_1)^{-1/2}$; then $\langle J_2x, J_2y\rangle = \langle x, y\rangle$ for every $x, y \in H_A$. We have $J_2(H_A) = J(H_A)$ and $J_2^*J_2 = 1$. Let $z \in H_A$ be an arbitrary element. Then

$$z = J_2J_2^*z + (z - J_2J_2^*z), \quad J_2J_2^*z \in J_2(H_A)$$

and

$$J_2^*(z - J_2J_2^*z) = J_2^*z - (J_2^*J_2)J_2^*z = J_2^*z - J_2^*z = 0,$$

so $(z - J_2J_2^*z) \in \mathrm{Ker}\, J_2^*$, but

$$\begin{aligned} x \in \mathrm{Ker}\, J_2^* \quad &\Leftrightarrow \quad \forall y : \quad \langle J_2^*x, y\rangle = 0 \quad \Leftrightarrow \\ &\Leftrightarrow \quad \forall y : \quad \langle x, J_2y\rangle = 0 \quad \Leftrightarrow \quad x \in J_2(H_A)^{\perp}. \end{aligned}$$

Hence $J_2J_2^*z \in J_2(H_A)$, $(z - J_2J_2^*) \in J_2(H_A)^\perp$, and

$$H_A = J_2(H_A) \widehat{\bigoplus} J_2(H_A)^\perp = J(H_A) \widehat{\bigoplus} J(H_A)^\perp.$$

So, if $J(H_A)^\perp = 0$, then J is an isomorphism. Let $x \in J(H_A)^\perp$, then $x \in J^*(H_A)^\perp$, so $\forall y: \ \langle x, J^*y\rangle = 0$ or $\forall y: \ \langle Jx, y\rangle = 0$, and $x \in \operatorname{Ker} J$. But J is an injection, and so, $x = 0$. ∎

4.5. Lemma. *Let $F : M \to H_A$ be an injection admitting a conjugate. Then*

$$F(M) \widehat{\bigoplus} F(M)^\perp = H_A.$$

Proof. We can assume by the stabilization theorem that $M = H_A^1 \cong H_A$. Then $F^*F : H_A^1 \to H_A^1$ is a self adjoint operator. Let $\|x\| = 1$, then

$$\|Fx\|^2 = \| \langle Fx, Fx\rangle \| \geqslant c^2$$

by injectivity and

$$\|F^*Fx\| = \|F^*Fx\| \, \|x\| \geqslant \| \langle F^*Fx, x\rangle \| = \| \langle Fx, Fx\rangle \| \geqslant c^2.$$

So $F^*F : H_A^1 \to H_A^1$ is a self adjoint injection and it is an isomorphism by the previous lemma. Moreover, $F^*F \geqslant 0$, and so, $(F^*F)^{-1/2}$ can be defined. Hence $U = F(F^*F)^{-1/2} : M \to H_A$ (which is an injection with $U(M) = F(M)$) is well defined. We have $U^*U = \mathrm{Id}_M$. Let $z \in H_A$ be an arbitrary element. Then

$$z = UU^*z + (z - UU^*z),$$

$$U^*(z - UU^*z) = U^*z - (U^*U)U^*z = U^*z - U^*z = 0.$$

Since $y \in \operatorname{Ker} U^* \ \Leftrightarrow \ \langle U^*y, x\rangle = 0 \forall x \ \Leftrightarrow \langle y, Ux\rangle = 0 \forall x \ \Leftrightarrow y \perp \operatorname{Im} U$ we get

$$U^*Uz \in \operatorname{Im} U = \operatorname{Im} F, \quad (z - UU^*z) \in (\operatorname{Im} U)^\perp.$$

The proof is finished because z is an arbitrary element. ∎

4.6. Lemma. *Let $H_A = M \oplus N$, $p : H_A \to M$ be a projection, N be a finitely generated projective module. Then $M \widehat{\oplus} M^\perp = H_A$ if and only if p admits a conjugate.*

Proof. If there exists p^*, then there exists $(1-p)^* = 1 - p^*$, so by [11] $\operatorname{Ker}(1-p) = M$ is the kernel of a self adjoint projection.

To prove the converse statement let us start from the case where N is a free module and let us prove first that $H_A = N^\perp + M^\perp$. By the Kasparov stabilization theorem we can assume that

$$N = \operatorname{span}_A \langle e_1, \ldots, e_n \rangle, \quad N^\perp = \operatorname{span}_A \langle e_{n+1}, e_{n+2}, \ldots \rangle .$$

Let g_i be the image of e_i by the projection of N on $M^\perp$:

$$e_1 = f_1 + g_1, \ldots, e_n = f_n + g_n, \quad f_i \in M,\ g_i \in M^\perp .$$

This projection is an isomorphism of A-modules $N \cong M^\perp$, so the elements $g_1, \ldots, g_n$ are free generators and $\langle g_k, g_k \rangle > 0_A$. Hence, if

$$f_k = \sum_{k=1}^{\infty} f_k^i e_i, \quad \text{then} \quad e_k - f_k^k e_k = \sum_{i \neq k} f_k^i e_i + g_k .$$

On the other hand

$$1 = \langle e_k, e_k \rangle = \langle f_k, f_k \rangle + \langle g_k, g_k \rangle , \quad 1 - (f_k^k)(f_k^k)^* \geqslant \langle g_k, g_k \rangle > 0.$$

Then by 2.1 the element $1 - f_k^k$ is invertible in A,

$$e_k = \frac{1}{1 - f_k^k} \left(\sum_{i \neq k} f_k^i e_i + g_k \right) \in N^\perp + M^\perp \quad (k = 1, \ldots, n),$$

so, $N^\perp + M^\perp = H_A$. Let $x \in N^\perp \cap M^\perp$. Every $y \in H_A = M \oplus N$ has the form $y = m + n$, so $\langle x, y \rangle = \langle x, m \rangle + \langle x, n \rangle = 0$, in particular, $\langle x, x \rangle = 0$ and $x = 0$. Hence, $H_A = N^\perp \oplus M^\perp$. Let us consider

$$q = \begin{cases} 1 & \text{on } N^\perp, \\ 0 & \text{on } M^\perp. \end{cases}$$

It is a bounded projection because $H_A = N^\perp \oplus M^\perp$. Let $x + y \in M \oplus N$, $x_1 + y_1 \in N^\perp \oplus M^\perp$. Then

$$\langle p(x+y), x_1 + y_1 \rangle = \langle x, x_1 + y_1 \rangle = \langle x, x_1 \rangle ,$$
$$\langle x + y, q(x_1 + y_1) \rangle = \langle x + y, x_1 \rangle = \langle x, x_1 \rangle .$$

Hence, there exists $p^* = q$.

To prove the general case let $\tilde{H}_A = H_A \widehat{\bigoplus} \tilde{N}$ with $N \widehat{\bigoplus} \tilde{N}$ a free module. Then, by the previous case,

$$M \widehat{\bigoplus} \tilde{M} = \tilde{H}_A,$$

$$M \widehat{\bigoplus} (M^\perp \widehat{\bigoplus} \tilde{N}) = H_A \widehat{\bigoplus} \tilde{N},$$

$$M \widehat{\bigoplus} M^\perp = H_A. \blacksquare$$

4.7. Theorem. *In the decomposition in the definition of A-Fredholm operator we can always assume M_0 and M_1 admitting an orthogonal complement. More precisely, there exists a decomposition for F*

$$\begin{pmatrix} F_3 & 0 \\ 0 & F_4 \end{pmatrix} : H_A = V_0 \oplus W_0 \to V_1 \oplus W_1 = H_A,$$

such that $V_0^\perp \widehat{\bigoplus} V_0 = H_A$, $V_1^\perp \widehat{\bigoplus} V_1 = H_A$, or (by the previous lemma it is just the same) such that the projections

$$p_0 : V_0 \oplus W_0 \to V_1, \quad p_1 : V_1 \oplus W_1 \to V_1$$

admit conjugates.

Proof. Let $W_0 = N_0$, $V_0 = W_0^\perp$. This orthogonal complement exists by [4], and $F|_{W_0^\perp}$ is an isomorphism. Indeed, if $x_n \in W_0^\perp$, then let $x_n = x_1^n + x_2^n$, $x_1^n \in M_0$, $x_2^n \in W_0$, $\|x_n\| = 1$. Let us assume that $\|Fx_n\| \to 0$. Then $\|Fx_1^n + Fx_2^n\| \to 0$, and, since $Fx_1^n \in V_1$, $Fx_2^n \in W_1$, $V_1 \oplus W_1 = H_A$, then this means that $\|Fx_1^n\| \to 0$ and $\|Fx_2^n\| \to 0$, and, since F_1 is an isomorphism, then $\|x_1^n\| \to 0$. If $a_1, \ldots, a_s$ are the generators of $W_0 = N_0$, then

$$0 = \langle x_n, a_j \rangle = \langle x_1^n, a_j \rangle + \langle x_2^n, a_j \rangle,$$

$$\| \langle x_2^n, a_j \rangle \| = \| \langle x_1^n, a_j \rangle \| \leqslant \|x_1^n\| \, \|a_j\| \to 0 \quad (n \to \infty)$$

for any $j = 1, \ldots, s$. Hence, since $x_2^n \in N$, we have $x_2^n \to 0 \quad (n \to \infty)$ and $x_n = x_1^n + x_2^n \to 0$, but this contradicts the equality $\|x_n\| = 1$. This contradiction shows that $F|_{W_0^\perp}$ is an isomorphism.

Let $V_1 = F(V_0)$. Since $W_0 = N_0$, we can assume that $W_1 = N_1$. Indeed, any $y \in H_A$ has the form $y = m_1 + n_1 = F(m_0) + n_1$, where $m_1 \in M_1$, $n_1 \in$

N_1, $m_0 \in M_0$. On the other hand, $m_0 = v_0 + n_0$, where $v_0 \in V_0$, $n_0 \in W_0 = N_0$, and

$$y = F(v_0 + n_0) + n_1 = F(v_0) + (F(n_0) + n_1) \in V_1 + N_1.$$

Hence, $H_A = V_1 + W_1$.

Let $y \in V_1 \cap W_1 = V_1 \cap N_1$, so that $n_1 = y = F(v_0)$, $n_1 \in N_1$, $v_0 \in V_0$. Let us decompose $v_0 + n_0$, where $m_0 \in M_0$, $n_0 \in N_0$. Then

$$n_1 = F(m_0) + F(n_0),$$
$$F(m_0) = n_1 - F(n_0), \quad F(m_0) \in M_1, \quad n_1 - F(n_0) \in N_1.$$

Hence $F(m_0) = 0$, $\;n_1 - F(n_0) = 0$, and since $F : M_0 \cong M_1$, then $m_0 = 0$. We have $v_0 \in V_0 = N_0^{\perp}$ and hence,

$$0 = \langle v_0, n_0 \rangle = \langle m_0 + n_0, n_0 \rangle = \langle n_0, n_0 \rangle, \quad n_0 = 0.$$

So, $v_0 = m_0 + n_0 = 0$, $y = F(v_0) = 0$. Hence $V_1 \cap W_1 = 0$ and $H_A = V_1 \oplus W_1$.

By 4.5 V_1 has an orthogonal complement $V_1^{\perp}$, $V_1 \widehat{\bigoplus} V_1^{\perp} = H_A$, and this completes the proof. ■

4.8. Remark. If we do not restrict the operator F to admit a conjugate, we can assert that there exists a decomposition

$$F : N_0^{\perp} \oplus N_0 \to M_1 \oplus L_n,$$

where $L_n = \operatorname{span}_A(e_1, \dots, e_n)$, but M_1 may have no orthogonal complement. This result was proved in [6].

5. Lefschetz numbers with values in $HC_0(A)$

5.1. Definition. Let $\{e_1, e_2, \dots\}$ be an A-orthobasis of $H_A = l_2(A)$ (the Hilbert module over A) with A-inner product $(\ ,\ )$. Let $S \in \operatorname{End}_A^* H_A$ (the A-linear endomorphisms of H_A admitting an adjoint) and $S(e_i) = 0$ $(i > k)$. We define the trace of S by

$$t(S, \{e_i\}, k) = \sum_{i=1}^{\infty} f((Se_i, e_i)) = \sum_{i=1}^{k} f(S_i^i),$$

where $f : A \to A/[A, A] = HC_0(A)$, $\|S_j^i\|$ is the matrix of S with respect to $\{e_i\}$, $S_j^i \in A$.

5.2. Lemma. *$t(S,\{e_i\},k) = t(S,\{e_i\},l) := t(S,\{e_i\})$ for $l \geqslant k$.*

The proofs of this lemma and the other statements of this Section can be found in [18].

5.3. Lemma. *Let S, $\{e_i\}$, k be as in 5.1 and $\{h_j\}$ a new A-basis of H_A (in general non-orthogonal). Then the series*

$$\sum_{r=1}^{\infty} f((S_h)_r^r)$$

converges to $t(S,\{e_i\})$, where $(S_h)_r^p$ are the matrix elements of S with respect to $\{h_i\}$.

Let us note that a basis of H_A is a system of elements $\{h_i\}$, such that $h_i = Be_i$, where $B \in \mathrm{GL}^*$ (automorphisms admitting a conjugate). The matrix of S with respect to the $\{h_i\}$ is the matrix of $B^{-1}SB$ with the respect to $\{e_i\}$, i.e., $(S_h)_j^i = (B^{-1}SB)_j^i = \langle B^{-1}SBe_i, e_j\rangle$.

So we can give instead of 5.1 the following correct definition.

5.4. Definition. Let $S \in \mathrm{End}_A^* H_A$, M and N Hilbert submodules of H_A, N finitely generated, $H_A = M \oplus N$, $S|_M = 0$. For an arbitrary basis $\{e_i\}$ we define

$$t(S) = \sum_{i=1}^{\infty} f(S_i^i).$$

5.5. Lemma. *Let M, N, S be as in 5.4, and $\tilde{N}$ be a countably generated Hilbert A-module, $\tilde{H}_A = H_A \widehat{\bigoplus} \tilde{N} \cong H_A$,*

$$\tilde{S} = \begin{pmatrix} S & 0 \\ 0 & 0 \end{pmatrix} : H_A \widehat{\bigoplus} \tilde{N} \to H_A \widehat{\bigoplus} \tilde{N}.$$

Then $t(S) = t(\tilde{S})$.

5.6. Lemma. *Let M, N, S be as in 5.4, $M \cong H_A$, $N = \bar{N} \oplus \bar{\bar{N}}$, $S|_{\bar{\bar{N}}} = 0$. Then*

$$t(S) = t(pSp),$$

where $p : M \oplus \bar{N} \oplus \bar{\bar{N}} \to M \oplus \bar{N}$ is a projection, and the sum on the right is in the space $M \oplus \bar{N} \cong H_A$. Let us notice, that if we denote by

$$q : M \oplus N \to M, \quad p_1 : N \to \bar{N}$$

the projections, then they admit conjugates. Hence, the projection $p = q + p_1(1-q)$ admits one, too.

5.7. Corollary. *If in 5.5 $M \oplus \bar{N}$ is orthogonal to $\bar{\bar{N}}$, and $\{h_i\}$ is an A-orthobasis of $M \oplus \bar{N}$, then*

$$t(S) = \sum_{i=1}^{\infty} f(\langle Sh_i, h_i \rangle). \blacksquare$$

Definition. Let $F : H_A \to H_A$ be an A-Fredholm operator (admitting an adjoint),

$$\begin{pmatrix} F_1 & 0 \\ 0 & F_2 \end{pmatrix} : H_A = M_0 \oplus N_0 \to M_1 \oplus N_1 = H_A \qquad (D)$$

a corresponding decomposition, restricted to satisfy the condition as in 4.7 (we always will assume this without specification). Let S_0 and S_1 be operators from $\mathrm{End}^*_A H_A$, such that the diagram

$$\begin{array}{ccc} H_A & \xrightarrow{F} & H_A \\ {\scriptstyle S_0}\downarrow & & \downarrow{\scriptstyle S_1} \\ H_A & \xrightarrow{F} & H_A. \end{array}$$

commutes. Let

$$\tilde{S}_0 = \begin{cases} 0 \text{ on } M_0, \\ S_0 \text{ on } N_0, \end{cases} \qquad \tilde{S}_1 = \begin{cases} 0 \text{ on } M_1, \\ S_1 \text{ on } N_1. \end{cases}$$

Let us define

$$L(F, S, D) = t(\tilde{S}_0) - t(\tilde{S}_1).$$

5.9. Lemma. *Let*

$$H_A = M_0 \oplus N_0 \to M_1 \oplus N_1 = H_A, \qquad (D)$$

$$H_A = \tilde{M}_0 \oplus N_0 \to \tilde{M}_1 \oplus N_1 = H_A \qquad (\tilde{D})$$

then

$$L(F, S, D) = L(F, S, \tilde{D}).$$

5.10. Lemma. *Let*

$$H_A = (M_0 \oplus N_0) \oplus K_0 \to (M_1 \oplus N_1) \oplus K_1 = H_A, \quad (D_1)$$

$$H_A = M_0 \oplus (N_0 \oplus K_0) \to M_1 \oplus (N_1 \oplus K_1) = H_A \quad (D_2)$$

be two decompositions for F. Then $L(F, S, D_1) = L(F, S, D_2)$.

5.11. Lemma. *Let*

$$H_A = M_0 \oplus N_0 \to M_1 \oplus N_1 = H_A \qquad (D)$$

and

$$H_A = \bar{M}_0 \oplus \bar{N}_0 \to \bar{M}_1 \oplus \bar{N}_1 = H_A \qquad (\bar{D})$$

be two decompositions for F. Then $L(F, S, D) = L(F, S, \bar{D})$. So L does not depend on D and we denote it by $L(F, S)$.

5.12. Remark. By the stabilization theorem and Lemma 5.5, we can define $L(F, S)$ for any countably generated Hilbert A-module instead of H_A.

5.13. Definition. Let $T = \{T_i\}$ be an endomorphism of an A-elliptic complex E:

$$\begin{array}{ccccccc} 0 & \longrightarrow & \Gamma(E_0) & \xrightarrow{d_0} & \Gamma(E_1) & \longrightarrow & \dots \\ & & \downarrow T_0 & & \downarrow T_1 & & , \\ 0 & \longrightarrow & \Gamma(E_0) & \xrightarrow{d_0} & \Gamma(E_1) & \longrightarrow & \dots \end{array}$$

$$T_{i+1} d_i = d_i T_i, \quad T_i \in \operatorname{End}^*_A \Gamma(E_i).$$

Assume the following

5.14. Condition. Sobolev products in $\Gamma(E)$ can be chosen in such a way that

$$T_i d_i^* = d_i^* T_{i+1}.$$

We take $E_{ev} = \oplus E_{2i}$, $E_{od} = \oplus E_{2i+1}$,

$$F = d + d^* : \Gamma(E_{ev}) \to \Gamma(E_{od}).$$

Then F is an A-Fredholm operator and the diagram stated below commutes, where

$$S_0 = \oplus T_{2i}, \qquad S_1 = \oplus T_{2i+1}.$$

$$\begin{array}{ccc} \Gamma(E_{ev}) & \xrightarrow{F} & \Gamma(E_{od}) \\ {\scriptstyle S_0}\downarrow & & \downarrow{\scriptstyle S_1} \\ \Gamma(E_{ev}) & \xrightarrow{F} & \Gamma(E_{od}). \end{array}$$

We define *the Lefschetz number of the second type* as

$$L_0(E, T, m) = L(F, S) \in HC_0(A),$$

where m denotes the dependence on inner products (via d^*).

5.15. Lemma. *Let $T = T_g$, $g \in G$ as in §2. Then the condition 5.14 is fulfilled.*

5.16. Theorem. *If $T = T_g$, $g \in G$, then*

$$L_0(E, T_g, m_G) = \tilde{\mathrm{Ch}}_0^0(L_1(g, E)),$$

where Ch_0^0 is the Chern character

$$\mathrm{Ch}_0^0 : K_0(A) \to HC_0(A)$$

(see [3, 7, 8]), and

$$\tilde{\mathrm{Ch}}_0^0(a \otimes z) = \mathrm{Ch}_0^0(a)z, \qquad z \in \mathbb{C}.$$

In particular, L_0 does not depend on m_G.

Proof. We have

$$L_1(g, E) = \mathrm{ind}_{G,A}^X([\sigma(E)])(g) = \mathrm{ind}_{G,A}^X(F)(g).$$

Let

$$M_o \oplus N_0 \to M_1 \oplus N_1 \qquad (D)$$

be a decomposition for F. Then by 2.8 and [15]

$$N_0 = \bigoplus_{k=1}^{K} V_k \otimes P_k, \qquad N_1 = \bigoplus_{l=1}^{L} W_l \otimes Q_l,$$

where V_k and W_l are $\mathbb{C}$-vector spaces of irreducible representations of G, P_k and Q_l are G-trivial projective finitely generated A-modules. Then (representations are unitary)

$$\mathrm{ind}_{G,A}^X(F) = \sum_{k=1}^{K} [P_k] \otimes \chi(V_k) - \sum_{l=1}^{L} [Q_l] \otimes \chi(W_l)$$

and

$$L_1(g, E) = \sum_{k=1}^{K} [P_k] \otimes \mathrm{Trace}(g|V_k) - \sum_{l=1}^{L} [Q_l] \otimes \mathrm{Trace}(g|W_l). \tag{2}$$

The end of the proof see in [18]. ■

6. Lefschetz numbers with values in $HC_{2l}(A)$

Let W^*A be the universal enveloping von Neumann algebra of the algebra A with the norm topology. Let U be a unitary operator in the Hilbert module A^n. Then

$$U = \int_{S^1} e^{i\varphi}\, dP(\varphi), \tag{3}$$

where $P(\varphi)$ is the projection valued measure valued in the space of matrices $M(n, W^*A)$, and the integral converges with respect to the norm. Let us associate with the integral sum

$$\sum_k e^{i\varphi_k} P(E_k)$$

the following class of the cyclic homology $HC_{2l}(M(n, W^*A))$:

$$\sum_k P(E_k) \otimes \ldots \otimes P(E_k) \cdot e^{i\varphi_k}.$$

Passing to the limit we get the following element

$$\tilde{T}U = \int_{S^1} e^{i\varphi}\, d(P \otimes \ldots \otimes P)(\varphi) \in HC_{2l}(M(n, W^*A)).$$

Then we define

$$T(U) = \mathrm{Tr}^n_* \tilde{T}U \in HC_{2l}(W^*A).$$

6.1. Lemma. *Let $J : M = A^m \to N = A^n$ be an isomorphism, $U_M : M \to M$, $U_N : N \to N$ be A-unitary operators and $JU_M = U_N J$. Then*

$$T(U_M) = T(U_N).$$

Proof. If

$$U_M = \int_{S^1} e^{i\varphi}\, dP(\varphi),$$

then

$$U_N = JU_M J^{-1} = \int_{S^1} e^{i\varphi}\, dJPJ^{-1}(\varphi).$$

To verify the equality $T(U_M) = T(U_N)$ it is sufficient to verify that

$$\mathrm{Tr}^m_* \left[\sum_k P(E_k) \otimes \ldots \otimes P(E_k) \cdot e^{i\varphi_k} \right] =$$

$$= \mathrm{Tr}^n_* \left[\sum_k JP(E_k)J^{-1} \otimes \ldots \otimes JP(E_k)J^{-1} \cdot e^{i\varphi_k} \right] \in HC_{2l}(W^*A),$$

but this follows from well-definedness of the Chern character

$$\mathrm{Ch}^0_{2l} : K_0(B) \to HC_{2l}(B)$$

(see [3, 8]). ■

Let now U be equal to U_g, i.e. an operator representing $g \in G$. Then (3) turns to be the sum associated with the decomposition from 2.8 and [15]

$$A^n \cong \bigoplus_{k=1}^{M} Q_k \otimes V_k,$$

where $V_k \cong \mathbb{C}^{L_k}$, and Q_k are projective A-modules of finite type. Then

$$U_g \left(\sum_{k=1}^{M} x_k \otimes v_k \right) = \sum_{k=1}^{M} x_k \otimes u^k_g v_k = \sum_{k=1}^{M} \sum_{l=1}^{L_k} x_k \otimes e^{i\varphi^k_l} v^l_k f_l,$$

where $f_1, \ldots, f_{L_k}$ is the diagonalizing basis for u^k_g; $v_k = \sum v^l_k f_l$. Then we can define

$$\tau(U_g) = \sum_{k=1}^{M} \sum_{l=1}^{L_k} \mathrm{Ch}^0_{2l}[P_k] \cdot \mathrm{Trace}(u^k_g) \in HC_{2l}(A). \tag{4}$$

We have $T(U_g) = i_*(\tau(U_g))$, where $i : A \to W^*A$.

A similar technique can be developed for a projective module N instead of A^n. For this purpose we take $N = q(A^n)$,

$$U \oplus 1 : A^n \cong N \oplus (1-q)A^n \to N \oplus (1-q)A^n \cong A^n,$$

$$\tilde{T}U = \int_{S^1} e^{i\varphi}\, d(qPq \otimes \ldots \otimes qPq)(\varphi).$$

The well-definedness is an immediate consequence of Lemma 6.1.

Let us consider a G-invariant A-elliptic complex (E, d), and let the Sobolev A-products be chosen invariant, so that $T_g = U_g$ are unitary operators (see §3).

6.2. Lemma. *We can choose a decomposition for the A-Fredholm operator*

$$F = d + d^* : \Gamma(E_{ev}) \to \Gamma(E_{od}),$$

$$F : M_0 \oplus \tilde{N}_0 \to M_1 \oplus \tilde{N}_1, \quad F : M_0 \cong M_1,$$

such that

$$\begin{array}{ll} \tilde{N}_0 = \oplus_i N_{2i}, & N_{2i} \subset \Gamma(E_{2i}), \\ \tilde{N}_1 = \oplus_i N_{2i+1}, & N_{2i+1} \subset \Gamma(E_{2i+1}), \end{array}$$

where N_m are projective invariant modules.

Proof. Let us assume that the complex consists of operators of the degree m, so $F = d + d^*$ is an A-Fredholm operator in the spaces $H^m(E_{ev}) \to H^0(E_{od})$. We can choose the basis in $H^m(E_{ev})$ (or the decomposition into modules P_j in $l_2(P)$) in such a way that $e_{ms+j} \in \Gamma(E_{2j})$, where $E_0, E_2, \ldots, E_{2j}, \ldots, E_{2m}$ are all non-zero terms of the complex, $s \in \mathbb{N}$, $j = 0, \ldots, m$ (and in a similar way for P_j). As usual, without loss of generality we can assume that

$$\tilde{N}_0 = \mathrm{span}_A(e_1, \ldots, e_{n_0}), \quad M_0 = \mathrm{span}_A(e_{n_0+1}, e_{n_0+2}, \ldots),$$

and $M_1 = F(M_0)$ has in $H^0(E_{od})$ the A-orthogonal complement $M_1^{\perp}$. Then for every $x \in M_1$, $y \in \tilde{N}_0$

$$\langle x, Fy \rangle = \langle Fx, y \rangle_0, \tag{5}$$

where the first brackets mean the pairing of a functional and an element. So, $F(\tilde{N}_0) \subset M_1^{\perp}$ and taking $\tilde{N}_1 = M_1^{\perp}$, we get a decomposition $F : M_0 \oplus \tilde{N}_0 \to M_1 \oplus \tilde{N}_1$.

Let

$$y = y_1 + y_3 + \cdots + y_{2m+1} \in \tilde{N}_1 \subset H^0(E_{od}), \quad y_{2j+1} \in H^0(E_{2j+1}),$$

and

$$x = x_0 + x_2 + \cdots + x_{2m} \in M_0 \subset H^m(E_{ev}), \quad x_{2j} \in H^m(E_{2j}).$$

Then $\langle Fx, y \rangle_0 = 0$, where

$$\begin{array}{ccccccc} Fx & = & d^* x_0 & + & \sum_{i=1}^m (dx_{2i-2} + d^* x_{2i}) & + & dx_{2m} \in \\ & \in & 0 & \oplus & \oplus_{i=1}^m H^0(E_{2i+1}) & \oplus & 0. \end{array}$$

Since (E, d) is a complex, $d^2 = 0$ and

$$\langle du, d^* v \rangle = \langle d^2 u, v \rangle = 0,$$

so

$$\begin{array}{ll} \langle y_{2j+1}, dx_{2j} \rangle = 0, & \langle y_{2j+1}, d^* x 2j + 2 \rangle = 0 \quad (j = 0, 1, \ldots, m) \\ \langle y_{2j+1}, dx \rangle = 0, & \langle y_{2j+1}, d^* x \rangle = 0. \end{array}$$

Hence $e_{2j+1} \in F(M_0)^{\perp} = M_1^{\perp} = \tilde{N}_1$, and

$$\tilde{N}_1 = \oplus_i (\tilde{N}_1 \cap \Gamma(E_{2i+1})) = \oplus_i N_{2i+1}.$$

■

6.3. Definition. *The Lefschetz number* L_{2l} we define as

$$L_{2l}(E, U_g, m_G) = \sum_i (-1)^i \tau(U_g|N_i) \in HC_{2l}(A),$$

where m_G denotes the dependence on inner products (via d^*).

Remark. For more general situations we hope to use T instead of τ.

6.4. Lemma. *The definition of* L_{2l} *is correct, i.e. this number does not depend on the choice of decompositions in Lemma 6.2.*

Proof. For any two decompositions we can by use of projection (as in [13, 15]) replace $\tilde{N}_0$ by a module inside $\mathrm{span}_A(e_1, \ldots, e_n)$ for a sufficiently great n (we use the notation of Lemma 6.2). By 6.1 $\tau(U_g|N_i)$ does not change under this replacement. So we can assume that we have to compare the decomposition as in 6.2 and the decomposition

$$F : \bar{M}_0 \oplus \bar{\tilde{N}}_0 \to \bar{M}_1 \oplus \bar{\tilde{N}}_1,$$

$$\bar{\tilde{N}}_0 = \oplus_i \bar{N}_{2i}, \quad \bar{N}_{2i} \subset N_{2i} \subset \Gamma(E_{2i}),$$

$$\bar{\tilde{N}}_1 = \oplus_i \bar{N}_{2i+1}, \quad \bar{N}_{2i+1} \subset \Gamma(E_{2i}).$$

Hence by (5), $\bar{N}_{2i+1} \subset N_{2i+1}$. Let $K_i = (\bar{N}_i)^{\perp}_{N_i}$. Then $F : K_{2i} \cong K_{2i+1}$ and by Lemma 6.1 we get $\tau(U_g|K_{2i}) = \tau(U_g|K_{2i+1})$. Hence

$$\sum_i (-1)^i \tau(U_g|N_i) = \sum_i (-1)^i \left(\tau(U_g|\bar{N}_i) + \tau(U_g|K_i)\right) =$$

$$= \sum_i (-1)^i \left(\tau(U_g|\bar{N}_i)\right). \quad \blacksquare$$

6.5. Theorem. *Let* $\tilde{\mathrm{Ch}}^0_{2l}(a \otimes z) = \mathrm{Ch}^0_{2l}(a) \cdot z$, *where* $z \in \mathbb{C}$. *Then*

$$L_{2l}(E, U_g, m_G) = \tilde{\mathrm{Ch}}^0_{2l}(L_1(g, E)),$$

in particular, L_{2l} *does not depend on* m_G.

Proof. We get the statement immediately from (2) and (4). $\blacksquare$

References

1. Atiyah M. F. and Segal G. B., *The index of elliptic operators. II.*, Ann. of Math. (2) **87** (1958), 531–545.
2. Barut A. O., Rączka R., *Theory of group representations and applications*, PWN – Polish Scientific Publishers, Warszawa, 1977.
3. Connes A., *Non-commutative differential geometry*, Publ. Math. IHES **62** (1985), 41–144.
4. Dupré M. J., Fillmore P. A., *Triviality theorems for Hilbert modules*, Topics in modern operator theory, 5th International conference on operator theory. Timisoara and Herculane (Romania), June 2–12, 1980, Birkhäuser Verlag, Basel-Boston-Stuttgart, 1981, pp. 71–79.
5. Friedrich T., *Vorlesungen über K-Theorie*, Teubner, Leipzig, 1987.
6. Irmatov A., *On a new topology in the space of Fredholm operators*, Annals of Global Analysis and Geometry **7** no. 2 (1989), 93–106.
7. Karoubi M., *Homologie cyclique des groupes et des algèbres*, C. R. Acad. Sci. Paris, Série 1 **297** (1983), 381–384.
8. Karoubi M., *Homologie cyclique et K-théorie algébrique. I.*, C. R. Acad. Sci. Paris. Série 1 **297** no. 8 (1983), 447–450.
9. Kasparov G. G., *Topological invariants of elliptic operators. I: K-homology*, Izv. Akad. Nauk SSSR Ser. Mat. **39** (1975), 796–838 (in Russian); (English transl.: Math. USSR – Izv. **9** (1975), 751–792).
10. Kasparov G. G., *Hilbert C^*-modules: theorems of Stinespring and Voiculescu*, J. Operator Theory **4** (1980), 133–150.
11. Mishchenko A. S., *Banach algebras, pseudo differential operators and their application to K-theory*, Usp. Mat. Nauk **34** no. 6 (1979), 67–79 (in Russian); (English transl.: Russ. Math. Surv. **34**, No. 6 (1979), 77–91).
12. Mishchenko A. S., *Representations of compact groups in Hilbert modules over C^*-algebras*, Trudy Mat. Inst. Steklov **166** (1984), 161–176 (in Russian); (English transl.: Proc. Steklov Inst. Math., 1986, No. 1 (166)).
13. Mishchenko A. S., Fomenko A. T., *The index of elliptic operators over C^*-algebras*, Izv. Akad. Nauk SSSR. Ser. Mat. **43** no. 4 (1979), 831–859 (in Russian); (English transl.: Math. USSR – Izv., 1980, V. 15, 87–112.).
14. Troitsky E. V., *The contractibility of the complete general linear group of the C^*-Hilbert module $l_2(A)$*, Func. Anal. i Pril. **20** no. 4 (1986), 58–64 (in Russian); (English transl.: Funct. Anal. Appl., 1986, V. 20, No. 4, 301–307).
15. Troitsky E. V., *The index of equivariant elliptic operators over C^*-algebras*, Ann. Global Anal. Geom. **5** no. 1 (1987), 3–22.
16. Troitsky E. V., *An exact K-cohomological C^*-index formula. I. The Thom isomorphism and the topological index*, Vestnik Mosc. Univ. Ser. 1. Mat. Meh. no. 2 (1988), 83–85 (in Russian); (English transl.: Moscow Univ. Math. Bull., 1988, v. 43, No. 2, 57–60).
17. Troitsky E. V., *An exact K-cohomological C^*-index formula. II: The index theorem and its applications*, Usp. Mat. Nauk. **44** no. 1, 213–214 (in Russian); (English transl.:

Russian Math. Surv., 1989, V. 44, No. 1, 259–261).

18. Troitsky E. V., *Lefschetz numbers of C^*-complexes*, Algebraic Topology, Poznań 1989 (S. Jackowski, B. Oliver, and K. Pawałowski, eds.), Lect. Notes in Math., vol. 1474, Springer-Verlag, Berlin, Heidelberg, and New York, 1991, pp. 193–206.

Chair of Higher Geometry and Topology, Dept. of Mech. and Math., Moscow State University, Moscow, 119899, Russia

email: **troitsky@mech.math.msu.su**

Assembly

Michael Weiss and Bruce Williams

Abstract. The goal of assembly is to approximate homotopy invariant functors from spaces to spectra by homotopy invariant and excisive functors from spaces to spectra. We show that there exists a *best* approximation, characterized by a universal property.

1. The Ordinary Assembly Map

We adopt a very category theoretic point of view in describing assembly maps. It has been formulated explicitly by Quinn in the appendix to [Q], and more implicitly in Quinn's thesis, in [QGF], in [And] and in articles of Waldhausen, e.g. [Wa1], [Wa2]. See also [QAB]. From this point of view, the goal of assembly is: Given a homotopy invariant functor F from spaces to spectra, to approximate F from the left by an *excisive* homotopy invariant functor $F^{\%}$.

In this section, all spaces are homotopy equivalent to CW-spaces, all pairs of spaces are homotopy equivalent to CW-pairs, and all spectra are CW-spectra.

A functor $\boldsymbol{F}$ from spaces to spectra is *homotopy invariant* if it takes homotopy equivalences to homotopy equivalences. A homotopy invariant $\boldsymbol{F}$ is *excisive* if $\boldsymbol{F}(\emptyset)$ is contractible and if $\boldsymbol{F}$ preserves homotopy pushout squares (alias homotopy cocartesian squares, see [Go1], [Go2]). The excision condition implies that $\boldsymbol{F}$ preserves finite coproducts, up to homotopy equivalence. Call $\boldsymbol{F}$ *strongly excisive* if it preserves arbitrary coproducts, up to homotopy equivalence.

If $\boldsymbol{F}$ is strongly excisive, then the functor $\pi_*\boldsymbol{F}$ from spaces to graded abelian groups is a generalized homology theory—it has Mayer–Vietoris sequences, and satisfies the strong wedge axiom. Conversely, homotopy theorists know that any generalized homology theory satisfying the strong wedge axiom is isomorphic to one of the form $\pi_*\boldsymbol{F}$ where $\boldsymbol{F}(X) = X_+ \wedge \boldsymbol{Y}$ and $\boldsymbol{Y}$ is a fixed spectrum. Such an $\boldsymbol{F}$ is of course strongly excisive.

1991 *Mathematics Subject Classification.* Primary 55P65; Secondary 55N20, 55P42, 19D10, 19E20.

1.1. Theorem. *For any homotopy invariant $\mathbf{F}$ from spaces to spectra, there exist a strongly excisive (and homotopy invariant) $\mathbf{F}^{\%}$ from spaces to spectra and a natural transformation*

$$\alpha = \alpha_{\mathbf{F}} : \mathbf{F}^{\%} \longrightarrow \mathbf{F}$$

such that $\alpha : \mathbf{F}^{\%}(*) \to \mathbf{F}(*)$ is a homotopy equivalence. Moreover, $\mathbf{F}^{\%}$ and $\alpha_{\mathbf{F}}$ can be made to depend functorially on $\mathbf{F}$.

Preliminaries. We are going to use homotopy colimits in the proof. Here is a description: Let Z be a functor from a small category $\mathcal{C}$ to the category of spaces. For $n \geq 0$ let $[n]$ be the ordered set $\{0, 1, \ldots, n\}$; we view this as a category, with exactly one morphism from i to j whenever $i \leq j$, and no morphism from i to j if $i > j$. The homotopy colimit of Z, denoted hocolim Z, is the geometric realization of the simplicial space

$$n \mapsto \coprod_{G:[n]\to\mathcal{C}} Z(G(0))$$

where the coproduct must be taken over all covariant functors G from $[n]$ to $\mathcal{C}$. We hope the face and degeneracy maps are obvious. See [BK] for more details. It is often convenient to use informal notation for a homotopy colimit, e.g.

$$\operatorname*{hocolim}_{C \text{ in } \mathcal{C}} Z(C)$$

instead of hocolim Z. This is particularly true when the values of the functor have "names" and the functor as such has not been named.

A special case of special interest: When $Z(C)$ is a point for every C in $\mathcal{C}$, then clearly hocolim Z is the classifying space of $\mathcal{C}$. (We shall also say: the *nerve* of $\mathcal{C}$; strictly speaking, the nerve of $\mathcal{C}$ is a simplicial set, and the classifying space of $\mathcal{C}$ is the geometric realization of the nerve of $\mathcal{C}$.) More generally, when Z is a constant functor, then hocolim Z is the product of the classifying space of $\mathcal{C}$ with the constant value of Z. In some examples below, $\mathcal{C}$ is the category of faces of an incomplete simplicial set ; then the classifying space of $\mathcal{C}$ is the barycentric subdivision of the incomplete simplicial set. (An *incomplete* simplicial set is a simplicial set without degeneracy operators.)

In general, a key property of homotopy colimits is their homotopy invariance. Suppose that $f : Z \to Z'$ is a natural transformation between functors from $\mathcal{C}$ to spaces. If f_C from $Z(C)$ to $Z'(C)$ is a homotopy equivalence for every C in $\mathcal{C}$, then f_* from hocolim Z to hocolim Z' is a homotopy equivalence.

Variations: The above formula for hocolim Z remains meaningful when Z is a functor from $\mathcal{C}$ to spaces or spectra. Bear in mind that the geometric realization of a simplicial pointed space or simplicial spectrum $[n] \mapsto X_n$ is given by a formula of type $\left(\coprod_n \Delta^n_+ \wedge X_n\right) / \sim$ where $\sim$ stands for the usual relations.

First proof of 1.1. For a space X, let simp(X) be the category whose objects are maps $\Delta^n \to X$ where $n \geq 0$, and whose morphisms are commutative triangles

$$\begin{array}{ccc} \Delta^m & \xrightarrow{f_*} & \Delta^n \\ & \searrow \quad \swarrow & \\ & X & \end{array}$$

where f_* is the map induced by a monotone injection f from $\{0, 1, \ldots, m\}$ to $\{0, 1, \ldots, n\}$. Let $\boldsymbol{F}_X$ from simp(X) to spectra be the covariant functor sending $g : \Delta^n \to X$ to $\boldsymbol{F}(\Delta^n)$, and let

$$\boldsymbol{F}^{\%}(X) := \text{hocolim}\, \boldsymbol{F}_X\,.$$

For each $g : \Delta^n \to X$ in simp(X) we have $g_* : \boldsymbol{F}(\Delta^n) \to \boldsymbol{F}(X)$. Letting g vary, we regard this as a natural transformation from $\boldsymbol{F}_X$ to the constant functor with value $\boldsymbol{F}(X)$. It induces

$$\alpha : \boldsymbol{F}^{\%}(X) \longrightarrow \boldsymbol{F}(X)\,.$$

Clearly α is a homotopy equivalence when X is a point. For arbitrary X, and $g : \Delta^n \to X$ in simp(X), we have the map $\Delta^n \to *$ which induces $\boldsymbol{F}(\Delta^n) \to \boldsymbol{F}(*)$, a homotopy equivalence. We regard this as a natural transformation from $\boldsymbol{F}_X$ to the constant functor with value $\boldsymbol{F}(*)$; by the homotopy invariance of homotopy colimits, the induced map of homotopy colimits is a homotopy equivalence

$$\boldsymbol{F}^{\%}(X) \longrightarrow |\text{simp}(X)|_+ \wedge \boldsymbol{F}(*)\,.$$

It is an exercise to show that $|\,\text{simp}(X)| \simeq X$. Thus $\boldsymbol{F}^{\%}(X)$ is related to $X_+ \wedge \boldsymbol{F}(*)$ by a chain of natural homotopy equivalences. □

Second proof of 1.1. We compose $\boldsymbol{F}$ with the geometric realization functor from incomplete simplicial sets to spaces, and henceforth assume that $\boldsymbol{F}$ is a functor from incomplete simplicial sets to spaces. For an incomplete simplicial set X, we define simp(X) much as before: objects are the simplicial maps $\Delta^n \to X$, for arbitrary n. (These are in bijection with the simplices of X.) We define $\boldsymbol{F}_X$ from simp(X) to spectra much as before. We let

$$\boldsymbol{F}^{\%}(X) = \operatorname{hocolim} \boldsymbol{F}_X$$

as before, and we define $\alpha : \boldsymbol{F}^{\%}(X) \to \boldsymbol{F}(X)$ as before. Then we observe that $\boldsymbol{F}^{\%}(X)$ has a natural filtration:

$$\boldsymbol{F}^{\%}(X) = \bigcup_k \boldsymbol{F}^{\%}(X^k)$$

where X^k is the k–skeleton. Applying the homotopy invariance of $\boldsymbol{F}$ to the constant map from a simplex to a point, one finds that

$$\boldsymbol{F}^{\%}(X^k)/\boldsymbol{F}^{\%}(X^{k-1}) \simeq \bigvee_z \mathbb{S}^k \wedge \boldsymbol{F}(*)$$

where z runs over the k–simplices of X. Hence the natural filtration of $\boldsymbol{F}^{\%}(X)$ leads to a spectral sequence converging to the homotopy groups of $\boldsymbol{F}^{\%}(X)$, with

$$E^2_{p,q} = H_p(X; \pi_q \boldsymbol{F}(*))$$

as E^2–term. But if the E^2–term is already homotopy invariant, then so is the E^∞–term, which implies the homotopy invariance of $\boldsymbol{F}^{\%}$. Also, we see that $\alpha : \boldsymbol{F}^{\%}(X) \to \boldsymbol{F}(X)$ is a homotopy equivalence for $X = *$. Further, we see that the functor

$$X \mapsto \boldsymbol{F}^{\%}(X^k)/\boldsymbol{F}^{\%}(X^{k-1})$$

takes squares of simplicial sets of the form

$$\begin{array}{ccc} X_1 \cap Y_2 & \longrightarrow & X_1 \\ \downarrow & & \downarrow \\ X_2 & \longrightarrow & X_1 \cup X_2 \end{array}$$

to homotopy pushout squares, and preserves arbitrary coproducts (up to homotopy equivalence). Using induction on k, we conclude that the functors

$$X \mapsto \boldsymbol{F}^{\%}(X^k)$$

have these properties, too ; then $\boldsymbol{F}^{\%}$ itself has these properties. Together with homotopy invariance this implies that $\boldsymbol{F}^{\%}$ is strongly excisive. □

1.2. Observation. *$\boldsymbol{F}^{\%}(X)$ is naturally homotopy equivalent to $X_+ \wedge \boldsymbol{F}(*)$.*

This is clear from the first proof of 1.1. We have not included it in Theorem 1.1 because it does not generalize well, as we shall see. In fact, our first proof does not generalize well ; that is why we have a second proof.

1.3. Observation. *If $\boldsymbol{F}$ is already excisive, then $\alpha : \boldsymbol{F}^{\%}(X) \to \boldsymbol{F}(X)$ is a homotopy equivalence for any X which is homotopy equivalent to a compact CW–space. If $\boldsymbol{F}$ is strongly excisive, then α is a homotopy equivalence for all X.*

Proof. By arguments going back to Eilenberg and Steenrod it is sufficient to verify that α is a homotopy equivalence for $X = *$. □

We want to show that $\alpha = \alpha_{\boldsymbol{F}}$ is the "universal" approximation (from the left) of $\boldsymbol{F}$ by a strongly excisive homotopy invariant functor. Suppose therefore that

$$\beta : \boldsymbol{E} \longrightarrow \boldsymbol{F}$$

is another natural transformation with strongly excisive and homotopy invariant $\boldsymbol{E}$. The commutative square

$$\begin{array}{ccc} \boldsymbol{E}^{\%} & \xrightarrow{\alpha_{\boldsymbol{E}}} & \boldsymbol{E} \\ \downarrow{\scriptstyle \beta^{\%}} & & \downarrow{\scriptstyle \beta} \\ \boldsymbol{F}^{\%} & \xrightarrow{\alpha_{\boldsymbol{F}}} & \boldsymbol{F} \end{array}$$

in which the upper horizontal arrow is a homotopy equivalence by 1.3, shows that β essentially factors through $\alpha_{\boldsymbol{F}}$. Note that if $\beta : \boldsymbol{E}(*) \to \boldsymbol{F}(*)$ happens to be a homotopy equivalence, then $\alpha_{\boldsymbol{E}} : \boldsymbol{E}^{\%}(X) \to \boldsymbol{E}(X)$ and $\beta^{\%} : \boldsymbol{E}^{\%}(X) \to \boldsymbol{F}^{\%}(X)$ are homotopy equivalences for all X, by the usual Eilenberg–Steenrod arguments.

Applications. Carlsson and Pedersen [CaPe] have used this "universal" approximation property to identify their *forget control* map with the assembly map for linear algebraic K-theory. Similarly Rosenberg [Ro] has used the "universal" approximation property to identify the Kasparov index map β with the assembly map in L-theory after localizing at odd primes. Ranicki has a construction of an assembly map for homotopy invariant functors from simplicial complexes to spectra [Ra, 12.19]. His construction may be identified with the one above by universality.

In many applications to geometry, assembly is the passage from local to global. For example, the *normal invariant* of a surgery problem $f : M \to N$

(with closed n–manifolds M and N, where $n \geq 5$, and some bundle data which we suppress) is an element in $\pi_n \boldsymbol{F}^{\%}(N)$, where $\boldsymbol{F}$ is the functor taking a space X to the L-theory spectrum $\boldsymbol{L}(\mathbb{Z}\pi_1(X))$ (details below). The normal invariant vanishes if and only if the surgery problem is bordant to another surgery problem $f_1 : M_1 \to N$ where f_1 is a *homeomorphism.* The image of the normal invariant under assembly is the *surgery obstruction* ; it vanishes if and only if the surgery problem is bordant to another surgery problem $f_1 : M_1 \to N$ where f_1 is a *homotopy equivalence.*

For another illustration, we mention the Whitehead torsion of a homotopy equivalence $f : X \to Y$ between compact euclidean neighborhood retracts. This is an element in the cokernel of $\alpha_* : \pi_1 \boldsymbol{F}^{\%}(Y) \to \pi_1 \boldsymbol{F}(Y)$, where $\boldsymbol{F}$ is the functor taking Y to the algebraic K-theory spectrum $\boldsymbol{K}(\mathbb{Z}\pi_1(Y))$ (details below). The torsion of f depends only the homotopy class of f, and it vanishes when f is a *homeomorphism.* This is of course the topological invariance of Whitehead torsion, due to Chapman. See [Ch] and [RaYa].

2. Examples

2.1. Linear K-theory

Recall that Quillen has defined a functor $\boldsymbol{K}$: *Exact* $\to$ *Spectra* where *Exact* is the category of exact categories. Alternatively, one can note that an exact category $\mathcal{M}$ determines a category with cofibrations and weak equivalences in the sense of Waldhausen by letting the cofibrations be the admissible monomorphisms and letting the isomorphisms be the weak equivalences. Then Waldhausen's $S_\bullet$ construction yields a functor $\boldsymbol{K}$ which is naturally homotopy equivalent to Quillen's $\boldsymbol{K}$. Let *Spaces*$_*$ be the category of spaces homotopy equivalent to CW-spaces which are equipped with nondegenerate base points. Then $\boldsymbol{K}(\mathbb{Z}\pi_1(X,*))$ is a functor from *Spaces*$_*$ to *Spectra.* In order to apply the construction of the assembly map from section 1 we have to show that this functor factors through the functor *Spaces*$_*$ $\to$ *Spaces* which forgets basepoints. The point of view is due to Quinn [QA], but the language we use is that of Lück and tom Dieck, [Lü, ch. II], [tD]. See also [Mitch].

Following a suggestion of MacLane [MaL], we use the word *ringoid* to mean a small category in which all morphism sets come equipped with an abelian group structure, and composition of morphisms is bilinear. Notice that a ringoid with one object is just a ring.

Any small category $\mathcal{C}$ gives rise to a ringoid $\mathbb{Z}\mathcal{C}$ having the same objects as $\mathcal{C}$. The set of morphisms from x_0 to x_1 in $\mathbb{Z}\mathcal{C}$ is the free abelian group generated by the set of morphisms from x_0 to x_1 in $\mathcal{C}$.

In particular, taking $\mathcal{C}$ to be the fundamental *groupoid* $\pi_1(X)$ of a space X, as in [Spa], we obtain a ringoid $\mathbb{Z}\pi_1(X)$. Objects in $\mathbb{Z}\pi_1(X)$ are points of X, and a morphism from y_0 to y_1 is a finite formal linear combination $\Sigma n_g \cdot g$, where the g are path classes beginning in y_0 and ending in y_1, and the n_g are integers.

Let $\mathcal{R}$ be a ringoid. A *left $\mathcal{R}$-module* is a covariant functor from $\mathcal{R}$ to abelian groups which is homomorphic on morphism sets; a *right $\mathcal{R}$-module* is a left $\mathcal{R}^{\mathrm{op}}$-module. A left $\mathcal{R}$-module is *free on one generator* if it is representable (that is, isomorphic to a morphism functor $\hom(x,-)$ for some object x in $\mathcal{R}$). It is *finitely generated free* if it is isomorphic to a finite direct sum of representable ones, and just *free* if it is isomorphic to an arbitrary direct sum of representable ones. It is *projective* if it is a direct summand of a free one, and *finitely generated projective* if it is a direct summand of a f. g. free one.

Left $\mathcal{R}$-modules form an abelian category in which the morphisms are natural transformations. Exercise for the reader: prove that a left $\mathcal{R}$-module P is projective if and only if any $\mathcal{R}$-module epimorphism with target P splits. The subcategory $\mathcal{P}_{\mathcal{R}}$ of finitely generated projective modules is then an exact category. For a space X, let $\boldsymbol{K}(X) = \boldsymbol{K}(\mathcal{P}_{\mathcal{R}})$ where $\mathcal{R} = \mathbb{Z}\pi_1(X)$. Since a homotopy equivalence between spaces induces an equivalence between their fundamental groupoids, our functor $\boldsymbol{K}$ is a homotopy functor and section 1 yields an assembly map for linear algebraic K-theory.

2.2. A-theory

Since Waldhausen has shown that his functor $X \mapsto \boldsymbol{A}(X)$ is a homotopy functor [Wa1, Prop. 2.1.7] we can directly apply Section 1 to get an assembly map for A-theory. (We use boldface notation, $\boldsymbol{A}(X)$, for the spectrum associated with the infinite loop space $A(X)$.)

2.3. L-theory

Recall that Ranicki [Ra, Ex. 13.6] [Ra, Ex. 1.3] has defined functors

$$\mathbb{L}_\bullet : \{\text{additive categories with chain duality}\} \longrightarrow \textit{Spectra}\ ,$$
$$\{\text{rings with involution}\} \longrightarrow \{\text{additive categories with chain duality}\}\ .$$

We write $\boldsymbol{L}$ for the first functor, rather than $\mathbb{L}_\bullet$, to be consistent. The second functor sends a ring R with involution j to the triple $(\mathcal{P}_R, T, e)$ where

- $\mathcal{P}_R$ is the category of f.g. projective left R-modules;
- T is the functor $\mathcal{P}_R \to \mathcal{P}_R$ which sends a module M to $\hom_R(M, R)$ where the involution j is used to convert this right R-module to a left R-module; and

- e is the inverse to the natural equivalence $\eta : \mathrm{id} \to T^2$ that maps a module M to $T^2(M)$ by taking the adjoint of the pairing

$$\hom_R(M, R) \times M \to R$$

which maps (f, m) to $j(f(m))$.

If X is a space with base point $*$, then $Z\pi_1(X, *)$ is equipped with the standard involution that takes an element $g \in \pi_1(X, *)$ to g^{-1}. Thus we again get a functor *Spaces*$_* \to$ *Spectra* which we have to factor through the forgetful functor *Spaces*$_* \to$ *Spaces*.

A *ringoid with involution* is a ringoid $\mathcal{R}$ together with a ringoid isomorphism

$$j : \mathcal{R} \longrightarrow \mathcal{R}^{\mathrm{op}}$$

such that the composite functor $\mathcal{R} \xrightarrow{j} \mathcal{R}^{\mathrm{op}} \xrightarrow{j^{\mathrm{op}}} \mathcal{R}$ is the identity. Notice that a ringoid with involution, with one object, is just a ring with involution.

For any space X, the ringoid $\mathbb{Z}\pi_1(X)$ has a standard involution. The involution is trivial on objects, and maps $\sum n_g g : x_0 \longrightarrow x_1$ (a typical morphism) to

$$\sum n_g g^{-1} : x_1 \longrightarrow x_0.$$

Thus we are done if we can show Ranicki's functor

$$\{\text{rings with involution}\} \to \{\text{additive categories with chain duality}\}$$

factors through the category of ringoids with involution.

Henceforth we assume the ringoid $\mathcal{R}$ comes equipped with an involution j. Then a left $\mathcal{R}$-module P can also be regarded as a right $\mathcal{R}$-module P^t (compose with $j^{-1} = j^{\mathrm{op}}$). Similarly a right $\mathcal{R}$-module P can also be regarded as a left $\mathcal{R}$-module.

Notice that for any object x in $\mathcal{R}$, the functor $\hom_{\mathcal{R}}(x, -)$ is a left $\mathcal{R}$-module, and $\hom_{\mathcal{R}}(-, x)$ is a right $\mathcal{R}$-module. For any two left $\mathcal{R}$-modules, M and N, let $\mathrm{HOM}_{\mathcal{R}}(M, N)$ be the abelian group of natural transformations from M to N.

For any left $\mathcal{R}$-module M, consider the contravariant functor from $\mathcal{R}$ to abelian groups which sends an object x to $\mathrm{HOM}_{\mathcal{R}}(M, \hom_{\mathcal{R}}(x, -))$. We let $T(M)$ be the left module obtained by using j to make this functor covariant. Notice that if $M = \hom_{\mathcal{R}}(y, -)$, then the Yoneda lemma implies $T(M)$ is just $\hom_{\mathcal{R}}(-, y)$ converted into a left module via j. Explicitly, $T(M)(x) \cong \hom(j(x), y) \cong \hom(j(y), x)$. Thus T sends f.g. free modules to f.g. free modules and f.g. projective modules to f.g. projective modules.

Fix a left $\mathcal{R}$-module M. Then for any pair of objects x and y in $\mathcal{R}$, we get the pairing

$$\mathrm{HOM}_{\mathcal{R}}(M, \mathrm{hom}_{\mathcal{R}}(y, -)) \times M(x) \to \mathrm{hom}_{\mathcal{R}}(x, y)$$

which sends (f, m) to $j(f(m))$. The adjoints of these pairings determine a natural transformation from the identity functor to T^2. If we restrict this natural transformation to the category of f.g. projective modules it is a natural equivalence, and we let e be the inverse natural transformation. Then $(\mathcal{P}_{\mathcal{R}}, T, e)$ is an additive category with (0-dimensional) chain duality and we are done.

2.4. The Novikov Conjecture

The Novikov conjecture, for a homotopy invariant functor $\boldsymbol{F}$ from spaces to spectra and a discrete group π, is the hypothesis that

$$\alpha_* : \pi_* \boldsymbol{F}^{\%}(B\pi) \otimes \mathbb{Q} \longrightarrow \pi_* \boldsymbol{F}(B\pi) \otimes \mathbb{Q}$$

is injective. It was originally formulated by Novikov for the L-theory functor, 2.3 above, and for all groups. The L-theory Novikov conjecture has been verified for many groups with a finite dimensional classifying space. See [RaNo] for details. Bökstedt, Hsiang and Madsen [BHM] proved the Novikov conjecture for the algebraic K-theory functor, 2.1 above, and all groups π such that $H_i(B\pi; \mathbb{Z})$ is finitely generated for all i.

3. Easy Variations

3.1. Variation. We can still do assembly when the functor $\boldsymbol{F}$ is defined on the category of spaces *over* a reference space B. (For example, B could be BG, the classifying space for stable spherical fibrations.) By abuse of notation, a map between spaces over B is a *homotopy equivalence* if it becomes a homotopy equivalence when the reference maps to B are omitted. A square of spaces over B is a *homotopy pushout square* if it becomes a homotopy pushout square when the reference maps are omitted. We call $\boldsymbol{F}$ homotopy invariant if it takes homotopy equivalences (over B) to homotopy equivalences. We call a homotopy invariant $\boldsymbol{F}$ *excisive* if it takes the empty set to a contractible spectrum and if it takes homotopy pushout squares (over B) to homotopy pushout squares. We call it *strongly excisive* if in addition it preserves arbitrary coproducts up to homotopy equivalence. — For any homotopy invariant $\boldsymbol{F}$ defined on spaces over B we have

$$\alpha : \boldsymbol{F}^{\%} \longrightarrow \boldsymbol{F},$$

natural in $\boldsymbol{F}$, where $\boldsymbol{F}^{\%}$ is homotopy invariant, strongly excisive and

$$\alpha : \boldsymbol{F}^{\%}(* \hookrightarrow B) \quad \longrightarrow \quad \boldsymbol{F}(* \hookrightarrow B)$$

is a homotopy equivalence for any point $*$ in B. If $\boldsymbol{F}$ is already strongly excisive, then α is a homotopy equivalence for all spaces over B. Prove this using the methods developed in the second proof of 1.1.

Example: Classical twisted L-theory. Let $B = K(\mathbb{Z}/2, 1)$. Then a map $X \to B$ determines a double covering $w : X^\natural \to X$. Unfortunately w does not, as one might expect, determine an involution on the ringoid $\mathbb{Z}\pi_1(X)$. But it does determine an involution on an equivalent category $\mathbb{Z}^w\pi_1(X)$. The objects of $\mathbb{Z}^w\pi_1(X)$ are the points of $X^\natural$, not X ; a morphism from x_0 to x_1 in $\mathbb{Z}^w\pi_1(X)$ is the same as a morphism from $w(x_0)$ to $w(x_1)$ in $\mathbb{Z}\pi_1(X)$. The involution is trivial on objects, and maps $\sum n_g g : x_0 \longrightarrow x_1$ (a typical morphism) to

$$\sum \text{sign}(g) \cdot n_g g^{-1} : x_1 \longrightarrow x_0,$$

where the *sign* of a path class g from $w(x_0)$ to $w(x_1)$ is $+1$ if g lifts to a path class from x_0 to x_1 in $X^\natural$, and -1 otherwise. — Refining 2.3 we let $\boldsymbol{L}(X{\to}B)$ be the L-theory spectrum of the ringoid with involution $\mathbb{Z}^w\pi_1(X)$.

Example: Tate Cohomology and the Ξ transformation. Let $B = BG$, the classifying space for stable spherical fibrations. Any map $X \to B$ determines an action of $\mathbb{Z}/2$ on a spectrum $\boldsymbol{A}(X{\to}B)$ which is homotopy equivalent to Waldhausen's A-theory spectrum $\boldsymbol{A}(X)$. See [Vog3] and [WW2]. Thus we can consider the functor sending X to the Tate cohomology spectrum

$$\widehat{\boldsymbol{H}}^{\blacktriangledown}(\mathbb{Z}/2; \boldsymbol{A}(X{\to}B))$$

(see [WW2] for details). In [WW2] we construct a natural transformation

$$\Xi : \boldsymbol{L}(X \xrightarrow{c_1} B_1) \quad \longrightarrow \quad \widehat{\boldsymbol{H}}^{\blacktriangledown}(\mathbb{Z}/2; \boldsymbol{A}(X \xrightarrow{c} B))$$

where c_1 is the composition of c with the Postnikov projection $B \to B_1 = K(\mathbb{Z}/2, 1)$. Together with the appropriate assembly maps, Ξ is used to study automorphisms of manifolds. See [WW1], [WW3] for the manifolds.

3.2. Example. Let $\mathcal{G}$ be a topological group with classifying space $B\mathcal{G}$, and suppose that $\mathcal{G}$ acts on a spectrum $\boldsymbol{T}$. For a space over $B\mathcal{G}$, say $f : X \to B\mathcal{G}$, let X^f be the pullback of

$$X \xrightarrow{f} B\mathcal{G} \leftarrow E\mathcal{G}.$$

The functor from spaces over $B\mathcal{G}$ to spectra given by

$$(f : X \to B\mathcal{G}) \quad \mapsto \quad X_+^f \wedge_{\mathcal{G}} \boldsymbol{T}$$

is strongly excisive. (The example is "typical", but we shall not go into details.)

3.3. Variation. There is a variant of assembly which applies to functors defined on *pairs* of spaces. Let $\boldsymbol{F}$ be such a functor, from pairs (X, Y) to spectra. We call $\boldsymbol{F}$ *homotopy invariant* if it takes homotopy equivalences of pairs to homotopy equivalences. We call a homotopy invariant $\boldsymbol{F}$ *excisive* if it takes the empty pair to a contractible spectrum, and if it takes homotopy pushout squares of pairs to homotopy pushout squares. (A square of pairs

$$\begin{array}{ccc} (X_1, Y_1) & \longrightarrow & (X_2, Y_2) \\ \downarrow & & \downarrow \\ (X_3, Y_3) & \longrightarrow & (X_4, Y_4) \end{array}$$

is a homotopy pushout square if the two squares made from the X_i and the Y_i, respectively, are homotopy pushout squares.) Finally $\boldsymbol{F}$ is *strongly excisive* if it is excisive and respects arbitrary coproducts, up to homotopy equivalence. — For any homotopy invariant $\boldsymbol{F}$ from pairs of spaces to spectra, there exist a strongly excisive (and homotopy invariant) $\boldsymbol{F}^{\%}$ from pairs of spaces to spectra and a natural transformation

$$\alpha = \alpha_{\boldsymbol{F}} : \boldsymbol{F}^{\%} \longrightarrow \boldsymbol{F}$$

such that

$$\alpha : \boldsymbol{F}^{\%}(*, \emptyset) \to \boldsymbol{F}(*, \emptyset), \qquad \alpha : \boldsymbol{F}^{\%}(*, *) \longrightarrow \boldsymbol{F}(*, *)$$

are homotopy equivalences. Moreover, $\boldsymbol{F}^{\%}$ and $\alpha_{\boldsymbol{F}}$ can be made to depend functorially on $\boldsymbol{F}$. If $\boldsymbol{F}$ is already strongly excisive, then α is a homotopy equivalence for every pair (X, Y). Here is a brief description of $\boldsymbol{F}^{\%}$: For a pair (X, Y) we have $\mathrm{simp}(Y) \subset \mathrm{simp}(X)$, and we define $\boldsymbol{F}^{\%}(X, Y)$ as the homotopy pushout (double mapping cylinder) of

$$\operatorname*{hocolim}_{g:\Delta^n \to X} \boldsymbol{F}(\Delta^n, \emptyset) \leftarrow \operatorname*{hocolim}_{g:\Delta^n \to Y} \boldsymbol{F}(\Delta^n, \emptyset) \to \operatorname*{hocolim}_{g:\Delta^n \to Y} \boldsymbol{F}(\Delta^n, \Delta^n)$$

where the homotopy colimits are to be taken over $\mathrm{simp}(X)$, $\mathrm{simp}(Y)$ and $\mathrm{simp}(Y)$, respectively.

3.4. Remark. Let $\boldsymbol{T}$ be a spectrum ; then the functor

$$X \mapsto X_+ \wedge \boldsymbol{T}$$

is homotopy invariant and strongly excisive. Any homotopy invariant and strongly excisive functor $\boldsymbol{F}$ from spaces to spectra has this form, up to a chain of natural homotopy equivalences (observations 1.2 and 1.3). The appropriate $\boldsymbol{T}$ is of course $\boldsymbol{F}(*)$. Next, let $f : \boldsymbol{T}_1 \to \boldsymbol{T}_2$ be a map of spectra. Then the functor

$$(X, Y) \mapsto \text{ homotopy pushout of } \left(Y_+ \wedge \boldsymbol{T}_2 \xleftarrow{f_*} Y_+ \wedge \boldsymbol{T}_1 \hookrightarrow X_+ \wedge \boldsymbol{T}_1\right)$$

is strongly excisive. Any strongly excisive functor $\boldsymbol{F}$ from pairs of spaces to spectra has this form, up to a chain of natural homotopy equivalences. The appropriate $\boldsymbol{T}_1$ is $\boldsymbol{F}(*, \emptyset)$, the appropriate $\boldsymbol{T}_2$ is $\boldsymbol{F}(*, *)$, and the appropriate f is induced by the inclusion of $(*, \emptyset)$ in $(*, *)$.

It follows that a strongly excisive $\boldsymbol{F}$ defined on pairs need not take every collapse map $(X, Y) \to (X/Y, *)$ to a homotopy equivalence. It does, however, if $\boldsymbol{F}(*, *)$ is contractible ; then $\boldsymbol{F}$ has the form $(X, Y) \mapsto (X/Y) \wedge \boldsymbol{F}(*, \emptyset)$ up to a chain of natural homotopy equivalences.

Equivariant versions of assembly are currently being developed by J. Davis and W. Lück [DaLü].

4. Assembly with Control

For the purposes of this section, a *control space* is a pair of spaces $(\bar{X}, X)$ where $\bar{X}$ is compact Hausdorff, X is open dense in $\bar{X}$, and X is an ENR. Informally, the set $\bar{X} \smallsetminus X$ is the *singular set*, whereas X is the *nonsingular set*. A *morphism* of control spaces is a continuous map of pairs $f : (\bar{X}, X) \to (\bar{Y}, Y)$ such that $f^{-1}(Y) = X$.

It seems that the use of *control* in topology began with Connell and Hollingsworth [CoHo]. For a survey of applications until 1986, see [QLA]. Through the influence of [Q], controlled topology led to *bounded algebra* and *controlled algebra*, [QA], [PW1], [PW2], [ACFP], and a plethora of functors from control spaces to spectra. Most of these have some homotopy invariance properties, i. e., they take homotopy equivalences to homotopy equivalences ; some of them also have excision properties [PW1], [PW2] [Vog1], [Vog2]. For applications, see also [CaPe] and [DWW], and many others.

Our goal here is roughly the following. Suppose that $\boldsymbol{F}$ is a homotopy invariant functor (details follow) from control spaces to spectra. We want to construct another functor $\boldsymbol{F}^{\%}$ from control spaces to spectra, homotopy invariant and excisive (details follow), and a natural transformation

$$\alpha : \boldsymbol{F}^{\%}(\bar{X}, X) \longrightarrow \boldsymbol{F}(\bar{X}, X)$$

which is a homotopy equivalence for $(\bar{X}, X) = (*, *)$. Moreover we would like to say that $\boldsymbol{F}^{\%}(\bar{X}, X)$ is related to $X^{\bullet} \wedge \boldsymbol{F}(*, *)$ by a chain of (weak) homotopy equivalences. Here $X^{\bullet}$ is the one–point compactification, usually not homotopy equivalent to a CW–space, so that $X^{\bullet} \wedge \boldsymbol{F}(*, *)$ is usually not homotopy equivalent to a CW–spectrum. (Hence we must allow *weak* homotopy equivalences in the chain.)

4.1. Terminology. Two morphisms $f_0, f_1 : (\bar{X}, X) \to (\bar{Y}, Y)$ between control spaces are *homotopic* if they agree on $\bar{X} \smallsetminus X$ and if they extend to a continuous one–parameter family of morphisms $f_t : (\bar{X}, X) \to (\bar{Y}, Y)$, where $0 \le t \le 1$, and all f_t agree on $\bar{X} \smallsetminus X$. A morphism $f : (\bar{X}, X) \to (\bar{Y}, Y)$ is a *homotopy equivalence* if there exists another morphism $g : (\bar{Y}, Y) \to (\bar{X}, X)$ such that gf and fg are homotopic to the identity. Note that a homotopy equivalence restricts to a *homeomorphism* of the singular sets.

A commutative square in the category of control spaces is a *homotopy pushout square* if the underlying square of nonsingular sets is a proper homotopy pushout square (details follow) in the category of locally compact spaces. *Details*: Recall that a map between locally compact spaces is *proper* if it extends to a continuous map between their one–point compactifications. A commutative square of locally compact spaces and proper maps

$$\begin{array}{ccc} X_1 & \longrightarrow & X_2 \\ \downarrow & & \downarrow \\ X_3 & \longrightarrow & X_4 \end{array}$$

is a *proper homotopy pushout square* if the resulting proper map from the homotopy pushout of $X_3 \leftarrow X_1 \rightarrow X_2$ to X_4 is a proper homotopy equivalence.

4.2. More terminology. A covariant functor $\boldsymbol{F}$ from control spaces to CW–spectra is *homotopy invariant* if it takes homotopy equivalences to homotopy equivalences. A homotopy invariant $\boldsymbol{F}$ is *excisive* if it takes homotopy pushout squares of control spaces to homotopy pushout squares of spectra, and $\boldsymbol{F}(\emptyset, \emptyset)$ is contractible.

Suppose that $\boldsymbol{F}$ is homotopy invariant and excisive, and let $(\bar{X}, X)$ be a control space with *discrete* but possibly infinite X. For any $y \in X$, we have a homotopy equivalence

$$\boldsymbol{F}(\bar{X}\smallsetminus y, X\smallsetminus y) \vee \boldsymbol{F}(y,y) \longrightarrow \boldsymbol{F}(\bar{X},X)$$

by excision, and hence a projection $\boldsymbol{F}(\bar{X},X) \to \boldsymbol{F}(y,y)$, well defined up to homotopy. We call $\boldsymbol{F}$ *pro-excisive* if these projection maps induce an isomorphism

$$\pi_n\boldsymbol{F}(\bar{X},X) \longrightarrow \prod_{y\in X} \pi_n\boldsymbol{F}(y,y) \qquad (n\in\mathbb{Z}).$$

In the following example, let $(\text{—})_{CW}$ be the standard CW approximation procedure replacing arbitrary spectra by CW–spectra. In detail, if $\boldsymbol{Y} = \{Y_n \mid n \in \mathbb{Z}\}$ is a spectrum with structure maps $\Sigma Y_n \to Y_{n+1}$, then the geometric realizations of the singular simplicial sets of the Y_n form a CW–spectrum $(\boldsymbol{Y})_{CW}$. Note that the functor π_* does not distinguish between $\boldsymbol{Y}$ and $(\boldsymbol{Y})_{CW}$.

Example. The functor $(\bar{X},X) \mapsto (X^\bullet \wedge \boldsymbol{S}^0)_{CW}$ is homotopy invariant and pro–excisive. Here $\boldsymbol{S}^0$ is the sphere spectrum. *Proof:* Transversality and Thom–Pontryagin construction lead to an interpretation of $\pi_n(X^\bullet \wedge \boldsymbol{S}^0) = \pi_n^s(X^\bullet)$ as the bordism group of stably framed smooth n–manifolds equipped with a *proper* map to X. This in turn leads to Mayer–Vietoris sequences from homotopy pushout squares of control spaces. Excision follows, and then pro–excision is clear. *Warning:* Be sure to use the correct topology on $X^\bullet$. Note that X could be any ENR, such as the universal cover of a wedge of two circles, or a countably infinite discrete set.

4.3. Proposition. *Suppose that $\boldsymbol{Y}$ is a CW–spectrum. Suppose also that $\boldsymbol{Y}$ is an Ω–spectrum (details below), or the suspension spectrum of a CW–space. Then the functor $(\bar{X},X) \mapsto (X^\bullet \wedge \boldsymbol{Y})_{CW}$ is homotopy invariant and pro–excisive.*

Proof. First suppose that $\boldsymbol{Y}$ is a suspension spectrum $\Sigma^\infty Y_0$. If the CW–space Y_0 is finite–dimensional, then we can use the preceding example and induction on the dimension of Y_0 to prove that $(\bar{X},X) \mapsto (X^\bullet \wedge \boldsymbol{Y})_{CW}$ is homotopy invariant and pro–excisive. If Y_0 has infinite dimension, we reduce to the finite dimensional case by observing that

$$\pi_n(X^\bullet \wedge \boldsymbol{Y}) \cong \pi_n^s(X^\bullet \wedge Y_0) \cong \pi_n^s(X^\bullet \wedge Y_0^{n+1})$$

where Y_0^{n+1} is the $(n+1)$–skeleton of Y_0.

Now suppose that $\boldsymbol{Y}$ is an arbitrary CW–spectrum. Then

$$\pi_n(X^\bullet \wedge \boldsymbol{Y}) := \operatorname*{colim}_k \pi_{n+k}(X^\bullet \wedge Y_k) \cong \operatorname*{colim}_k \pi_{n+k}(X^\bullet \wedge \Sigma^\infty Y_k).$$

Using the suspension spectrum case of 4.3, which we have established, we deduce immediately that the functor $(\bar{X}, X) \mapsto (X^\bullet \wedge \boldsymbol{Y})_{CW}$ is homotopy invariant and excisive. Furthermore, for a control space $(\bar{X}, X)$ with discrete X, we have

$$\pi_n(X^\bullet \wedge \boldsymbol{Y}) = \operatorname*{colim}_k \pi_{n+k}(X^\bullet \wedge Y_k) \cong \operatorname*{colim}_k \prod_{x\in X} \pi_{n+k}(Y_k).$$

Here we want to exchange direct limit and product to get

$$\prod_{x\in X} \operatorname*{colim}_k \pi_{n+k}(Y_k) \cong \prod_{x\in X} \pi_n(\boldsymbol{Y}).$$

In general this is not permitted. But it is clearly permitted if $\boldsymbol{Y}$ is an Ω–spectrum—the adjoints of the structure maps $\Sigma Y_k \to Y_{k+1}$ are homotopy equivalences $Y_k \to \Omega Y_{k+1}$. □

4.4. Theorem. *Suppose that $\boldsymbol{F}$ is a homotopy invariant functor from control spaces to CW–spectra. Suppose also that $\boldsymbol{F}$ behaves like a pro–excisive functor on the category of control spaces $(\bar{X}, X)$ with discrete X (details follow). Then there exists a pro–excisive functor $\boldsymbol{F}^\%$ from control spaces to CW–spectra, and a natural transformation $\alpha = \alpha_{\boldsymbol{F}} : \boldsymbol{F}^\% \to \boldsymbol{F}$ such that*

$$\alpha : \boldsymbol{F}^\%(*, *) \to \boldsymbol{F}(*, *)$$

is a homotopy equivalence. The construction can be made natural in $\boldsymbol{F}$.

Details. The extra hypothesis on $\boldsymbol{F}$ means that $\boldsymbol{F}$ takes a homotopy pushout square of control spaces with *discrete* nonsingular sets to a homotopy pushout square of spectra, and that, for any $(\bar{X}, X)$ with discrete X, the homomorphisms

$$\pi_n \boldsymbol{F}(\bar{X}, X) \longrightarrow \prod_{y\in X} \pi_n \boldsymbol{F}(y, y)$$

(defined as in 4.2) are isomorphisms. Carlsson [Car] has shown that functors of type "controlled algebraic K-theory" satisfy this condition. (Carlsson seems to have been the first to realize that this requires proof.)

4.5. Construction. The following *teardrop construction* will be needed in the proof of 4.4. Let $f : X \to Y$ be a proper map of ENR's, where Y is the nonsingular set of a control space $(\bar{Y}, Y)$. We note that the diagram of control spaces

$$(\bar{Y}, Y) \xrightarrow{\text{collapse}} (Y^{\bullet}, Y) \xleftarrow{f} (X^{\bullet}, X)$$

has a limit (=pullback) in the category of control spaces ; its nonsingular set is canonically identified with X, and we denote it by $(\bar{X}, X)$.

4.6. Notation. Suppose that X is the geometric realization of an incomplete simplicial set (simplicial set without degeneracies). Then

- X_n is the set of n–simplices in X.
- X^n is the n–skeleton.
- For each monotone injection $f : [m] \to [n]$, we write X_f to mean $\Delta^m \times X_n$. There is a *characteristic map* from X_f to X, via $\Delta^n \times X_n$. Note that this depends on f, not just on m and n. When f equals $\mathrm{id} : [n] \to [n]$, we write $X_{[n]}$ instead of X_f.

Proof of 4.4. Let $\mathcal{C}$ be the category of all control spaces. A key observation is that $\boldsymbol{F}$ is sufficiently determined by its restriction to a certain subcategory $\mathcal{C}'$, which we now describe. An object in $\mathcal{C}'$ is a control space $(\bar{X}, X)$ where X is the geometric realization of an incomplete simplicial set. Then X is a CW–space, and we require additionally that X have *small cells*, which means the following: For every $z \in \bar{X} \smallsetminus X$ and neighbourhood U of z in $\bar{X}$, there exists another neighborhood W of z in $\bar{X}$ such that any (open) cell of X intersecting W is contained in U. Note also that since X is an ENR, the underlying incomplete simplicial set must be locally finite, finite dimensional and countably generated. A morphism in $\mathcal{C}'$, say from $(\bar{X}, X)$ to $(\bar{Y}, Y)$, is a morphism of control spaces whose restriction to nonsingular sets is given by a simplicial map. Note that any finite diagram (=finitely generated simplicial subset of the nerve) in $\mathcal{C}'$ has a colimit.

The standard way to attempt recovery of a functor from its restriction to a subcategory is by *Kan extension*, here: *homotopy Kan extension.* Hence the following claim: for every $(\bar{Y}, Y)$ in $\mathcal{C}$, the canonical map

$$\operatorname*{hocolim}_{\substack{(\bar{X},X)\to(\bar{Y},Y) \\ (\bar{X},X) \text{ in } \mathcal{C}'}} \boldsymbol{F}(\bar{X}, X) \quad \longrightarrow \quad \boldsymbol{F}(\bar{Y}, Y)$$

is a homotopy equivalence. The homotopy colimit is taken over the category whose objects are objects in $\mathcal{C}'$ with a reference morphism to $(\bar{Y}, Y)$, and whose morphisms are morphisms in $\mathcal{C}'$, over $(\bar{Y}, Y)$. We denote this category by $(\mathcal{C}' \downarrow (\bar{Y}, Y))$.

To prove this claim, we observe that the canonical map in question is a natural transformation of functors in the variable $(\bar{Y}, Y)$. Since every $(\bar{Y}, Y)$ in $\mathcal{C}$ is a retract of some object $(\bar{X}, X)$ in $\mathcal{C}'$ (with a retraction morphism $(\bar{X}, X) \to (\bar{X}, X)$ which need not belong to $\mathcal{C}'$), it is enough to check the claim when $(\bar{Y}, Y)$ is already in $\mathcal{C}'$. Since any finite diagram in $(\mathcal{C}' \downarrow (\bar{Y}, Y))$ has a colimit, we have, almost from the definition,

$$\operatorname*{colim}_{(\bar{X},X)\to(\bar{Y},Y)} \pi_* \boldsymbol{F}(\bar{X}, X) \xrightarrow{\cong} \pi_*\big(\operatorname*{hocolim}_{(\bar{X},X)\to(\bar{Y},Y)} \boldsymbol{F}(\bar{X}, X)\big).$$

Hence our claim is proved if we can show that the canonical homomorphism

$$\operatorname*{colim}_{(\bar{X},X)\to(\bar{Y},Y)} \pi_* \boldsymbol{F}(\bar{X}, X) \longrightarrow \boldsymbol{F}(\bar{Y}, Y)$$

is an isomorphism. But this is obvious. We conclude that homotopy invariant functors on $\mathcal{C}$ are sufficiently determined by, and can be recovered from, their restriction to $\mathcal{C}'$. From now on we regard 4.4 as a statement about functors on $\mathcal{C}'$.

For $(\bar{X}, X)$ in $\mathcal{C}'$ and a monotone injection $f : [m] \to [n]$, we have the characteristic map $X_f \to X$ which we can use to compactify X_f (teardrop). This compactification is understood in the following definition:

$$\boldsymbol{F}^{\%}(\bar{X}, X) := \operatorname*{hocolim}_{f} \boldsymbol{F}(\bar{X}_f, X_f).$$

The homotopy colimit is taken over the category whose objects are monotone injections $f : [m] \to [n]$, with arbitrary $m, n \geq 0$; a morphism from f to g is a commutative square of monotone injections

$$\begin{array}{ccc} [m] & \xrightarrow{f} & [n] \\ \downarrow & & \uparrow \\ [p] & \xrightarrow{g} & [q]. \end{array}$$

We can now proceed as in the second proof of 1.1. The filtration of X by skeletons X^k leads to a filtration of $\boldsymbol{F}^{\%}(\bar{X}, X)$ by subspectra $\boldsymbol{F}^{\%}(\bar{X}^k, X^k)$. Here another teardrop construction is understood. By inspection,

$$\boldsymbol{F}^{\%}(\bar{X}^k, X^k)/\boldsymbol{F}^{\%}(\bar{X}^{k-1}, X^{k-1}) \simeq \mathbb{S}^k \wedge \boldsymbol{F}(\bar{X}_{[k]}, X_{[k]}).$$

From our extra hypothesis on $\boldsymbol{F}$, we then get isomorphisms

$$\pi_n\big(\boldsymbol{F}^{\%}(\bar{X}^k, X^k)/\boldsymbol{F}^{\%}(\bar{X}^{k-1}, X^{k-1})\big) \quad \simeq \quad \prod_{x \in X_k} \pi_{n-k}\boldsymbol{F}(*,*)$$

for $n \in \mathbb{Z}$, and this shows immediately that $\boldsymbol{F}^{\%}$ is homotopy invariant and excisive, and even pro–excisive. (Imitate the second proof of 1.1 ; use homology with locally finite coefficients to describe the E^2–term of the appropriate spectral sequence converging to $\pi_*\boldsymbol{F}^{\%}(\bar{X}, X)$.) Finally the assembly map

$$\alpha : \boldsymbol{F}^{\%}(\bar{X}, X) \longrightarrow \boldsymbol{F}(\bar{X}, X)$$

is obvious, and it is an isomorphism when $(\bar{X}, X) = (*,*)$. □

4.7. Observation. *If $\boldsymbol{F}$ in 4.4. is already pro–excisive, then the assembly α from $\boldsymbol{F}^{\%}(\bar{X}, X)$ to $\boldsymbol{F}(\bar{X}, X)$ is a homotopy equivalence for every $(\bar{X}, X)$.*

Proof. Fix $\boldsymbol{F}$, homotopy invariant and pro–excisive. We lose nothing by restricting $\boldsymbol{F}$ to $\mathcal{C}'$ (see proof of 4.4). When $(\bar{X}, X) = (*,*)$, the assembly α is an isomorphism by 4.4. By pro–excision, assembly is then an isomorphism for any $(\bar{X}, X)$ where X is discrete. For arbitrary $(\bar{X}, X)$ in $\mathcal{C}'$, we can argue by induction on skeletons: X is the strict and homotopy pushout of a diagram

$$X^{k-1} \leftarrow \partial\Delta^k \times X_k \xrightarrow{\subset} \Delta^k \times X_k\,.$$

Each of the spaces in this diagram has a canonical (teardrop) compactification ; two of the spaces in the diagram have dimension $< k$, the third is homotopy equivalent (with control) to a discrete space. Note that we use the condition on small cells at this point. □

4.8. Corollary. *If $\boldsymbol{F}$ in 4.4. is pro–excisive then there exists a chain of natural weak homotopy equivalences*

$$\boldsymbol{F}(\bar{X}, X) \simeq \ldots \simeq X^{\bullet} \wedge \boldsymbol{F}(*,*)_{\Omega}$$

where $\boldsymbol{F}(*,*)_{\Omega}$ is an Ω–spectrum envelope of $\boldsymbol{F}(*,*)$.

Proof. We may restrict to $\mathcal{C}'$. We may also assume $\boldsymbol{F}$ is a functor from control spaces to CW–Ω–spectra. Here it is understood that the morphisms in the category of CW–Ω–spectra are *functions*, not *maps*, in the language of [Ad, III§2]. Reason for making this technical assumption: the category of

CW–Ω–spectra has arbitrary and well-behaved products whereas the category of CW–spectra does not. Writing (≃ and ≃) for weak homotopy equivalences going in the direction indicated, we have

$$
\begin{aligned}
\boldsymbol{F}(\bar{X},X) \;\langle\simeq\; & \boldsymbol{F}^{\%}(\bar{X},X) \\
= \; & \operatorname*{hocolim}_{f} \boldsymbol{F}(\bar{X}_f, X_f) \\
\simeq\rangle \; & \operatorname*{hocolim}_{f:[m]\to[n]} \boldsymbol{F}(\bar{X}_n, X_n) \\
\simeq\rangle \; & \operatorname*{hocolim}_{f:[m]\to[n]} \prod_{y\in X_n} \text{cofiber } [\boldsymbol{F}(\bar{X}_n \smallsetminus y, X_n \smallsetminus y) \longrightarrow \boldsymbol{F}(\bar{X}_n, X_n)] \\
\langle\simeq \; & \operatorname*{hocolim}_{f:[m]\to[n]} \prod_{y\in X_n} \boldsymbol{F}(y,y) \\
\cong \; & \operatorname*{hocolim}_{f:[m]\to[n]} \prod_{y\in X_n} \boldsymbol{F}(*,*)\,.
\end{aligned}
$$

The first ≃⟩ is induced by the projections $p_f : X_f \to X_n$, for f from $[m]$ to $[n]$, where X_n must be compactified in such a way that p_f extends to a morphism of control spaces restricting to a homeomorphism of the singular sets. Again, this uses the small cells condition. The second of the weak homotopy equivalences labelled ⟨≃ is an inclusion, and it is a weak homotopy equivalence by excision.

We conclude that a homotopy invariant and pro–excisive functor $\boldsymbol{F}$ on $\mathcal{C}'$ is determined, up to a chain of weak homotopy equivalences, by what it does to the control space $(*,*)$. Hence such an $\boldsymbol{F}$ is related by a chain of natural weak homotopy equivalences to the functor

$$(\bar{X},X) \mapsto X^{\bullet} \wedge \boldsymbol{F}(*,*)_{\Omega}$$

whose CW–approximation is homotopy invariant and pro–excisive by 4.3.

REFERENCES

[ACFP]: D.R. Anderson, F.X. Connolly, S. Ferry and E.K. Pedersen, *Algebraic K-theory with continuous control at infinity*, J. Pure and Appl. Algebra **94** (1994), 25–47.

[Ad]: J.F. Adams, *Stable homotopy and generalised homology*, Chicago Lectures in Math., Univ. of Chicago Press, Chicago and London, 1974.

[And]: D.W. Anderson, *Chain functors and homology theories.*, Proc. 1971 Seattle Algebraic Topology Symposium, Lect. Notes in Math., vol. 249, Springer-Verlag, Berlin and New York, 1971, pp. 1–12.

[BHM]: M. Bökstedt, W.-C. Hsiang and I. Madsen, *The cyclotomic trace and algebraic K-theory of spaces*, Invent. Math. (1993), 465–540.

[BK]: A.K. Bousfield and D.M. Kan, *Homotopy limits, completions, and localizations*, Lect. Notes in Math., vol. 304, Springer-Verlag, Berlin and New York, 1972.

[Car]: G. Carlsson, *Bounded K-theory and the assembly map in algebraic K-theory*, these proceedings.

[CaPe]: G. Carlsson and E.K. Pedersen, *Controlled algebra and the Novikov Conjectures for K- and L-theory*, Topology (to appear).

[Ch]: T.A. Chapman, *The topological invariance of Whitehead torsion*, Amer. J. Math. **96** (1974), 488–497.

[CoHo]: E.H. Connell and J. Hollingsworth, *Geometric groups and Whitehead torsion*, Trans. Amer. Math. Soc. **140** (1969), 161–181.

[DaLü]: J.F. Davis and W. Lück, *Spaces over a category, assembly maps, and applications of the isomorphism conjecture in K- and L-theory*, in preparation.

[DWW]: W. Dwyer, M. Weiss and B.Williams, *A parametrized index theorem for the algebraic K-theory Euler class*, Preprint, Feb. 1995, 63 pp.

[Go1]: T.G. Goodwillie, *Calculus I: The First Derivative of Pseudoisotopy Theory*, *K*-Theory **4** (1990), 1–27.

[Go2]: T.G. Goodwillie, *Calculus II: Analytic Functors*, *K*-Theory **5** (1992), 295–332.

[Lü]: W. Lück, *Transformation Groups and Algebraic K-Theory*, Lect. Notes in Math., vol. 1408, Springer-Verlag, Berlin and New York, 1989.

[MaL]: S. MacLane, *Categories for the Working Mathematician*, Graduate Texts in Math., vol. 5, Springer-Verlag, Berlin and New York, 1971.

[Mitch]: B. Mitchell, *Rings with several objects*, Adv. in Math. **8** (1972), 1–161.

[PW1]: E.K. Pedersen and C. Weibel, *A nonconnective delooping of algebraic K-theory*, Proc. 1983 Rutgers Conf. on Algebraic Topology, Lecture Notes in Math., vol. 1126, Springer-Verlag, Berlin and New York, 1985, pp. 166–181.

[PW2]: E.K. Pedersen and C. Weibel, *K-theory homology of spaces*, Proc. 1986 Int. Conf. Algebraic Topology, Arcata, Lecture Notes in Math., vol. 1370, Springer-Verlag, Berlin and New York, 1989.

[Q]: F. Quinn, *Ends of maps II*, Invent. Math. **68** (1982), 353–424.

[QA]: F. Quinn, *Geometric algebra*, Proc. 1983 Rutgers Conf. on Algebraic Topology, Lecture Notes in Math., vol. 1126, Springer-Verlag, Berlin and New York, 1985, pp. 182–198.

[QAB]: F. Quinn, *Assembly maps in bordism-type theories*, these proceedings.

[QGF]: F. Quinn, *A geometric formulation of surgery*, Topology of Manifolds, Proceedings 1969 Georgia Topology Conf., Markham Press, Chicago, 1970, pp. 500–511.

[QLA]: F. Quinn, *Local algebraic topology*, Notices Amer. Math. Soc. **33** (1986), 895–899.

[Ra]: A. Ranicki, *Algebraic L-theory and Topological Manifolds*, Cambridge Tracts in Math., Cambridge Univ. Press, Cambridge, 1992.

[RaNo]: A. Ranicki, *On the Novikov Conjecture*, these proceedings.

[RaYa]: A. Ranicki and M. Yamasaki, *Controlled K-theory*, Topology and its Appl. **61** (1995), 1–59.

[Ro]: J. Rosenberg, *Analytic Novikov for topologists*, these proceedings.

[Spa]: E. Spanier, *Algebraic Topology*, McGraw-Hill, New York, 1966.

[tD]: T. tom Dieck, *Transformation Groups*, Studies in Math., vol. 8, De Gruyter, Berlin and New York, 1987.

[Vog1]: W. Vogell, *Algebraic K-theory of spaces, with bounded control*, Acta Math. **165** (1990), 161–187.

[Vog2]: W. Vogell, *Continuously controlled A-Theory*, Preprint, SFB 343 at Bielefeld University, Germany, 1992, 11 pp.

[Vog3]: W. Vogell, *The involution in the algebraic K-theory of spaces*, Proc. 1983 Rutgers Conf. on Algebraic Topology, Lecture Notes in Math., vol. 1126, Springer-Verlag, Berlin and New York, 1985, pp. 277–317.

[Wa1]: F. Waldhausen, *Algebraic K-theory of Spaces*, Proc. 1983 Rutgers Conf. on Algebraic Topology, Lecture Notes in Math., vol. 1126, Springer-Verlag, Berlin and New York, 1985, pp. 318–419.

[Wa2]: F. Waldhausen, *Algebraic K-theory of spaces, concordance, and stable homotopy theory*, Proc. 1983 Princeton Conf. on Alg. Topology and Alg. K-Theory, Annals of Math. Studies, vol. 113, Princeton Univ. Press, Princeton, 1987, pp. 392–417.

[WW1]: M. Weiss and B. Williams, *Automorphisms of Manifolds and Algebraic K-Theory: I*, K-Theory **1** (1988), 575–626.

[WW2]: M. Weiss and B. Williams, *Automorphisms of Manifolds and Algebraic K-Theory: II*, J. Pure and Appl. Algebra **62** (1989), 47–107.

[WW3]: M. Weiss and B. Williams, *Automorphisms of Manifolds and Algebraic K-Theory: Finale*, Preprint, 1994, 62pp.

DEPT. OF MATHEMATICS, UNIV. OF MICHIGAN, ANN ARBOR, MI 48109-1003, USA

email: `msweiss@math.lsa.umich.edu`

DEPT. OF MATHEMATICS, UNIV. OF NOTRE DAME, NOTRE DAME, IN 46556, USA

email: `bruce@bruce.math.nd.edu`

Pro–Excisive Functors

Michael Weiss and Bruce Williams

ABSTRACT. We classify homotopy invariant and *pro–excisive* functors F from Euclidean neighbourhood retracts to spectra. The pro–excision axiom ensures that $\pi_* F$ is a generalized "locally finite" homology theory.

0. Introduction

Index theorems usually involve some form of homology, as a receptacle for the *symbol*. Proofs of index theorems often involve some form of *locally finite homology* as well. Locally finite homology is so useful in this connection because it has some contravariant features in addition to the usual covariant ones. Thus, a proper map $f : X \to Y$ (details below) between locally compact spaces induces a map f_* in locally finite homology, going in the same direction ; but an inclusion j of an open subset $V \subset X$ induces a *wrong way* map j^* from the locally finite homology of X to that of V. In the case of singular locally finite homology, this is clear from the definition

$$H_*^{lf}(X) := \lim_{U} H_*(X, U)$$

where the inverse limit is taken over all $U \subset X$ with compact complement. (Note the excision property $H_*(X, U) \cong H_*(V, V \cap U)$ which applies when $X \smallsetminus U$ is compact and contained in V.)

In practice, when locally finite homology makes its appearance in the proof of an index theorem, or elsewhere, it may not be immediately recognizable as such. If the contravariant features are not needed, then theorem 1.2 below, essentially quoted from [WWA], should solve the problem. A recognition problem of this sort with a similar solution appears in the work on the Novikov conjecture of [CaPe]. (To see the similarity, note for example that a reduced Steenrod homology theory applied to one-point compactifications of locally compact subsets of some $\mathbb{R}^n$ makes a perfectly good locally finite homology theory.) If the contravariant features *are* relevant, then theorem 1.2 is not good enough and instead theorem 2.1 below should be used.

1991 *Mathematics Subject Classification.* Primary 55N20; Secondary 55N07, 55P42, 19D10, 57Q10.

We have used it in [DWW], in the proof of a parametrized index theorem for the algebraic K-theory *Euler class.* The proof is outlined below in §3. We emphasize that this proof is not analytic. It is part of a topological reply to the paper by Bismut and Lott [BiLo] and to the "Lott challenge" which asked for an explanation in topological language of a Riemann–Roch theorem for flat vector bundles [BiLo, Thm. 0.1].

While this may not be directly related to the Novikov conjecture, the reader should realize that index theorems relating Euler class and Euler characteristic can often be strengthened by keeping track of Poincaré duality. For example, the Euler characteristic of a closed smooth oriented manifold M^{4k} may be regarded for simplicity as an element in the topological K-group $K_0^{\mathrm{top}}(*) \cong \mathbb{Z}$. Keeping track of the Poincaré duality, one obtains much more: an element in

$$K_0^{\mathrm{top}}(B\mathbb{Z}/2) \cong \mathbb{Z} \oplus \widehat{\mathbb{Z}}_2$$

[AtSe] whose first component is the Euler characteristic of M, and whose second component is the difference of Euler characteristic and signature divided by two. The relationship between Euler class and $\mathcal{L}$–class is similar. A strengthened index theorem along these lines, parametrized and with plenty of algebraic K-theory instead of topological K-theory, is already implicit in [WW3] and may be more explicit in the next revision.

1. Excision and Proper Maps

Recall that a map $f : X \to Y$ between locally compact spaces is *proper* if it extends to a continuous pointed map $f^\bullet : X^\bullet \to Y^\bullet$ between their one-point compactifications. Note also that, in general, not every pointed continuous map $X^\bullet \to Y^\bullet$ is of the form $f^\bullet$ for a proper $f : X \to Y$. Let $\mathcal{E}$ be the category of ENR's (euclidean neighborhood retracts), with *proper* maps as morphisms. Let $\mathcal{E}^\bullet$ be the larger category whose objects are the ENR's, and where a morphism from $X \to Y$ is a continuous pointed map $X^\bullet \to Y^\bullet$. The goal is to characterize functors of the form

$$(*) \qquad X \mapsto X^\bullet \wedge \boldsymbol{Y} ,$$

where $\boldsymbol{Y}$ is a CW–spectrum and an Ω–spectrum, by their homotopy invariance and excision properties. (We call $\boldsymbol{Y}$ an *Ω-spectrum* if the adjoints of the structure maps $\Sigma Y_n \to Y_{n+1}$ are homotopy equivalences $Y_n \to \Omega Y_{n+1}$.) Reason for setting this goal: $\pi_*(X^\bullet \wedge \boldsymbol{Y})$, as a functor in the variable X, has all the properties one expects from a locally finite homology theory—details below, just before Thm. 1.3. We may view (*) as a functor from $\mathcal{E}$ to spectra, or as a functor from $\mathcal{E}^\bullet$ to spectra, so the task is twofold.

1.1. Terminology. A commutative square of locally compact spaces and proper maps

$$\begin{array}{ccc} X_1 & \longrightarrow & X_2 \\ \downarrow & & \downarrow \\ X_3 & \longrightarrow & X_4 \end{array}$$

is a *proper homotopy pushout square* if the resulting proper map from the homotopy pushout of $X_3 \leftarrow X_1 \rightarrow X_2$ to X_4 is a proper homotopy equivalence. A covariant functor $\boldsymbol{F}$ from $\mathcal{E}$ to CW–spectra is *homotopy invariant* if it takes proper homotopy equivalences to homotopy equivalences. A homotopy invariant $\boldsymbol{F}$ is *excisive* if it takes homotopy pushout squares in $\mathcal{E}$ to homotopy pushout squares of spectra, and $\boldsymbol{F}(\emptyset)$ is contractible.

Suppose that $\boldsymbol{F}$ is homotopy invariant and excisive, and suppose that X in $\mathcal{E}$ is *discrete.* For any $y \in X$, we have a homotopy equivalence

$$\boldsymbol{F}(X \smallsetminus y) \vee \boldsymbol{F}(y) \longrightarrow \boldsymbol{F}(X)$$

by excision, and hence a projection $\boldsymbol{F}(X) \rightarrow \boldsymbol{F}(y)$, well defined up to homotopy. We call $\boldsymbol{F}$ *pro-excisive* if these projection maps induce an isomorphism

$$\pi_n \boldsymbol{F}(X) \longrightarrow \prod_{y \in X} \pi_n \boldsymbol{F}(y) \qquad (n \in \mathbb{Z}).$$

Example. Let $\boldsymbol{Y}$ be a CW–spectrum. Assume also that $\boldsymbol{Y}$ is an Ω–spectrum, or that $\boldsymbol{Y}$ is the suspension spectrum of a CW–space. Then the functor taking X to the standard CW–approximation of $X^\bullet \wedge \boldsymbol{Y}$ is a homotopy invariant and pro–excisive functor from ENR's to CW–spectra. See [WWA, 4.3] for the proof. *Warning:* Be sure to use the correct topology on $X^\bullet$. Note that X can be a countably infinite discrete set, or the universal cover of a wedge of two circles, or worse. *Illustration:* Suppose $\boldsymbol{Y}$ is the sphere spectrum. Then $\pi_n(X^\bullet \wedge \boldsymbol{Y}) \cong \pi_n^s(X^\bullet)$ and this can be interpreted via transversality as the bordism group of framed manifolds equipped with a *proper map* to X. The excision properties follow easily from this interpretation. In the following theorem we invoke a functorial construction which associates to each CW–spectrum $\boldsymbol{Y}$ a CW–Ω–spectrum $\boldsymbol{Y}_\Omega$ and a homotopy equivalence $\boldsymbol{Y} \rightarrow \boldsymbol{Y}_\Omega$.

1.2. Theorem. *If $\boldsymbol{F}$ from $\mathcal{E}$ to CW–spectra is homotopy invariant and pro–excisive, then there exists a chain of natural weak homotopy equivalences*

$$\boldsymbol{F}(X) \simeq \ldots \simeq X^\bullet \wedge \boldsymbol{F}(*)_\Omega .$$

Proof. This is contained in Cor. 4.8 of [WWA], which is about homotopy invariant and pro–excisive functors from *control spaces* to spectra. A control space is a pair $(\bar{X}, X)$ where $\bar{X}$ is compact, X is open dense in $\bar{X}$, and X is an ENR. A morphism of control spaces, $f : (\bar{X}_1, X_1) \to (\bar{X}_2, X_2)$, is a map of pairs such that $f^{-1}(X_2) = X_1$. Writing $\mathcal{C}$ for the category of control spaces, we see that $\mathcal{E}$ is a *retract* of $\mathcal{C}$ via

$$X \mapsto (X^\bullet, X) \qquad\qquad (\bar{X}, X) \mapsto X\,.$$

Hence any homotopy invariant and strongly excisive $\boldsymbol{F}$ on $\mathcal{E}$ determines one on $\mathcal{C}$, and this is covered by Thm. 4.4 of [WWA]. □

2. Excision and One-Point Compactification

Here we are interested in functors from $\mathcal{E}^\bullet$, the "enlarged" category of ENR's, to spectra. Such a functor will be called *homotopy invariant* and *pro–excisive* if its restriction to $\mathcal{E}$ has these properties. Before proving any theorems about pro–excisive functors on $\mathcal{E}^\bullet$, we elucidate the structure of $\mathcal{E}^\bullet$. Note that every diagram of ENR's of the form

$$(*) \qquad\qquad X \supset V \xrightarrow{g} Y$$

where V is open in X and g is proper, gives rise to a continuous pointed map $X^\bullet \to Y^\bullet$ which agrees with g on V and maps the complement of V to the base point in $Y^\bullet$. Clearly every continuous pointed map $X^\bullet \to Y^\bullet$ arises in this way, for unique V and g, so that (*) may be regarded as the description of a typical morphism in $\mathcal{E}^\bullet$. If it happens that $V = X$ in (*), then the morphism under consideration is in $\mathcal{E}$. If it happens that $g = \mathrm{id}$, then we must think of the morphism as some kind of *reversed inclusion*. We see from (*) that every morphism in $\mathcal{E}^\bullet$ can be written in the form gh, where g is in $\mathcal{E}$ and h is a reversed inclusion of an open subset. This decomposition is unique.

A functor $\boldsymbol{F}$ from $\mathcal{E}^\bullet$ to CW–spectra is therefore a gadget which to every X in $\mathcal{E}^\bullet$ associates a CW–spectrum $\boldsymbol{F}(X)$, to every proper map $g : X \to Y$ a map $\boldsymbol{F}(X) \to \boldsymbol{F}(Y)$, and to every inclusion $V \subset X$ of an open subset, a wrong way map $\boldsymbol{F}(X) \to \boldsymbol{F}(V)$ which we think of as a *restriction map*. Certain associativity relations and identity relations must hold—after all, a functor is a functor.

Similarly, a natural transformation $\tau : \boldsymbol{F}_1 \to \boldsymbol{F}_2$ between functors from $\mathcal{E}^\bullet$ to spectra is a gadget which to every X in $\mathcal{E}^\bullet$ associates a map $\boldsymbol{F}_1(X) \to$

$F_2(X)$ such that the diagrams

$$\begin{array}{ccc} F_1(X) & \xrightarrow{\tau} & F_2(X) \\ \downarrow g_* & & \downarrow g_* \\ F_1(Y) & \xrightarrow{\tau} & F_2(Y) \end{array} \qquad \text{and} \qquad \begin{array}{ccc} F_1(X) & \xrightarrow{\tau} & F_2(X) \\ \downarrow \text{restriction} & & \downarrow \text{restriction} \\ F_1(V) & \xrightarrow{\tau} & F_2(V) \end{array}$$

commute, for every proper $g : X \to Y$ and every inclusion of an open subset $V \subset X$. In particular, the natural weak homotopy equivalences in the following theorem 2.1. are gadgets of this type. Thus 2.1 is *not* a formal consequence of 1.2.

2.1. Theorem. *If F from $\mathcal{E}^\bullet$ to CW–spectra is homotopy invariant and pro–excisive, then there exists a chain of natural weak homotopy equivalences*

$$F(X) \simeq \dots \simeq X^\bullet \wedge F(*)_\Omega .$$

Proof. Let $\mathcal{E}_1^\bullet$ be the full subcategory of $\mathcal{E}^\bullet$ consisting of those objects which are geometric realizations of simplicial sets. Let $\mathcal{E}_2^\bullet$ be the full subcategory of $\mathcal{E}^\bullet$ consisting of the objects X such that $X^\bullet$ is homeomorphic to the geometric realization of a pointed finitely generated simplicial set. Note that $\mathcal{E}_2^\bullet$ is equivalent to the category of finitely generated pointed simplicial sets, where the morphisms are the pointed continuous maps between geometric realizations. Let $\iota_1 : \mathcal{E}_1^\bullet \to \mathcal{E}^\bullet$ and $\iota_2 : \mathcal{E}_2^\bullet \to \mathcal{E}^\bullet$ be the inclusion functors. Our strategy is to show that $F\iota_2$ determines $F\iota_1$, up to a natural chain of weak homotopy equivalences. It is comparatively easy to analyze $F\iota_2$ and to recover F from $F\iota_1$, up to a natural chain of weak homotopy equivalences.

We shall pretend that $\mathcal{E}^\bullet$, $\mathcal{E}_1^\bullet$ and $\mathcal{E}_2^\bullet$ are *small* categories ; the truth is of course that they are *equivalent* to small categories. We can also throw in a few technical assumptions about F, as follows. Each spectrum $F(X)$ is made up of pointed simplicial sets $F_n(X)$ and simplicial maps from $\Sigma F_n(X)$ to $F_{n+1}(X)$, for $n \in \mathbb{Z}$. For each morphism $X \to X'$ in $\mathcal{E}^\bullet$, the induced map $F(X) \to F(X')$ is in fact a *function* [Ad, III§2], given by compatible simplicial maps $F_n(X) \to F_n(X')$ for all $n \in \mathbb{Z}$. Finally, each $F(X)$ is an Ω–spectrum. These assumptions facilitate the definition of homotopy limits.

Step 1. For Z in $\mathcal{E}_1^\bullet$ we have the canonical map

$$F\iota_1(Z) \longrightarrow \operatorname*{holim}_K F\iota_2(Z \smallsetminus K) . \tag{**}$$

where K runs over all closed subsets of Z such that $Z \smallsetminus K$ is in $\mathcal{E}_2^\bullet$ and $Z \smallsetminus K$ has *compact* closure in Z. We want to think of (**) as a natural transformation between *functors* on $\mathcal{E}_1^\bullet$, in the variable Z. Thus if Z_1 and Z_2 are in $\mathcal{E}_1^\bullet$, and $f : Z_1^\bullet \to Z_2^\bullet$ is a continuous pointed map, and $L \subset Z_2$ is closed, and the closure of $Z_2 \smallsetminus L$ in Z_2 is compact, we let $f^*(L)$ be the inverse image of L under f, minus the base point. Then we have

$$\operatorname*{holim}_K \boldsymbol{F}\iota_2(Z_1 \smallsetminus K) \longrightarrow \operatorname*{holim}_L \boldsymbol{F}\iota_2(Z_1 \smallsetminus f^*L) \longrightarrow \operatorname*{holim}_L \boldsymbol{F}\iota_2(Z_2 \smallsetminus L)\,.$$

Step 2. We shall verify that the codomain of (**) has certain excision properties. From a strict pushout square of (realized) simplicial sets, all in $\mathcal{E}_1^\bullet$, and simplicial maps

$$(***)\qquad \begin{array}{ccc} Z_1 & \xrightarrow{\ \subset\ } & Z_2 \\ \downarrow & & \downarrow \\ Z_3 & \xrightarrow{\ \subset\ } & Z_4 \end{array}$$

and a cofinite simplicial subset $K \subset Z_4$, we obtain another pushout square

$$\begin{array}{ccc} K_1 & \longrightarrow & K_2 \\ \downarrow & & \downarrow \\ K_3 & \longrightarrow & K_4 \end{array}$$

where K_i is the inverse image of K in Z_i. Then the spaces $Z_i \smallsetminus K_i$ form a proper homotopy pushout square, so that the spectra $\boldsymbol{F}\iota_2(Z_i \smallsetminus K_i)$ form a homotopy pushout square, alias homotopy pullback square. Passing to homotopy limits and noting that a homotopy limit of homotopy pullback squares is a homotopy pullback square, we see that the codomain of (**) does indeed have certain excision properties: it takes (***) to a homotopy pullback square alias homotopy pushout square. (We have taken the liberty to "prune" the indexing categories for the homotopy limits involved. This is justified by [DwKa, 9.3].)

In particular, the filtration of an arbitrary Z in $\mathcal{E}_1^\bullet$ by skeletons leads to a filtration of domain and codomain of (**), and by excision and inspection the induced maps of filtration quotients are weak homotopy equivalences. Hence (**) is a weak homotopy equivalence, by induction on the dimension of Z.

Step 3. The functor $\boldsymbol{F}\iota_2$ on $\mathcal{E}_2^\bullet$ is homotopy invariant and excisive in the following sense. For any Z in $\mathcal{E}_2^\bullet$, the map

$$\boldsymbol{F}\iota_2(Z \times [0,1]) \to \boldsymbol{F}\iota_2(Z)$$

induced by projection is a homotopy equivalence. The square of spectra

$$\begin{array}{ccc} \boldsymbol{F}\iota_2(Z_1 \cap Z_2) & \longrightarrow & \boldsymbol{F}\iota_2(Z_1) \\ \downarrow & & \downarrow \\ \boldsymbol{F}\iota_2(Z_2) & \longrightarrow & \boldsymbol{F}\iota(Z_1 \cup Z_2) \end{array}$$

is a weak homotopy pushout square provided $Z_1^\bullet, Z_2^\bullet$ are geometric realizations of pointed simplicial subsets of a finitely generated pointed simplicial set. Further, $\boldsymbol{F}\iota_2(\emptyset)$ is weakly contractible. This follows directly from our hypotheses on $\boldsymbol{F}$.

Step 4. The functor $\boldsymbol{F}\iota_2$ is related to $Z \mapsto Z^\bullet \wedge \boldsymbol{F}(*)$ by a chain of natural weak homotopy equivalences. The argument follows the lines of [WWA, §1]. For Z in $\mathcal{E}_2^\bullet$ let simp$(Z^\bullet)$ be the category whose objects are maps $\Delta^n \to Z^\bullet$ and whose morphisms are linear maps $f_* : \Delta^m \to \Delta^n$, over $Z^\bullet$, induced by some monotone f from $\{0, 1, \ldots, m\}$ to $\{0, 1, \ldots, n\}$. Let $\boldsymbol{F}_{Z^\bullet}$ be the functor from simp$(Z^\bullet)$ to spectra taking $g : \Delta^n \to Z^\bullet$ to $\boldsymbol{F}(\Delta^n)$, and let

$$\boldsymbol{F}^\%(Z^\bullet) := \operatorname{hocolim} \boldsymbol{F}_{Z^\bullet}\,.$$

Each $g : \Delta^n \to Z^\bullet$ in simp$(Z^\bullet)$ is a morphism $\Delta^n \to Z$ in $\mathcal{E}_2^\bullet$ which induces g_* from $\boldsymbol{F}\iota_2(\Delta^n) = \boldsymbol{F}(\Delta^n)$ to $\boldsymbol{F}\iota_2(Z)$. Collecting all these, we have the *assembly*

$$\alpha : \boldsymbol{F}^\%(Z^\bullet) \longrightarrow \boldsymbol{F}\iota_2(Z)\,.$$

The domain of α is homotopy invariant and excisive like $\boldsymbol{F}\iota_2$, but $\boldsymbol{F}^\%(\emptyset^\bullet)$ need not be contractible. However, the map of vertical (homotopy) cofibers in

$$\begin{array}{ccc} \boldsymbol{F}^\%(\emptyset^\bullet) & \xrightarrow{\ \alpha\ } & \boldsymbol{F}\iota_2(\emptyset) \\ \downarrow & & \downarrow \\ \boldsymbol{F}^\%(Z^\bullet) & \xrightarrow{\ \alpha\ } & \boldsymbol{F}\iota_2(Z) \end{array}$$

is a natural transformation between homotopy invariant and excisive functors in the variable Z, and it is clearly a homotopy equivalence when Z is a point and when Z is empty. By Eilenberg–Steenrod arguments, it is always a homotopy equivalence. Further, $\boldsymbol{F}^\%(Z^\bullet)$ can be related to $Z_+^\bullet \wedge \boldsymbol{F}(*)$ as in [WWA, §1].

Step 5. We must verify that $\boldsymbol{F}$ can be recovered from $\boldsymbol{F}\iota_1$. For every X in $\mathcal{E}^\bullet$, the map

$$\boldsymbol{F}(X) \longrightarrow \operatorname*{hocolim}_{Z \to X} \boldsymbol{F}(Z)$$

is a homotopy equivalence. The hocolim is taken over the category with objects $Z \to X$, where Z is in $\mathcal{E}_1^\bullet$. The claim is obvious for X in $\mathcal{E}_1^\bullet$. The general case follows because every X in $\mathcal{E}^\bullet$ is a retract of some object in $\mathcal{E}_1^\bullet$. □

Remark. Theorems 1.3 and 2.1 are reminiscent of uniqueness and existence theorems for generalized *Steenrod homology theories*, [KKS], [EH], [M]. For variations and applications see [CaPe].

3. An Example

For the purposes of this section, a *control space* is a pair of spaces $(\bar{Y}, Y)$ where $\bar{Y}$ is *locally* compact Hausdorff, Y is open dense in $\bar{Y}$, and Y is an ENR. Informally, the set $\bar{Y} \smallsetminus Y$ is the *singular set*, whereas Y is the *nonsingular set*. A *morphism* of control spaces is a continuous *proper* map of pairs $f : (\bar{Y}, Y) \to (\bar{Z}, Z)$ such that $f^{-1}(Z) = Y$. Note that we are less restrictive here than in [WWA, §4] because we allow $\bar{Y}$ to be noncompact. In any case these ideas come from [ACFP].

Fix a control space $(\bar{Y}, Y)$. By a *geometric module* on Y we mean a free abelian group B with an (internal) direct sum decomposition

$$B = \bigoplus_{x \in Y} B_x$$

where each B_x is finitely generated, and the set $\{x \in Y \mid B_x \neq 0\}$ is closed and discrete in Y. Given two geometric modules B and B' on Y, a *controlled homomorphism* $f : B \to B'$ is a group homomorphism, with components $f_y^x : B_x \to B_y$ say, subject to the following condition. For any $z \in \bar{Y} \smallsetminus Y$ and any neighborhood V of z in $\bar{Y}$, there exists a smaller neighborhood W of z in $\bar{Y}$ such that $f_y^x = 0$ and $f_x^y = 0$ whenever $x \in W$ and $y \notin V$. Clearly the composition of two geometric homomorphisms $B \to B'$, $B' \to B''$ is a geometric homomorphism $B \to B''$.

For a geometric module B on Y and a neighborhood U of $\bar{Y} \smallsetminus Y$ in Y, we let B^U be the geometric submodule of B which is the direct sum of the B_x for $x \in U$. Given B and B', as before, a *germ* of controlled homomorphisms from B to B' is an equivalence class of pairs $(U, f : B^U \to B')$. Here $(U, f : B^U \to B')$ and $(W, g : B^W \to B')$ are *equivalent* if f and g agree on $B^{U \cap W}$.

A controlled homomorphism germ as above is *invertible* if it is an isomorphism in the germ category. Geometric modules on Y and invertible germs of controlled homomorphisms between them form a symmetric monoidal category. With this we can associate a K-theory spectrum, using the construction of [Se], say. Since it depends ultimately on $(\bar{Y}, Y)$, we denote it by $\boldsymbol{E}(\bar{Y}, Y)$. It is clear that $\boldsymbol{E}(\bar{Y}, Y)$ is covariantly functorial in $(\bar{Y}, Y)$. Finally we put

$$\boldsymbol{F}(X) := \boldsymbol{E}(X\times[0,1], X\times[0,1))$$

so that $\boldsymbol{F}$ is a functor from ENR's and proper maps to spectra. It is well known [ACFP] that $\boldsymbol{F}$ is homotopy invariant and excisive on the category of *compact* polyhedra and piecewise linear maps. It follows immediately that $\boldsymbol{F}$ is also homotopy invariant and excisive on the category of compact ENR's, since every compact ENR is a retract of a compact polyhedron. Using the result of [Ca], in addition to arguments proving excision as in [ACFP] or [Vog], one can verify that $\boldsymbol{F}$ is in fact homotopy invariant and pro–excisive on the category $\mathcal{E}$ of all ENR's and their proper maps. We hope to give more details elsewhere.

The functor $\boldsymbol{F}$ has an extension to $\mathcal{E}^\bullet$, the "enlarged" category of ENR's. To understand this extension, recall the canonical decomposition of morphisms $X \to Y$ in $\mathcal{E}^\bullet$ as reversed inclusion $X \supset V$ of an open subset, followed by a proper map $g : V \to Y$. We know already how the proper map $g : V \to Y$ induces $g_* : \boldsymbol{F}(V) \to \boldsymbol{F}(Y)$, so that our task now is to produce a *restriction map* $\boldsymbol{F}(X) \to \boldsymbol{F}(V)$ which we may (pre–)compose with g_*. Now the word *restriction* almost gives it away. Recall that $\boldsymbol{F}(X)$ was constructed as the K-theory of a certain symmetric monoidal category whose objects are the geometric modules on $X\times[0,1)$. We can indeed *restrict* a geometric module B on $X\times[0,1)$ to $V\times[0,1)$ by discarding all the B_z for $z \notin V\times[0,1)$. Similarly, if $f : B \to B'$ is a morphism of geometric modules on $X\times[0,1)$, we restrict by discarding all the f_z^y where $y \notin V\times[0,1)$ or $z \notin V\times[0,1)$.

Now it is important to realize that restriction, as we have defined it, is *not* a functor from the category of geometric modules on $X\times[0,1)$ to the category of geometric modules on $V\times[0,1)$. It does not respect composition of controlled homomorphisms. However, restriction is compatible with passage to controlled homomorphism germs, and after passage to germs restriction does respect composition. This is easily verified. Hence we have enough of a restriction functor to get an induced restriction map $\boldsymbol{F}(X) \to \boldsymbol{F}(V)$.

It is known [ACFP] that $\boldsymbol{F}(*) = \boldsymbol{E}([0,1],[0,1)) \simeq \mathbb{S}^1 \wedge \boldsymbol{K}(\mathbb{Z})$ where $\boldsymbol{K}(\mathbb{Z})$ is the algebraic K-theory spectrum of $\mathbb{Z}$. Therefore, by theorem 2.1, there exists a chain of natural weak homotopy equivalences

$$(**) \qquad \boldsymbol{F}(X) \quad \simeq \quad \cdots \quad \simeq \quad X^\bullet \wedge \mathbb{S}^1 \wedge \boldsymbol{K}(\mathbb{Z})$$

for X in $\mathcal{E}^\bullet$. We stress once again that each of the natural homotopy equivalences in the chain is natural for arbitrary morphisms in $\mathcal{E}^\bullet$, not just those in the subcategory $\mathcal{E}$.

The functor $\boldsymbol{F}$ has a nonlinear version which is used in [DWW] to state and prove an index theorem. We proceed to explain how, keeping the linear $\boldsymbol{F}$ for simplicity. A key fact is that $\boldsymbol{F}$ comes equipped with a rule which selects for each X in $\mathcal{E}^\bullet$ a point $\langle\!\langle X\rangle\!\rangle$ in the infinite loop space $\Omega^{\infty+1}\boldsymbol{F}(X)$, the *microcharacteristic* of X. It is a refined sort of Euler characteristic. For example, when X is compact and connected, then the component of $\langle\!\langle X\rangle\!\rangle$ in

$$\pi_0(\Omega^{\infty+1}\boldsymbol{F}(X)) \cong \mathbb{Z}$$

(see (**) above) is the "usual" Euler characteristic of X. Note however that the microcharacteristic is a point, not a connected component. Microcharacteristics enjoy some naturality. Namely, bending the truth just a little, we may say that for any open subset $V \subset X$, the restriction $\Omega^{\infty+1}\boldsymbol{F}(X) \to \Omega^{\infty+1}\boldsymbol{F}(V)$ (explained earlier) takes $\langle\!\langle X\rangle\!\rangle$ to $\langle\!\langle V\rangle\!\rangle$.

Let $\gamma : E \to X$ be a fiber bundle whose fibers E_x are homeomorphic to $\mathbb{R}^n$. We can make another fibration on X, the *Euler fibration* of γ, with infinite loop space fiber $\Omega^{\infty+1}\boldsymbol{F}(E_x)$ over $x \in X$. Then $x \mapsto \langle\!\langle E_x\rangle\!\rangle$ determines a section of the new fibration, which we call the *Euler section.* It is a refined sort of Euler class. Note that we have omitted a number of serious technical points (what is the topology on the total space of the Euler fibration ; why is the Euler section continuous).

Digression. The geometric significance of the Euler section is clearer in the *nonlinear* set-up: when $\dim(X) < (4n/3)-5$, say, the structure group of γ can be reduced from $\mathrm{TOP}(\mathbb{R}^n)$ to $\mathrm{TOP}(\mathbb{R}^{n-1})$ if and only if the (nonlinear) Euler section of γ is nullhomotopic. *End of digression.*

Let M^n be a closed topological manifold with tangent bundle $TM \to M$. The tangent bundle is a fiber bundle with distinguished "zero" section and with fibers homeomorphic to $\mathbb{R}^n$; we can sufficiently characterize it by assuming that it comes with an *exponential map* $\exp : TM \to M$ which is left inverse to the zero section and embeds each fiber. There is a map $\wp$ from $\Omega^{\infty+1}\boldsymbol{F}(M)$ to the space of sections of the Euler fibration of the tangent bundle which takes $z \in \Omega^{\infty+1}\boldsymbol{F}(M)$ to the section

$$x \mapsto \mathrm{res}(z) \in \Omega^{\infty+1}\boldsymbol{F}(\exp(T_xM)) \cong \Omega^{\infty+1}\boldsymbol{F}(T_xM)$$

where res means restriction $\boldsymbol{F}(M) \to \boldsymbol{F}(\exp(T_xM))$. At this point it is necessary to know what the restriction maps do, not just that they exist. This is what we have theorem 2.1 for. It follows quite easily that $\wp$ is a version of *Poincaré duality.* In particular, $\wp$ is a homotopy equivalence. By the naturality property of microcharacteristics, $\wp$ takes the microcharacteristic $\langle\langle M\rangle\rangle$ to the Euler section of the tangent bundle. Hence we have proved a version of Heinz Hopf's index theorem: *the Poincaré dual of the Euler characteristic of M is the Euler class of M.* This works quite well for families, and then the advantage of working with *algebraic K-theory* as opposed to working with the group K_0 becomes apparent.

References

[ACFP]: D.R. Anderson, F.X. Connolly, S. Ferry and E.K. Pedersen, *Algebraic K-theory with continuous control at infinity*, J. Pure and Appl. Algebra **94** (1994), 25–47.

[Ad]: J.F.Adams, *Stable homotopy and generalised homology*, Chicago Lectures in Math., Univ. of Chicago Press, Chicago and London, 1974.

[AtSe]: M. Atiyah and G. Segal, *Equivariant K-theory and Completion*, J. Diff. Geometry **3** (1969), 1–18.

[BiLo]: J.-M. Bismut and J. Lott, *Flat vector bundles, direct images and higher real analytic torsion*, J. Amer. Math. Soc. **8** (1995), 291–363.

[Ca]: G. Carlsson, *On the algebraic K-theory of infinite product categories*, *K*-Theory (to appear).

[CaPe]: G. Carlsson and E.K. Pedersen, *Controlled algebra and the Novikov conjectures for K- and L-theory*, Topology (to appear).

[DwKa]: W. Dwyer and D. Kan, *A classification theorem for diagrams of simplicial sets*, Topology **23** (1984), 139–155.

[DWW]: W. Dwyer, M. Weiss and B. Williams, *A parametrized index theorem for the algebraic K-theory Euler class*, Preprint, Feb. 1995, 63 pp.

[EH]: D.A. Edwards and H.M. Hastings, *Čech and Steenrod Homotopy Theories with Applications to Geometric Topology*, Lecture Notes in Math., vol. 542, Springer-Verlag, Berlin and New York, 1976.

[KKS]: D.S. Kahn, J. Kaminker and C. Schochet, *Generalized homology theories on compact metric spaces*, Michigan Math. J. **24** (1977), 203–224.

[M]: J. Milnor, *On the Steenrod Homology Theory*, these proceedings.

[Se]: G.B. Segal, *Categories and Cohomology Theories*, Topology **13** (1974), 293–312.

[Vog]: W. Vogell, *Algebraic K-theory of spaces, with bounded control*, Acta Math. **165** (1990), 161–187.

[WWA]: M. Weiss and B. Williams, *Assembly*, these proceedings.

[WW3]: M. Weiss and B. Williams, *Automorphisms of Manifolds and Algebraic K-Theory: Finale*, Preprint, 1994, 62pp.

Dept. of Mathematics, Univ. of Michigan, Ann Arbor, MI 48109-1003, USA

email: msweiss@math.lsa.umich.edu

DEPT. OF MATHEMATICS, UNIV. OF NOTRE DAME, NOTRE DAME, IN 46556, USA
email: bruce@bruce.math.nd.edu

Printed in the United States
By Bookmasters